ISBN 978-1-5282-0650-1
PIBN 10024765

BOTANICAL ABSTRACTS

A monthly serial furnishing abstracts and citations of publications in the international field of botany in its broadest sense.

VOLUME V

AUGUST–SEPTEMBER, 1920

PUBLISHED MONTHLY UNDER THE DIRECTION OF

THE BOARD OF CONTROL OF BOTANICAL ABSTRACTS, INC.

A democratically constituted organization, with members representing many societies interested in plants.

BALTIMORE, U. S. A.

WILLIAMS & WILKINS COMPANY

1920

THE SOCIETIES NOW REPRESENTED

AND

THE MEMBERS OF THE BOARD OF CONTROL

(The Executive Committee for 1920 are indicated by asterisks)

American Association for the Advancement of Science, Section G.

*B. E. LIVINGSTON, Johns Hopkins University, Baltimore, Maryland.

A. F. BLAKESLEE, Station for Experimental Evolution, Cold Spring Harbor, Long Island, New York.

Botanical Society of America, General Section.

B. M. DAVIS, University of Michigan, Ann Arbor, Michigan.

*R. A. HARPER, Columbia University, New York City.

Botanical Society of America, Physiology Section.

B. M. DUGGAR, Missouri Botanical Garden, St. Louis, Missouri.

W. J. V. OSTERHOUT, Harvard University, Cambridge, Massachusetts.

Botanical Society of America, Systematic Section.

J. H. BARNHART, New York Botanical Garden, Bronx Park, New York City.

A. S. HITCHCOCK, U. S. Bureau of Plant Industry, Washington, D. C.

American Society of Naturalists.

J. A. HARRIS, Station for Experimental Evolution, Cold Spring Harbor, Long Island, New York.

E. M. EAST, Harvard University, Bussey Institution, Forest Hills, Boston, Massachusetts.

Ecological Society of America.

FORREST SHREVE, Desert Laboratory, Carnegie Institution, Tucson, Arizona.

*GEO. H. NICHOLS, Yale University, New Haven, Connecticut.

Paleontological Society of America.

E. W. BERRY, Johns Hopkins University, Baltimore, Maryland.

F. H. KNOWLTON, U. S. National Museum, Washington, D. C.

American Society of Agronomy.

C. A. MOOERS, University of Tennessee, Knoxville, Tennessee.

E. G. MONTGOMERY, Cornell University, Ithaca, New York.

Society for Horticultural Science.

*E. J. KRAUS, University of Wisconsin, Madison, Wisconsin.

W. A. MCCUE, Delaware Agricultural Experiment Station, Newark, Delaware.

American Phytopathological Society.

*DONALD REDDICK (*Chairman of the Board*), Cornell University, Ithaca, New York.

C. L. SHEAR, U. S. Bureau of Plant Industry, Washington, D. C.

Society of American Foresters.

J. S. ILLICK, State Forest Academy, Mount Alto, Pennsylvania.

BARRINGTON MOORE, American Museum of Natural History, New York City.

American Conference of Pharmaceutical Faculties.

HENRY KRAEMER, University of Michigan, Ann Arbor, Michigan.

WORTLEY F. RUDD, Medical College, Richmond, Virginia.

Royal Society of Canada.

No elections.

At large.

W. A. ORTON, U. S. Bureau of Plant Industry, Washington, D. C.

CONTENTS

BOTANICAL ABSTRACTS

A monthly serial furnishing abstracts and citations of publications in the international field of botany in its broadest sense.

UNDER THE DIRECTION OF

THE BOARD OF CONTROL OF BOTANICAL ABSTRACTS, INC.

BURTON E. LIVINGSTON, Editor-in-Chief
The Johns Hopkins University, Baltimore, Maryland

Vol. V — AUGUST, 1920 — No. 1

ENTRIES 1-1085

AGRONOMY

C. V. PIPER, *Editor*

MARY R. BURR, *Assistant Editor*

1. ANONYMOUS. **Electricity in agriculture.** Sci. Amer. Supplem. 88: 269. 1919.

2. ANONYMOUS. **The value of lupins in the cultivation of poor, light land.** Sci. Amer. Supplem. 88: 265. 1919. [Abstract of paper read by A. W. OLDERSHAW before Agricultural Section, British Assoc. Adv. Sci. Reprinted, *Ibid.* 88: 321. 1919.

3. ANONYMOUS. **Rispentypen des Hafers.** [Types of oat panicles.] Illustrierte Landw. Zeitg. 39: 87. *Fig. 68–72.* 1919.—This article is taken from the book entitled "Der Hafer" by Adolph Zade: Jena, 1918. Five different types of panicles are described and illustrated: 1. Stiff or vertical panicle. 2. Loose or hanging panicle. 3. Bushy panicle. 4. Spreading or open panicle. 5. Flag-shaped panicle.—*John W. Roberts.*

4. ANONYMOUS. **Kartoffelanbauversuche in der Schweiz.** [Potato culture experiments in Switzerland.] Illustrierte Landw. Zeitg. 39: 97–98. 1919.—Two portions of a field were planted to potatoes. In one portion the cut surface of the tubers was placed downward, in the other it was placed upward. Each portion of the field was divided into four plats according to the portion of the tuber used in planting: 1. "Kopfe." 2. Tubers cut into halves longitudinally. 3. Entire tubers. 4. Eyes cut out from tubers. For each plat, the weight of the seed potato, the total crop, and the proportion of weight of seed potato to weight of yield are given. The position of the cut surface made no difference in the yield. There was little difference in the yields from plats 1, 2, and 3; a good yield was had from all three. In proportion to the weight of the material planted, the yield of plat 4 was the highest of all, but the yield was not sufficient to make proper use of the ground. Experiments to determine proper plant spacing are also given.—*John W. Roberts.*

5. ANONYMOUS. **Seed importation act defined.** Seed World. 6^{12}: 20. 1919.

6. ANTHONY, STEPHEN, AND HARRY V. HARLAN. **Germination of barley pollen.** Jour. Agric. Res. 18: 525–536. *Pl. 60–61.* 1920.—See Bot. Absts. 5, Entry 949.

7. BARBER, C. A. **The effect of salinity on the growth of sugar cane.** International Sugar Jour. 22: 17–18. 1920.—From experiments carried on at the cane breeding station at Coimbatore it was found that common salt in the soil seriously affects the sprouting of sugar canes; the color of the leaves is rarely good; and the growth is stunted.—*E. Koch.*

8. Becker, Josef. **Versuche zur Unterscheidung landwirtschaftl. Sämereien und Futtermittel mit Hilfe der Serumreaktion. [Serum reaction an aid in the determination of agricultural seeds and feeds.]** Fühl. Landw. Zeit. 67: 114–120. 1918.—An antiserum, produced by inoculating into animals (rabbits) a certain albumen, possesses the power of causing precipitation of the substance used for inoculation. By means of such a serum reaction it is possible to clearly distinguish between various agricultural seeds and feeds and easily detect adulterations. In preparing the material for inoculation the seeds are ground into a fine powder, extracted with a 10 per cent sodium chlorid solution, the extract filtered and the protein precipitated with ammonium sulphate. The precipitate is filtered, washed and dried. Before being used the dried powder is dissolved in a physiological salt solution—5 grams of the powder in 100 cc. of solution. Of course, it must also be borne in mind that the serum is in many cases specific only when used in the proper dilution.—*Ernst Artschwager.*

9. Brown, W. H., and A. F. Fischer. **Philippine forest products as sources of paper pulp.** Forest. Bur. Philippine Islands Bull. 16. *13 p. Pl. 1.* (1918) 1919.—See Bot. Absts. 5, Entry 161.

10. Bussy, P. **Étude agricole des terres de la Cochinchine. [An agricultural study of the soils of Cochinchina.]** Bull. Agric. Inst. Sci. Saigon 2: 1–11. 1920.

11. Chalmers, D. F. **Report on the operations of the Department of Agriculture, Burma, 1919.** *15 p.* 1919.—The annual report of the Director of Agriculture for Burma, giving the results of development and testing of improved varieties of crop plants, commonly cultivated in Burma. Pebyugale, a variety of *Phaseolus lunatus*, condemned for export purposes on account of its hydrocyanide content, is found to contain a negligible amount of the poison.—*Winfield Dudgeon.*

12. Chevalier, A. **Culture et valeur alimentaire des principales légumeneuses tropicales. [Culture and food value of the principal tropical legumes.]** Bull. Agric. Inst. Sci. Saigon 1: 330–340. 1919.—A general discussion of the commonly cultivated species of the genera *Soja, Arachis, Mucuna, Phaseolus, Vigna,* etc.—*E. D. Merrill.*

13. Chittenden, E. J. **The effect of "place" on yield of crops.** Jour. Roy. Hortic. Soc. 44: 72–74. *Fig. 20, 21.* 1919.—This is a report of a comparison of yields of outside and inside rows of potatoes planted in plots in which the yields averaged 100 for the former to 72 for the latter.—*J. K. Shaw.*

14. Christianson, C. **General consideration of peat problems.** Jour. Amer. Peat Soc. 13: 7–9. 1920.—Peat and peat lands are valuable for both agricultural and industrial purposes. Working out the details of the utilization of peat lands for agricultural and fuel purposes, constitutes the peat problem.—*G. B. Rigg.*

15. Clouston, D. **The selection of rice on the Raipur Experimental Farm.** Agric. and Co-op. Gaz. [India] 15[1]: 5–9. 1919.—See Bot. Absts. 4, Entry 543.

16. Collens, A. E., and others. **Sugar-cane experiments in the Leeward Islands. Report on experiments conducted in Antigua and St. Kitts-Nevis in the season 1916–17 and 1917–18, Part 1.** Imperial Department of Agriculture, Barbados. 1919.—In Antigua the experiments were carried on at nine different stations of varying soil conditions. The varieties which have given the best results as plant canes over a long period of experimentation are B. 4596, Sealy Seedling, B. 6308, B. 1528 and B. 3922. B. 3412 tops the list in the experiments with ratoons over a period of 16 years. In the Colony of St. Kitts-Nevis, B. 6308 heads the list of plant canes for 1916–17. In 1917–18, Ba. 6032 is first, followed very closely by B. 6308 and B. H. 10(12). As ratoons, A. 2 and B. 1528 head the lists respectively.—*J. S. Dash.*

17. Connor, S. D. **Agricultural value of Indiana peat and necessary fertilizers.** Jour. Amer. Peat Soc. 13: 13-17. 1920.—Indiana contains several hundred thousand acres of peat and muck soils, mostly neutral, but some acid. If properly drained and fertilized these soils are capable of producing large and profitable crops. Ordinary crops on neutral peat soils respond to potash fertilization; on acid ones to lime and phosphate.—*G. B. Rigg.*

18. Dunbar, B. A., and E. R. Binnewies. **Proso millet investigations—analysis of the oil— a characteristic alcohol.** Jour. Amer. Chem. Soc. 42: 658-666. 1920.

19. Elayda, I. **A preliminary report on the acclimatization of alfalfa.** Philippine Agric. 8: 70-76. *1 pl.* 1919.

20. Ellis, J. H. **The stage of maturity of cutting wheat when affected with black stem rust.** Agric. Gaz. Canada 6: 971. 1919.—Experiments conducted at the Manitoba Agricultural College show that, contrary to popular notion, wheat attacked by rust should not be cut on the green side. Two fields of badly rusted Marquis wheat were divided into seven plots each. Seven stages of maturity starting with the late milk stage were examined in relation to weight and quality of grain yield. Premature cutting resulted in a brighter color of the grain but decreased yield. Cutting when the grain was firm showed the greatest weight per bushel and greatest yields. Grain cut in the "late" milk stage gave 56 pounds per bushel and that cut in the "firm" stage 59 pounds per bushel.—*O. W. Dynes.*

21. Francis, T. C. **Tobacco-growing in Cuba.** Sci. Amer. Supplem. 88: 304-305. *6 fig.* 1919.

22. Garner, W. W., and H. A. Allard. **Effect of the relative length of day and night and other factors of the environment on growth and reproduction in plants.** Jour. Agric. Res. 18: 553-605. *Pl. 64-79, 35 fig.* 1920.—The duration of the daily period of illumination was found to be a factor of the first importance in the growth and development of plants, particularly with respect to sexual reproduction. At Washington, D. C., during the summer months a number of species and varieties were subjected to continuous daily periods of solar illumination of 5, 7 and 12 hours' duration, by placing the different series of test plants in a dark chamber at 3, 4 and 6 o'clock, p.m., respectively, and returning them to the open at 10, 9 and 6 a.m., respectively, on the following morning. In certain cases the daily exposure consisted of two periods, daylight at 10 a.m. and 2 p.m. to dark, 4 hours of darkness at midday thus intervening. The control plants were fully exposed throughout the entire day. *Soja max, Nicotiana tabacum, Aster linariifolius, Mikania scandens, Phaseolus vulgaris, Ambrosia artemisiifolia, Raphanus sativus, Daucus carota, Lactuca sativa, Brassica oleracea, Hibiscus moscheutos, Viola fimbriatula, Solidago juncea,* were used. In all species tested the rate of growth was proportional to the duration of the daily exposure to light. The length of the vegetative period (germination to flowering stage) was shortened, lengthened or not affected, depending on the species and variety. The time required for ripening of fruit was markedly reduced. Under the artificially shortened daily illumination the duration of the vegetative period of early, medium, late, and very late maturing varieties of soy beans was only 21 to 28 days while the respective periods of the controls were 26, 62, 73, 110 days. All varieties thus behaved as early maturing ones. Similarly, the vegetative period of *Aster linariifolius* was reduced from 122 to 36 days and that of Maryland Mammoth tobacco was reduced from 155 to 60 days while Connecticut Broadleaf tobacco was not materially affected. A variety of *Phaseolus vulgaris* from the tropics attained the flowering stage in 28 days under the shortened exposures as against 109 days required by the controls, and the corresponding periods for *Ambrosia artemisiifolia* were 27 and 85 days. *Mikania scandens, Raphanus sativus* and *Hibiscus moscheutos,* on the other hand, were unable to flower under the reduced light exposures. Two daily exposures with 4 hours' darkness intervening had little effect on time of flowering. By suitably controlling the duration of the daily illumination soy beans, aster and ragweed were induced to complete two vegetative and reproductive cycles in one season. The relation of the seasonal length of day to the natural distribution of plants and to practical

crop production are discussed. The above results showing the significance of the length of day in sexual reproduction were confirmed by the use of incandescent electric lights to lengthen the normal daily illumination period during the winter months. Under suitable exposures *Fagopyrum vulgare*, *Spinacea oleracea* and other plants assumed the ever-blooming type of development. Although the plants of buckwheat showed general similarity in behavior under the normal illumination of the short winter days, the individuals growing under the influence of the lengthened illumination period manifested striking differences among themselves in time of flowering and in size attained. Under controlled conditions differences in water supply and light intensity were without effect on the time of flowering of soy beans. It is tentatively concluded that: Sexual reproduction can be attained by the plant only when it is exposed to a specifically favorable length of day (the requirements in this particular varying widely with the species and variety), and exposure to a length of day unfavorable to reproduction but favorable to growth tends to produce gigantism or indefinite continuation of vegetative development, while exposure to a length of day favorable alike to sexual reproduction and to vegetative development extends the period of sexual reproduction and tends to induce the "ever-bearing" type of fruiting. The term *photoperiodism* is suggested to designate the phenomena disclosed. A bibliography is appended.—*W. W. Garner.*

23. Hawtrey, S. H. C. **Notes on a few useful plants and home industries of Paraguay.** South African Jour. Indust. 3: 35–41. 1920.

24. Helyar, J. P. **Report of the Department of Seed Analysis.** New Jersey Agric. Exp. Sta. Ann. Rept. 1918: 93–97. 1919.—Gives a summarization of the tests for field crop seeds, vegetable seeds and corn.—*Mel. T. Cook.*

25. Hendry, G. W. **Mariout barley with a brief discussion of barley culture in California.** California Agric. Exp. Sta. Bull. 312: 57–109. *Fig. 19.* 1919.—A brief history of Mariout barley is given, including an account of its introduction into the United States. The bulletin is devoted mainly to a discussion of the practical aspects of barley culture in California. The moisture and soil requirements, methods of preparing the soil and seeding, methods of harvesting the crop and comparative yields in different states are discussed.—*W. P. Kelley.*

26. Hepner, Frank E. **Wyoming forage plants and their chemical composition.** Wyoming Agric. Exp. Sta. Ann. Rept. 28 (1917–18): 117–128. 1918.—This paper consists of two parts. Part I deals with the relation of the soil to the nitrogen content of high altitude plants. In earlier work done at this station (Wyoming Agric. Exp. Sta. Bulls. 65, 70, 76, and 87) it was discovered that the native plants were richer in nitrogen than those of the same species grown in the more humid climates of lower altitudes, and later investigations developed the fact that there was a tendency for the nitrogen content to increase with the altitude. In an attempt to find out whether the cause of this increase might not be found in the higher nitrogen content of the soil at higher altitudes, 54 samples of 33 different species of grasses, sedges and rushes were collected at different altitudes and at the same time the soils on which they grew were sampled. These were analyzed and the results are given in tabular form. These results appear to show that the increase of nitrogen in the plants at higher elevation is not so marked as the earlier work would indicate, although the statements made in the earlier bulletins were generally true. Regarding the question as to whether the soils of high altitudes are richer in nitrogen than those of lower elevations, the conclusion is that although nitrogen in the soil is practically the sole source of the nitrogen in the plant, and that the quantity present doubtless exerts a considerable influence on the amount taken up by the plant, still the abundance of nitrogen found in high altitude grasses is not due entirely, if at all, to the greater amount of nitrogen, either total or nitrate, in the soils, nor is it due to excessive quantities of any other soil constituent. Part 2 gives the complete proximate analyses of some of the forage plants including those dealt with in the previous paper. They are all Grasses, Sedges, and Rushes, including *Agropyron occidentale* Scribn.; *Agropyron pseudo-*

repens Scribn. & Smith; *Agropyron tenerum* Vasey; *Agrostis alba* L; *Beckmannia erucaeformis* (L) Host; *Bouteloua oligostachya* (Nutt.) Torr.; *Bromus inermis* Leyss; *Bromus porteri* (Coult.) Nash; *Carex aristata* R. Br.; *Carex festiva ebenea* (Rydb.) A. Nels.; *Carex nebrascensis* Dew; *Carex scopulorum* Holm; *Carex siccata* Dew; *Carex utriculata* Boott.; *Carex variabilis* Bailey; *Deschampsia caespitosa* (L.) Beauv.; *Eleocharis palustris* L.; *Elymus macounii* Vasey; *Glyceria grandis* Wats.; *Hordeum jubatum* L; *Juncus balticus* L; *Juncus longistylis* Torr.; *Juncus nodosus* L; *Juncus mertensianus* Bong; *Juncus richardsonianus* R. & S.; *Phleum alpinum* L; *Phleum pratense* L; *Poa reflexa* Vasey & Scribn.; *Poa nevadensis* Vasey; *Puccinellia airoides* (Nutt.) Wats & Coult.; *Scirpus americanus* Pers.; *Sporobolus airoides* Torr.; *Sporobolus brevifolius* (Nutt.) Scribn.; *Trisetum subspicatum* Beauv.—*James P. Poole.*

27. Hillman, F. H., and Helen M. Henry. **Identification of seed of Italian alfalfa and red clover.** Seed World 7[3]: 15. 1920.—Studies made in the Federal Seed Laboratory of the United States Department of Agriculture indicated that it is possible for the expert seed analyst to identify with reasonable certainty alfalfa and red clover seed grown in Italy, when the seed is represented by samples of sufficient size. The six kinds of incidental seeds peculiar to the Italian strains constitute the basis of identification, namely: *Hedysarum coronarium*, *Galega sp.*, probably *G. officinalis*, *Trifolium supinum*, *Cephalaria transylvanica* of the Dipsacaceae, a species of Phalaris closely allied to *Phalaris canariensis*, and an undetermined species of Valerianella very similar to *V. dentata*.—*M. T. Munn.*

28. Hiltner, Lorenz. **Vermehrte Futtergewinnung aus der heimischen Pflanzenwelt. 1. Teil. Die Gewinnung von Futter auf dem akerland. II. Teil. Wald, Heide und Moor als Futterquellen. Die Verwertung der Wasser- und Sumpfpflanzen. Futtergewinnung aus Gemüse—Obst-, Wein- und Hopfengarten. [Increased forage production from the native flora. Pt. 1. Obtaining of cattle feed from the farm. Pt. 2. Forest, meadow and moor as sources of cattle feed. The use of aquatic and swamp plants as cattle feed, etc.]** Stuttgart, 1917–1918.—The first part of Hiltner's book was written in the spring of 1917 and is perhaps best described to American agronomists by saying that it is comparable in subject-matter and manner of treatment to a high-grade station or Department bulletin on forage and fodder crops, with special reference to war conditions. The 84 pages of this publication are devoted to a discussion of forage products grown on the fields, both cultivated plants and weeds. Under each of the more important crops the author gives the composition in terms of the percentage of protein, fat, and nitrogen-free extract, discusses methods of culture, fertilizers, and the best methods of utilizing the feed, whether green, ensiled, or as dried feed. In the second part, written in the spring of 1918, the author discusses fodder that may be secured from woodland, moorland, or other waste lands, water and swamp plants, feeds from the waste of gardens, orchards, vineyards, and hop fields. And finally, in an appendix the author discusses the methods of treating straw to make it a desirable feed.—In 1913 Germany imported a total of one million tons (of 1000 kg. each) of food stuffs for farm animals. This had a value of 43.3 marks per head of large live stock (Hauptgrossvieh), while the value of food imported for human consumption was valued at 26.66 marks per capita. A large part of the imports too consisted of protein and fat-rich foods. The object of Hiltner, therefore, is to point out how German farmers may increase their output of forage by producing more per acre or by utilizing weeds and other plants not commonly used, and waste products. Much of the advice given the German farmer would be inapplicable to American conditions because of the considerable amount of hand labor involved. The saving of waste products by laborious processes may be necessary under certain conditions, but would certainly not appeal to American farmers.—The author frankly points out that while many plants not commonly used may be fed, these will in most cases serve only as roughage, and have not the protein or fat content to make them valuable as substitutes for imported concentrates.—The discussion in part I falls under five heads: 1. Legumes and clovers. 2. Potatoes. 3. Sugar beets, mangels, swedes, carrots. 4. Miscellaneous forage plants. 5. Weeds.—The cultivation of legumes is urged but nothing new is brought out. Most emphasis is placed on potatoes and sugar beets. Before the war 12 per cent of the arable land in Germany was devoted to pota-

toes and 40 per cent of the crop was fed to animals. Besides the tubers the herbage, cut just as the tubers ripen can be used as hay or ensilage. Miscellaneous information is given on various minor forage plants and weeds with a view to the more general utilisation of everything edible.—In part II food stuffs to be secured from trees, shrubs, water and swamp plants and from various water products are discussed.—The use of forest tree foliage and twigs is especially urged and there is an alphabetical list of species under which are given the essential items of information for each species.—Wood, chemically treated, was being used in 1918 but apparently not as yet very largely or successfully. The author refers hopefully however to many plans underway. In an appendix the treatment of straw with caustic soda is discussed.—*A. J. Pieters.*

29. Himber, F. C. **Flour and mill feed prices.** North Dakota Agric. Exp. Sta. Special Bull. 15: 360–368. 1919.—A questionnaire sent to flour mills in North Dakota secured wholesale flour prices at a date when federal supervision of milling was in force and thereafter. Comparative profits on flour and mill feeds are discussed.—*L. R. Waldron.*

30. Holmes Smith, E. **Flax cultivation.** South African Jour. Indust. 2: 1153–1159. 1919.

31. Jabs, Asmus. **Einiges über unsere Torfmoore. [Notes on our peat bogs.]** Naturwissenschaften 7: 491–495. 1919.—The agricultural use of peat lands in Germany as well as the industrial uses of peat are discussed in the light of post-war conditions.—*Orton L. Clark.*

32. Jones, James W. **Beet top silage and other by-products of the sugar beet.** U. S. Dept. Agric. Farmers Bull. 1095. *24 p. Fig. 1–12.* 1919.

33. Kaiser, Paul. **Der Stachelginster. [Prickly broom. (*Ulex europaeus.*)]** Illustrierte Landw. Zeitg. 39: 38. 1919.

34. Kidd, Franklin. **Laboratory experiments on the sprouting of potatoes in various gas mixtures. [Nitrogen, oxygen, and carbon dioxide.]** New Phytol. 18: 248–252. 1919.—See Bot. Absts. 5, Entry 960.

35. Kling, Max. **Die Kriegsfuttermittel. [War live-stock food.]** Stuttgart, 1918.—This is essentially a handy compendium of information regarding the various feeds on the German market in 1918 or which might be produced by the farmer. In general it covers the same ground as Hiltner but without the cultural directions and with the data on the composition of the various substances more conveniently arranged. In many cases only the trade name and chemical composition of the substance is given. References to sources of chemical data are given, and as a rule there are one or two, rarely three analyses.—Besides prepared feeds there are data on all sorts of major and minor forage crops, trees and shrubs, weeds, swamp plants, vegetable and animal wastes. Preparations from chemically treated wood and straw are discussed and some directions given.—*A. J. Pieters.*

36. Kondo, M. **Ueber Nachreife und Keimung verschieden reifer Reiskörner (Oryza sativa). [After-ripening and germination of rice seeds in various stages of maturity.]** Ber. Ohara Inst. Landw. Forsch. 1: 361–387. 1918.—Grains in the "milk stage" are capable of germination, though the percentage germinating is small. However, if they are kept 15 days in dry storage, or 30 days in moist storage, they will germinate well. The "yellow-ripe" grains germinate sparingly, but if kept for 3 months they will germinate as well as fully ripe grains. The "fully-ripe" grains germinate at once, but germinate better if kept for a month after harvesting. The "dead-ripe" grains germinate immediately after harvesting and need no after ripening.—The after-ripening process is rapidly accomplished, if the rice seeds are kept in a dry condition, but is delayed under moist conditions. Seeds ripened under moist conditions germinate better, however, than those ripened under dry conditions. It is unnecessary to keep the seeds in the panicles.—The germination of freshly harvested,

unripe seeds is hastened after drying in the sun.—The riper the seeds and the further the after-ripening has progressed, the more quickly they germinate and the higher the percentage of germination and the better the seedlings they produce.—Abnormal seedlings often appear "Milk-ripe" grains often produce radicles but no plumules. Fully ripe grains often produce plumules but no radicles.—*H. S. Reed.*

37. Kondo, M. **Ueber die in der Landwirtschaft Japans gebrauchten Samen. [Seeds used in Japanese agriculture.]** Ber. Ohara Inst. Landw. Forsch. 1: 261–324. *17 fig.* 1918.—An account of the morphological characters of certain seeds and their seedlings. Discusses such features as the external appearance of the seed, color, size, weight, anatomical structure of the seed coat, embryo, and seedling.—Seeds of the following plants are so described: *Raphanus sativus*, *Solanum Melongena*, *Cucurbita moschata* var. *Toonas Makino*, *Lagenaria vulgaris*, *Benincasa cerifera*, *Citrullus vulgaris*, *Luffa cylindrica*, *Momordica charantia*, *Cucumis melo*, *Cucumis sativus*.—Literature cited.—*H. S. Reed.*

38. Kulkarni, M. L. **Further experiments and improvements in the method of planting sugar cane and further study of the position of seed in the ground while planting.** Agric. Jour. India 14: 791–796. *Pl. 29–32.* 1919.—Sugar cane cuttings with one bud, planted with the bud pointing upward, sprouted 82 per cent and averaged 5.1 pounds per cane as compared with 50 per cent sprouting and 4.3 pounds per cane where cuttings with three buds were planted with the buds pointed sideways. The yield of crude sugar was about 25 per cent greater from the single bud plantings. Results from placing maize, cotton and jack beans with the seeds pointing upwards, sideways and downwards are given. In all cases seeds pointed upwards gave the poorest results. The author attributes poor stands and sickly plants to indiscriminate placing of seeds, or of buds where cuttings are used in planting.—*J. J. Skinner.*

39. Maceda, F. N. **Selection in soy beans.** Philippine Agric. 8: 92–98. 1919.

40. Menual, Paul, and C. T. Dowell. **Cyanogenesis in sudan grass: A modification of the Francis-Connell method of determining hydrocyanic acid.**—Jour. Agric. Res. 18: 447–450. 1920.—Sudan grass [*Andropogon sorghum Sudanensis*] is found to contain about one-third as much hydrocyanic acid as is found in grain sorghums. The quantity is greatest in the young plant and decreases rapidly as the plant matures. There is more acid in the plant in the morning than in the afternoon.—*D. Reddick.*

41. Miévelle, R. **Essais des culture du blé au Tran-ninh. [Experiments in cultivating wheat in Tran-ninh.]** Bull. Agric. Inst. Sci. Saigon 1: 364–369. 1919.

42. Molegode, W. **Transplanting of paddy.** Tropic. Agriculturist 52: 199–200. 1919. —Results of many experiments on the effect of transplanting rice are given which show an increase of 33⅓ to 220 per cent in yield. Figures are also given to show that in all recorded tests the increased yield and the seed saved by transplanting more than equalled the extra cost incurred by the operation.—*R. G. Wiggans.*

43. Mooers, C. A. **Planting rates and spacing for corn under southern conditions.** Jour. Amer. Soc. Agron. 12: 1–22. 1920.—In general the small and short seasoned varieties require thicker planting than the large long-seasoned varieties. Experimental results indicate a close relationship between the best rate of planting for grain production and a definite yield of grain per plant. To approximate the proper stand of corn a simple equation may be used as follows: $N = \frac{56Y}{F}$. In this equation N stands for the number of stalks per acre, Y for the expectancy or approximate production in bushels per acre of the field in question under average seasonal conditions and F is the standard varietal factor or the average weight of grain per plant in pounds at the best rate of planting as determined experimentally for the variety in question. In the spacing experiments it was concluded that the best results in practice will probably be attained with a width of row which permits the satisfactory use of tillage implements but allows the determined number of stalks to be as widely spaced as possible.—*F. M. Schertz.*

44. Moulton, R. H. **Kudzu, the latest forage plant.** Sci. Amer. Supplem. 88: 364–365. *5 fig.* 1919.—Descriptive of a rapid-growing perennial plant, rich in protein, starch and sugar, which it is asserted gives promise of becoming one of the leading sources of wealth in certain sections of the U. S., especially in some of the southern states.—*Chas. H. Otis.*

45. Mundy, H. G., and J. A. T. Walters. **Rotation experiments. 1913:1919.** Rhodesia Agric. Jour. 16: 513–520. 1919.

46. Nagel, ——. **Kartoffellagerungsversuche. [Potato storage experiments.]** Illustrierte Landw. Zeitg. 39: 6. 1919.—Contrary to the results of Noffe, who found that potatoes lost the least starch when stored in a cool, dry, but well lighted place, the author's experiments resulted in the least loss of both starch and sugar in potatoes stored in a cool, dry, but dark place. Tables showing the percentages of loss under different conditions are given.—*John W. Roberts.*

47. Oldershaw, A. W. **The value of lupins in the cultivation of poor, light land.** Jour. Ministry Agric. Great Britain 26: 982–991. *Fig. 1–3.* 1920.—The value of the cultivation of lupins (Blue and yellow, *Lupinus luteus*) as a means of improving and reclaiming poor light land is not sufficiently appreciated. Lupins grow with surprising luxuriance upon poor, blowing sand, which will grow practically nothing else but rye. The effect of a crop of lupins upon the succeeding crop is really astonishing. Information is given on the sowing, harvesting and utilisation of lupins and on the removal of the possible poisonous properties from lupins.—*M. B. McKay.*

48. Parnell, F. R. **Experimental error in variety tests with rice.** Agric. Jour. India 14: 747–757. 1919.—Experimental errors in field work under Indian conditions are given and data presented. The probable error of long, narrow field plots (20 × 250 lks.) is much less than square plots.—*J. J. Skinner.*

49. Perez, P. F., Manuel A. Suárez, Manuel F. Grau, and Antonio García Villa. **Experiencias en el cultivo del tabaco. [Experiments in the cultivation of tobacco.]** Revist. Agric. Com. y Trab. 2: 484–488. 1919.—This is the report of a commission appointed by the Secretary of Agriculture to report on the results of experiments with tobacco obtained by Francisco B. Cruz. The experiments involve the comparison of tobacco grown without shade, shaded by palm leaves and shaded with cheese cloth. Tobacco produced under shade was declared most desirable for the American market. The yield produced under cheese cloth was largest.—*F. M. Blodgett.*

50. Pescott, E. E. **Excursion to Nobelius's nursery, Emerald.** Victorian Nat. 36: 9, 124, 125. Jan. 8, 1920.—Paper read before the Field Naturalists Club of Victoria, Australia. The paper is a popular account of an excursion taken to the tree-nursery of Messrs. C. A. Nobelius and Sons at Emerald. Uncultivated plants which attracted especial attention were noted including *Erica arborea*; *Ranunculus repens* the English buttercup, which has become naturalised; and *Chiloglottis* the Green Bird Orchid, a clump of which was found in the top of a tree fern. The feature of the nursery, however, was the establishment of the flax industry, many acres of land being devoted to the culture of the New Zealand Flax, *Phormium tenax.* A flax mill has been installed. The flax plants are ready to cut at three years old, and subsequently every three years for an indefinite period. The leaves are graded by throwing a bundle of them upright in a sunken cask. The different lengths are withdrawn and assembled in three grades. They are then scutched, the freed fiber washed, dried and bleached and the fiber is ready for baling and despatch to the rope mills. A ton of fiber is obtained from seven tons of leaves, whereas in New Zealand eight to ten tons of leaves are required to produce one ton of fiber. In New Zealand the flax grows best in swamps, while all of Mr. Nobelius' was hill grown. The local fiber is of superior quality—and graded "special" at the rope mills.—*F. Detmers.*

51. Plymen, F. J. **Nitrate of soda as a manure for cotton.** Agric. and Co-op. Gaz. [India] 15[7]: 10–11. 1919.—Nitrate of soda is strongly recommended as a fertiliser for cotton. Methods for application and instructions for storage are given.—*Winfield Dudgeon.*

52. Ponsdomenech, J. **Elementos químicos necesarios a un terreno para caña.** [Fertilizer necessary for sugar cane.] Revist. Agric. Com. y Trab. 2: 489–493. 1919.

53. Powers, W. L. **The improvement of wild meadow and tule land.** Jour. Amer. Peat Soc. 13: 18–25. 1920. Oregon has about 500,000 acres of such land. There are two soil types—peat and silt loam. Its crop production can be greatly increased by regulating the water supply by drainage and irrigation.—*G. B. Rigg.*

54. Richey, Frederick D. **Formaldehyde treatment of seed corn.** Jour. Amer. Soc. Agron. 12: 39–43. 1920.—Seed corn was treated with solutions of 5, 15 and 25 cc. of formaldehyde per liter. The weakest solution did not materially affect the vitality of the seed while the 15-cc. solution was injurious, as evidenced by the germination and development in sand. The treatment with 5 cc. per liter was markedly injurious. Fungus development was best checked by soaking the seed in a solution (5 cc. HCHO in 9.95 cc. of water) and "fuming" the seed for 2–24 hours. This treatment did not interfere with the normal development of corn seedlings in water culture.—*F. M. Schertz.*

55. Rindl, M. **Vegetable fats and oils. I.** South African Jour. Indust. 3: 14–23. 1920.

56. Robson, W. **Cotton experiments.** Report on the Agricultural Department, Montserrat, 1917–18: 3–12. Imperial Department of Agriculture, Barbados, 1919.—Full account is given of the breeding and selection work with this crop done by the Agricultural Department.—*J. S. Dash.*

57. Roemer, Th. **Die technik der Sortenprüfung.** [The technique of variety testing.] Illustrierte Landw. Zeitg. 39: 35–36. 1919.—As a result of experiments to determine the best experimental technique in variety tests, the author considers the following as important factors: (1) weather (2) kind of fruit (3) size of plats (4) shape of plats (5) number of replicate plats (6) number of plats for comparison (7) situation of the plats with regard to one another (8) treatment at harvest time. The field for the experiments should be carefully selected. There should be at least six replicates of each plat. Care should be taken to give each plat proper cultivation. The author also discusses the things to be considered in determining the quality of the yield. Among these are size of grain, susceptibility to fungous attack, and ability of the seeds to germinate.—*John W. Roberts.*

58. Rosenfeld, A. H. **Kavangire: Porto Rico's Mosaic Disease-Resisting Cane.** Internat. Sugar Jour. 22: 26–33. 1920.—An account of the history and behaviour of Kavangire in the Argentine is presented.—From investigations carried on for the purpose of combating the mosaic or mottling disease of sugar cane in Porto Rico, it was found that of 20 imported varieties there was one Japanese variety (Kavangire) which proved to be immune. This cane was obtained from the National Agricultural School in Tucuman, which in turn obtained the variety from the Experiment Station in Campinas, Brazil. When tried out at the Tucuman Sugar Experiment Station, it showed on first germination remarkable vigor, dark color, high agricultural production, fair juice if left for late cropping, and extreme resistance to fungous disease and attacks of boring insects.—It is a typically thin Japanese bamboo type of cane, identical with the Uba variety of Natal and bears no relation to the Cavangerie which is a large soft red cane with faint black stripes. Experiments were continued with the variety under the name of Kavangire and a consignment of this variety was sent to the Federal Experiment Station at Mayaguez, Porto Rico.—Being resistant to root disease, borer and stem rot, and to frost, it requires less replanting than other varieties which reduces cost of production. Experiments at Tucuman with Kavangire in comparison with native striped and purple canes (Cheribon) show that the yield of cane per hectare as second, third, and fourth

year stubble of Kavangire is in each case much greater than that of the local cane. One crop of plant and four of stubble gives an average yield of cane and sugar per hectare for Kavangire of three times that of the local striped cane.—The objections to this type of cane can be controlled and if the Kavangire turns out to be the only variety in Porto Rico immune to the mottling disease, it will be adopted as the staple cane of the Island.—*E. Koch.*

59. Russell, E. J. **Report on the proposed electrolytic treatment of seeds (Wolfryn Process) before sowing.** Jour. Ministry Agric. Great Britain 26: 971-981. 1920.—Tests made chiefly with wheat, oats, and barley to determine the value of the electrolytic treatment of seeds before sowing gave uncertain results, with occasionally an increase, sometimes no influence, and at other times a reduction in yield. At present the treatment should be looked upon as an adventure which may or may not prove profitable.—*M. B. McKay.*

60. Schander, R. **Beobachtungen und Versuche über Kartoffeln und Kartoffelkrankheiten im Sommer 1917. [Observations and investigations of potatoes and potato diseases in 1917.]** Fühl. Landw. Zeit. 67: 204-226. *1 fig.* 1918.—In general, uncut tubers are to be preferred to cut tubers for seed. The practice of permitting the cut surfaces of seed potatoes to dry before planting seems to be inferior to direct planting; at least the yields are higher in the latter case. Spacing the plants 30 to 40 cm. apart in the row with the rows 50 to 60 cm. wide gives the highest net yields. In light soils the distance may be decreased while in heavy soils it may safely be increased. Varieties with red skin, notably variety Wohltman, produced a number of tubers which were of a light color and contained red stripes. No explanation for this phenomenon has been given. The extreme dryness of the summer of 1917 delayed, and, in the early varieties, prevented the occurrence of late blight. On examination of the tubers, however, it was found that many were covered with mycelium of *Phytophthora infestans.* After all, is the fungus carried on the tubers and does it from them enter the stems and foliage? The stems and foliage seem to be least resistant to the fungus between the time of flowering and maturity. The best way to combat the fungus is to grow varieties which, at the time of the appearance of the fungus, are but little affected.—*Ernst Artschwager.*

61. Shepherd, F. R. **Cotton experiments.** Report on the Agricultural Department, St. Kitts-Nevis, 1917-18: 7-14. Imperial Department of Agriculture, Barbados. 1919.—Details given relating to selection work with cotton in the Colony; bolling and flowering curves are included.—*J. S. Dash.*

62. Stokes, Fred. **The food value of vegetables.** Jour. Roy. Hortic. Soc. 44: 21-30. 1919.—See Bot. Absts. 5, Entry 1857.

63. Störmer, ——. **Keimungshemmungen bei blauen Lupinen. [A case of arrested germination in blue lupines.]** Illustrierte Landw. Zeitg. 39: 12. 1919.—The seeds of the 1918 crop of blue lupines gave a germination percentage of only 24. However, a high percentage of germination (89 to 92 per cent) was obtained after treatment with concentrated sulphuric acid for 15 minutes. followed by a thorough washing with water and then drying.—*John W. Roberts.*

64. Störmer, ——. **Die Anwending von schwefelsäuren Ammoniak und Kalkstickstoff als Kopfdügung zu Winterroggen. [The use of ammonium sulphate and calcium nitrate as the principal fertilizers for winter rye.]** Illustrierte Landw. Zeitg. 39: 73-74, 83-84. 1919.

65. Taylor, H. W. **Tobacco culture. Harvesting and curing.** Rhodesia Agric. Jour. 16: 521-530. *6 fig.* 1919.

66. Trueman, J. M. **Fourteenth Annual Report of the Nova Scotia Agricultural College and Farm. Part 2—Report of J. M. Trueman, Professor of Agriculture and Farm Superintendent.** Prov. of Nova Scotia Ann. Rept. Secretary Agric. 1918: 26-50. 1919.

67. Vendrell, Ernesto. **Estudio sobre los abonos verdes en rotacion con las demás plantas cultivadas en Cuba.** [Green manures in the rotation.] Revist. Agric. Com. y Trab. 2: 553–556. 1919.

68. Vieillard, P. **Notes sur le fonctionnement de quelques services de recherches agricoles de Java.** [Notes on the functions of certain services of agricultural research in Java.] Bull. Agric. Inst. Sci. Saigon 1: 353-358. 1919.

69. Waldron, L. R., and John C. Thysell. **Report of the Dickinson Sub-station for the years 1914 to 1918 inclusive.** North Dakota Agric. Exp. Sta. Bull. 131. *84 p. 19 fig.* 1919.—Authors not jointly responsible. Yields are given for wheat, oats, barley, emmer, flax, maize, potatoes, and certain forage crops for the years indicated and for earlier years for certain crops. Also tables are presented showing the effect of the previous crop treatment and cultural treatment upon the succeeding crop, especially upon the wheat crop. Weather data are presented.—*L. R. Waldron.*

70. Westover, H. I., and Samuel Garver. **A cheap and convenient experimental silo.** Jour. Amer. Soc. Agron. 12: 69–72. 1920.—Experiments conducted at Redfield, S. Dakota, showed that nearly all of the common plants can be preserved as silage which is readily eaten by cattle. Motor oil barrels were used as experimental silos.—*F. M. Schertz.*

71. Wilson, J., and F. J. Chittenden. **Some further experiments with potatoes.** Jour. Roy. Hortic. Soc. 44: 83–88. 1919.—*I. Effect of spacing on yield.* In 1917 nine different spacings were used. In 1918 more spacings, namely sixteen, were used ranging from 9 to 18 inches between plants in the row. For spacings used in 1918 they reiterate their conclusions drawn in 1917 as follows: "(1) The greater the space given to the individual plant the greater the yield of that individual is likely to be. (2) The greater the number of plants on a given area the greater the yield from that area will be." In spacing the other important factors besides yield that must be given due consideration are "relative quantity of seed required," "convenience in cultivating among and earthing up the plants and the need of circulation of air as a preventative of disease."—*II. Effect of different origin on yield of potatoes.* The author is of the opinion that locality alone is not a guarantee of seed potatoes of high producing value. Other factors besides immaturity of seed potatoes at time of planting may be important. Emphasis is laid upon the importance of uniform condition of temperature and moisture in the soil during the growing and maturation periods.—*H. A. Jones.*

BIBLIOGRAPHY, BIOGRAPHY AND HISTORY

Lincoln W. Riddle, *Editor*

72. Anonymous. **Ethel Sargant. (1863–1918.)** New Phytol. 18: 120-128. *2 fig.* 1919. —This is an obituary account of Miss Sargant, with a critical appreciation of her botanical work. A bibliography of her papers is appended.—*I. F. Lewis.*

73. Anonymous. **Introduction of the sugar-cane into the West Indies.** Agric. News [Barbados] 18: 242. 1919.—Information given is based principally on what is known of the life and voyages of Christopher Columbus, and it appears that sugar-cane was not indigenous to the West Indies but that it was introduced by Columbus on his second voyage about 1493. —*J. S. Dash.*

74. Barber, C. A. **Reminiscences of sugar cane work in India.** International Sugar Jour. 21: 390–395. 1919.—An historical account of the difficulty of cane growing in India due to faulty methods of cultivation and an attack of *Colletotrichum falcatum* is presented. Barber worked out a system for cultivation and discovered resistant varieties which when introduced to the cultivators made cane growing successful.—*E. Koch.*

75. BONNIER, G. **Notice sur Viviand-Morel.** Rev. Gén. Bot. **31**: 5-9. 1919.—A brief sketch of M. Viviand-Morel (1843-1915), a French taxonomist whose researches dealt chiefly with the problem of elementary species.—*L. W. Sharp.*

76. CHODAT, R. **Casimir De Candolle, 1836-1918. [Avec un portrait.]** Arch. Sci. Phys. Nat. Genéve v: **1**: 5-28. 1919.—Anne Casimir De Candolle was born in Geneva, Feb. 20, 1836, the son of Alph. De Candolle. He received a thorough training in physics, mathematics and chemistry in Paris under the direction of Berthelot. He then visited London where he remained for some time with the mycologist Berkeley. England became to him a second home; there he married the daughter of a fellow countryman and there his four children were born. De Candolle's botanical contributions were varied, including collaboration with his distinguished father on the Prodromus; but his love for the physical sciences led him mainly into the newer physiological fields of his day, and it was in these fields that he did his best work. De Candolle's strong human sympathies and great versatility won many close friends, and his death is widely lamented. One son, M. Augustin, continues the botanical labors of the family De Candolle, a race of outstanding botanists.—*J. H. Faull.*

77. FARLOW, W. G., ROLAND THAXTER, AND L. H. BAILEY. **George Francis Atkinson.** Amer. Jour. Bot. **6**: 301-302. 1919.—A sketch of the life and work of Professor Atkinson.—*E. W. Sinnott.*

78. FITZPATRICK, HARRY M. **George Francis Atkinson.** Science **49**: 371-372. 1919.—An appreciation of Professor Atkinson as a teacher, investigator and friend, together with a brief résumé of his life and work.—*A. H. Chivers.*

79. FITZPATRICK, HARRY M. **Publications of George Francis Atkinson.** Amer. Jour. Bot. **6**: 303-308. 1919.—A compilation of 178 titles of Professor Atkinson's papers, arranged in chronological order.—*E. W. Sinnott.*

80. FRIEDEL, J. **Notice sur Charles-Louis Gatin.** Rev. Gén. Bot. **31**: 65-74. *Portrait.* 1919.—An account of the work of Charles-Louis Gatin (1877-1916), a French botanist who fell at Douaumont. In Algiers and at the Sorbonne he carried out a number of important researches on the anatomy and physiology of germination in palms and certain other monocotyledonous families. A list of his 51 papers is given.—*L. W. Sharp.*

81. HAMILTON, A. G. **List of papers and books on, or containing references to, the pollination of Australian plants.** Australian Nat. **4**: 81-86. 1919.

82. JANVRIN, C. E. **The scientific writings of Thomas J. Burrill.** Trans. Illinois Hortic. Soc. **51**: 195-201. 1918.—A complete bibliography of the scientific publications of this pioneer botanist is given. The first paper was in 1869 and the last in 1917. Most of the papers dealt with some phase of plant pathology.—*H. W. Anderson.*

83. KROK, TH. O. B. **En sällsynt botanisk skrift. [A rare botanical publication.]** Bot. Notiser **1919**: 165-166. 1919.—In the Royal Library at Stockholm, there is found a little publication of 31 unnumbered pages in small 8vo, entitled: "Catalogus plantarum Tám in excultis quam incultis locis prope Aboam superiori aestate masci abservatarum. In gratiam Philo-Botanicorum concinnatus. Ab Elia Til-Landz. Maij 1673, Aboae-Excusus à Petro Hansonio." This is the only copy now known in existence. It contains the enumeration of 496 plants, wild and cultivated. A second edition was published in Åbo 1683, enumerating 536 plants. Til-Landz was born in 1640. His original name was Tillander, but after having been saved from a shipwreck, he changed it to Til-Landz, which means "on land." Linnaeus named Tillandsia of the Family Bromeliaceae after him.—*P. A. Rydberg.*

84. MANGIN, L. **Paul Hariot (1854-1917). Notice nécrologique. [Obituary notice.]** Bull. Soc. Path. Veg. France **5**: 65-70. [With portrait.] 1918. [Issued April 1919.]—The subject of this notice was the son of a pharmacist and was trained in the same profession. His

first botanical work was in connection with an expedition to Cape Horn. Upon his return to Paris, he became associated with Van Tieghem in the Natural History Museum. He was chiefly interested in the algae and fungi. Later he gave special attention to the rusts, and became one of the founders of the Plant Pathological Society of France. At the time of his death, he was curator of the Cryptogamic Herbarium at the Jardin des Plantes. [See also next following Entry, 85.]—*C. L. Shear.*

85. Mangin, L. **Paul Hariot (1854–1917). Notice nécrologique.** [Obituary notice.] Bull. Trimest. Soc. Mycol. France 35: 4–11. 1919.—See also next preceding Entry, 84.

86. Mitra, Sarat Chandra. **On the use of the swallow-worts in the ritual, sorcery, and leechcraft of the Hindus and the Pre-Islamitic Arabs.** Jour. Bihar and Orissa Research Society [Patna] 4: 191–213, 351–356. 1918.—Treats of religious beliefs and ritualistic practices with reference to *Calotropis gigantea* and *C. procera.*—*B. Laufer.*

87. [Nordstedt, C. T. O.] [Swedish rev. of: Gertz, O. **Christopher Rostii Herbarium Vivum i Lund.**] Bot. Notiser 1918: 214. 1918.—A notice of a Pre-Linnean herbarium found in the University Library at Lund, Sweden. It has the title: "*Herbarium vivum de anno* 1610," and contains 372 plants. It became the property of the University in 1687.—*P. A. Rydberg.*

88. Ostenfeld, C. H. **Botanikeren Johan Lange.** [John Lange, the botanist.] Bot. Tidsskr. 36: 175–181. 1918.—Address on the occasion of the commemoration of the birth of John Lange, author of the handbook of the Danish flora. This took place on March 20, 1918. —*A. L. Bakke.*

89. Pammel, L. H. **Recent literature on fungous diseases of plants.** Rept. Iowa State Hortic. Soc. 53: 185–225. 1918.—Contains abstracts of recent literature on fungous diseases of plants under the following heads, diseases of apple, pear or quince; diseases of the potato; tomato diseases; root crops and vegetable diseases; diseases of forest trees; miscellaneous diseases of fruits; miscellaneous fungicides; diseases of cereal and forage crops; systematic papers, biographical and historical. Under the last topics are given a review of Whetzel's History of Phytopathology, and notices of R. H. Pearson, H. S. Coe, Geo. F. Atkinson, V. M. Spalding, Byron D. Halsted and P. H. Mell.—*L. H. Pammel.*

90. Roberts, H. F. **The founders of the art of breeding. I.** Jour. Heredity 10: 99–106. *4 fig.* 1919.—An historical discussion of the investigations and writings of the founders of the art of breeding. It is shown that sex was recognised in the date palm by the Babylonians and Assyrians but was forgotten. The Greek writers, Aristotle, Pliny and Theophrastus, commented upon the supposed nature of sex in plants, but it remained for Camerer, professor of Natural Philosophy in the University of Tübingen in 1694, to discover by actual experiment that pollination is indispensable to seed production. The article closes with a bibliography of the early publications. [See also next following Entry, 91.]—*M. J. Dorsey.*

91. Roberts, H. F. **The founders of the art of breeding. II.** Jour. Heredity 10: 147–152. *1 fig.* 1919.—The second article describing the work of the early hybridists. Koelreuter published a series of articles from 1761 to 1766 in which he records the results of 136 experiments in crossing plants. To Koelreuter belongs the credit of having produced in 1760 the first plant hybrid—a cross between *Nicotiana paniculata* and *N. rustica.* He also experimented with other plants. The author points out, however, that Thomas Fairchild, an Englishman, crossed two kinds of pinks 41 years previous to the experiments of Koelreuter, and that Richard Bradley, who wrote of the experiments of Fairchild, had, two years before this (1717), removed the anthers from twelve tulips in a remote corner of the garden and found that they produced no seeds, while some four hundred others in another section of the garden produced seeds freely. Still others experimented with sex in plants before the work of Koelreuter. In 1739 James Logan, governor of Pennsylvania, found that when isolated corn plants

were detasseled, or the ears covered before pollination, no seeds developed. He showed the direct relation of the tassels to seed production by cutting the tassels off of a portion of the ear before pollination, in which case he found that that portion from which the tassels were cut bore no grains. Philip Miller repeated the experiments of Bradley in 1741. In 1750 Gleditsch published a learned account of his experiments in the palm. A pistillate palm some eighty years old had never fruited but when pollinated with "male" pollen bore fruit, the seeds of which germinated in 1751. Thus between the time of Camerarius and Koelreuter a number of experimenters were investigating sex in plants, but these experiments appeared to have had but little influence upon the scientific thought of their day. Following these experiments Sprengel (1750–1816) first showed the extent of insect pollination. In the early 19th century the work of Andrew Knight and William Herbert in England and Gärtner in Germany is outstanding. The author shows that there were many breaks in the trend of thought regarding sex in plants up to the time of the publication of Mendel's papers in 1866. [See also next preceding Entry, 90.]—*M. J. Dorsey.*

92. Romell, L. **Svamplitteratur, särskilt för studium av hymenomyceter (hattsvampar).** [Mycological literature, especially for the study of the hymenomycetes (cap fungi).] Svensk. Bot. Tidskr. [Stockholm] 13: 110–112. 1919.—See Bot. Absts. 5, Entry 680.

93. Rosenvinge, L. Kolderup. **Jacob Severin Deichmann Branth.** Bot. Tidsskr. 36: 213–218. 1918.—A biographical sketch of Branth, the well known student of the lichens of Denmark.—*A. L. Bakke.*

94. Shear, C. L., and Neil E. Stevens. **The mycological work of Moses Ashley Curtis.** Mycologia 11: 181–201. 1919.—The life and work of Curtis as revealed mainly through his correspondence is presented in a thorough manner. He was not only a mycologist but also a student of flowering plants and lichens. He collected lichens at the suggestion of Tuckerman (1845), and then turned his attention to the fungi (1846). In 1848 appeared his first mycological paper, in which he acknowledges indebtedness to Berkeley for assistance in its preparation. From 1846 to 1872 he corresponded with Berkeley, exchanging notes and specimens of fungi and thus making possible the important mycological contributions which appeared under their joint authorship. Curtis's original herbarium now forms part of the Farlow Herbarium of Harvard University. Among other institutions which are known to have collections of Curtis's fungi are the Royal Botanical Garden, Kew, England; the U. S. Department of Agriculture, the New York State Museum, and the University of Nebraska. —*H. R. Rosen.*

95. Stevens, N. E. **Two southern botanists and the Civil War.** Sci. Monthly 9: 157–166. 1919.—Rev. M. A. Curtis and H. W. Ravenel were distinguished for their contributions to botany, especially in the field of mycology. The letters of these two botanists to each other and to others are quoted and commented upon. In those days as well as in the world war just ending, the botanist placed his knowledge at the disposal of his country.—*L. Pace.*

96. Whetzel, H. H. **George Francis Atkinson.** Bot. Gaz. 67: 366–368. *Fig.* 1919.—A biographical sketch.

BOTANICAL EDUCATION

C. Stuart Gager, *Editor*

Alfred Gundersen, *Assistant Editor*

97. A[damson], R. S. **The quadrat method.** [Rev. of: Weaver, J. E. **The quadrat method in teaching ecology.** Plant World 21: 267–283. *7 fig.* 1918.] Jour. Ecol. 7: 216. 1919.

98. Anonymous. [Rev. of: Bower, F. O. **Botany of the living plant.** Macmillan and Co.: New York, 1919.] New Phytol. 18: 259–261. 1919.

99. Anonymous. [Rev. of: Cork, M. T. **Applied economic botany.** *261 p., 148 fig.* J. B. Lippincott: Philadelphia, 1919.] Amer. Bot. 25: 116–117. Aug., 1919.—"One of the first books to indicate an approaching change in the subject matter of plant studies."—*Reviewer.*

100. Anonymous. [Rev. of: Ellis, G. S. M. **Applied botany.** *viii + 248 p. 67 fig. 2 maps.* Hodder & Stoughton. "One of the new teaching series of practical text-books."] Jour. Botany 58: 93–94. 1920.

101. Bancroft, Wilder T. [Rev. of: Buisson, Ferdinand, and Frederick E. Farrington. **French educational ideals of today.** 21 × 14 cm., *xii + 326 p.* Yonkers-on-Hudson: World Book Company, 1919. $2.25.] Jour. Phys. Chem. 24: 80. 1920.—"It is a good book and an interesting one" but the title is misleading for "it does not help the university teacher with his problems and never was intended to."—*H. E. Pulling.*

102. Boulger, G. S. [Rev. of: Martin, John N. **Botany for agricultural students.** *x + 585 p.*] Jour. Botany 58: 29–30. 1920.

103. Buckman, H. C. **The teaching of elementary soils.** Jour. Amer. Soc. Agron. 12: 55–57. 1920.—The paper discusses the placing of soil science on a sound theoretical pedagogical basis.—*F. M. Schertz.*

104. Clute, Willard N. **Plant names and their meanings.** Amer. Bot. 25: 122–129. 1919.—The derivation of scientific and vernacular names of the Ranunculaceae discussed.—*W. N. Clute.*

105. Davis, Bradley M. **Introductory courses in botany.** School Sci. Math. 20: 52–56. Jan., 1920.—Outline No. 7. Structure and function, breeding, economic plants, plant communities. Activities and structure showing adaptation emphasized. Outline No. 8. Parts of seed plants, the cell, functions, life histories, plant families, evolution. Emphasis on philosophical aspects. Outline No. 9. History of botany, soil, root, transpiration, photosynthesis, respiration, growth, reproduction. Classification. Emphasis on functions. Outline No. 10. Structure and function of tissues 3 weeks, reproduction 3 weeks, survey of plants: thallophytes 4 weeks, higher plants 3 weeks. [See also next following Entry, 106.]—*A. Gundersen.*

106. Davis, Bradley M. **Introductory courses in botany IV.** School Sci. Math. 20: 352–360. April, 1920.—Outline No. 11. Water relations of plants, nutrition, growth, seeds. Dependent plants. Principal groups of independent plants, industries, plant geography.—No. 12. Seed plant, composite flowers, herbarium of autumn flowers, weeds, pollination, seeds, trees, fall gardens. Algae, bacteria, etc.—No. 13. Plant as a whole. Seeds, fruits, bacteria, yeast, algae and main groups. Last forestry, gardening, orcharding.—No. 14. Nasturtium or Bouncing Bet and composite. Weeds, fruits, bulbs, bacteria, algae, etc., ending with leaves and flowers.—No. 15. Morphology of common plants, physiology, commercial products. Trees, soils, wild flowers, weeds. Decorative planting, plant breeding, seeds, ecology, the cell, algae, fungi, field trips.—No. 16. Algae, bacteria, fungi, gymnosperms, plant physiology, water relations, soils, monocotyledons and dicotyledons, roots, fertilisation, budding, fertilizers, weeds, visits to farms. [See also next preceding Entry, 105.]—*A. Gundersen.*

107. Giles, J. K. **Corn club lessons.** Georgia State Coll. Agric. Bull. 193. *20 p., 3 fig.* 1920.—Contains ten lessons for the Corn Club boys, as follows: No. 1, History of corn (*Zea Mays*); No. 2, Fall preparation; No. 3, Preparation of the seed bed; No. 4, Seed corn; No. 5, Planting; No. 6, Cultivation; No. 7, Selection of seed corn; No. 8, Grow legumes in your corn; No. 9, Selecting exhibits—score card; No. 10, Diseases and insect pests.—*T. H. McHatton.*

108. Prain, David, and others. **Report of the Committee on the Royal Botanic Society.** Royal Bot. Soc. London Quarterly Summary and Meteorological Readings 2: 4-8. Oct., 1919. —The committee was appointed by Lord Ernle to inquire and report what steps should be taken to render the work of the Royal Botanical Society of London as useful as possible from the scientific and educational point of view. The committee recommends the establishment of 1. A school of economic botany; 2. A research institute with special reference to plant physiology; 3. A center for teaching horticulture; 4. Courses in school gardening especially for teachers. The report continues with suggestions for buildings and equipment to cost about £5,500 and the organization of a staff involving an annual budget of £3,000-£3,500 (= pre-war, say £2,000-£2,250). It is also suggested that the new institute should cooperate with local colleges and botany schools by supplying material for teaching and research. [See also abst. from London Times, in Science 51: 58. 1920.]—*C. S. Gager.*

109. Randall, J. L. **Gardening as a part of city education.** Nat. Study Rev. 16: 95-97. 1920.—There is an imperative demand for a new education. The school directed home garden is the most economic form of gardening for small cities and the suburbs of larger cities. In congested parts of large cities school or vacant lot gardens must be substituted. Teachers may receive information from United States School Garden Army, Bureau of Education, Washington, D. C.—*A. Gundersen.*

110. Shaw, Ellen Eddy. **Efficiency aids to garden work.** Nat. Study Rev. 16: 89-94. 1920.—Suggestions to garden teachers in children's work on ways of preparing children for their outdoor work, and on methods of planning and planting a garden, where children have individual plots. The use of the older boys and girls as junior assistant teachers is recommended. Hints for registration of children and keeping of garden crop records.—*A. Gundersen.*

111. Smith, Arthur. **A lesson on soil formation and its bacteria.** Gard. Chron. Amer. 24: 409-410. 1920.

112. Smith, R. S. **Introductory courses in soils.** Jour. Amer. Soc. Agron. 12: 58-60. 1920.—The paper states in broad terms a tentative outline of the general purpose to be attained by an introductory soils course.—*F. M. Schertz.*

113. Stevens, F. L. **Practical botany.** [Rev. of: (1) Cook, M. T. **Applied economic botany.** *261 p., 142 fig.* J. B. Lippincott: Philadelphia, 1919 (see Bot. Absts. 3, Entry 491); (2) Martin, J. N. **Botany for agricultural students.** *585 p., 488 fig.* John Wiley and Sons: New York, 1919 (see Bot. Absts. 3, Entry 2165).] Bot. Gaz. 68: 307-308. 1919.—Cook's work is "written in attractive style, and the material is well-selected, and is a commendable effort to differentiate secondary-school botany from university botany. The numerous half-tones are of unusually good quality." In Martin's work "the presentation is botanical rather than agricultural. The line drawings are not as well done or as accurate as they should be, and the illustrations in general are in contrast with the excellent presswork and the easy and pleasing style of presentation."—*H. C. Cowles.*

114. Trelease, Sam F. **Laboratory exercises in agricultural botany.** College Cooperative Co., Inc.: College of Agriculture, Los Baños, P. I. April, 1919.—Contains 109 pages covering directions for laboratory study for agricultural students as follows: Part I. Physiological Plant Anatomy, including general characteristics of the plant, seed, plant cell, root, stem, leaf, flower, fruit; Part II. Systematic Botany, including I. Primitive organisms (*Bacteria, Cyanophyceae, Flagellata, Myxomycetes, Diatomeae*), II. Plants (*Algae, Fungi, Bryophyta, Spermatophyta*). The guide has been prepared for use with Copeland's "The first year of Botany," a multigraphed text in use at the College of Agriculture, Los Baños.—*C. S. Gager.*

115. Waller, A. E. **Xenia.** School Sci. Math. 19: 150–157. Feb., 1919.—Historical and popular account of xenia, from both a genetic and cytological standpoint. Several illustrations of xenia given, and simple demonstration experiments with maize characters, of instructional value, suggested. [See also Bot. Absts. 5, Entry 496.]—*Orland E. White.*

CYTOLOGY

Gilbert M. Smith, *Editor*
George S. Bryan, *Assistant Editor*

116. Bobilioff, W. **De inwendige bouw der schorselementen ven Hevea brasiliensis. [The structure of cell elements in the bark of Hevea brasiliensis.]** Arch. Rubbercult. Nederlandsch-Indië 3: 222–231. 1919.—See Bot. Absts. 5, Entry 546.

117. Carter, Nellie. **The cytology of the Cladophoraceae.** Ann. Botany 33: 467–478. *1 pl., 2 fig.* 1919.—The chloroplast in *Cladophora*, *Chaetomorpha* and *Rhizoclonium* consists of a parietal film lining the cell wall and often more or less reticulated. Pyrenoids are very numerous and scattered in both the peripheral and internal parts of the chloroplast. The nuclei are confined almost invariably to the chloroplast, not being found as a general rule in the colorless cytoplasm. During mitosis the nucleus of *Rhizoclonium* and *Cladophora* is characterised by the formation of a long thin spireme, which gives rise to very numerous chromosomes. After the migration of the chromosomes to the opposite poles of the spindle the daughter nuclei are separated by constriction of the spindle in the region of the equator.—*G. S. Bryan.*

118. Carter, Nellie. **On the cytology of two species of Characiopsis.** New Phytol. 18: 177–186. *3 fig.* 1919.—*Characiopsis saccata* n. sp. and *Ch. Naegelii* (A. Br.) Lemm. are treated. The cytological features of the vegetative cells were found to differ in important respects in the two species. Zoögonidangia were not found. The cytology of *Characium angustum* is also described, in which the regular successive cleavage of the protoplast contrasts strongly with the progressive cleavage found in *Ch. Sieboldii* by Smith.—*I. F. Lewis.*

119. Chambers, Robert. **Changes in protoplasmic consistency and their relation to cell division.** Jour. Gen. Physiol. 2: 49–68. 1919.—The author has continued his microdissection studies with dividing eggs of *Arbacia* and *Asterias*. Periodic changes in the consistency of the egg cytoplasm after fertilization and during cleavage are described. It is shown that the development of the amphiaster is associated with the formation of two semisolid masses within the more fluid egg substance. After the cleavage furrow has completed the separation of the two blastomeres, the semisolid masses revert to a more fluid state. By various treatments the formation of a cleavage furrow may be prevented following which the egg reverts to a single, spherical, semifluid mass with two nuclei. An egg mutilated in its semisolid state may revert to a more fluid state in which case the furrow becomes obliterated, the nuclei tend to more to positions which may assure symmetry in aster formation and a new cleavage furrow is developed, or the cleavage furrow may persist until cleavage is completed, cutting off non-nucleated segments.—*O. F. Curtis.*

120. Coulter, M. C. **A new conception of sex.** [Rev. of: Jones, W. N. **On the nature of fertilization and sex.** New Phytol. 17: 167–188. 1918. (See Bot. Absts. 3, Entry 637.)] Bot. Gaz. 68: 68–69. 1919.

121. Gatenby, J. Bronte. **Identification of intracellular structures.** Jour. Roy. Microsc. Soc. London 2: 93–119. *14 fig.* 1919.—The author tries to show certain results in practical histo-chemistry from the cytologist's point of view. Every animal cell is composed of the following fairly sharply marked bodies; nucleus, cytoplasm and centrosome. The cytoplasm is composed of (1) protoplasmic or living inclusions such as mitrochondria, Golgi apparatus and possibly other less numerous enigmatic protoplasmic granules; (2) deutoplasmic

inclusions (dead) containing yolk, fat or oil, glycogen or starch, and pigment when not united with mitochondria; (3) ground protoplasm or cytoplasm (living). This classification is particularly true of embryonic or indifferent cells and other cells containing many secondary formations derived from various sources in the differentiation of the cell. He also gives the nomenclature of cell division, saying that every cell undergoes the process of karyokinesis which involves the division of the chromatin; dictyokinesis which involves the division of the Golgi apparatus; chondrokinesis, the division of the mitochondria. All three processes are preceded by the division of the centrosome, which is possibly stimulated to divide by the nucleus and is therefore called "centrokinesis." He describes at length the various inclusions of the cells emphasizing their morphological distinctions, their chemical constitution, and also tabulates the chemical and staining tests for these cytoplasmic and deutoplasmic inclusions. Formal metallic methods for detecting cell inclusions have a future before them. The chromeosmium tetroxide fixatives at present give the best results, but great improvement in the manufacture of microscopic lenses is necessary.—*Julia Moesel Haber.*

122. Levine, Michael. **Life history and sexuality of Basidiomycetes.** [Rev. of: Bensaude, Mathilde. **Recherches sur le cycle évolutif et la sexualité chez les Basidiomycètes.** *156 p., 13 pl., 30 fig.* Nemours, 1918. (See Bot. Absts. 3, Entry 347.)] Bot. Gaz. 68: 67–68. 1919.

123. Mirande, Marcel. **Sur la formation cytologique de l'amidon et de l'huile dans l'oogone des Chara.** [Formation of starch and oil in the egg of Chara.] Compt. rend. Acad. Sci. Paris 168: 528–529. 1919.—The cytoplasm of the young egg of *Chara* is crowded with mitochondria. Numerous clear vesicles appear, which enlarge greatly, forcing the mitochondria into dark staining lines around the clear areas. Starch grains appear in the vesicles and the result in the mature egg is a "mitochondrial pseudo-parenchyma" in which the starch grains are embedded. The mitochondria are the primordia of amyloplasts.—Oil appears in the young egg as minute droplets, which increase in size as the egg matures. In the older stages the drops occur in the meshes of the "mitochondrial pseudo-parenchyma." They are not the products of special mitochondria, and may be secreted by the amylogenes themselves. —*F. B. Wann.*

124. Molisch, Hans. **Das Plasmamosaik in den Raphidenzellen der Orchideen Haemaria und Anoectochilus.** [Plasma mosaic in raphid cells of the orchids Haemaria and Anoectochilus.] Sitzungsber. K. Akad. Wiss. Wien (Math.-Nat. Kl.) 126: 231–242. *Pl. 1.* 1917.

125. Putterill, Victor Armsby. **Notes on the morphology and life history of Uromyces Aloes Cke.** South African Jour. Sci. 15: 656–662. *Pl. 22–23, fig. 1–6.* 1919.—See Bot. Absts. 4, Entry 1153.

126. Small, James. **The origin and development of the Compositae. Miscellaneous topics.** New Phytol. 18: 129–176. *Fig. 64–78.* 1919.—See Bot. Absts. 5, Entry 720.

127. Stålfelt, M. G. **Über die Schwankungen in der Zellteilungsfrequenz bei den Wurzeln von Pisum sativum.** [Variations in the frequency of cell division in the roots of Pisum sativum.] Svensk. Bot. Tidskr. [Stockholm] 13: 61–70. 1919.—See Bot. Absts. 5, Entry 945.

FOREST BOTANY AND FORESTRY

Raphael Zon, *Editor*

J. V. Hofmann, *Assistant Editor*

128. Agan, Joseph E. **Brazilian fibers.** Bull. Pan-American Union 50: 394–404. *4 pl.* 1920.—Seven fibers of importance are discussed briefly. These are "Piassava," from the bark of the palms *Attalea funifera* Mart. and *Leopoldina piassaba* Wall. This fiber is now used in the United States for the manufacture of snow sweepers for street cars. "Piteira"

is obtained from the leaves of *Fourcroya gigantea* Vent. "Aramine" or "Guaxima Roxa," from the trunk of *Urena lobata* L., is used in making bags. *Hibiscus canabinus* L. furnishes another fiber of value for manufacturing bags. *Sida rhombifolia* L. and *S. cordifolia* L. furnish good fiber, but the wild plants are small with crooked branches. "Gravata" (*Ananas sagenaria* Schult.) and "Gravata de Gaucho" (*Bromelia karatas* L.) are also common. The possibilities of growing and of using these fiber plants are discussed.—*G. R. Bisby.*

129. Andrews, Eliza F. **Oddities in tree stems.** Amer. Forest. 25: 1476–1478. *7 fig.* 1919.

130. Anonymous. **"Black bean" or "Moreton Bay chestnut."** Australian Forest. Jour. 2: 14, 19. 1919.—A brief account of the silvical characteristics of *Castanospermum australe* A. Cunn.—*C. F. Korstian.*

131. Anonymous. **Blackboy and its commercial uses.** Australian Forest. Jour. 2: 178. 1919.—A brief note on *Xanthorrhoe preissii* of Western Australia. This species yields a resinous powder which, when heated, forms lumps known locally as "blackboy gum," from which glucose, treacle, scents, alcohol, picric acid and certain tar products, and from these latter again two dyes have been obtained.—*C. F. Korstian.*

132. Anonymous. **A complete wood preserving plant mounted on cars.** Sci. Amer. Supplem. 88: 332–333. *4 fig.* 1919. [From the *Railway Age.*]

133. Anonymous. **Gathering chicle gum for American gum chewers.** Sci. Amer. Supplem. 88: 172. *3 fig.* 1919.—Describes the process of obtaining chicle gum from the naseberry (*Achras sapota*), a tree of Central and tropical South America.—*Chas. H. Otis.*

134. Anonymous. **Grass tree fibre.** Australian Forest. Jour. 2: 175. 1919.—A brief note on the kingia grass tree which at present is used mainly in manufacturing coarse brooms and brushes, but which is believed to possess qualities making it suitable for insulating material for freezing works.—*C. F. Korstian.*

135. Anonymous. **Hints on storing timber to prevent decay.** Sci. Amer. 120: 359–360. 1919.

136. Anonymous. **Kiln drying oak for vehicles.** Sci. Amer. 120: 343. 1919.

137. Anonymous. **Laboratory tests in built-up wood.** Sci. Amer. 121: 606. 1919.

138. Anonymous. **"Napoleon willow" dying.** Amer. Forest. 24: 1414. *1 fig.* 1919.

139. Anonymous. **New uses for balsa wood.** Sci. Amer. 121: 559. 1919.

140. Anonymous. **Preparing cork for shipment.** Sci. Amer. Supplem. 88: 200–201. *3 fig.* 1919.

141. Anonymous. **Steaming of vehicle stock during kiln drying.** Sci. Amer. 120: 360. 1919.

142. Anonymous. **Valuable wandoo.** Australian Forest. Jour. 2: 213. 1919.—A brief note on characteristics of *Eucalyptus redunca.*—*C. F. Korstian.*

143. Anonymous. **Western Australian tuart.** Australian Forest. Jour. 2: 174–175. 1919. —A note on the characteristics of *Eucalyptus gomphocophala.* *C. F. Korstian.*

144. Anonymous. **What are naval stores?** Sci. Amer. 121: 328. 1919.

145. ANONYMOUS. **Holztrocknung durch kalte Luft. [The drying of wood by means of cold air.]** Naturwissenschaften 7: 353. 1919.—A review of an article appearing in the Quarterly Journal of Forestry.—*Orton L. Clark.*

146. ANONYMOUS. **Un bon exemple à suivre. [A good example to follow.]** Bull. Trimest. Soc. Forest. Franche-Comté et Belfort 13: 55-56. 1919.—The city council of Épinal on May 3, 1919, adopted a resolution urging that the fines for forest trespass provided by Article 192 of the Code forestier be increased and that the penalty of imprisonment be restored, at least to the extent of making it optional in the case of habitual offenders. The example set by Épinal should be widely followed and every effort made to secure legislation which will more adequately protect the forests, particularly in the vicinity of cities.—*S. T. Dana.*

147. ANONYMOUS. **Ce que valent les chênes sur pied. [Oak stumpage values.]** Bull. Trimest. Soc. Forest. Franche-Comté et Belfort 13: 53-55. 1919.—Stumpage prices of oak timber in eastern France have approximately doubled since 1916, while the prices of many other commodities are three or even four times what they were before the war. Taking into account the decreased purchasing power of money, oak stumpage, in spite of the apparent increase in price, is worth relatively less than it was a few years ago. Owners of timber of good quality would therefore do well to hold it for the further increase in price which is sure to take place.—*S. T. Dana.*

148. ANONYMOUS. **La forêt de Haguenau (étude d'un forestier française. [A study of the forest of Haguenau.]** Bull. Trimest. Soc. Forest. Franche-Comté et Belfort 13: 117-146. 1919.—The historic forest of Haguenau, owned jointly by the State and the city of Haguenau, comprises an almost unbroken expanse of 13,699 hectares in northern Alsace between the Rhine and the Vosges. It is situated on a practically level plain with a heavy, impermeable clay subsoil, generally overlain with a mixture of sand and clay in varying proportions. The area as a whole is cold, poorly drained, and in spots marshy. The continuity of the forest, which has decreased comparatively little in size since the middle ages, is doubtless due to the fact that the soil is in general unsuitable for cultivation. Injuries from frost, snow-break, and windfall are not uncommon and are at times severe. There is also more or less damage from animals (chiefly deer), insects (chiefly May beetles), various fungi, and, rarely, fire. Scotch pine forms 50 per cent of the stand, oak 30 per cent, hornbeam 8 per cent, and beech 6 per cent. Scotch pine grows rapidly up to 70 or 80 years of age, and ordinarily reaches maturity at about 120 years, with a height of from 28 to 30 metres and a diameter of 60 centimeters. It accommodates itself to all except the most marshy sites; is ordinarily rather poorly formed, but produces wood of excellent quality; and forms rather open stands which at maturity seldom have more than 200 trees per hectare. Seed years occur annually after 50 years of age with particuarly heavy crops every 3 or 4 years. Oak, which formerly occupied a much more important place in the forest, thrives best in the alluvial soils along stream bottoms and produces a fine-grained wood which is much sought after, particularly for ship-building. Although it often attains a much greater age, it ordinarily matures at from 150 to 180 years with a height of from 25 to 30 meters and a diameter of 70 centimeters to 1 meter. Seed crops, which are much less frequent than formerly, occur at intervals of approximately 7 years, with full crops not oftener than once in 50 years. Hornbeam is of little value except as a filler and is often more or less of a weed tree. Beech was formerly much more abundant than at present, but has been increasing in importance again since 1870 because of its frequent use by the Germans for underplanting with pine and oak. Herbaceous vegetation is generally abundant, some times to the extent of interfering with reproduction, and local residents derive a considerable revenue from the abundant crops of whortleberry. The forest is more or less burdened with rights of use, most of which date back to time immemorial, and considerable damage has been done to the soil by the constant removal of the hardwood leaf litter. Transportation facilities and markets are good.—Prior to the seventeenth century, the forest of Haguenau appears to have been regarded as chiefly valuable for pasturage. The first real attempts at forest regulation were made in 1695, and it was not until 1845 that a complete

and systematic plan of forest management was put into effect. This plan was followed until after the Franco-Prussian war, when, in 1874, it was revised by the German foresters. The latter completely reorganized the division of the forest into blocks, compartments, and subcompartments; determined on the management of the entire area as high forest (nearly 7 per cent had been handled by the French as coppice under standards); fixed the rotation for Scotch pine at 70 to 120 years, and for oak at 160 years; and arranged the cutting series so as to progress against the direction of the prevailing winds. Natural reproduction by the shelterwood system, which was almost uniformly used by the French, was at first employed by the Germans as well, but was gradually abandoned in favor of artificial reproduction. During the last years of German management Scotch pine was reproduced almost entirely by direct seeding in strips, supplemented when necessary by planting; while oak was reproduced chiefly by the planting of 3-year-old transplants, and occasionally by direct seeding in strips. Thinnings were practised every 7 to 10 years, frequent and moderate thinnings being preferred to less frequent and heavier ones. In the judgment of the French foresters the Germans tended to favor too dense a stocking, both at the establishment of the stands and later. Underplanting of beech, chiefly to improve soil conditions, was common, wild seedlings generally being used for the purpose. A few of the best trees (from 15 to 25 per hectare) were nearly always reserved at the final cutting for the production of large-sized material. The practice of selling stumpage, which had been followed by the French, was superseded under German management by logging by the forest administration. The net revenue from the forest increased from 44 francs per hectare in the period from 1889 to 1900 to 57 francs in 1912–1914 and to 120 francs in 1915–1918. The recent war led to the turpentining by the Germans of the Scotch pine. The total cut remained about the same but the proportion of pine increased while that of oak decreased. Thinnings were neglected, stock accumulated in the nurseries, and the regeneration of cut-over areas did not keep pace with the cuttings. On the whole, however, the war did not seriously interfere with the management of the forest which is still in good condition.—*S. T. Dana.*

149. Anonymous. **Historique d'une coupe.** [History of a cutting area.] Bull. Trimest. Soc. Forest. Franche-Comté et Belfort 13: 51–53. 1919.—In 1844 steps were taken to convert a cutting area of 7.23 hectares, chiefly oak with a little beech, in the communal forest of Corravillers on the borders of the Vosges, into coppice under standards. Since 1844 there have been three cuttings of standards at regular intervals of 25 years. The records show that the yields in fuel and bark secured from these successive cuttings have remained approximately constant. The transformation of the stand from pure coppice into coppice under standards has therefore been accomplished without loss in current yield, and the timber contained in the boles of the standards represents clear gain. As a result of the transformation the money value of the yield has increased from 460 to 680 francs per hectare. Still better results would have been obtained in a more moderate climate and a more fertile soil than that of the Vosges. —*S. T. Dana.*

150. Anonymous. **Notre domaine forestier et la guerre. (Extrait du Bulletin d'informations du G. Q. G.)** [Our forest domain and the war.] Bull. Trimest. Soc. Forest. Franche-Comté et Belfort 13: 43–46. 1919.—The forest area of 600,000 hectares included in that part of France lying in the war zone suffered severely both as a result of battle and of its extensive exploitation by the French themselves and more particuarly by the Germans. The latter not only used wood lavishly in the zone of operations but shipped considerable quantities back to Germany in order to save their own resources and to cripple France, which in 1913 imported 177,000,000 francs' worth of wood, for the post-war competition. Direct damages to the forests in the war zone are estimated roughly to amount to 1,400,000,000 francs, and indirect damages to 260,000,000 francs; while the forests in other parts of France also suffered serious damage because of the tremendous consumption necessitated by the war and by lack of tonnage. While the forests are recovering, France should meet its needs for wood, which are still great, by utilising part of the enormous reserves offered by its colonies. The German possessions in the Kamerun, one of the most richly forested countries in Africa, will offer partial compensation for the devastation of the French forests caused by the war.—*S. T. Dana.*

151. ANONYMOUS. **Wattle and wattle growing.** Australian Forest. Jour. 3: 45-46. 1920.—A note on the growing of various species of acacia and the products of the destructive distillation of black wattle wood.—*C. F. Korstian.*

152. ARIAS, BERNARDO. **Un sustituto del corcho.** [A substitute for cork.] Revist. Agric. Com. y Trab. 2: 493-497. *3 fig.* 1919.—In this article attention is called to the tree *Ochroma lagopus* Sw. as a native tree valuable for planting because of the lightness of its wood, its rapid growth, medicinal properties and the wool or fiber in its fruits.—*F. M. Blodgett.*

153. BADOUX, H. **Die Waldreservationen in der Schweiz.** [Forest reserves in Switzerland.] Schweiz. Zeitsch. Forstwesen 71: 2-4. 1920.—The policy for acquiring national forests was approved in 1906, and in 1910 three forest reserves were approved involving a total area of about 50 hectares. These areas were in effect leased by the government for periods of 25 and 60 years. The policy of the continuation of the forests was left to be determined when the period of lease expires. Some areas were paid up for the entire term, and others are paid by annual installments.—*J. V. Hofmann.*

154. BAILEY, W. A. **Artificial regeneration in sal forests.** Indian Forester 45: 519-521. 1919.—Coppice overtops planted stock after cuttings in sal forests. To prevent this planting is now made about five years in advance of the opening of the stand giving the planted stock an opportunity to develop and become dominant at the start.—*E. N. Munns.*

155. BARBEY, A. **Les forêts suisse pendant la guerre.** [The Swiss forests during the war.] Bull. Trimest. Soc. Forest. Franche-Comté et Belfort 13: 46-51. 1919.—Administration of the 982,000 hectares of forest lands in Switzerland, one-fourth of the total area of the country, is decentralized. Cantonal forests comprise 4 per cent of the forest area, communal forests 67 per cent, and private forests 29 per cent. There are no national forests, and the national forest service employs only 17 professional foresters. It contributes, however, to the salaries of the cantonal forest officers; supervises the use made of subsidies granted to the cantons; administers the federal forest law; provides technical instruction at the forest school at Zürich; and directs the forest experiment station.—At the outbreak of the war construction was automatically arrested and cutting materially decreased. After ten or twelve months, however, the foreign demand for timber and the native demand for wood fuel (due to the scarcity of coal), resulted in a steadily increasing cut. In 1916 wood exports, which before the war had been from 40,000,000 to 50,000,000 francs a year less than wood imports, exceeded the latter by 68,000,000 francs. The increased cut was accompanied by increased prices, fuel doubling and timber trebling in value in three years or less. Little or no overcutting took place in the public forests, but was more or less marked in the private forests, where advantage was taken of the extraordinary demand to improve the stands by the removal of many old reserves which before the war could not be marketed profitably. Strict supervision was exercised over all cuttings, a federal decree in 1917 requiring a permit for all cuttings of 20 cubic meters or more and fixing a fine of from 10 to 40 francs per cubic meter for all cuttings made without a permit. Moreover, measures were taken to maintain and if possible to increase the future productivity of the forest. For instance, in the Canton of Vaud, the number of inspectors was increased so that the average area under the supervision of each was reduced from 7,300 to 4,000 hectares. This example should be followed by other cantons as a means of increasing production and of rendering Switzerland independent of foreign supplies. An increase of only 1.1 cubic meters per hectare in the annual growth of the 600,000 hectares of communal forests would be sufficient to wipe out the present deficit of 700,000 cubic meters, but this can hardly be expected as long as the average area under the supervision of a technical forester remains as high as 8,570 hectares.—*S. T. Dana.*

156. BEESON, C. F. C. **Food plants of Indian forest insects. Part IV.** Indian Forester 45: 488-495. 1919.—A continuation of previous work. Forty-four species of three families are listed with the plants attacked by each.—*E. N. Munns.*

157. Berry, James B. **Wood famine imminent.** Georgia State Coll. Agric. Bull. 187. *4 p., 4 fig.* 1920.—This bulletin notes that the acme of wood production was reached in Georgia in 1909, with the cutting of a billion board feet. Since then there has been a gradual falling off in production.—*T. H. McHatton.*

158. Biolley, H. **Betrachtungen über die Wirtschafts-Einrichtung der Waldungen in der Schweiz. (Bemerkungen zu den Studien des Herrn. Dr. Ph. Flury.) [Observations concerning improvement of forest management in Switzerland. Remarks on Dr. Ph. Flury's studies translated from the Journal of Forestry of Perret, Couvêt.]** Schweiz. Zeitschr. Forstwesen 71: 37–49. 1920.—Forestry is divided into two groups, one based on practical experience and the other on biological principles. Emphasis is placed on the fundamental biological studies to be used as a basis for all forest practice. The practical concerns itself too much with the present production, and one part of a forest may be left unproductive due to over maturity while another is exploited during its growing period. Among the first essentials for improvement are definite forest boundaries, compartments; definite volume and growth tables and cutting cycles based on accurate local growth figures. The relation of density of stand and increment must be correlated with cutting periods in order to secure continuous production. The principal points recommended for the improvement of the forest are: every acre must reach its maximum production; production as influenced by stand, site, species, etc., must be determined locally; species to be used and care required; improvement for regulation only should be reduced to a minimum. All changes in forest management should be based on thorough scientific research.—*J. V. Hofmann.*

159. Bontrager, W. E. **What shade and ornamental trees shall we plant?** Monthly Bull. Ohio Agric. Exp. Sta. 5: 35–41. *5 pl.* 1920.—See Bot. Absts. 5, Entry 1798.

160. Bouvet, Schaeffer, and others. **Congrès de 1919. [Congress of 1919.]** Bull. Trimest. Soc. Forest. Franche-Comté et Belfort 13: 72–109. 1919.—The first meeting of the Society since the outbreak of the war was held at Strassburg, August 3 to 6, 1919. In connection with the rejoicing over the recovery of the "lost provinces," attention was called to the flattering comments regarding French methods of forest management in Alsace-Lorraine which were made by German foresters after the war of 1870. Field trips were made to the forests of Haguenau, Hoh-Koenigsburg, Sainte-Odile, Hohwold, Haslach and Nideck, brief descriptions of the character and management of which are given.—*S. T. Dana.*

161. Brown, W. H., and A. F. Fischer. **Philippine forest products as sources of paper pulp.** Forest. Bur. Philippine Islands Bull. 16: *13 p. Pl. 1.* 1918. (1919).—A general consideration of the bamboos, coarse grasses such as *Imperata exaltata* and *Saccharum spontaneum*, various fiber plants, and some trees as potential sources of paper pulp.—*E. D. Merrill.*

162. Brown, W. H., and A. F. Fischer. **Philippine mangrove swamps.** Forest. Bur. Philippine Islands Bull. 17: 1–132. *47 pl.* 1918.—A general consideration of the mangrove swamps, their constituent species, and economic products. Keys and descriptions are given to all species, as well as local names, etc. The illustrations, chiefly photographic, are excellent. In addition to general mangrove scenes each individual species is illustrated. The economic discussion includes data on stand, cultivation, firewood, tanbark and dyes, with a discussion of the nipa palm and its uses.—*E. D. Merrill.*

163. Brown, W. H., and A. F. Fischer. **Philippine bamboos.** Forest. Bur. Philippine Islands Bull. 15. *32 p. Pl. 1–33.* 1918.—See Bot. Absts. 5, Entry 1015.

164. Brunnhofer, A. **Berufsfragen. [Questions of professional forestry.]** Schweiz. Zeitschr. Forstwesen 71: 4–6. 1920.—A discussion of the relation of technical and commercial forestry. A separation of the two phases is condemned on the basis that the technical forester must be familiar with the commercial phases in order to practice his profession intelligently, and the commercial man must take technical forestry into consideration in utili-

zation and harvesting, otherwise the scientific phase, which aims at continuous production, will be defeated. For these reasons a forester in either field must have a good knowledge of the other field, and the best interests of forestry will be served by keeping the two phases combined and making up the deficiency of men by reducing the areas under each forester and furnishing him with an assistant.—*J. V. Hofmann.*

165. Burkill, I. H. **The composition of a piece of well-drained Singapore secondary jungle thirty years old.** Gardens' Bull. Straits Settlements 2: 145–157. 1919.—See Bot. Absts. 4, Entry 280.

166. Burrow, Gordon. **Reproduction of cypress pine.** Australian Forest. Jour. 2: 91–92. 1919.—A note on the factors governing the reproduction of this species. The author is convinced that a good seeding season and a good growing season are co-essentials. A good seed crop is dependent upon sufficient precipitation to set and nourish the young cones and bring them to maturity. Drought, rabbits, and fire are serious enemies of young reproduction.—*C. F. Korstian.*

167. Champion, H. G. **Observations on some effects of fires in the chir (Pinus longifolia) forests of the West Almora Division.** Indian Forester 45: 353–364. *1 pl.* 1919.—Examinations of burned areas after a fire show damage cannot be estimated until several months later. Insects for some unknown reason did not appear in large numbers after fire in mature stands though death continues afterward, which may be due to a destructive fungus. Damage by fire may be as much due to heat-killing as flame itself. In young trees damage bears an inverse ratio to height, the smaller the tree the greater the loss. On reproduction, fire appears to have a beneficial effect, probably due to reduced competition, food or soil water. Fire in mixed stands operates to thin out the chir and increase oaks and other trees.—*E. N. Munns.*

168. Chapman, H. H. **A program for private forestry.** Amer. Forest. 25: 1405–1406. 1919.

169. Claudy, C. H. **Economic tree murder. How we are denuding our forests to supply Europe while she is conserving her own timber.** Sci. Amer. 121: 132. 145. 1919.

170. Cook, O. F. **Olneya beans.** Jour. Heredity 10: 321–331. *Fig. 13–17.* 1919.—See Bot. Absts. 4, Entry 549.

171. Cremata, Merlino. **Algo sobre nuestros bosques.** [Forest preservation.] Revist. Agric. Com. y Trab. 2: 610–611. 1919. An article of forest conditions in Cuba and on forest preservation.—*F. M. Blodgett.*

172. Crevost, C., and C. Lemarie. **Plantes et produits filamenteux et textiles de l'Indochine.** [Fiber- and textile-producing plants of Indo-China.] Bull. Econ. Indochine 22: 813–837. *Pl. 2.* 1919.—See Bot. Absts. 5, Entry 1122.

173. Dana, S. T. **National forests and the water supply.** Amer. Forest. 25: 1507–1522. *33 fig.* 1919.

174. Danielsson, Uno. **Naturskydd i Södra Kalmar län** [Protection of natural beauty in southern Kalmar (Sweden).] Skogen 6: 17–22. *5 fig.* 1919.

175. Darnell-Smith, G. P. **Dry rot in timber.** Australian Forest. Jour. 2: 314–316. 1919.—A brief discussion of the characters of some dry rot fungi and measures for their control. Creosote and tar are effective, but their odor and color restrict their use. Boric acid and magnesium fluosilicate are strongly recommended. Wood-preserving oil, prepared from kerosene shale, is effective if the ventilation is good.—*C. F. Korstian.*

176. Darvey, Mason. **Forest tree planting in Nelson District.** New Zealand Jour. Agric. 19: 297-299. 1919.—It is believed that *Pinus insignis* and several species of Eucalyptus may be planted on land costing about $50 an acre as a very profitable long term investment.—*N. J. Giddings.*

177. Dawkins, C. G. E. **Yemané (Gmelina arborea) in Upper Burma.** Indian Forester 45: 505-519. 1919.—The results of trials to introduce the yemané into the forests of Burma are given. Three methods have been tried; broadcast sowing, dibbling and field planting. Notes on the growth of plantations made are given.—*E. N. Munns.*

178. De Jong, A. W. K. **Tapproeven bij Hevea brasiliensis. [Tapping experiments on Hevea brasiliensis.]** Arch. Rubbercult. Nederlandsch-Indië 3: 277-278. 1919.—Tapping a quarter, a third or half the circumference of the tree with one left hand cut gave the following results:

	Proportion of the rubber yields for		
	¼ of the C.	⅓ of the C.	½ of the C.
For the first area tapped	100	117	140
For the second area tapped	100	116	135.5
For the third area tapped	100	109.5	100
For the three areas tapped	100	114	122

—*W. E. Cake.*

179. Demorlaine, J. **La nécessité d'un service forestier d'armée sous l'ancien régime. [The need for an army forest service.]** Rev. Eaux et Forêts 57: 229-230. 1919.—Duhamel du Montceau, in 1764, in his "Exploitation des Bois," pointed out the need of attaching forest officers to the engineers crops of the army in order to prevent the serious damage done to the forests when the timber and other forest products needed by the army were secured by ordinary soldiers without technical supervision. The need of an army forest service of this sort has been strikingly demonstrated by the great war. Such a service should be autonomous, with the same standing as the Engineer or Quartermaster Corps, and should direct the formation, management, instruction, and organization of companies of mobilized foresters.—*S. T. Dana.*

180. Descombes, Paul. **Installation d'expériences prolongées sur le ruissellement. [Protracted experiments upon stream-flow.]** Mém. Soc. Sci. Phys. Nat. Bordeaux VII, 2: 17-35. *2 fig.* 1918.—The author gives a brief résumé of methods adopted by L'Association Centrale pour l'Aménagement des Montagnes in studying the relations between precipitation and stream-flow in the drainage basin of the Arises. An apparatus for automatically gauging and recording changes in stream level is described. Data are presented to indicate a correlation between changes in the flow of the Ariège (1896-1910) and the sylvo-pastoral conditions in its drainage basin.—*I. W. Bailey.*

181. Descombes, Paul. **Le reboisement et le développement economique de la France. [Reforestation and the economic development of France.** Mém. Soc. Sci. Phys. Nat. Bordeaux VII, 2: 103-217. *2 fig.* 1918.—Deforestation and over-grazing in the uplands of France prevent an extensive substitution of waterpower for coal and are considered to be responsible for the depopulation and degradation of these regions. Reforestation and other remedial projects for improving the range have been combated by the mountaineers, who fear curtailment of their herds and flocks. L'Association Centrale pour l Aménagement des Montagnes has conducted a series of extensive experiments to prove that it is possible to prevent overgrazing and to reforest the mountains without reducing the live stock of the mountaineers. This is done by excluding from the alpine pastures migratory herds and flocks from the lowlands. In considering measures for reforestation of both uplands and lowlands the author

devotes considerable attention to a discussion of the status of French forests and the reforestation movement during the nineteenth century, and quotes various legislative enactments at length. The paper contains much statistical information.—*I. W. Bailey.*

182. De Vries, O. **Over de bruikbaarheid van instrumenten als metrolac en latexometer voor het bepalen van het rubbergehalte van de latex. [On the use of hydrometers (metrolac and latexometer) to determine the rubber content of latex.]** Arch. Rubbercult. Nederlandsch-Indië 3: 207–221. 1919.—Very large differences may occur between the real rubber content of Hevea latex as determined by actual coagulation and the figures obtained from the hydrometric specific gravity readings. The metrolac and latexometer are constructed for a special case, perhaps an original latex of 37½ per cent rubber content and 0.9775 specific gravity or some other combination near there, when the specific gravity of the original serum varies from 1.022. When such a latex is diluted with water the reading of the instrument is correct, but for latices of other composition the rubber content cannot be determined by these instruments. In general on the estates in Java the results obtained by hydrometric readings are too low, usually giving values between 70 and 80 per cent of the real content.—*W. E. Cake.*

183. De Vries, O. **Verband tusschen het soortelijk gewicht van latex en serum en het rubbergehalte van de latex. [The relation between the specific gravity of latex and serum and the rubber content of latex.]** Arch. Rubbercult. Nederlandsch-Indië 3: 183–206. 1919.—The relation between the specific gravity of Hevea latex and its rubber content was determined in the following five cases: (1) continued tapping after a period of rest, (2) light or heavy tapping systems, (3) pollarding, which also acts as a "heavier stress," (4) periods of rest and shallow tapping, and (5) individual trees. In all cases the results are the same, showing that the specific gravity is inversely proportional to the rubber content. The actual specific gravity of the latex is determined by the proportion of the rubber and serum (i.e., the rubber content of the latex) and only to a small extent by the specific gravity of the serum which remains nearly constant.—*W. E. Cake.*

184. De Vries, O., and W. Spoon. **Variabiliteit van plantage-rubber. [Variability in plantation-rubber.]** Arch. Rubbercult. Nederlandsch-Indië 3: 246–276. 1919.—Data from the Central rubber station comparing the tensile strength, slope, rate of cure, and viscosity of smoked sheet and crepe rubber for the years 1917 and 1918. The principal causes for deviation and variability in properties are pointed out.—*W. E. Cake.*

185. Essig, E. O. **New hosts of oak-root fungus in Humboldt County.** Monthly Bull. Comm. Hortic. California 8: 79–80. 1919.—See Bot. Absts. 4, Entry 1170.

186. F[oster], J. H. [Rev. of: Rankin, W. Howard. **Manual of tree diseases.** *398 p.* Macmillan Co.: New York, 1918.] Jour. Forest. 17: 321. 1919.

187. Geete, Erik. **Ur timmersaxens historia. [From the history of the timber "grab hook."]** Skogen 6: 23–25. *3 fig.* 1919.

188. Gellatly, F. M. **Investigatory work needed: relation of commonwealth to states.** Australian Forest. Jour. 2: 137–139. 1919—The more important benefits to be derived from a forest products laboratory are discussed. Urgent need is voiced for research along the following lines: (1) tests of pulping and paper-making qualities of indigenous woods and materials, (2) distillation tests to determine the tar oil, gas, acid and other properties of commercial value in indigenous woods, (3) investigation of the chemical and commercial properties of gums, kinos, resins, and saps.—*C. F. Korstian.*

189. Grinndal, Th. **Tidig eller sen skogssådd? [Early or late forest sowing?]** Skogen 6: 124–127. 1919.

190. Gupta, B. L. **New Indian species of forest importance.** Indian Forester 45: 388–392. 1919.—A continuation of previous work (*Ibid.* 43: 132. 1917). The present list includes 48 species recently described from India, bringing the total forest species up to 393.—*E. N. Munns.*

191. HAINES, H. H. **Indian species of Carissa.** Indian Forester 45: 375–388. *Pl 17–20, fig. 1–7.* 1919.

192. HALL, CUTHBERT. **On a new species or form of Eucalyptus.** Proc. Linnean Soc. New South Wales 43: 747–749. *Pl. 75.* 1918.

193. HECK, G. E. **Splintering of airplane woods.** Sci. Amer. Supplem. 88: 68–69. *4 fig* 1919.

194. HEIM, A. L. **Airplane propeller manufacture.** Sci. Amer. Supplem. 88: 162. 1919. —Considers problems of manufacture which have been or need to be studied.—*Chas. H. Otis.*

195. HOFFMAN. **Ist die Vergesellschaftung im Forstbetriebe möglich? [Is socialization of forest industry practicable?)** Forstwiss. Centralbl. 41: 210–226. 1919.—Most socialists agree that forest industries of Germany should be socialized, in order to avoid danger of monopoly, to insure continuity of employment and of supplies of forest products, and to insure maximum sustained production at lowest cost. Methods suggested are State ownership, either by purchase or confiscation, syndicalization, or division of large holdings. State ownership is unnecessary because the State already owns a sufficient proportion of the forests to prevent monopoly, and undesirable because of the probable decrease in efficiency due to bureaucratic inertia and political influences. Moreover, it is financially impossible. Syndicalization is not desirable because the nature of the business is not adapted to this form of management. Division of holdings is contrary to the requirements of efficient forest production, and unnecessary anyway because there are few very large holdings. The best way for a democratic state to control forest production is by use of its powers of taxation. The forest law should require that all forest tracts of more than 100 hectares be managed according to a working plan, under technical supervision. Beyond this, the owner should have entire freedom of action. Owners of smaller tracts should form cooperative bodies or looser associations, in order to be able to take steps toward more efficient management. The State should supervise the activities of these associations.—*W. N. Sparhawk.*

196. HORNE, W. T. **Oak-fungous, oak-root fungus disease, fungus root-rot, toadstool root-rot or mushroom root-rot.** Monthly Bull. Comm. Hortic. California 8: 64–68. *Fig. 36–39.* 1919.—See Bot. Absts. 4, Entry 1176.

197. HUBUALT, E. **L'après guerre dans les îles britanniques: projets de reconstitution forestière. [Forest reconstruction in Great Britain.]** [Rev. of: **Final Report of Forestry Subcommittee, Reconstruction Committee, Ministry of Reconstruction.** *106 p.* 1918.] Rev. Eaux et Forêts 57: 213–228. *1 fig.* 1919.—The critical situation in which Great Britain found itself during the war as a result of totally inadequate native wood supplies has led to the formulation by a specially appointed committee of a comprehensive forestation program, intended to decrease materially Great Britain's present dependence on other countries and to provide a reserve capable, in case of war, of meeting for three years all its needs for wood at a rate of cutting five times as great as the normal annual consumption. The program contemplates the establishment in 80 years of 717,000 hectares of coniferous plantations, chiefly Scotch pine, European larch, Douglas fir, Sitka Spruce, Norway Spruce, and western red cedar. Two-thirds of this area, or 478,000 hectares, will be forested during the first 40 years, and 101,000 hectares during the first ten years. Of this latter area, the state will itself acquire, either by purchase or lease, and plant 60,000 hectares; it will associate itself with communities and individuals in the cooperative planting and management of 10,000 hectares; and through the granting of subsidies of one kind or another it will encourage the forestation of 10,000 hectares by communities and individuals. The remaining 21,000 hectares are to be secured through the voluntary or forced reforestation by their owners of areas cut clear during the war. In addition the reforestation during this period of 4,000 hectares of hardwoods (and eventually of 8,000 hectares) is contemplated. The committee proposes certain reductions in forest taxes and in freight rates for forest products, the systematic training of both

higher and lower forest officers, and the establishment of adequately equipped forest experiment stations. The carrying out of this program, the cost of which during the first 10 years is estimated at 84,162,000 francs, is to be entrusted to an independent forest commission consisting of three salaried and three non-salaried members, and having attached to it three subcommissioners, ten or eleven divisional officers, and fifty or fifty-five forest officers. From the French point of view the most characteristic feature of the program is the fact that particular care is taken to prevent the state, in spite of the important part played by it, from encroaching on the rights of private owners, and to encourage, rather than to force, coöperation on the part of the latter.—*S. T. Dana.*

198. ILLICK, J. S. **When trees grow.** Canadian Forest. Jour. 15: 351–354. 1919.—A series of studies carried out for several years involving daily measurements on 200 trees during the growing season lead to conclusions that: (a) Trees grow almost twice as fast during the night as during the day; (b) The growing season for white pine and Norway spruce, in Pennsylvania at least, is ended by July 1st; (c) Such knowledge is of high utility in choosing season for planting trees.—*H. C. Belyea.*

199. IWAKI, TAKANORI. **Microscopical distinctions of some Japanese coniferous woods.** [Article in Japanese.] Bot. Mag. Tôkyô 32: 187–198, 219–237. 1918.—See Bot. Absts. 4, Entry 1299.

200. JAUFFRAT, AIMÉ. **La détermination des bois de deux Dalbergia de Madagascar, d'après les caractères de leurs matières colorantes.** [Identification of wood of Dalbergia by staining reactions.] Compt. Rend. Acad. Sci. Paris 168: 693–694. 1919.—See Bot. Absts. 5, Entry 565.

201. JOLLY, N. W. **The importance of the wood pulp industry to Australian forests.** Australian Forest. Jour. 2: 9. 1919.—The possibility of Australia manufacturing wood pulp from its own forests is discussed. The author advocates the utilisation of hardwood and *Pinus insignis* saplings and poles for wood pulp as a means of utilizing waste or of rendering thinnings profitable.—*C. F. Korstian.*

202. JONES, J. **Shea butter tree.** Imperial Department of Agriculture for the West Indies. Report on the Agricultural Department, Dominica, 1918–19: 3. 1919.—Nuts from Dominica examined at the Imperial Institute, London, were found to contain 44 per cent. of fat, a somewhat lower percentage than that contained in West African nuts.—*J. S. Dash.*

203. KHAN, A. HAFIZ. **Red wood of Himalayan spruce (Picea morinda).** Indian Forester 45: 496–498. *1 pl.* 1919.—The water absorptive capacity of the red wood which occurs in the heart of *Picea morinda* is less than that of the white wood, while it is at the same time heavier, volume for volume, than white wood. Both colored woods are lighter than water. —*E. N. Munns.*

204. KOEHLER, A. **Selecting wood for airplanes.** Sci. Amer. Supplem. 88: 148–149. *5 fig.* 1919.

205. LANTES, ADELAIDE. **El alamo.** [The pipal tree.] Revist. Agric. Com. y Trab. 2: 612–613. *3 fig.* 1919.

206. LA TOUCHE, T. H. D. **The submerged forest at Bombay.** Rec. Geol. Surv. India 49: 214–219. *Pl. 17–19.* 1919.—During excavations in Bombay harbor in 1878 a submerged forest with many stumps in situ was found over an area of 30 acres. The trees were embedded in stiff blue clay 6 to 20 feet thick, resting on decomposed basaltic rock, and covered with 4 to 5 feet of harbor silt. The deepest stumps were rooted 33 feet below the present mean high tide. Most of the wood was identified as *Acacia catechu*, but two apparently drift logs were teak (*Tectona grandis*). In 1910 excavations on an adjacent area disclosed more stumps,

some rooted 40 feet below high tide. The conclusion is that there has been gradual depression of a forested rocky coastal plain, forming quiet lagoons in which the trees became embedded in the clay; then a tilting movement brought in the open sea, and *Teredo* bored the trunks, causing them to break off at the clay surface.—*Winfield Dudgeon.*

207. Lindberg, Ferd. **Då skogen snoar in. [When the forest is snowed in.]** Skogen 6: 128–132. *4 fig.* 1919.

208. Maas, J. G. J. A. **Gewijzigde methode voor veldproeven met Hevea. [Other methods for field experimentation with Hevea.]** Arch. Rubbercult. Nederlandsch-Indië 3: 233–237. 1919.—In this article the author sets forth a plan for the elimination of error due to the personal factor of the tapper in field experiments with *Hevea*. His plan is to have the tapping rows and collecting rows perpendicular to each other, so that each tapper taps a part of the trees of each collecting task.—*W. E. Cake.*

209. Maas, J. G. J. A. **Nog eenige kiemproeven met Hevea-zaden. [Some more germination trials with Hevea seed.]** Arch. Rubbercult. Nederlandsch-Indië 3: 237–243. 1919.—In preserving *Hevea* seed the packing material must be moist and not air tight. When *Hevea* seeds are to be preserved for longer than one month the packing material should be moistened every 3 or 4 weeks. At a temperature of 4 to 8°C. the seeds will stand a drier and more air-tight package better than at ordinary temperatures. Air-tight packages however cause them to lose their germinating power quickly. Merely ensilaging *Hevea* seed in the ground seems to be good for preserving the seeds on an estate for a short period like a month. Treatment with water at about 50°C. resulted in increased germination energy, and a slightly improved germination. Sprinkling with warm water at 45°C. increased the rapidity of germination a little but had practically no effect on the germination per cent.—*W. E. Cake.*

210. Mackay, H. **Conifers in Victoria.** Australian Forest. Jour. 2: 265–267. 1919.—Summary of a paper on "Coniferous plantations in Southeastern Australia," read before the first Inter-State Conference on Forestry, embodying the experience of that State in the establishment of exotic conifers over a period of 34 years. Thirteen conifers indigenous to North America are found in the list.—*C. F. Korstian.*

211. Mackay, H. **Treatment of indigenous hardwoods.** Australian Forest. Jour. 2: 19–20. 1919.—Extract from a paper read before the first Interstate Conference on Forestry at Sydney, November, 1911, in which the silvicultural management of eucalyptus forests is briefly discussed. Wherever the standing crop is fairly uniform in age and size, a clear cutting in sections, leaving, in addition to seed trees, only trees fit for piles and girders, is advocated.—*C. F. Korstian.*

212. Madelin, J. **Les cèdres du Liban. [The cedars of Lebanon.]** Rev. Eaux et Forêts 57: 275–276. 1919.—The cedars of Lebanon, formerly regarded by the natives as divine beings in tree form, flourish only at El-Herzé at an altitude of over 2200 meters. Some of them are over a hundred feet high and the largest is 3 feet in diameter. The few trees which still survive have suffered severely at the hands of tourists and should be protected from further damage.—*S. T. Dana.*

213. Main, J. M. **Eden and its timber resources.** Australian Forest. Jour. 3: 48–49. 1920.—A note on the forest resources adjacent to the town of Eden on the South Coast of Australia with a list of the principal timber species of eucalyptus and their uses.—*C. F. Korstian.*

214. Martin, Percy F. **Great forests of South America.** Canadian Forest. Jour. 15: 264–266. 1919.—Four types of timber are recognized: small scrubby forests of dry temperate or sub-tropical regions; good forests of Antarctic beech and a few conifers of temperate regions in the Andes; the fresh and salt-water swamps of mangroves and species with soft woods; the tropical rain forest of a great variety of hardwoods.—*E. N. Munns.*

215. Massias, J. **Les forêts de Grèce.** [The forests of Greece.] Rev. Eaux et Forêts 57: 237-247. 1919.—Prior to 1913 the forest area of Greece, excluding areas once forested but now devastated, amounted to some 800,000 hectares, or about 12 per cent of the total area of the country. Including the new provinces added by the war, the total forest area is about 13 per cent. Approximately 50 per cent belongs to the State, 20 per cent to convents and communes, and 30 per cent to private owners. Aleppo pine constitutes 35 per cent of the stands, Cephalonian fir 25 per cent, and various oaks 20 per cent. The value of the forest products harvested annually, including timber, fuel, charcoal, resin, forage, and other minor products, amounts to about 3,300,000 francs, of which nearly one-half is fuel.—All forests, both public and private, are theoretically subject to a forest regime in the department of Agriculture, but lack of personnel makes this control ineffective. Even in the State forests there are no real plans of management. These, as well as certain private forests, are heavily burdened with various rights of use which have resulted in serious damage, particularly through the unrestricted grazing of sheep and goats. The forests themselves are not subject to a land tax, but forest products (with certain exceptions, the most important of which is fuel harvested by the peasants for their own use) are taxed at varying rates according to the nature of the product and the character of the ownership. Recent laws aim to secure better fire protection, the reforestation of denuded lands, the codification and revision of existing rights of user, and improved management of all forest lands, both public and private. There are two schools for the training of guards and rangers and one (at Athens) for the training of higher forest officers.—*S. T. Dana.*

216. Mattoon, Wilbur R. **Making woodlands profitable in the Southern States.** U. S. Dept. Agric. Farmers Bull. 1071. *38 p. 55 fig.* 1920.

217. Mattoon, Wilbur R. **Treating fence posts on farm.** Louisiana State Univ. Div. Agric. Exp. Circ. 37. *20 p. 11 fig.* 1920.—Fence posts treated with creosote and set in the ground at Calhoun, Louisiana, in 1908 were examined after 10 years. Of the black gum posts, 97 per cent were sound; cypress, 96 per cent; tupelo gum, 88 per cent; sweet gum, 87 per cent; sap pine, 73 per cent; bay, 68 per cent. Methods of treating posts are also discussed. —*C. W. Edgerton.*

218. Miller, Robert B. **The wood of Machaerium Whitfordii.** Bull. Torrey Bot. Club 47: 73-79. *8 fig.* 1920.—A study is made of the wood of *Machaerium Whitfordii* Macbride, which came from Colombia. Color, density and other gross characters are given; it is related to the true rosewoods and is of commercial importance. It is diffuse porous, usually has uniseriate rays, storied arrangement of elements, small half-bordered pits between vessels and ray cells, and sieve-like perforations of pit membrane. Wood parenchyma is diffuse, paratracheal, and on the face of the summer wood.—*P. A. Munz.*

219. Morrison, W. G. **Natural afforestation in a New Zealand mountain area.** Australian Forest. Jour. 2: 380-384. 1919.—The first installment of a discussion treating the merits of natural regeneration by seed with particular reference to the indigenous forests of the Hanmer area. It is contended that natural regeneration ought to be accomplished at less than one-tenth the cost of relatively cheap planting methods. [See also next following Entry, 220.]—*C. F. Korstian.*

220. Morrison, W. G. **Natural afforestation in a New Zealand mountain area.** Australian Forest. Jour. 3: 23-25. 55-58. 1920.—A continuation and final installment of an article, the first part of which has been abstracted. The spontaneous reproduction of exotic shelter plantations on the Hanmer Plains is described. *Pinus radiata*, *P. pinaster*, *Betula alba*, *Quercus pedunculata* and *Larix europea* were found reproducing themselves from seed at rates varying from several hundred to tens of thousands per acre depending on the species, the distance from seed trees and site conditions. The mean annual rainfall for the years 1905 to 1918 is approximately 48 inches, which is well above the safety limit for successful planta-

tions. The author cites evidence to show that natural afforestation of the high country is feasible but suggests that on the more accessible waste areas it be augmented by artificial afforestation as now practiced. [See also next preceding Entry, 219.]—*C. F. Korstian.*

221. Nordstedt, C. T. O. [Swedish rev. of: Heribert-Nilsson, N. **Experimentelle Studien uber Variabilität, Spaltung, Artbildung und Evolution in der Gattung Salix. [Experimental studies on variability, segregation, speciation and evolution in the genus Salix.]** Lunds Universitets Arsskr. N. F. (Avd. 2.) 14[28]: 1–145. *65 fig.* 1918.] Bot. Notiser 1919: 39–40. 1919.

222. Pearson, R. S. **Note on the mechanical strength and seasoning properties of Shorea robusta timber.** Indian For. Rec. 7: 120–145. 1919.—The results of tests on sal for transverse strain, compression, shearing and hardness are given in detail on timber felled at different times of the year, from different localities, and from trees of different origin. Data is also presented on the rate of seasoning of woods obtained under the same conditions as those described above.—*E. N. Munns.*

223. Petch, T. **The effect of time intervals in rubber tapping.** Dept. Agric. Ceylon Bull. 42. *8 p.* 1919.

224. Pierre, L. **Note sur l'Isonandra Krantziana (arbre à Gutta-Percha de la Cochinchine et du Cambodge). [Note on Isonandra Krantziana, a gutta percha tree of Cochinchina and Cambodia.]** Bull. Agric. Inst. Sci. Saigon 2: 33–40. 1920.—A report on the economic possibilities of the above species, this one probably being the form described by Pierre as *Dichopsis Krantziana.*—*E. D. Merrill.*

225. Raux, Marcel. **Une devise de politique forestière. [A motto of forest policy.]** Rev. Eaux et Forêts 57: 248–254. 261–274. 1919.—A comprehensive forest policy should include both a far-sighted administrative program and legislation necessary to make this program effective. The essence of such a policy can be expressed by the simple motto, "To create and to conserve." The State should take the lead in creating, not by the purchase of private lands already forested, but by the acquisition and reforestation, chiefly with native conifers, of lands now uncultivated or abandoned. These plantations, scattered throughout the country, would not only prove profitable financially, but would prove more effective in stimulating similar work on the part of other owners than any amount of literary propaganda. Reforestation by communities should be further encouraged by State loans, and the resulting plantations should be subject to the forest regime. Private owners and forestry societies should be given free advice and other assistance by the State, and plantations established by them should be granted liberal exemptions from taxation until they reach a certain height.—The conservation of privately owned forests, which constitute more than two-thirds of the forest area of France, is a matter of very real public concern and should therefore be undertaken by the State. Supervision of cuttings in such forests should be exercised by the State, without charge to the owner; while clear cuttings in protection forests should be prohibited, and in other forests should be followed by reforestation. As to clearings, legislation should be enacted providing that the forest area of France must not be diminished; prohibiting the clearing of all stands in the zone of protection forests; and requiring a permit from the Minister of Agriculture for the clearing of all stands outside of this zone. These measures would require an increased forest personnel, which could be secured in part by relieving forest officers of their duties as fish wardens. Supervision of private cuttings should also be facilitated by commissioning private forest guards as forest officers. Finally, conservation should be promoted by giving forest owners, both public and private, more adequate protection against trespass by increased penalties.—*S. T. Dana.*

226. Reynard, J. **Les arbres de la paix. [Trees of peace.]** Bull. Trimest. Soc. Forest Franche-Comté et Belfort 13: 111–112. 1919.—Trees should be widely planted as the simplest and most practical means of commemorating the peace treaty of Versailles. Better than anything else they serve to bind father to son, dead to living, generation to generation.—*S. T. Dana.*

227. Romell, Lars-Gunnar. **Sammanväxning och naturympning. [Growing together and natural grafting.]** Skogen 6: 133–141. *4 fig.* 1919.

228. Rumbold, Caroline. **The injection of chemicals into chestnut trees.** Amer. Jour. Bot. 7: 1–20. *7 fig.* 1920.—See Bot. Absts. 5, Entry 964.

229. Scheidter, Franz. **Das Tannensterben im Frankenwalde. [Death of firs in the Frankenwald.]** Naturw. Zeitschr. Forst- u. Landw. 17: 69–90. 1919.—The dying of firs in the State-owned Frankenwald, and also to a lesser extent in other middle-European forests, which has become gradually and only in recent years of alarming extent, is described in great detail. After dissertating upon various theories which have been advanced by other investigators, especially Neger, the writer states it as his own opinion that insects and fungi (the Hallimasch most commonly), are only secondary causes, and that the fundamental difficulty arises from the improper silvicultural system followed in the State forests. In these the effort seems always to have been to grow fir, and spruce-fir mixtures, in even-aged stands, whereas privately-owned forests, under similar conditions, are usually handled as all-aged or selection forests, a plan which is better adapted to fir. The opinion is advanced, and is backed by much evidence, that the rapid loss of fir in the Frankenwald is due primarily to crowding when the even-aged stands attain a certain age or density, being particularly marked where fir must compete with the broader-crowned spruce. In any event, in such stands, the lower limbs are lost very rapidly, and in the opinion of the writer, the small crown remaining at the top of the tree is then unable to draw to itself sufficient moisture for existence. The older needles die, then the growing tip succumbs, and death of the entire tree soon follows. Often, before death occurs, there is a vigorous production of "water-sprouts" on the lower portion of the stem. The evil is augmented by drought years, and by snow-damage and windfall which, by opening the canopy, apparently encourage the production of these "water-sprouts" and also cause drying of the soil, the growth of grass, etc. A horde of insects, and some of the most destructive fungi, attack the weakened trees, and of course hasten death and contribute to the aggregate losses. The suggested remedy is a system of management which will give the fir more ample space for its late development and maturing. This the selection system would appear to do.—*C. G. Bates.*

230. Schotte, Gunnar. **Meddelanden från Svenska Skogsvårdsforeningen.—Protokoll; fört vid Svenska Skogsvårdsföreningens årsmöte i Stockholm den 14 mars, 1919. [Proceedings at the annual meeting of the Swedish forestry association, Stockholm, March 14, 1919.]** Skogen 6: 217–224. 1919.

231. Secrest, Edmund. **Salient features of a forestry policy for Ohio.** Monthly Bull. Ohio Agric. Exp. Sta. 5: 15–19. 1920.—The depletion of forests cannot be permitted longer to escape public attention. Private ownership has failed to provide for renewal of forests after cutting. The effect of such a policy is very marked in small communities where certain phases of the lumbering industry have been the chief source of income. A state forestry policy is proposed whereby non-agricultural or idle lands may be purchased for reforestation purposes. Ohio has 500,000 acres of such land which should come under public ownership, or state or municipal custody. To encourage private owners to reforest waste lands the state should establish nurseries where planting stock could be obtained at the cost of production. —*R. C. Thomas.*

232. Show, S. B. **Climate and forest fires in northern California.** Jour. Forestry 17: 965–979. 1919.—Relationships existing between fire and climate have long been recognised by foresters but not before studied intensively. The moisture content of the forest litter is a prime consideration as to both ignition and rate of spread of fire. Litter dries out exceedingly fast under summer conditions and when it contains 8 per cent or less moisture, burns readily. Over this amount fire will not spread. Litter moisture is affected by climatic conditions, being driest on south slopes and the most moist on north slopes and at high elevations. Litter behaves like soil as regards hydroscopic moisture, taking up as much as 6 per cent of

its own weight.—The rate of spread of fires is best measured by perimeter rather than by area or distance, and is governed largely by wind velocity. This speed varies as the square of the wind velocity.—*E. N. Munns.*

233. Shull, C. A. **Curing timber.** [Rev. of: Stone, Herbert. **The ascent of the sap and the drying of timber.** Quart. Jour. Forest. 12: 261–266. 1918.] Bot. Gaz. 68: 310. 1919.—The author's suggestion may be sound on the practical side, but his "assumptions as to the movement of sap in trees will not meet with favor among plant physiologists. It is hard to imagine a conception more at variance with experimental results of physiological studies."

234. Sim, T. R. **South African rubber. I.** South African Jour. Indust. 2: 1127–1137. *5 pl.* 1919.

235. Sim, T. R. **South African rubber. II.** South African Jour. Indust. 3: 24–34. 1920.

236. Society of American Foresters, Committee for the Application of Forestry. **Forest devastation: a national danger and a plan to meet it.** Jour. Forest. 17: 911–945. 1919.—A detailed and comprehensive program of action is outlined. Blame is placed on the lumber industry and economic development for the state of affairs at present. To correct the evils which now exist, plans for constructive legislation are offered including the purchase and control of forest lands and production, the establishment of forest insurance agencies and forest loan banks, and state coöperation in securing tax and fire-prevention reforms. A minority report of the committee is also presented.—*E. N. Munns.*

237. Starte, H. W. **Reservation of standards in strips and checks in exploitation.** Indian Forester 45: 414–416. *1 fig.* 1919.—A system of parallel strips in cutting in coppice with standards has been worked out to prevent the tendency towards overcutting, and frauds by operators.—*E. N. Munns.*

✓ 238. Stevens, J. L. **Blackboy and its commercial uses.** Australian Forest Jour. 2: 201–202. 1919.—The outside portions of the blackboy or grass tree are reported to yield very fine drying oils and turpentine substitutes suitable for the manufacture of paints and varnishes. The acidic liquors obtained in the distillation process contain large quantities of acetic acid, methyl alcohol and tannin extract, while the gas is of high calorific value and purity, being free from sulphur and nitrogen compounds.—*C. F. Korstian.*

239. Taylor, A. A. **California's redwood park.** Amer. Forestry 25: 1446–1450. *4 fig.* 1919.

240. Tiemann, H. D. **Kiln-drying specifications for airplane lumber.** Sci. Amer. Suplem. 88: 104. *3 fig.* 1919.

241. Trägårdh, Ivar. **Några allmänna men hittills föga uppmärksammade barkborrar och deras gångsystem.** [Some common but hitherto little known bark beetles and their galleries.] Skogen 6: 237–246. *Pl. 1–7.* 1919.

242. Vernet, G. **Précautions à prendre dans l'enfumage du caoutchouc (Incendies-stickage).** [Precautions to be taken in smoking rubber.] Bull. Agric. Inst. Sci. Saigon 1: 362–364. 1919.

243. von Fankhauser, F. **Zur Kenntnis der Lärche.** [A larch study.] Schweiz. Zeitschr. Forstwesen 70: 188–194. *3 fig.* 1919.—The natural range of the species is taken as the area over which natural reproduction occurs, although good growth may be secured in other regions by artificial reproduction. Soil moisture is emphasized as the principal factor that limits the distribution of larch. Other writers have attributed depth and character of soil as important limiting factors, but the occurrence of larch on all types of soil and its distribution, limited only by elevation and exposure, are taken as conclusive evidence that soil texture

and depth are important only in so far as these qualities affect soil moisture. Variations of the root systems and the development of deep tap roots are influenced more by depth of water table than by character of soil. Transpiration is also an important factor. DR. F. VON HÖNEL's experiments, which he conducted in 1879 with 21 species, showed that the amount of water transpired to produce 100 grams dry weight of leaves in various species was as follows: Larch, 115 L., Ash, 98 L., Beech, 86 L., Birch, 85 L., Spruce, 21 L., Pine, 10 L. The service-berry was the only species that transpired more than the larch. KIRCHNER describes the anatomy of the larch needle as being especially adapted for aeration by the arrangement of the cells length-wise in the needles, and the cell walls joined only at the corners. Air spaces about the size of the cells occur between each two layers of cells. The thin cuticle of the needle is also a factor. Excessive transpiration indicates the necessity of an abundant supply of water. The shedding of leaves in the winter is a habit necessitated by the excessive transpiration. In periods of severe drought the needles turn yellow, and part of them may fall to conserve moisture. The tree, however, recovers readily and new leaves develop, whereas other conifers die. Specific cases were noted during the severe drought of 1911. The dense parabolic crowns formed on good moist soil and the open neiloid crowns formed on drier sites are so different that a division of species based on this character has been advocated. Competition of larch with other species is largely controlled by the supply of available water. The fir and the spruce spread their lateral roots near the surface and, to a large extent, prevent surface water from reaching the deeper soil in which the larch roots usually occur. Where the larch successfully competes with other species it is due to sub-irrigation of the area with water from other areas.—*J. V. Hofmann.*

244. VON KUNZ, I. **Zwanzigjährige forstliche Betätigung eines Laien. [Twenty years' forestry experience of a layman.]** Schweiz. Zeitschr. Forstwesen **70**: 195–200. 1919.—The author is a chemist whose interest in forestry prompted him to purchase a forest meadow of two hectares and plant it to tree seedlings. Spruce, fir, pine, larch, beech, oak, hornbeam and elm were used. The plantation was very successful, and at the age of twenty years the conifers formed a complete ground cover where they were spaced 1.25 m. by 1.25 m. The pines had begun to clear, but the spruce branches were still all green.—*J. V. Hofmann.*

245. VON SEELEN, D. **Der Wald als Bruder des Feldes. [The interdependence of forest and farm.]** Zeitschr. Forst- u. Jagdw. **51**: 308–315. 1919.—A plea for more thorough use of German forest resources. A policy is outlined to accomplish this end. The war, and its results, has made it necessary for Germany to adopt a broader policy of forest management. The former rather restrictive policy resulted in much waste of such natural resources as forage and nut crops within the forests, owing to the fact that grazing animals were apt to cause damage to reproduction. The author argues, however, that through proper regulation such damage can be minimized. Free use and administrative use policies are also outlined. Article, on whole, is an answer to an opponent to this broader concept of a forest policy.—*Hermann Krauch.*

246. WAHLGREN, A. **Skogen och människan i förhistorisk tid. [The forest and man in prehistoric times.]** Skogen **6**: 1–8, 65–68, 229–236. 1919.

247. WALKER, R. S. **The Paulownia tomentosa tree.** Amer. Forest. **25**: 1485–1486. *3 fig.* 1919.

248. WATT, A. S. **On the causes of failure of natural regeneration in British oakwoods.** Jour. Ecol. **7**: 173–203. 1919.

249. WEIR, JAMES E., AND ERNEST E. HUBERT. **The influence of thinning on western hemlock and grand fir infected with Echinodontium tinctorium.** Jour. Forest. **17**: 21–35. 1919.—See Bot. Absts. 3, Entry 574.

250. Welo, L. A. **Emergency seasoning of Sitka spruce.** Sci. Amer. Supplem. 87: 404-405. *2 fig.* 1919.

251. Wood, B. R. **Note on proposed system for regeneration of sal forests.** Indian Forester 45: 403-413. 1919.—Changes in the management of sal forests are not believed essential and strip cutting is not feasible. Suggestions are made to study the growth and the relation of forest and fire to the regeneration of sal.—*E. N. Munns.*

252. Zimmer, Walter J. **Regeneration of forests.** Australian Forest. Jour. 2: 75-76. 1919.—A brief discussion of the suitability of the coppice method of regeneration to the eucalyptus forests of Australia, which sucker very freely.—*C. F. Korstian.*

GENETICS

George H. Shull, *Editor*

James P. Kelly, *Assistant Editor*

253. Abidin, J. **Pferdezucht und Pferderassen im osmanischen Reich. [Horse breeding and horse breeds in the Osmanian country.]** Flugschr. Deutsch. Ges. Züchtungsk. 1918: 31. *47 fig.* 1918.

254. Åkerman, Å. **Växternas kölddöd och frosthärdighet. Föredrag vid Sveriges Utsädesforenings extra möte under Landtbruksveckan 1919. [Winter killing and frost-resistance of plants. A paper read at a special meeting of the Swedish Seed-Grain Association during the "Farmers Week," 1919.]** Sveriges Utsädesförenings Tidskrift 29: 61-85. *4 fig.* 1919.—Detailed exposition of different theories to explain killing of plants by cooling. According to experiments of Lidfors and others on the importance of sugar in protecting plants against cold, it is supposable that hereditary differences in frost-resistance in different kinds of plants might possibly depend on hereditary differences in sugar content. Author also has been able to show that for wheat a parallelism seems to exist between sugar content and hardiness against cold, in such way that plants which are more resistant to frost contain more sugar than plants less resistant to frost.—In the following table four kinds of wheat are arranged in order of their resistance against cold, beginning with the least resistant:

VARIETY	DRY SUBSTANCE IN PER CENT OF FRESH WEIGHT	SUGAR IN PER CENT OF FRESH WEIGHT
Smaavete II	23.2	13.3
Solvete	23.8	14.8
Thulevete	24.7	17.1
Lantvete	26.0	19.6

The quantity of sugar varies much during different periods; but the sugar-curves are rather nearly parallel for the different sorts of wheat.—*K. V. Ossian Dahlgren.*

255. Allendorf and Ehrenberg. **Die Aufgaben des Sonderausschusses für Zuckerrübenbau. [Special problems of sugar-beet breeding.]** Mitt. Deutsch. Landw. Ges. 1919: 531-534. 1919.—See Bot. Absts. 5, Entry 259.

256. Amend F. **Untersuchungen über flämischen Roggen unter besonderer Berücksichtigung des veredelten flämischen Landroggens und seiner Züchtung. [Investigations on Flemish rye with special reference to improved varieties and their breeding.]** Landw. Jahrbüch. 52: 614-669. 1919.—See Bot. Absts. 5, Entry 260.

257. Anonymous. The improvement of agricultural crops by selection and hybridization. Scot. Jour. Agric. 2: 10–20. 1919.—Substance of address delivered to Glasgow and West Scotland Agricultural Discussion Society by T. Anderson, Director of the Board's Seed Testing Station. Mass selection, pure line selection, hybridisation, and Mendelism in relation to crop improvement are discussed. Emphasis is placed on value of pure seed stocks to the farmer.—*R. J. Garber.*

258. Anonymous. Report of the work of the plant breeding division for 1919. Jour. Dept. Agric. Ireland 20: 102–107. 1920.

259. Anonymous. [German rev. of: Allendorf and Ehrenberg. Die Aufgaben des Sonderausschusses für Zuckerrübenbau. (Special problems of sugar-beet breeding.) Mitt. Deutsch. Landw. Ges. 1919: 531–534. 1919.] Zeitschr. Pflanzenzücht. 7: 112. Dec., 1919.

260. Anonymous. [German rev. of: Amend, F. Untersuchungen über flämischen Roggen unter besonderer Berücksichtigung des veredelten flämischen Landroggens und seiner Züchtung. (Investigations on Flemish rye with special reference to improved varieties and their breeding.) Landw. Jahrbüch. 52: 614–669. 1919.] Zeitschr. Pflanzenzücht. 7: 112. Dec., 1919.

261. Anonymous. [German rev. of: Barker, E. Heredity studies in the morning-glory (Ipomoea purpurea). New York Cornell Agric. Exp. Sta. Bull. 392. *39 p., 3 pl.* 1917. (See Bot. Absts. 1, Entry 1164.)] Zeitschr. Pflanzenzücht. 7: 113. Dec., 1919.

262. Anonymous. [German rev. of: Baur, Erwin. Über Selbststerilität und über Kreuzungsversuche einer selbstfertilen und einer selbststerilen Art in der Gattung Antirrhinum. (On self-sterility and crossing experiments with a self-fertile and self-sterile species in the genus Antirrhinum.] Zeitschr. indukt. Abstamm. Vererb. 21: 48–52. May, 1919. (See Bot. Absts. 3, Entry 2082.)] Zeitschr. Pflanzenzücht. 7: 114. Dec., 1919.

263. Anonymous. [German rev. of: Becking, L. G. M. Baas. Over Limietverhoudingen in Mendelsche populaties. (Limiting proportions in Mendelian populations.) Genetica 1: 443–456. *4 fig.* Sept. 1919. (See Bot. Absts. 3, Entry 2086.)] Zeitschr. Pflanzenzücht. 7: 113. Dec., 1919.

264. Anonymous. [German rev. of: Emerson, R. A. A fifth pair of factors, Aa, for aleurone color in maize, and its relation to the Cc and Rr pairs. Cornell Univ. Agric. Exp. Sta. Mem. 16: 231-289. *Fig. 71.* Nov., 1918. (See Bot. Absts. 1, Entry 877.)] Zeitschr. Pflanzenzücht. 7: 115. Dec., 1919.

265. Anonymous. [German rev. of: Fraser, Allan Cameron. The inheritance of the weak awn in certain Avena crosses and its relation to other characters of the oat grain. Cornell Univ. Agric. Exp. Sta. Mem. 23: 635–676. June, 1919.] Zeitschr. Pflanzenzücht. 7: 116–117. Dec., 1919.—See also Bot. Absts. 5, Entry 292.

266. Anonymous. [German rev. of: Freeman, G. F. Linked quantitative characters in wheat crosses. Amer. Nat. 51: 683–689. 1917.] Zeitschr. Pflanzenzücht. 7: 116. Dec., 1919.

267. Anonymous. [German rev. of: Frölich, G. Die Umzüchtung von Wintergetreide in Sommergetreide. (The breeding of winter cereals into spring cereals.) Friedrichswerther Monatsber. 9: 27–30. 1919.] Zeitschr. Pflanzenzücht. 7: 118. Dec., 1919.—See also Bot. Absts. 5, Entry 284.

268. Anonymous. [German rev. of: Frölich, G. Die Beeinflussung der Kornschwere durch Auslese bei der Züchtung der Ackerbohne. (The influencing of grain-weight by selection in the breeding of field beans.) Friedrichswerther Monatsber. 9: 7–8, 17–20. 1919.] Zeitschr. Pflanzenzücht. 7: 117–118. Dec., 1919.

269. Anonymous. [German rev. of: Fruwirth, C. **Die gegenwärtige Organisation der Pflanzenzüchtung in Deutschland und in Österreich-Ungarn. (The present organization of plant breeding in Germany and Austria.)** Nachricht. Deutsch. Landw. Ges. Österreich. 1919: 35-39. 1919.] Zeitschr. Pflanzenzücht. 7: 118. Dec., 1919.

270. Anonymous. [German rev. of: Fruwirth, C., Dr. Th. Roemer, Dr. E. von Tschermak. **Handbuch der landwirtschaftlichen Pflanzenzüchtung. 4. Die Züchtung der vier Hauptgetreidearten und der Zuckerrübe. (Handbook of agricultural plant breeding. 4. Breeding of the four chief cereals and the sugar beet).** *8vo., xv+504 p., 42 fig.* Paul Parey: Berlin, 1918.] Zeitschr. Pflanzenzücht. 7: 145. Dec., 1919.

271. Anonymous. [German rev. of: Gassner, S. **Beiträge zur physiologischen Charakteristik sommer- und winterannueller Gewächse, insbesondere der Getreidepflanzen. (Contribution to the physiological characteristics of summer and winter annuals with special reference to the cereals.)** Zeitschr. Bot. 10: 417-480. *7 pl., 2 fig.* 1918.] Zeitschr. Pflanzenzücht. 7: 118-120. Dec., 1919.

272. Anonymous. [German rev. of: Hansen, W. **Einiges über Rübenzucht. (Something about beet-breeding.)** Landw. Zeitung 1919.] Zeitschr. Pflanzenzücht. 7: 120. Dec., 1919.

273. Anonymous. [German rev. of: Jones, D. F. **Natural cross-pollination in the tomato.** Science 43: 509-510. 1916.] Zeitschr. Pflanzenzücht. 7: 120. Dec., 1919.

274. Anonymous. [German rev. of: Jones, D. F. **Linkage in Lycopersicum.** Amer. Nat. 51: 608-621. 1917.] Zeitschr. Pflanzenzücht. 7: 120-121. Dec., 1919.

275. Anonymous. [German rev. of: Jones, D. F. **Dominance of linked factors as a means of accounting for heterosis.** Genetics 2: 466-479. *1 fig.* 1917. See Bot. Absts. 1, Entry 1245.] Zeitschr. Pflanzenzücht. 7: 121. Dec., 1919.

276. Anonymous. [German rev. of: Jones, D. F. **The effect of inbreeding and crossbreeding upon development.** Connecticut Agric. Exp. Sta. Bull. 207. *100 p., 12 pl.* New Haven, 1918. (See Bot. Absts. 2, Entry 34; 3, Entry 988.)] Zeitschr. Pflanzenzücht. 7: 122. Dec., 1919.

277. Anonymous. [German rev. of: Jones, Donald F. **Bearing of heterosis upon double fertilization.** Bot. Gaz. 65: 324-333. April, 1918. (See Bot. Absts. 1, Entry 228.)] Zeitschr. Pflanzenzücht. 7: 121-122. Dec., 1919.

278. Anonymous. [German rev. of: Kajanus, Birger. **Genetische Papaver-Notizen. (Genetical notes on Papaver.)** Bot. Notiser 1919: 99-102. 1919. (See Bot. Absts. 3, Entry 2145.)] Zeitschr. Pflanzenzücht. 7: 123. Dec., 1919.

279. Anonymous. [German rev. of: Kajanus, B. **Genetische Studien über die Blüten von Papaver somniferum L. (Genetical studies on the flowers of Papaver somniferum L.)** Arkiv Bot. K. Svensk. Vetenskapsakad. 15: 1-87. *3 pl.* 1919. (See Bot. Absts. 3, Entry 2147.)] Zeitschr. Pflanzenzücht. 7: 123-125. Dec., 1919.

280. Anonymous. [German rev. of: Kajanus, Birger. **Über eine konstant gelbbunte Pisum-Rasse. (On a constantly yellow-variegated variety of Pisum.)** Bot. Notiser 1918: 83-84. 1918. (See Bot. Absts. 3, Entry 2146.)] Zeitschr. Pflanzenzücht. 7: 125. Dec., 1919.

281. Anonymous. [German rev. of: Kajanus, B., and S. O. Berg. **Pisum-Kreuzungen. (Pea-crosses.)** Arkiv Bot. K. Svensk. Vetenskapsakad. 15: 1-18. 1919. (See Bot. Absts. 3, Entry 2148.)] Zeitschr. Pflanzenzücht. 7: 125-126. Dec., 1919.

282. Anonymous. [German rev. of: Kalt, B., and A. Schulz. **Über Rückschlags individuen mit Spelzweizeneigenschaften bei Nacktweizen der Emmerreihe des Weizens. (Concerning reversionary individuals with characters of the Spelt type in the naked wheat of the Emmer series.)** Ber. Deutsch. Bot. Ges. 36: 669–671. 1918. (See Bot. Absts. 4, Entry 624.)] Zeitschr. Pflanzenzücht. 7: 126. Dec., 1919.

283. Anonymous. [German rev. of: Kiessling, L. **Die Leistung der Wintergerste und deren züchterische Beeinflüssung. (The performance of winter barley and its modification by breeding.)** Illustr. Landw. Zeitg. 1919: 310–311. 1919.] Zeitschr. Pflanzenzücht. 7: 126. Dec., 1919.

284. Anonymous. [German rev. of: Killer, J. **Über die Umzüchtung reiner Linien von Winterweizen in Sommerweizen. (Concerning the changing over of pure lines of winter wheat into spring wheat.)** Jour. Landw. 67: 59–62. 1919.] Zeitschr. Pflanzenzücht. 7: 126. Dec., 1919.—See also Bot. Absts. 5, Entry 267.

285. Anonymous. [German rev. of: Küster, E. **Über Mosaikpanaschierung und vergleichbare Erscheinungen. (Mosaic variegation and comparable phenomena.)** Ber. Deutsch. Bot. Ges. 36: 54–61. 1918. (See Bot. Absts. 3, Entry 265.)] Zeitschr. Pflanzenzücht. 7: 126. Dec., 1919.

286. Anonymous. **Origin of maize.** [Rev, of: Kuwada, Y. **Die Chromosomenzahl von Zea Mays L. Ein Beitrag zur Hypothese der Individualität der Chromosomen und zur Frage über die Herkunft von Zea Mays L. (The chromosome number of Zea Mays L. A contribution to the hypothesis of the individuality of chromosomes and to the problem of the origin of Zea Mays L.)** Jour. Coll. Sci. Imperial Univ. Tôkyô 39: 1–148. *2 pl., 4 fig.* Aug., 1919. (See Bot. Absts. 4, Entry 643.)] Gard. Chron. 67: 114. Mar. 6, 1920.

287. Anonymous. [German rev. of: Lehmann, Ernst. **Über die Selbststerilität von Veronica syriaca. (On the self-sterility of Veronica syriaca.)** Zeitschr. indukt. Abstamm. Vererb. 21: 1–47. *1 fig.* May, 1919.] (See Bot. Absts. 3, Entry 2159.)] Zeitschr. Pflanzenzücht. 7: 127. Dec., 1919.

288. Anonymous. [German rev. of: Lindstrom, E. **Linkage in maize: aleurone and chlorophyll factors.** Amer. Nat. 51: 225–237. 1917.] Zeitschr. Pflanzenzücht. 7: 127. Dec., 1919.

289. Anonymous. [German rev. of: Lindstrom, E. W. **Chlorophyll inheritance in maize.** Cornell Univ. Agric. Exp. Sta. Mem. 13: 1–68. *5 colored pl.* Aug., 1918. (See Bot. Absts. 1, Entry 484.)] Zeitschr. Pflanzenzücht. 7: 127–129. Dec., 1919.

290. Anonymous. [German rev. of: Love, H. H., and W. T. Craig. **Methods used and results obtained in cereal investigations at the Cornell Station.** Jour. Amer. Soc. Agron. 10: 145–157. *1 pl., 1 fig.* April, 1918. (See Bot. Absts. 3, Entry 2163.)] Zeitschr. Pflanzenzücht. 7: 129–130. Dec., 1919.

291. Anonymous. [German rev. of: Love, H. H., and W. T. Craig. **The relation between color and other characters in certain Avena crosses.** Amer. Nat. 52: 369–383. Aug.–Sept., 1918. (See Bot. Absts. 1, Entry 914.)] Zeitschr. Pflanzenzücht. 7: 130–131. Dec., 1919.

292. Anonymous. [German rev. of: Love, H. H., and A. C. Fraser. **The inheritance of the weak awn in certain Avena crosses.** Amer. Nat. 51: 481–493. *2 fig.* 1917. (See Bot. Absts. 1, Entry 1263.)] Zeitschr. Pflanzenzücht. 7: 129. Dec., 1919.—See also Bot. Absts. 4, Entry 265.

293. Anonymous. [German rev. of: Love, H. H., and G. P. McRostie. **The inheritance of hullessness in oat hybrids.** Amer. Nat. 53: 5–32. *7 fig.* Jan.–Feb., 1919. (See Bot. Absts. 1, Entry 1264; 2, Entry 420.)] Zeitschr. Pflanzenzücht. 7: 131–132. Dec., 1919.

294. Anonymous. [German rev. of: Meunissier, A. **Expériences génétiques faites à Verrières. (Genetical experiments made at Verrière.)** Bull. Soc. Nation. Acclimat. France 1918: 1–31. 1918. (See Bot. Absts. 4, Entry 677.)] Zeitschr. Pflanzenzücht. 7: 132–134. Dec., 1919.

295. Anonymous. [German rev. of: Nilsson-Ehle, H. **Untersuchungen über Speltoidmutationen beim Weizen. (Experiments on speltoid mutations in wheat.)** Bot. Notiser 1917: 305–329. *1 fig.* 1917.] Zeitschr. Pflanzenzücht. 7: 134. Dec., 1919.

296. Anonymous. [German rev. of: Oberstein, O. **Über das Vorkommen echter Knospenvariationen bei pommerschen und anderen Kartoffelsorten. (Occurrence of true bud variation in Pommeranian and other varieties of potato.** Deutsch. Landw. Presse 1919: 560–561. *1 pl.* 1919.] Zeitschr. Pflanzenzücht. 7: 135. Dec., 1919.

297. Anonymous. [German rev. of: Rasmuson, Hans. **Zur Genetik der Blütenfarben von Tropaeolum majus. (On the genetics of the flower colors of Tropaeolum majus.)** Bot. Notiser 1918: 253–259. Nov., 1918. (See Bot. Absts. 3, Entry 2180.)] Zeitschr. Pflanzenzücht. 7: 135. Dec., 1919.

298. Anonymous. [German rev. of: Rasmuson, Hans. **Über eine Petunia-Kreuzung. (On a Petunia cross.)** Bot. Notiser 1918: 287–294. 1918. (See Bot. Absts. 3, Entry 2181.)] Zeitschr. Pflanzenzücht. 7: 135–136. Dec., 1919.

299. Anonymous. [German rev. of: Roemer, Th. **Über Lupinenzüchtung. (On lupine breeding.)** Deutsch. Landw. Presse 1919: 174–175. 1919.] Zeitschr. Pflanzenzücht. 7: 136. Dec., 1919.

300. Anonymous. [German rev. of: Schmidt, J. **Investigations on hops. X. On the aroma in plants raised by crossing.** Compt. Rend. Trav. Lab. Carlsberg 11: 330–332. 1917. (See Bot. Absts. 1, Entry 1290.)] Zeitschr. Pflanzenzücht. 7: 136. Dec., 1919.

301. Anonymous. [German rev. of: Schmidt, J. **Investigations on hops (Humulus lupulus). XI. Can different clones be characterized by the number of marginal teeth in the leaves?** Compt. Rend. Trav. Lab. Carlsberg 14: 1–23. *8 fig.* 1918. (See Bot. Absts. 3, Entry 2192.)] Zeitschr. Pflanzenzücht. 7: 136–137. Dec., 1919.

302. Anonymous. [German rev. of: Schmidt, Johs. **La valeur de l'individu a titre de générateur appréciée suivant la méthode du croisement diallèle. (Individual potency appraised by the method of diallel crossing.)** Compt. Rend. Trav. Lab. Carlsberg 14: 1–33. 1919.] Zeitschr. Pflanzenzücht. 7: 136. Dec., 1919.

303. Anonymous. [German rev. of: Schmidt, Johannes. **Der Zeugungswert des Individuums beurteilt nach dem Verfahren kreuzweiser Paarung. (Individual potency, based on experiences in cross-matings.)** *8vo., 40 p.* Gustav Fischer: Jena. 1919. (See Bot. Absts. 3, Entry 2190.)] Zeitschr. Pflanzenzücht. 7: 145–146. Dec., 1919.

304. Anonymous. [German rev. of: Siegel, W. **Das Recht des Gemüsezüchters. (The right of the vegetable breeder.)** *8vo.* Frick: Wien, 1919.] Zeitschr. Pflanzenzücht. 7: 146. Dec., 1919.

305. Anonymous. [German rev. of: Sirks, M. J. **Stérilité, auto-inconceptibilité et differentiation sexuelle physiologique. (Sterility, self-incompatibility and physiological differentiation of the sexes.)** Arch. Néerland. (Sci. Ser.) III, 1917: 205–234. 1917.] Zeitschr. Pflanzenzücht. 7: 137. Dec., 1919.

306. Anonymous. [German rev. of: Snell, K. **Farbenänderung der Kartoffelblüte und Saatenanerkennung. (Color changes of the potato blossom and the recognition of varieties.)** Der Kartoffelbau 1919: 1–3. 1919.] Zeitschr. Pflanzenzücht. 7: 137. Dec., 1919.

307. Anonymous. [German rev. of: Sommer, K. **Über Kartoffelzüchtung und vergleichende anbauversuche mit Neuzüchtungen auf der Domäne Ellischau. (Potato breeding and comparative cultural tests of new varieties on the Ellischau estate.)** Nachr. Deutsch. Landw. Ges. Österr. 1919: 190–193. 1919.] Zeitschr. Pflanzenzücht. 7: 138. Dec., 1919.

308. Anonymous. [German rev. of: Stahel, G. **Eerste verslag over de werkzamheden ten behoeve van de selectie van koffie en cacao. (First report on the effectiveness of selection in coffee and cacao.)** Dept. Landbouw in Suriname (Paramaribo) Bull. 36. *23 p.* 1919.] Zeitschr. Pflanzenzücht. 7: 138–139. Dec., 1919.

309. Anonymous [R.]. [German rev. of: (1) Stout, A. B. **Self- and cross-pollinations in Cichorium intybus with reference to sterility.** Mem. N. Y. Bot. Gard. 6: 333–454. 1916. (2) Idem. **Fertility in Cichorium intybus: The sporadic occurrence of self-fertile plants among the progeny of self-sterile plants.** Amer. Jour. Bot. 4: 375–395. *2 fig.* 1917. (3) Idem. **Fertility in Cichorium intybus: Self-compatibility and self-incompatibility among the offspring of self-fertile lines of descent.** Jour. Genetics 7: 71–103. Feb., 1918. (See Bot. Absts. 1, Entry 243.)] Zeitschr. Pflanzenzücht. 7: 139–140. Dec., 1919.

310. Anonymous. [German rev. of: Tammes, T. **Die Flachsblüte. (The flower of flax.)** Recueil Trav. Bot. Néerland. 15: 185–227. *22 fig.* 1918.] Zeitschr. Pflanzenzücht. 7: 140. Dec., 1919.

311. Anonymous. [German rev. of: Tjebbes, K., and H. N. Kooiman. **Erfelijkheidsonderzoekingen bij boonen. (Genetical experiments with beans.)** Genetica 1: 323–346. *1 colored pl.* 1919. (See Bot. Absts. 3, Entry 1041.)] Zeitschr. Pflanzenzücht. 7: 140–141. Dec., 1919.

312. Anonymous. [German rev. of: Urban, J. **Hochpolarisierende Rübe und ihre Nachkommenschaft. (High-polarizing beets and their progeny.)** Zeitschr. Zuckerindustr. Böhmen 42: 387–391. 1919.] Zeitschr. Pflanzenzücht. 7: 141–142. Dec., 1919.

313. Anonymous. [German rev. of: Volkart, A. **40. und 41. Jahresbericht. Schweizerische Samenuntersuchungs- und Versuchsanstalt in Oerlikon-Zürich. (40th and 41st Ann. Rept. Swiss seed control and experiment station in Oerlikon-Zürich.)** Land. Jahrb. Schweiz. 1919: 1–40. 1919.] Zeitschr. Pflanzenzücht. 7: 142. Dec., 1919.

314. Anonymous. [German rev. of: von Caron-Eldingen. **Physiologische Spaltungen ohne Mendelismus. (Physiological segregation without Mendelism.)** Deutsch. Landw. Presse 1919: 515–516. 1919.] Zeitschr. Pflanzenzücht. 7: 114–115. Dec., 1919.

315. Anonymous. [German rev. of: von Ubisch, G. **Gerstenkreuzungen. (Barley crosses.)** Landw. Jahrb. 53: 191–244. *3 pl., 23 fig.* 1919.] Zeitschr. Pflanzenzücht. 7: 141. Dec., 1919.

316. Anonymous. **Flugblatt der Ungarischen Gesellschaft für Rassenhygiene und Bevölkerungspolitik. [Circular of the Hungarian Society for race hygiene and colonization policy.]** Münchener Med. Wochenschr. 66: 76–77. 1919.

317. Anstead, R. D. **Improvement of coffee by seed selection and hybridization.** Agric. Jour. India 14: 639–644. 1919.—An address at the Coffee Planters' Conference at Mysore, India, July 1918. It is urged that the growers select high-yielding coffee trees for propagation in the belief that the present practice of raising nursery stock from "plantation run" seed is resulting in the deterioration of the varieties. It is suggested that facilities be provided to economic botanists for developing new varieties by hybridization. Author also reports that a Mr. Jackson has obtained a vigorous and disease-resistant hybrid which comes true from seed. [See Bot. Absts. 4, Entry 893.]—*J. H. Kempton.*

318. Arthur, J. M. [Rev. of: Folsom, Donald. **The influence of certain environmental conditions, especially water supply, upon form and structure in Ranunculus.** Physiol. Res. 2: 209–276. *24 fig.* Dec., 1918. (See Bot. Absts. 1, Entry 1484; 2, Entry 307.)] Bot. Gaz. 69: 271. Mar., 1920.

319. Bach, Siegfried. **Noch ein Bastardierungsversuch Pisum X Faba. [Another hybridization experiment, Pisum × Faba.)** Zeitschr. Pflanzenzücht. 7: 73–74. June 1919.—Of ten emasculated flowers of Victoria peas, seven were pollinated with *Vicia faba* pollen, while three were left unpollinated. All ten were bagged. After 48 hours, 3 of the pollinated flowers were fixed in Flemming's solution and imbedded in paraffin. Later sections stained with Heidenhain's haematoxylin showed only a few very short pollen-tubes and these in no case were observed penetrating the stigmatic surface. After 8 days, the remaining seven bagged flowers, both pollinated and unpollinated, were found to have developed to the same degree, small pods 1–2 cm. long 0.4 to 0.6 cm. wide with shriveled seed-"anlagen," and within another 10 days, these dried up and fell off. Results confirm Gärtner and von Tschermak. Seedless pods are parthenocarpic and formed without pollination. Inability of *Vicia faba* and *Pisum* to hybridize lies in lack of chemical stimuli to promote pollen-tube growth.—*Orland E. White.*

320. Bach, Siegfried. **Zur näheren Kenntnis der Faktoren der Anthozyanbildung bei Pisum. (To a more exact knowledge of the factors for the formation of anthocyan in Pisum.]** Zeitschr. Pflanzenzücht. 7: 64–65. June 1919.—Red F_1 heterozygote Pisum flowers from red-flowered × white-flowered (*ABaB*) and pink-flowered × white-flowered (*AbaB*) crosses are indistinguishable to the eye from those of the red-flowered homozygote (*ABAB*). Investigations of the concentration and other characteristics of anthocyan, demonstrated that anthocyan development, both qualitatively and quantitatively, is the same in all these genetic types. Comparisons of pink-flowered homozygous types (*AbAb*) with the above red-flowered types shows an anthocyan concentration difference of 2:1 in favor of the latter. Milton Bradley color scale showed color extracts from red-flowered types to be similar to "Violet red," and pink-flowered extracts to be "Violet red tint no. 1." Concludes that red-flower coloring matter differs from that of pink in having greater anthocyan concentration and in being a distinct kind of anthocyan. Factor *A* is more important in furnishing a basis for anthocyan formation than factor *B*, the latter acting as a modifying agent which changes the anthocyan of pink-flowers to that of a new type (red) with more violet in it. Names of pea varieties used are cited and methods of procedure are given in detail.—*Orland E. White.*

321. Bateson, W. **Dr. Kammerer's testimony to the inheritance of acquired characters.** Nature 103: 344–345. July 3, 1919.—Reply to Prof. MacBride (Nature, May 22), describing personal experiences which cast serious doubt upon veracity of Kammerer's claims of inheritance of acquired characters in salamanders.—*Merle C. Coulter.*

322. Baudouin, M. **Découverte d'un procède sûr pour reconnaitre le sexe des axis humains à tout âge. [Discovery of a process for the recognition of sex in the human axis at all ages.]** Compt. Rend. Acad. Sci. Paris 167: 652–663. 1918.

323. Baumann, E. **Zur Frage der Individual- und der Immunitätszüchtung bei der Kartoffel. [On individual selection and breeding for immunity in potatoes.]** Fühlings Landw. Zeitg. 1918: 246. 1918.

324. Baumann, E. **Beiträge zur Kenntnis der Rapspflanze und zur Züchtung des Rapses. [Contribution to a knowledge of the rape plant, and to the breeding of the rape.]** Zeitschr. Pflanzenzücht. 6: 139. *2 fig.* 1918.

325. Becker, J. **Vererbung gewisser Blütenmerkmale bei Papaver Rhoeas. [Inheritance of certain floral characters in Papaver Rhoeas.]** Zeitschr. Pflanzenzücht. 6: 215–221. *3 fig.* 1918.

326. BECKER, J. **Beiträge zur Züchtung der Kohlgewächse.** [Contribution to the breeding of the Brassicas.] Zeitschr. Pflanzenzücht. 7: 91–99. Dec., 1919.

327. BERGH, EBBE. **Studier över dövstumheten i Malmöhus län.** [Studies on deaf-dumbness in the district of Malmöhus, Sweden.] *185 × 250 mm., 199 p.* Stockholm, 1919.—Among deaf-and-dumbs there are a greater number of individuals with brown or black hair and brown eyes than among normal persons in Sweden. The author considers that this fact is caused by descent from immigrant darker types. He points out that there is scarcely any chance to restrain the consanguineal deaf-dumbness by legal directions.—*K. V. Ossian Dahlgren.*

328. BIGGAR, H. H. **The relation of certain ear characters to yield in corn.** Jour. Amer. Soc. Agron. 11: 230–234. 1919.—Relationship of four ear characters to yield has been measured for five varieties of maize. The ear characters chosen were weight, length, numbers of rows and shelling percentage. Data were obtained for a period of several years. It was found that ear length was the most consistent index of subsequent yield though the highest correlation coefficient found in the series was between weight and yield. The author concludes that these four ear characters are not closely enough associated with yield to be of value as a basis for selection.—*J. H. Kempton.*

329. BIXBY, W. G. **The butternut and the Japan walnut.** Amer. Nut Jour. 10: 76–79. 82, 83, *11 fig.* 1919.—Occurrence of rough-shelled walnuts on American-grown trees of the two Japanese species, *Juglans cordiformis* and *J. Sieboldiana*, is discussed, illustrated and convincingly explained as due to natural hybridisation between the above species and the closely related native American species *J. cinerea*. Reference is also made to the possibility of producing new superior hybrid varieties between these oriental and American species which can be grown throughout a greater range of latitude than these walnuts at present occupy. —*E. B. Babcock.*

330. BLAKESLEE, ALBERT F. **Sexuality in mucors.** Science 51: 375–382, 403–409. *4 fig.* April 16 and 23, 1920.—Mucors are divided into two groups as regards sexual reproduction: (1) homothallic or hermaphroditic forms, and (2) heterothallic or dioecious forms. The latter are by far the most abundant in nature.—Sexes of different dioecious species show an imperfect sexual reaction and produce gametes which, however, never fuse. By this "imperfect hybridization" reaction the sex of unmated dioecious races may be determined. In dioecious species there are two types of zygospore germination. In one case the spores in a germsporangium are all of same sex, but in the other the spores are of both sexes. Environmental factors have a direct influence on zygospore formation. Many "neutral" races have been found which give no sexual reaction *inter se* or with testers of other species. The apparent neutrality of such races may be due to lack of the peculiar environmental conditions necessary for expression of the sex which is actually present. All dioecious species investigated are sexually dimorphic. Author discusses gamete differentiation in mucors and its possible significance in relation to sex differentiation in higher forms.—*W. H. Eyster.*

331. BLISS, A. J. **Hybrid bearded Irises.** Gard. Chron. 67: 76, 88. Feb. 14, 21, 1920.—Older varieties of June-flowering bearded Irises may be referred to two main species, *pallida* and *variegata*, or combinations of the two. *Amoena* is a color variety of *variegata*, due to inhibiting factor for yellow or absence of factors for yellow present in *variegata*. *Neglecta* is *squalens* minus yellow. Several hundred crossings of *plicata* color type do not yield conclusive evidence of origin. Characteristic beard is carried through generations of transition seedlings in which it has disappeared along with *plicata* color characters, reappearing unaltered in succeeding individuals of plicata color type. *Plicata* crossed with *pallida* or *squalens-pallida* forms give *plicata* only. Crossed with *pallida* or *variegata* the *plicata* type disappears but when crossed with certain *neglectas* or *squalens*-carrying *plicata* the Mendelian ratio of one-half *plicatas* is obtained, suggesting that the *plicata* type has arisen as a mutation from

pallida by the dropping of a single factor or set of linked factors. Standards and falls of an *Iris* appear to be controlled, both in form and in color, by independent sets of linked factors. —*J. Marion Shull.*

332. Bornmüller, J. **Notizen zur Flora Unterfrankens nebst einigen Bemerkungen über Bastarde und eine neue Form von Polystichum lonchitis (L) Roth im Alpengebiet. [Observations on the flora of Unterfranken, with several remarks on hybrids and a new form of Polystichum lonchitis (L) Roth in the alpine region.** Beih. Biol. Centralbl. 36: 183–199. *1 pl.* 1918.—See Bot. Absts. 4, Entry 1704.

333. Boulenger, G. A. **Un cas intéressant de dimorphisme sexuel chez un serpent africain (Bothrolychus ater Günther). [An interesting case of sexual dimorphism in an African snake.]** Compt. Rend. Acad. Sci. Paris 168: 666–669. 1919.—See Bot. Absts. 5, Entry 1463.

334. Brandl, J. **Die direkte Anpassung und Vererbung der Pflanzen. [Direct adaptation and heredity in plants.]** Wiener Landw. Zeit. 68: 790. 1918.

335. Brehm, V. **Über geschlechtsbegrenzte Speziesmerkmale der Süsswasserorganismen und deren eventuelle experimentelle Aufklärung durch das Mendelsche Spaltungsgesetz. [On the sex-limited species-characters of freshwater organisms and their experimental explanation through the Mendelian law of segregation.]** Naturw. Wochenschr. 18: 4–8. 1919.

336. Bridges, C. B., and T. H. Morgan. **Contributions to the genetics of Drosophila melanogaster. II. The second chromosome group of mutant characters.** Carnegie Inst. Washington Publ. 278. *P. 123–304, 7 pl., 17 fig.* Washington, D. C. 1919.—39 mutant races with genes in "second chromosome" are described, paralleling treatment of sex-linked characters in Carnegie Publ. 237; more than 35 others, discovered since 1916, remain to be described. Most important genes, with loci, are:

0.0 Star	(*S*)	affects mainly eye-facets
15.4 Streak	(*k*)	affects mainly thorax pattern
29.0 Dachs	(*d*)	affects mainly venation and legs
46.5 Black	(*b*)	affects mainly body color
52.7 Purple	(*pr*)	affects mainly eye color
65.0 Vestigial	(*vg*)	affects mainly wings and halteres
73.5 Curved	(*c*)	affects mainly wing curvature
96.2 Plexus	(*px*)	affects mainly wing venation
105.1 Speck	(*sp*)	affects mainly axil of wing

Mutants are treated in chronological order of discovery; special attention is given to genetic methods employed, and tracing their development. Each mutant is fully described as to origin, stock, determination of chromosome and locus, reoccurrences, allelomorphs, modifiers, literature, and value as a genetic tool. General topics, discussed under mutants to which they apply, include: modifying factors, autosomal and balanced lethals, variations in crossing-over due to age, temperature, and specific genes, causes of inviability and methods of "balancing" inviability in experiments, coincidence and its bearing on map-distance, linkage method of analysis for multiple-gene cases, etc.—Most of the mutants are recessive, i.e., the heterozygote can not be distinguished from normal. Only five are dominant; at least four of these are lethal when homozygous, like most dominant mutations in *Drosophila*. Some (e.g., black, blistered, etc.) are partially dominant; i.e., the heterozygote is intermediate between homozygote and normal, but usually more like normal. Two of the genes (lethal T and lethal IIa) show their presence only by disturbance of expected ratios, since they have no visible effect when heterozygous, and kill all flies homozygous for them. Certain genes are "specific modifiers," i.e., they produce no effect except in the presence of certain other "main" genes; thus cream II, cream b, and pinkish, all dilute eosin (sex-linked) eye color but produce no visible effect on non-eosin flies; again, one or more second-chromosome genes reduce bristle number in dichaete (third chromosome), but not in non-dichaete, flies. Purple is a "disproportionate modifier" of vermilion, i.e., it modifies vermilion (sex-linked) more than it does normal eye color.—One series of multiple (quintuple) allelomorphs is described; vestigial, strap, antlered, nick, all affecting wings.—The method of construction of map of second chromosome is described in detail. The "second chromosome" was originally defined arbitrarily as "that chromosome which carries the gene for black and such other genes as may be found to be linked to black." Loci lying on the same side of black as does curved were considered "to the right" or in plus direction from black; those on the opposite side "to the left" or in minus direction. First distance mapped, black-purple, based on 48,931 flies, is 6.2 units (6.2 per cent crossing over), a distance small enough to exclude double crossing over. Other loci located by combining data from different crosses, corrected, where necessary, for double crossing-over, and weighted according to numbers and probable accuracy. Thus vestigial was located 18.5 units to right of black, curved 27.0. These four loci form central framework of chromosome. Dachs was next located at −17.5 (with reference to black) streak at −31.1, star at −46.5. Most important locus at right end is speck, at +58.6 from black. All other loci are located with reference to one or more of the foregoing. As star is of known loci, farthest to left, it is taken as zero point, and other loci renumbered accordingly. Present map of second chromosome, made in this way, with location of all genes treated, is given in text; also constructional map, showing method of construction.—Working map, subject to continuous changes, shows also value of each mutant. Value depends on constancy of character, separability from normal, viability, fertility, accuracy of mapping, and location at convenient distance from other important loci.—*C. R. Plunkett.*

337. Burt, B. C., and N. Haider. **Cawnpore-American cotton: An account of experiments in its improvement by pure-line selection and of field trials, 1913–1917.** Agric. Res. Inst. Pusa Bull. 88. *32 p., 10 pl., 1 fig.* 1919.

338. Carle E. **Sélection pédigrée appliquée a la variété local de riz Phung-tien. [Pedigree selection applied to the local rice variety known as Phung-tien.]** Bull. Agric. Inst. Sci. Saigon 2: 26–32. 1920.

339. Cohen-Stuart, C. P. **Erfelijkheidsleer in dienst der bestrijding van dierlijke vijanden. [Genetics and the production of animal foods.]** Teysmannia 1918: 37–48. 1918.

340. Coppola, Alfredo. **L'acrocefalosindattilia. Contributo allo studio delle disendocrinie congenite. [Acrocephalosyndactylism. A contribution to the study of congenital disendocriny.]** Revista di Patol. Nerv. e. Ment. 24: 283–339. *19 fig.* Dec. 1919.

341. Correns, C. **Fortsetzung der Versuche zur experimentellen Verschiebung des Geschlechtsverhältnisses. [Continuation of the attempt to experimentally shift the sex ratio.]** Sitzungsber. Preuss. Akad. Wiss. Berlin 1918: 1175–1200. *3 fig.* 1918.

342. Crozier, W. J. **Sex-correlated coloration in Chiton tuberculatus.** Amer. Nat. 54: 84–88. Jan.-Feb., 1920.—Foot, ctenidia and other soft parts of male are pale buff color. Corresponding parts in female are salmon-pink to orange-red, depending principally on state of maturity of ovary. Pigment belongs to carotin-like "lipochromes." Evidence shows that color difference cannot possibly help in sex recognition and must therefore be looked upon as a "metabolic accident."—*H. L. Ibsen.*

343. Dahlgren, K. V. Ossian. **Heterostylie innerhalb der Gattung Plumbago. [On the occurrence of heterostyly in the genus Plumbago.]** Svensk Bot. Tidskr. 12: 362–372. *8 fig.* 1918.—*Plumbago capensis* Thunb., *P. rosea* L. and *P. europaea* L. are heterostylous plants. The anthers in long-styled flowers are not placed so deeply in the tube as the stigma in brevistylous ones. Stigmas of the two types are very different both in size and form. The difference between the pollens of the two sorts of plants is however relatively slight. Among forty investigated herbarium specimens of *Plumbago europaea* 18 were short-styled and 22 long-styled, which indicates that the two types may exist in about equal numbers. Heterostyly seems to exist also in the genera *Ceratostigma* and *Vogelia.*—*K. V. Ossian Dahlgren.*

344. Danforth, C. H. **An hereditary complex in the domestic fowl.** Genetics 4: 587-596. *5 fig.* Nov., 1919.—Brachydactyly, syndactyly, and ptilopody (booting) are believed by the author to be the somatic expression of a single gene and data in support of this view are presented.—*H. D. Goodale.*

345. Davenport, C. B. **Influence of the male in the production of human twins.** Amer. Nat. 54: 122-129. Mar.-Apr., 1920.—Both the fathers and the mothers of twins are found to come from fraternities in which twins are about four times as frequent as in the population at large. If only the data involving uniovular twins be considered, the frequency of twins in the parental generation is twelve times that of the population at large, and is as high on the father's side as on the mother's. Uniovular twinning is directly hereditary through either parent as in the armadillo. It is tentatively suggested that biovular twinning is indicative of marked reproductive vigor and relative absence of lethal factors on both sides. Since data from comparative sources show that only a fraction of the eggs ovulated become fertilised and reach late embryonic stages, and since there is good evidence that a high percentage of originally twin pregnancies result in only a single viable foetus, the assumption seems justified that two-egg ovulations are relatively common in man, but that only a small part of such ovulations actually result in twins that are born and recorded as such.—*C. H. Danforth.*

346. Dawson, Andrew Ignatius. **Bacterial variations induced by changes in the composition of culture media.** Jour. Bact. 4: 133-148. Mar., 1919.—As test organism author used a long-cultivated strain of *Bacterium coli*. Preliminary test showed that maximum growth of this organism on meat extract agar was attained in 9 to 11 days. In order to determine effect on this organism of change in environment, so far as regards media, chemical analysis was made of 9-days growth collected from 8 different media. These media consisted of 2 per cent agar to which was added various combinations of peptone, meat extract, edestin, flour proteins, butter soap, glucose and glycerol. Varying proportions of these substances were used, and in most cases no more than two appeared in each medium in addition to the agar. One medium consisted of potato juice alone. Considerable variability occurred in the proportions of nearly all bacterial constituents as the result of growth on these different media.—Production of acid and gas in various carbohydrates was tested in litmus-carbohydrate-serum water after about 200 generations growth on each of the 8 different media. Marked variability occurred; on one medium the organism behaved precisely as a *B. coli-communior*, while on two others it possessed almost the type characteristics of a *B. coli-communis*.—Agglutinability of organisms grown on all 8 media were tested with sera obtained by injection into rabbits of bacteria grown on 4 of the media. Differences in agglutinability were observed easily as great as those frequently utilized to demonstrate the existence of different "strains" of the same basic organism.—Morphological changes accompanying growth on different media appeared to be relatively unimportant. [See Bot. Absts. 3, Entry 1237.]—*M. A. Barber.*

347. Dawson, J. A. **An experimental study of an amicronucleate Oxytricha. I. Study of the normal animal, with an account of cannibalism.** Jour. Exp. Zool. 29: 473-513. *2 pl., 3 fig.* Nov. 20, 1919.—Pedigreed cultures of *Oxytricha hymenostoma* carried 289 generations, then from November 17, 1917, to April 30, 1918, in small petri-dish mass cultures, revealed the absence of micronucleus during all phases of life-history of cultures. This amicronucleate race apparently can live indefinitely under favorable environmental conditions without conjugation, autogamy, endomixis. In state resembling syngamy (*a*) animals fused in pairs die or separate and reproduce with no signs of depression. (*b*) cannibalism occurs causing increased fission rate among progeny of cannibal for short time. [See also next following Entry, 348.]—*Austin R. Middleton.*

348. Dawson, J. A. **An experimental study of an amicronucleate Oxytricha. II. The formation of double animals or 'twins.'** Jour. Exp. Zool. 30: 129-157. *1 pl., 13 fig.* Jan 5, 1920.—Under conditions similar to those in which syngamy usually occurs is strong tendency for formation of double animals, "twins," by plastogamic dorsal fusion. Twins have all

morphological structures of two single animals, reproduce by transverse fission. Favorable environmental conditions necessary for continued existence of twins, i.e., do not survive in competition with single animals. Selection produced striking increase in percentage of twins in pedigreed culture from single twin animal. Division rate of twins similar to that of normal animals. Miscible condition of twin cytoplasm handed on to twin progeny but is quickly lost in single animals derived from twins, kept under identical environmental conditions. Under favorable environmental conditions twin strains breed indefinitely. Pairing, cannibalism, twin formation, occur among animals in similar physiological condition, these phenomena therefore interpreted as abortive attempts to undergo syngamy, failure due to amicronucleate condition. Inability to undergo syngamy has no effect on viability of race. [See also next preceding Entry, 347.]—*Austin R. Middleton.*

349. De Vries, H. **Phylogenetische und gruppenweise Artbildung. [Phylogenetic and group-wise species-formation.]** Flora 11–12 (Festschr. E. Stahl): 208–226. 1918.—Under the term "gruppenweise Artbildung" de Vries understands the formation of a species within a genus. There are also frequent transitions such as the reappearance of the same mutation within a species. For example, the occasional appearance of a peloric form of *Linaria vulgaris.* For the study of "group-wise" species formation the genus *Oenothera* offers excellent material. The mutations observed in this genus can be divided into general and special. The general mutations can be considered as parallel and taxonomic from the standpoint of the systematist, and as progressive and retrogressive from the standpoint of the geneticist. The parallel mutations appear in different species, as for example, the dwarfs which are produced every year by *Oe. biennis* and *Oe. Lamarckiana,* and the *sulfurea* form of *Oe. biennis* and *Oe. suaveolens.* Parallelism is not limited to species of one genus but goes beyond these limitations. For example, the cruciate form of sepals of *Epilobium hirsutum cruciatum,* and very rare mutations of *Oe. biennis cruciata.* As an example of taxonomic mutation de Vries cites the complete lack of petals in the mutant *Oe. suaveolens.*—The absence of petals is a species character of *Fuchsia macrantha* and *F. procumbens.* Examples of progressive mutations are those in which a double number of chromosomes occurs,—*gigas* forms. Among retrogressive mutations are *Oe. nanella, Oe. brevistylis* and *Oe. rubrinervis.* The half-mutants are those which are produced by the fusion of a recessive mutated gamete with a normal gamete, as the mutant *gigas.* In this form we have annually 2 to 3 per cent mutants of the dwarf form. The half-mutants, which can be isolated here, give 25 per cent plants of the *gigas* form, 50 per cent half-mutants and 25 per cent dwarfs. The first and third forms are constant. The half-mutants lead us to the group of special mutations. The first example cited by author is *Oe. grandiflora.* Two-thirds of the plants grown from seed are green and like the parent, and one-third consists of yellow-green weak forms which die if left in the open. About one-fourth of the seed are sterile. This phenomenon author explains in the following manner: *Oe. grandiflora* is a half-mutant which segregates into 25 per cent *ochracea* forms, 50 per cent half-mutant forms, and 25 per cent homozygous forms, the latter of which cannot be formed because the factor for *grandiflora* is united with a lethal factor. Parallel with this is also the appearance of *Oe. Lamarckiana* mut. *rubrinervis,* which segregates in *Oe. deserens* and *Oe. rubrinervis.* About half of the seeds of *Oe. Lamarckiana* are empty. This is explained by au hor in that *Oe. Lamarckiana* produces two kinds of gametes, the typical or *laeta,* and the *velutina.* Each gamete has a lethal factor which is closely linked with the character factor. Heterozygous combinations of these factors give good seeds which produce plants and homozygotic combinations give the sterile seeds. If one of the two lethal factors becomes "vital" the *laeta* or the *velutina* mutation appears. Finally he considers heterogamy, i.e., the phenomenon in which the direct and the reciprocal crosses are not the same. He assumes that the species which are crossed are half-mutations but that part of the pollen is lethal.—*M. Demerec.*

350. De Wilde, P. A. **Verwantschap en Erfelijkheid bij doofstomheid en retinitis pigmentosa. [Relationship and heredity in deaf-and-dumbness and retinitis pigmentosa.]** Dissertation, Amsterdam. 1919.—See also Bot. Absts. 4, Entry 520.

351. De Winiwarter. H. **Les mitoses de l'épithélium séminal du chat. [Mitoses of the seminal epithelium of the cat.]** Arch. Biol. 30: 1-87. *1 double pl. with 34 fig.* 1919.—Thirty-six chromosomes occur in oögonial cells, thirty-five in spermatogonial, the difference depending on the heterochromosomes. The thirty-four autosomes unite to form seventeen bivalents in the primary spermatocyte, the heterochromosome constituting an eighteenth element. Secondary spermatocytes have eighteen and seventeen chromosomes respectively, and these numbers are maintained in the spermatids and consequently in the spermatozoa, since the last division is an equation-division. The heterochromosome is not detectable in the spermatogonia but appears gradually in the telophase of the last spermatogonial division. It finally becomes visible as an elongated body, often curved or even sharply bent. It never appears double as does its homologue in the oöcyte. It is readily distinguished from the nucleolus, which is spherical and visible in spermatogonia as well as in the spermatocytes.—Author believes that his earlier counts in oögenesis, in which he and Saintmont recorded twelve chromosomes on the first maturation spindle and estimated twenty-four as the somatic number, were incorrect. He now thinks that the division figures were abnormal or that in fixation the chromosomes agglutinated.—Various authors have described a "monosome" in the germ-cells of the female cat but author is convinced that what they have regarded as a single body is the two heterochromosomes in juxtaposition.—The observational part of the paper is followed by twenty-six pages of discussion of the literature and of general aspects of the work.—*M. F. Guyer.*

352. Doblas, José Herrera. **Seleccion de semillas. [Seed selection.]** Bol. Assoc. Agric. España 11: 90-95. 1919.

353. Dodge, Raynal. **Aspidium cristatum × marginale and A. simulatum.** Amer. Fern Jour. 9: 73-80. 1919.—Extracts from letter written to C. H. Knowlton by Dodge in 1907 containing a detailed account of his discovery of the Massachusetts fern and the hybrid between the crested and marginal ferns.—*F. C. Anderson.*

354. Dresel, Kurt. **Inweifern gelten die Mendelschen Vererbungsgesetze in der menschlichen Pathologie? [To what extent do Mendelian laws of heredity hold in human pathology?]** Virchow's Arch. 224: 256-303. 1917.—In general, the so-called laws of heredity (e.g., the "law of filial regression") are not such in the strictest sense, but the Mendelian law does present a conception which is fundamental to the study of human heredity. Hereditary disease may be due to single dominant or recessive factors or to combinations of factors. Occasional departures from expected results seeming to show incomplete dominance are due to the chance absence from the germplasm of a second factor which is usually present in homozygous form and which is essential to the actual manifestation of the condition. Sex-linked inheritance is wholly in accord with Mendel's law and is the expression of a certain degree of affinity between the sex factor ("gamete") and the disease-favoring factor. Since the proportion of affected individuals and female carriers is believed frequently to be high in sex-linked inheritance, the occasional presence of two equally potent but independent factors is suggested. The essay, which received the "Schulze Preis," is illustrated by forty-seven graphic diagrams and several tables classifying human diseases on the basis of their behavior in heredity. There is a rather extensive bibilography.—*C. H. Danforth.*

355. Dreyer, Th. F. **A suggested mechanism for the inheritance of acquired characters.** South African Jour. Sci. 14: 272-277. 1918.

356. Drude, O. **Erfahrungen bei Kreuzungsversuchen mit Cucurbita Pepo. [Experiences in crossing experiments with Cucurbita Pepo.]** Ber. Deutsch. Bot. Ges. 35: 26-57. *1 pl.* 1918.

357. Dunn, L. C. **The sable varieties of mice.** Amer. Nat. 54: 247-261. *3 fig.* May-June, 1920.—Sable is a form of yellow mouse showing considerable dark pigment on dorsal and lateral aspects. Black and tan is an extreme type of this variety. Darkness of sables

and black and tans appears due to genetic causes transferable to non-yellow varieties. Cross between agouti (light) and black and tan (dark) gives F_1 sables and agoutis both intermediate. Further hybrid generations showed many light segregates both yellow and non-yellow, and fewer dark segregates. No *extreme* dark segregates found in yellow (black and tan) types, and few extreme dark non-yellow segregates. These latter proved not homozygous for darkening factors. Results indicate presence of genetic factors similar to those producing differences in size of rabbits. This similarity indicates unsuitableness of material for production of clear and analyzable results, rather than insoluble nature of problem. Correct interpretation of such differences must await combination of optimum of material and method.—*C. C. Little.*

358. Eaton, S. V. [Rev. of: Dorsey, M. J. **Relation of weather to fruitfulness in the plum.** Jour. Agric. Res. 17: 103–126. *Pl. 13–16, 1 fig.* June 16, 1919. (See Bot. Absts. 3, Entry 1478.)] Bot. Gaz. 69: 269. Mar., 1920.

359. Ebstein, A. **Zur Frage des Vorkommens von Kretinen und Albinos in Lehrbach im Harz. [On the occurrence of cretins and albinoes in Lehrbach in the Harz.]** Die Naturwissenschaften 6: 561–565. 1918.

360. Eisenberg, P. **Untersuchungen über die Variabilität der Bakterien. VII. Über die Variabilität des Schleimbildungsvermögens und der Gramfestigkeit. [Investigations on the variability of bacteria. VII. On the variability of the slime-building capacity and in Gram-reaction.]** Centralbl. Bakt. Parasitenk. 82: 401. 1918.

361. Everitt, P. F. **Quadrature coefficients for Sheppard's formula (c). Biom. Vol. 1:** p. 276. Biometrika 12: 283. Nov., 1919.—This table gives constants necessary for rapid calculation of the area of a curve, from equally spaced ordinates.—*John W. Gowen.*

362. Findlay, Wm. M. **The size of seed.** North Scotland Coll. Agric. Bull. 23. *15 p.* 1919.—See Bot. Absts. 3, Entry 1361.

363. Fischer, E. **Die Beziehungen zwischen Sexualität und Reproduktion im Pflanzenreich. [Relation between sexuality and reproduction in the vegetable kingdom.]** Mitteil Naturf. Ges. Bern. 1918: 1–4. 1918.

364. Fries, Rob. E. **Strödda iakttagelser över Bergianska Trädgardens gymnospermer. [Miscellaneous observations on gymnosperms in the Bergian garden.]** Acta Horti Bergiani [Stockholm] 6: 1–19. *1 pl., 1 fig.* 19—.—The original specimen of *Larix americana* Michx. f. *glauca* Beissn. is characterised by chlorocarpy. Color of needles is certainly in large part blue-green (*glauca*). Shoots with typical light-green color are to be seen here and there, however, which is also shown in a colored plate. The cause of this fact, suggesting chimera-phenomena, is not as yet explained. Of *Picea Engelmannii* (Parr) Engelm., author describes a *virgata* and a *prostrata* form, both belonging to the *glauca* type. Teratological formations in the strobiles of *Larix decidua* are described.—Report is given on the winter-resistance of different kinds of needle-trees. Different observations concerning the process of flowering are given and discussed. *Pinus cembra*, *Picea nigra* and *Abies arizonica* seem during the individual life to have a ♀ stage preceding the androgynous stage. In *Pinus ponderosa* var. *scopulorum*, *Picea omorica* and *Abies concolor*, on the contrary, a ♂ stage seems to precede the stage with both sexes.—*K. V. Ossian Dahlgren.*

365. Fritsch, K. **Floristische Notizen. Über Rumex Heimerlii Beck und einige andere angebliche Tripelbastarde aus der Gattung Rumex. [Floristic notes on Rumex Heimerlii Beck and several other supposed triple hybrids in the genus Rumex.]** Österr. Bot. Zeitg. 67: 249–252. 1918.

366. Frölich, G. **Abstammungs- und Inzuchtsforschungen. Dargestellt an der wichtigsten Blutlinie des weissen deutschen Edelschweines, Ammerländer Zucht. [Pedigree and inbreeding investigations. Represented in the most important bloodlines of improved white German swine, Ammerland breed.]** Kühn-Archiv 7: 52–129. *6 pl.* 1918.

367. Frölich, G. **Wichtigste Blutlinie des weissen deutschen Edelschweines, Ammerländer Zucht. [Most important blood-lines of improved white German swine, Ammerland breed.]** Deutsch. Landw. Presse. 46: 24. *12 fig.* 1919.

368. Frölich, G. **Die Beeinflussung der Kornschwere durch Auslese bei der Züchtung der Ackerbohne. [The influencing of grain-weight by selection in the breeding of field beans.]** Friedrichswerther Monatsber. 9: 7–8, 17–20. 1919.—See Bot. Absts. 5, Entry 268.

369. Frölich, G. **Die Umzüchtung von Wintergetreide in Sommergetreide. [The breeding of winter cereals into spring cereals.]** Friedrichswerther Monatsber. 9: 27–30. 1919.—See Bot. Absts. 5, Entry 267.

370. Frost, H. B. **Mutation in Matthiola.** Univ. California Publ. Agric. Sci. 2: 81–190. 1919.—Occurrence, characteristics and heredity of certain aberrant types of *Matthiola annua* Sweet are described. These aberrant forms resemble some of the "mutant" types produced by *Oenothera Lamarckiana.* It is highly probable that they are originally produced by mutation but it is uncertain whether aberrant individuals arise by immediate mutation or by segregation. Although the species is typically Mendelian with respect to various characters, yet individuals of the mutant types give erratic hereditary ratios suggestive of *Oenothera.* Six out of eight types studied have shown their heritability in progeny tests. Some of the types have been produced by many parents and in several pure lines isolated from the original commercial variety, "Snowflake."—Mutant types are in general inferior to Snowflake in vigor, fertility and various form and size characters. The early type is practically a smaller and earlier Snowflake and is probably due to a single dominant mutant factor. In five other types no true-breeding individuals have yet been found although it is known that in three of the types the mutant factor (or factors) is carried by both eggs and sperms; hence it appears that these mutant factors are imperfectly recessive for a lethal effect. Evidence is reported for linkage of three mutant factors with the factor pair for singleness and doubleness of flowers but selfing ratios suggest duplication of a chromosome (non-disjunction) as in *Oenothera lata.* Further study may help to explain the remarkable genetic behavior of *Oenothera* and *Citrus.*—*E. B. Babcock.*

371. Fruwirth, C. **Zum Verhalten der Bastardierung spontaner Variationen mit der Ausgangsform. [The hybridization of a spontaneous variation with the original form.]** Zeitschr. Pflanzenzücht. 7: 66–73. *2 fig.* June, 1919.—Author observed a spontaneous variation in color of seed coats of a spotted strain of narrow-leaved lupine (*Lupinus angustifolius*). This variation was a dilution of the color. It has bred true since 1911. Reciprocal hybrids were made between this dilute-colored form and the parent strain. In F_1 dilute color was dominant when maternal parent was dilute and recessive when the paternal parent was dilute. Segregation occurred in both hybrids in second and subsequent generations but behavior was very irregular.—*J. H. Kempton.*

372. Fruwirth, C. [German rev. of: Fruwirth, C. **Handbuch der landwirtschaftlichen Pflanzenzüchtung. II. Die Züchtung von Mais, Futterrüben und anderen Rüben, Oelpflanzen und Gräsern. (Handbook of agricultural plant breeding. II. The breeding of maize, fodder beets and other root-crops, oil plants and grasses.)** *3rd. ed., 262 p., 50 fig.* Paul Parey: Berlin, 1918.] Zeitschr. Pflanzenzücht. 7: 144–145. Dec., 1919.

373. Fruwirth, C. **Die gegenwärtige Organisation der Pflanzenzüchtung in Deutschland und in Österreich-Ungarn. [The present organization of plant breeding in Germany and Austro-Hungary.]** Nachricht. Deutsch. Landw. Ges. Österreich 1919: 35–39. 1919.—See Bot. Absts. 5, Entry 269.

374. Fruwirth, C., Dr. Th. Roemer, and Dr. E. von Tschermak. **Handbuch der landwirtschaftlichen Pflanzenzüchtung. 4. Die Züchtung der vier Hauptgetreidearten und der Zuckerrübe. [Handbook of agricultural plant breeding. 4. Breeding of the four chief cereals and the sugar beet.]** *3rd. ed., 8vo., xv + 504 p., 42 fig.* Paul Parey: Berlin, 1918.—See Bot. Absts. 5, Entry 270.

375. Gassner, S. **Beiträge zur physiologischen Charakteristik sommer- und winterannueller Gewächse, insbesondere der Getreidepflanzen. [Contribution to the physiological characteristics of summer and winter annuals with special reference to the cereals.]** Zeitschr. Bot. 10: 417–480. *7 pl., 2 fig.* 1918.—See Bot. Absts. 5, Entry 271.]

376. Gassul, R. **Nachtrag zu meiner Mitteilung über "Eine durch Generationen prävalierende symmetrische Fingerkontraktur." [Supplement to my contribution on a symmetrical contraction of the fingers prevailing through generations.]** Deutsch. Mediz. Wochenschr. 44: 1196–1197. 1918.—See Also Bot. Absts. 4, Entry 578, and next following Entry, 378.

377. Gassul, R. **Eine durch Generationen prävalierende symmetrische Fingurkontraktur. [A symmetrical contraction of the fingers prevailing through generations.]** Deutsch. Mediz. Wochenschr. 44: 1197–1198. *2 fig.* 1918.—In a family from Mecklenburg-Schwerin three successive generations have produced individuals with permanent bilateral hyperextension of the basal phalanges of the fourth and fifth fingers. [See also next preceding Entry, 377.]—*C. H. Danforth.*

378. Gatenby, J. Brontë. **The cytoplasmic inclusions of the germ-cells. VI. On the origin and probable constitution of the germ-cell determinant of Apanteles glomeratus, with a note on the secondary nuclei.** Quart. Jour. Microsc. Sci. 64: 133–153. *1 pl., 10 fig.* Jan., 1920.—Author describes his attempts to determine the composition and origin of the germ-cell determinant in the oöcytes of the parasitic hymenopteran, *Apanteles glomeratus.* He finds that it arises as a concentrated area at the posterior pole of the young oöcytes; that it is probably formed of albuminous material rather than of chromatin, fat, yolk, or glycogen; and that the secondary nuclei have no connection with it.—*R. W. Hegner.*

379. Gatenby, J. Brontë. [Rev. of: Thomson, J. Arthur. **Heredity.** *3rd. ed., ix + 627 p., 47 fig.* John Murray: London, 1919.] Sci. Prog. 14: 517. Jan., 1920.

380. Geisenheyner, L. **Über einigen Panaschierungen. [On some variegations.]** Verhandl. Bot. Ver. Prov. Brandenburg 59: 51–61. *3 fig.* 1918.

381. Goebel, K. **Zur Kenntnis der Zwergfarne. [To a knowledge of the dwarf ferns.]** Flora 11–12 (Festschr. Stahl): 268–281. *6 fig.* 1918.—Describes dwarf mutants (?) from two spp. of *Aspidium,* one sp. of *Drynaria* (tropical), and two spp. of *Platycerium.* Mutation has not yet been directly observed in culture. Dwarfs are characterized by smaller and fewer cells, smaller or fewer bundles (or both), fewer sori, sporangia, and spores. Describes parallel investigation of dwarf mutant from *Salvia protensis.*—*Merle C. Coulter.*

382. Goldsmith, William M. **A comparative study of the chromosomes of tiger beetles (Cicindelidae).** Jour. Morph. 32: 437–487. *Pl. 1–10.* 1919.—Five species of *Cicindela* were studied, all of which conform to one type in regard to chromosome number and spermatogenesis. The male has a "double odd chromosome," the female two, making the formulae 20 + Xx = 22♂, 20 + Xx + Xx = 24♀. In other Coleoptera two additional types are known, (1) 20 + X + Y = 22♂, 20 + 2X = 22♀; and (2) 18 + X = 19♂, 18 + 2X = 20♀. In *Cicindela* spermatogonia are in syncytial cysts; the spermatocyte growth period includes, in sequence, the usual diffuse, leptotene, synaptic (synizesis) and diplotene stages, giving rise to prophase bivalents. The Xx complex is a single compound body in first division, going undivided to one pole, giving two types of second spermatocytes. X separates from x in anaphase and both divide in second division. Spermatogonia each have one nucleolus, oogonia have two, corresponding to sex-chromosome relations. Early stages of oocyte growth period correspond in general to those of spermatocyte.—*Chas. W. Metz.*

483. Guyer, M. F., and E. A. Smith. **Studies on cytolysins. I. Some prenatal effects of lens antibodies.** Jour. Exp. Zool. 26: 65–82. 1918.—The lenses of freshly-killed rabbits were reduced to a pulp and diluted with normal salt solution, then injected into the peritoneal cavity of fowls. Serum obtained from such fowls, when injected into the blood-vascular system of pregnant rabbits, attacked the lenses of some of the uterine young, though without effect on the lenses of the mothers. The affected lenses were rendered opaque or liquid. Similar results were obtained in mice. The experiments demonstrate that specific structural modifications can be engendered in the young *in utero* by means of specifically sensitized sera. —*Bertram G. Smith.*

384. Haecker, V. **Vererbungsgeschichtliche Einzelfragen IV. Über die Vererbung extremer Eigenschaftsstufen. [Historical genetical problems IV. On the inheritance of extreme character-gradations.]** Zeitschr. indukt. Abstamm. Vererb. 21: 145–157. *2 fig.* Sept., 1919. —Various cases already in the literature are brought together in support of the following generalization: The extreme grades of a varying character will show agreement with the law of segregation, but the intermediate grades will not show such phenomena among themselves. The cases cited as evidence include height of peas, Mendel's short and tall vs. Bateson's dwarf and semi-dwarf; the relations of *Oenothera gigas* and *nanella* vs. those of *O. muricata* and *nanella*; stature in man (an original pedigree is given of one family involving a size cross); crosses between the blue and white varieties of flax, and between two species of somewhat different blues; much the same situation in *Veronica*; leaf color in Shull's *Lychnis*; spotting in mice and rats; and finally various examples from butterfly crosses. In attempting to provide some theoretical explanation to cover the situation in general, the multiple factor theory is found impossible without far-reaching supporting hypotheses. A *special* factor influencing the extreme plus and minus grades is not accepted because this phenomenon is so far-reaching in plants and animals, involving color as well as form, that there must be a *common* final cause behind all cases. Neither can the popular theory of linkage be called in to help without the special assumption that linkage is effective when strong concentrations of duplicate factors are present, and also when these factors are in greatly reduced numbers, but in the intermediate conditions the factors exhibit their independence. But this explanation is not satisfactory, and in the present state of the science the best one can do is to say that, "In continuously varying characters the extremes show a greater inclination to inheritable independence than do the intermediate grades." In other words, the germplasm determining the extreme grades is much more stable and independently heritable than that determining the intermediate grades. The article is concluded with a cursory discussion of the antagonistic relation between white and black with special reference to mosaic arrangements and to ontogenetic reversals; it is suggested that one condition of the germplasm may completely turn over into the other condition with proportional ease.—*E. C. MacDowell.*

385. Haecker, V. **Die Annahme einer erblichen Übertragung körperlicher Kriegsschäden. [The supposition of a hereditary transmission of physical war injuries.]** Arch. Frauenk. u. Eugenik. 4: 1. 1919.

386. Haecker, V. **Über Regelmässigkeiten im Auftreten erblicher Normaleigenschaften, Anomalien und Krankheiten beim Menschen. [On regularity in the occurrence of hereditary normal characteristics, anomalies and diseases in man.]** Mediz. Klinik. 14: 177. 1918.

387. Hammerlund, H. C. **Förädling av grönsaksväxter vid Weibullsholms Växtförädlingsanstalt. [Improvement of green vegetables at the station for plant improvement of Weibullsholm.]** *13 p., 7 fig.* W. Weibulls Illustrerade Årsbok (Landskrona) 15 (1920). 1919.—Gives an account of the results obtained and methods practised. Self-fertility has been found to be very unequal for different sorts of cabbage, and seems also to vary for other kinds of green vegetables. In parsnips self-fertility seems however in general to be very effective.—*K. V. Ossian Dahlgren.*

388. Hansen, W. **Einiges über Rübenzucht. [Something about beet-breeding.]** Landw. Zeitung 39: 154–156. 1919.—See Bot. Absts. 5, Entry 272.

389. H[arland], S. C. **A note on a peculiar type of rogue in Sea-Island cotton.** Agric. News [Barbados] 19: 29. 1920.—A distinct type characterised by great reduction in size of all the organs and nearly complete sterility, constitutes about 0.05 per cent of plants in fields of Sea Island cotton in St. Vincent. No viable pollen is produced and seeds are very rarely developed. Plants grown from two seeds borne on a "rogue" plant, representing therefore F_1 of cross with Sea Island, had all characters of latter. A self-fertilized strain of Sea Island, which had produced hitherto only normal plants, gave rise in fourth selfed generation to rogue plants in 4 out of 62 progenies, the average percentage of rogues having been 1.6.—*T. H. Kearney.*

390. Heribert-Nilsson, H. N. **Ett försök med urval inom pedigreesorter av havre.** [**An experiment with selection among pedigree-varieties of oats.**] *4 p.* W. Weibulls Illustrerade Årsbok (Landskrona) 15 (1920). 1919.—Of the Danish "Tystofte Gulhvid," by pedigree selection, a new and more productive variety "Weibull's Fortunahavre" was obtained. Here is of special interest that selection within the pedigree variety "Tystofte Gulhvid" has given such a surprisingly good result. This shows that the mother variety "Tystofte Gulhvid," must either not have been homogeneous, although secured by pedigree selection, or the original plant of "Fortuna" oats must represent a mutation. Under high humidity combined with high temperature author has observed that the oat flowers are able to open and, contrary to the usual rule, disperse their pollen. Cross-fertilization thus is not excluded in oats, which as a rule however is an autogamous plant. The author also considers as most probable that the individual used as mother plant had its genotype changed by a new combination.—*K. V. Ossian Dahlgren.*

391. Hoffmann, Hermann. **Geschlechtsbegrenzte Vererbung und manisch-depressives Irresein.** [**Sex-linked inheritance and manic-depressive insanity.**] Zeitschr. ges. Neurol. Psych. 49: 336–356. 1919.—Author reviews suggestion of Lenz that certain diseases represent dominant sex-linked characters and develops the theoretical expectations for this form of heredity. One of the critical requirements in these cases is that a father characterized by a dominant sex-linked trait should produce only normal sons and affected daughters. Lenz mentioned manic-depressive insanity as possible example of this type. Author finds that in general the heredity of the diathesis does conform approximately to theoretical expectations based on Lenz's hypothesis, but there are numerous exceptions. Twelve such exceptional family histories are presented in some detail. In these families where affected men have married presumably normal women there have been produced instead of all normal sons twenty-four affected and two normal, from which it is concluded that manic depressive insanity does not present an entirely satisfactory example of dominant sex-linked heredity.—*C. H. Danforth.*

392. Hopkins, L. S. **A crested form of the lady fern.** Amer. Fern Jour. 9: 86–88. *Pl. 4.* 1919.

393. Jehle, R. A., and others. **I. Control of cotton wilt. II. Control of cotton anthracnose and improvement of cotton.** Bull. North Carolina Dept. Agric. 41[1]: Supplem. 5–28. *Fig. 1–6 and 1–5.* 1920.—See Bot. Absts. 5, Entry 747.

394. Jelinek, Dr. **Nächste Aufgaben der Pflanzenzüchtung und der Sortenprüfung.** [**The next problems of plant breeding and variety testing.**] Zeitschr. Pflanzenzücht. 7: 83–90. Dec., 1919.

395. Kajanus, H. B. (1) **Weibullsholms Ambrosia-kokärt.** *1 p.* (2) **Weibulls Kolibri-fodervicker.** *2 p., 2 fig.* (3) **Weibulls Tardus-Hundäxing.** *2 p., 2 fig.* W. Weibulls Illustrerade Årsbok (Landskrona) 15 (1920). 1919.—New and productive sorts of *Pisum sativum*, of *Vicia sativa*, and of *Dactylis glomerata* are described: the last flowers about two weeks later than the common sorts.—*K. V. Ossian Dahlgren.*

396. Kammerer. **Geschlechtsbestimmung und Geschlectsverwandlung. Zwei gemeinverständliche Vorträge. [Sex determination and sex modification. Two popular lectures.** *96 p., 16 fig.* Perles: Wien, 1918.

397. Kammerer, K. **Mischling. [Hybrids.]** Ornith. Monatshefte. 43: 31–32. 1918.

398. Kammerer, Paul. **Das Gesetz der Serie. Eine Lehre von den Wiederholungen im Lebens- und im Weltgeschehen. [The law of series. A doctrine of the repetition in life- and world-phenomena.** *17 × 24.5 cm., 486 p., 8 pl., 26 fig.* Deutsche Verlang-Anstalt: Stuttgart, Berlin, 1919.

399. Kiessling, L. **Die Leistung der Wintergerste und deren züchterische Beeinflussung. [The performance of winter barley and its modification by breeding.]** Illustr. Landw. Zeitung 1919: 310–311. 1919.—See Bot. Absts. 5, Entry 283.

400. Klatt, B. **Vergleichende metrische und morphologische Grosshirnstudien an Wild- und Haushunden. [Comparative metrical and morphological studies on the cerebrum of wild and domesticated dogs.]** Sitzungsber. Ges. Naturf. Freunde. 1918: 35–55. 1918.

401. Klatt, B. **Experimentelle Untersuchungen über die Beeinflussbarkeit der Erbanlagen durch den Körper. [Experimental investigations on the modifiability of the hereditary factors through the soma.]** Sitzungsber. Ges. Naturf. Freunde. 1919: 39–45. 1919.

402. Knibbs, G. H. **The problems of population, food supply and migration.** Scientia 26: 485–495. 1919.—Popular mathematical paper showing that the present world's population increase is too rapid when compared with possibilities of increasing the food supply.—*E. M. East.*

403. Kottur, G. L. **An improved type of cotton for the southern Maratha country (Bombay Presidency, India).** Agric. Jour. India 14: 165–167. *1 pl.* 1919.

404. Kraus, and L. Kiessling. **Die Landsortenzüchtung in Bayern. [Breeding of local varieties in Bavaria.]** Deutsch. Landw. Presse 1918: 247. 1918.

405. Kroemer, K. **Das staatliche Rebenveredelungswesen in Preussen. [State grape-improvement project in Prussia.]** Landw. Jahrb. 51: 1–292. *8 pl., 43 fig.* 1918.

406. Kronacher, C. **Die deutscher Schweinezucht und Haltung nach dem Kriege. [German swine breeding and maintenance after the war.]** Flugschr. Deutsch. Ges. Züchtsk. 1918: 47. 1918.

407. Kronacher, C. **Beitrag zur "Erbfehler" Forschung in der Tierzucht mit besonderer Berücksichtigung des Rorens beim Pferde. [Contribution to investigation of hereditary defects in animal breeding, with special reference to "Rorens" in horses.]** Flugschr. Deutsch. Ges. Züchtungsk. 1918: 1–32. 1918.

408. Kronacher, C. **Allgemeine Tierzucht. Ein Lehr- u. Handbuch für Studierende u. Züchter. 4. Abteilung (Abschnitt VI des Gesamtwerkes): Die Züchtung. [General animal breeding. A text and handbook for students and breeders. 4th part (Section VI of the complete work): Breeding.** *8vo, 357 p.* Paul Parey: Berlin, 1919.]

409. Lenz, Fritz. **Über dominant-geschlechtsbegrenzte Vererbung und die Erblickkeit der Basedowdiathese. [Dominant sex-linked heredity and the inheritance of the Basedow diathesis.]** Arch. Rassen u. Gesellschaftsbiol. 13: 1–9. *5 fig.* 1918.—The fact that certain sex-linked traits are recessive carries with it the corollary that allelomorphic traits are sex-linked dominants. Biologically there is no essential difference between normal and disease-favoring determiners, and consequently dominant sex-linked diseases might be expected. Such diseases, instead of being very rare in the female, should be twice as frequent as in the male.

Affected females mated to normal males should produce in equal numbers both normal and affected sons and daughters while affected males mated to normal females should produce only normal sons and affected daughters. The incidence of several diseases of man, including Basedow's, approximate the expectations for dominant sex-linked traits. That they are such can not be stated with assurance till further data shall have been accumulated. It is the purpose of this paper to point out the possibility of dominant sex-linked traits and to indicate their expected mode of inheritance.—*C. H. Danforth.*

410. LILLIE, FRANK RATTRAY. **Problems of fertilization.** *13 × 19 cm., xii + 278 p., 19 fig.* Univ. Chicago Press: Chicago, 1919.—Author distinguishes two phases of fertilization, rejuvenescence, and combination of inheritance from two parents. Latter is only feature common to all cases of fertilization. Morphology of fertilization is described. Chromosome equivalence of egg and sperm is emphasized. Origin of centrosome in fertilized egg is regarded as physiological rather than morphological. There is no evidence that mitochrondria of sperm have any function in heredity. Pathological polyspermy strongly supports nuclear theory of heredity.—Behavior of sperm under various circumstances is described, especially in response to chemical stimuli, including those originating in egg. Agglutination of sperm is due to substance in sperm, which is specific in its action. Approach of sperm to egg is not due solely to random activity, nor to chemotactic orientation alone, but to combination of different types of behavior. Gametes must both be in definite condition before fertilization may occur, and that condition lasts variable time in different species. Sperm owes its power of fertilization to a substance, not to its motility, and this substance may also be responsible for agglutination. Egg also owes fertilization capacity to hypothetical substance (fertilizin). Fertilization is accompanied by increase in rate of oxidation, changes in permeability, changes in colloidal condition, and chemical alterations. Fertilization involves long series of events, some cortical, some internal, and process may be arrested in middle, making fertilization partial. Such incomplete activation of egg results sooner or later in arrest of development.—Tissue specificity in fertilization is demonstrated when spermatozoa fail to enter accessible cells other than ova. Species specificity is shown by hybrid fertilization in echinoderms, teleosts, and Amphibia, and by self-fertilization in various animals. Such hybridization experiments demonstrate some non-specific and some specific factors. Latter are found in cortical reactions of egg. If cortical barrier is passed by foreign sperm, fertilization proceeds normally. In plants, sterility is due to inhibition of growth of pollen tube, not to incompatibility of gametes, and in some cases sterility factors are known to be inherited. Specificity is doubtless due to chemical phenomenon, problem related to agglutination of sperms. Analogy with immunity reaction is pointed out, but with warning that these phenomena may be fundamentally unlike.—Activation involves two phases, cortical and internal. Agglutination of sperm to egg is first step in cortical phase, and is due to agglutinating substance (fertilizin). This substance is combined on entrance of one sperm, and egg does not react to other sperms. Author criticises Loeb's view that activation of egg is due to cortical cytolysis; discusses increase of oxidation, also gelation and liquefaction of cortical protoplasm, and electrical polarization. Internal phase of activation mainly relates to preparation for karyokinesis.—*A. Franklin Shull.*

411. LUNDBORG, H. **Befolkningsstudier i Norrbotten och nordliga Lappland särskildt inågra fjällbyar av Torne sjö. [The structure of population in Norrbotten and in the northeast part of Lappland, specially in some mountain villages near Lake Torne.]** Ord och Bild [Stockholm] 28: 641–648. *11 fig.* 1919.—Author describes how the Lapponians are going over to settle in houses and the social and race biological consequences of this change. Crossings between Swedes, Finlanders and Lapponians are not uncommon. The lowest and poorest part of the population includes as a rule Lapponians and half-blood Lapponians; the middle part are Finlanders; the upper portion consists of Swedes or Swede Finlanders. The younger a village is and the more westward up to the mountain it is situated, the more the Lapponians or Lapponian Finlandian elements dominate. The reason for this difference in the structure of population depends undoubtedly upon the race inequalities or differences in cultural qualification of the tribes in question.—*K. V. Ossian Dahlgren.*

412. LUNDBORG, H. **Olika folk och kulturer, sedda i rasbiologiskt ljus.—Internationell Politik. [Different peoples and cultures in race-biological light.]** *125 × 200 mm., 8 p.* Stockholm, 1919.—Author treats the consequences of (1) inter-marriages, (2) extreme mixing of races, (3) marriages within the same tribe (inter-marriages in its wide sense) and (4) race-mixings between related peoples.—*K. V. Ossian Dahlgren.*

413. LUNDBORG, H. **Om modern ärftlighetsforskning med särskild hänsyn till människan. [On modern inquiry into heredity with special consideration to mankind.]** Ord och Bild [Stockholm] 28: 186–196. *4 fig.* 1919.—Popular treatise.—*K. V. Ossian Dahlgren.*

414. LUNDBORG, H. **En svensk bondesläkts historia sedd i rasbiologisk belysning.—Svenska Sällskapets for Rashygien skriftserie II. [The history of a Swedish peasant family in eugenical light. No. II. of the papers of the Swedish Eugenical Association.** *138 × 215 mm., 40 p., 8 fig.* P. A. Norstedt & Söners Förlag: Stockholm, 1920.—Author first discusses genealogical investigation as a cultural subject. Especially in Sweden it might be possible to practise genealogical inquiries on a greater scale, because the registration of the inhabitants of Sweden since centuries ago is more complete than in any other country. The "hüsforhörsböcker" are especially important, because in these books on the same page are noted whole families. After a small chapter on "genealogical principles" the author proceeds to a popular description of his investigation on the Lister family. This family was extensively discussed in author's great work "Medizinisch-biologische Familieforschungen innerhalb eines 2232-köpfigen Bauergeschlechtes in Schweden," Jena 1913.—*K. V. Ossian Dahlgren.*

415. LYNCH, CLARA J. **An analysis of certain cases of intra-specific sterility.** Genetics 4: 501–533. *2 fig.* Nov., 1919.—Analysis of sterility in certain mutant races of *Drosophila melanogaster*. Fused is sex-linked recessive. Males are fertile with normal or heterozygous females; fused females produce no offspring when mated to fused males, only a few (and these all daughters) when mated to normal males. XXY fused females, mated to normal males, produce a few sons, but these are all non-disjunctional exceptions. Hence fused gene acts to prevent eggs from developing, but this action may be inhibited by its normal allelomorph, either before maturation (in heterozygous female) or after fertilization (in not-fused offspring of fused female). Rudimentary, another sex-linked recessive, acts in same way as fused, but not so completely, as rudimentary females produce a few rudimentary offspring. Morula, reduced bristle, dwarf (autosomal recessives) have sterile females and fertile males. Dibro (autosomal recessive) apparently sterile in both sexes. Cleft (sex-linked recessive) has sterile males, and females have never been obtained. In none of the cases studied was it possible to isolate a sterility gene independent of the mutant gene itself. Sterility is probably one of the effects of these mutant genes.—*A. H. Sturtevant.*

416. MACOUN, W. T. **Blight resistant potatoes.** Canadian Hortic. 42: 129–156. 1919.—See Bot. Absts. 3, Entry 1644.

417. MACBRIDE, E. W. **The inheritance of acquired characters.** Nature 103: 222. May 22, 1919.—Refers to recent work of KAMMERER published in Archiv für Entwicklungsmechanik, 1919, extending earlier experiments with *Alycetes*, the "mid-wife" toad. These normally pair on land, the horny patch on the hand of the male, characteristic of water-breeding Anura, being absent. KAMMERER had previously found that *Alycetes* subjected to a higher temperature, paired in water, and that the F_1 and F_2 generations developed the horny patch, even when returned to a terrestrial environment. It is now found that the patch persists in the F_4 generation.—MCBRIDE deprecates certain criticisms of the work of KAMMERER and is inclined to support the results as evidence toward the inheritance of acquired characters. He notes that arrangements for a repetition of the experiment in the Zoological Gardens, are being made, although a minimum of six years will be required.—Although author is inclined to challenge Mendelians in connection with the results achieved by KAMMERER, experiments with *Drosophila*, particularly where abnormal abdomen develops, are suggestive that a common explanation may underlie both phenomena.—*L. B. Walton.*

418. Meader, Percy D. **Variation in the diphtheria group.** Jour. Infect. Diseases 24: 145–157. 1919.—Author's material consisted of 25 different strains of the diphtheria bacillus, isolated, for the most part, from throats of persons infected with diphtheria during epidemic of the disease. Pure cultures were made of each strain by repeated plating on agar. From each pure culture a series of subcultures were made by plating dilutions so prepared that as far as possible each colony represented the progeny of a single organism. Repeated subcultures were made from selected colonies of each strain. Progeny of the various colonies were examined in 20 hour slant cultures on Loeffler's serum stained with Loeffler's methylene blue. The frequency of the various Wesbrook types of morphology were tabulated for the original type of each strain and for the progeny of each type. Employing as a criterion of variability in type the fact that the predominating types of morphology present in subcultures were different from those present in the original culture, the author found that of his 25 strains 8 showed morphologic variation, 4 may have varied only slightly, if at all, and 13 showed no reasonable indication of variation.—To determine fermentative variability, each of the 25 strains were compared with their descendants after the 5th and 10th platings as regards their power to produce acid in dextrose, lactose, maltose, dextrin, and saccharose. More than half of the cultures investigated varied after successive platings as regards their power to produce acid in carbohydrates.—Variability of virulence of the 25 strains was tested by means of the inoculation into guinea pigs of each original type and of its progeny after the 5th and 10th platings. Some strains gained virulence, some lost it and some remained constant in the course of successive platings. Variations in virulence were only in part correlated with morphologic types. Cultures containing granular forms were frequently non-virulent, while those which consisted of solid-staining forms for the greater part of their cultivation were consistently non-virulent.—From a biometric study of the fermentative reactions of members of the diphtheria group it appears that they constitute a genetically related group of organisms. In subcultures derived from one parent strain variations in morphology, in fermentative reactions and in virulence, occur, but the virulence of a strain is not correlated with its fermentative reactions nor closely correlated with its morphology.—*M. A. Barber.*

419. Meunissier, A. **De quelques idées sur la selection des légumes. [Some ideas on the selection of vegetables.]** Rev. Hortic. 91: 300–303. June, 1919.—See Bot. Absts. 5, Entry 1855.

420. Meves, G. **Eine neue Stütze für die Plastosomen theorie der Vererbung. [A new support for the plastosome theory of heredity.]** Anat. Anzeig. 50: 1918.

421. Molz, C. **Natürliche und künstliche Auslese zur Erzielung widerstandsfähiger Sorten. [Natural and artificial selection for the achievement of resistant varieties.]** Deutsch. Landw. Presse 1918: 19. 1918.

422. Morgan, Thomas Hunt. **The physical basis of heredity.** *14x21 cm., 300 p., 117 fig.* J. B. Lippincott Co.: Philadelphia, 1919.—A presentation of the modern factorial theory of heredity, comprising the phenomena of segregation, independent assortment, linkage and crossing over, the linear arrangement of the genes, interference, and the limitation of the linkage groups. Both the genetic evidence and the cytological are presented, and it is shown how the genetic phenomena are explained by the chromosome mechanism. On the basis of these principles an analysis is given of sex and sex-linked inheritance, non-disjunction, parthenogenesis and pure lines, cytoplasmic and maternal inheritance. There is a discussion of variation in linkage caused by hereditary factors and by environmental conditions. The chapter on "Variation in the number of the chromosomes and its relation to the totality of the genes" deals with triploidy and tetraploidy, and recent work indicating deficiency, duplication of factors in a chromosome, and transposition of factors from one chromosome to another. The chapter on mutation includes the explanation of pseudo-mutations by balanced lethals. In "The particulate theory of heredity and the nature of the gene" the author discusses the relation of the genetic factor or gene to somatic characters and to ontogeny.—*Alexander Weinstein.*

423. Morgan, T. H. **Contributions to the genetics of Drosophila melanogaster. IV. A demonstration of genes modifying the character "notch."** Carnegie Inst. Washington Publ. 278. *P. 343–388. 1 pl., 15 fig.* Washington, D. C. 1919.—Notch is a dominant sex-linked gene affecting wings, lethal when homozygous; consequently all notch flies are female and heterozygous. Mass selection in the direction of slight notching, carried out through 24 generations of *Drosophila melanogaster*, resulted in marked change in direction of selection. Extreme selected females, out-crossed to wild-type flies, gave ordinary notch in first generation, showing notch gene unmodified. Linkage relations demonstrated results of selection due to recessive modifying factor in second chromosome. Second experiment (19 generations) gave similar results; crosses showed effect due to same modifier in both cases.—A modification in opposite direction, called "short notch," appeared several times; outcrosses to wild flies gave ordinary notch. Linkage relations showed this modification due to recessive modifier in first chromosome.—Notch gene is always necessarily heterozygous, but all results show no "contamination" by its normal allelomorph. Other mutations, modifying wings in somewhat similar or different ways, were all located in other chromosomes or different loci in X chromosome, thus showing them independent of notch.—High sex-ratios (76:1 and 119:10), given by two notch females, were undoubtedly due to lethal mutation in not-notch X chromosome, as shown in other cases. Only those few sons having crossover X survive.—*C. R. Plunkett.*

424. Morgan, T. H., and C. B. Bridges. **Contributions to the genetics of Drosophila melanogaster. I. The origin of gynandromorphs.** Carnegie Inst. Washington Publ. 278. *122 p., 4 pl., 10 fig.* Washington, D. C. 1919.—The genetic situation in *Drosophila melanogaster* made possible experimental demonstration of causes of production of mosaics and gynandromorphs (sex-mosaics). Principal recent theories are: delayed fertilization of one cleavage nucleus (Boveri 1888); development from a supernumerary sperm (Morgan 1905); and chromosomal elimination, i.e., elimination of one X chromosome from one of daughter cells at an early embryonic division (Morgan 1914). Critical evidence is obtained when gynandromorphs are hybrids of known sex-linked characters, and also contain known autosomal characters. A number of such cases, all described in detail, all show male and female parts differ by sex-chromosome only. The elimination theory is only possible one in these cases, and covers all but very few gynandromorphs in *Drosophila*.—Gynandromorphs start as females; a striking preponderance of female parts is found, as expected on elimination theory. Starting as a male is theoretically possible, but not indicated in any known cases. Starting as XX female, the male parts will be XO, therefore sterile (as shown in primary non-disjunction); except in case of XXY (non-disjunctional) individuals, where male parts will be XY, fertile. All evidence from gynandromorphs with male abdomen and testes supports these predictions. —Earlier theories of gynandromorphs are critically considered. The only one besides elimination found necessary to employ, in a few cases, is the theory of bi-nucleated eggs. Doncaster has found such eggs in *Abraxas*.—Both gonads of same individual are always alike; which is expected if germ plasm of *Drosophila* arises from single cell, as in *Miastor*, *Chironomus*, *Calliphora*, and other flies.—Only one certain case was found of a somatic mosaic, i.e., one not involving sex-chromosome; may be accounted for by autosomal elimination or bi-nucleated egg. Rarity may be due to failure of autosomal elimination or to inviability of such flies.—Ten somatic mutations described are all males, of which nine look like known sex-linked characters. This is in accord with expectation, if mutation occurs in only one chromosome of a pair, as is highly probable; since visible sex-linked mutations are four times as frequent as all dominants. Mosaics in plants are discussed; somatic mutation or chromosome elimination the most probable explanations in most cases.—All known gynandromorphs of Drosophila are thoroughly treated as to parentage, description, and explanation, with figures and diagrams of chromosomes. The great majority are adequately explained by simple X elimination, including a number from XXY mothers. Many are approximately bilateral, others largely antero-posterior, some mainly female, a few mainly male, and a few very irregular. In all, the male and female parts and their characters are strictly self-determining. No region, however small, is interfered with by neighboring parts or action of the gonad. The few cases not explicable by simple elimination are most simply explained as binucleated

eggs; but on this view there should be as many autosomal mosaics as gynandromorphs of this type, which is not the case. An alternative explanation is non-disjunction, followed by either "somatic reduction" or double elimination in a cleavage division; no critical evidence to decide between these views.—Gynandromorphs in other animals are discussed at length. In bees, both Eugster and von Engelhardt gynandromorphs can be accounted for by chromosomal elimination, so far as the evidence goes. In moths, those cases where sex-linked factors furnish critical evidence can be explained by chromosome elimination; here the gynandromorphs start as males (ZZ). This explanation applies to two mosaics in *Abraxas*. Toyama's gynandromorphs in silk-worms can be explained as bi-nucleated eggs. Goldschmidt's mosaics in the gypsy moth can not be explained because there are no sex-linked factors involved.—In Crustacea, molluscs, and some worms (e.g., *Bonellia*) external conditions and age seem, in some cases, to be factors in determining sex; there may be genetic factors that determine sex under ordinary, or other, circumstances.—In birds, a few bilateral gynandromorphs are known. Internal secretions of the ovary are known to suppress male secondary sexual characters in most cases. Apparently particular differences, in some species, are not influenced.—In man and other mammals, cases of gynandromorphs are known. Mechanism of sex determination is the same as in *Drosophila*. Modification by hormones also possible. Freemartin caused by male sex-hormone, through common circulation, suppressing normal development of ovary (Lillie). Possibility is suggested that cancer may be conditioned by inherited gene or genes liable to frequent somatic mutation or chromosome aberrations.—*C. R. Plunkett.*

425. Mosséri, V. M. **Egyptian cottons: Their deterioration and means of remedying it.** Bull. Union Agric. Egypte 16: 53–79. 1918.—Supposed greater resistance to "pink boll worm" (*Pectinophora gossypiella*) of certain varieties of cotton in Egypt said to be due merely to greater precocity. In India, supposed home of this insect, however, native cottons appear really more resistant than introduced Egyptian cotton. Deterioration of varieties grown in Egypt believed to be caused by mixing of seed and by natural hybridization, rather than by any process of spontaneous degeneration. Three methods of procedure are suggested for improvement of Egyptian cotton crop: (1) "Mendelian synthesis" as practiced by Balls; (2) selection and roguing to increase uniformity of existing varieties; (3) isolation of desirable mutants which originate new varieties.—*T. H. Kearney.*

426. Myerson, Abraham. **Mental disease in families.** Mental Hygiene 3: 230–239. Apr., 1919.—Author used records of Taunton State Hospital from 1854 to 1916 covering 16,000 persons, of whom 1547 were related. He compared the marriage rate of four groups—alcoholic insanities, general paresis, dementia praecox and senile dementia. In the first three groups the percentage of married males was found to be less than for females, in the seniles the reverse was true. The dementia praecox group showed the lowest fertility as compared with the total population. He concludes that marriage acts as barrier to propagation of endogenous diseases, such as dementia praecox, but not against exogenous, such as syphilis.—The preponderance of insane women recorded may be accounted for on the theory that women transmit their mental peculiarities to their female children more than to their male, but there is a more obvious explanation. Since men migrate to other districts more than women, female descendants are more likely to appear in a given asylum. The data at this particular institution show the mother-daughter group to be the largest and sisters decidedly outnumber brothers.—Notwithstanding the numerous factors tending to discount the actual meaning of the figures, author considers it probable that descendants of insane who themselves become insane do so at an earlier age than their ancestors and are tending to reproduce themselves in smaller proportion.—With regard to the character of transmission his findings lead him to believe that (1) The paranoid type of psychosis gives either paranoid or dementia praecox. (2) Dementia praecox gives dementia praecox or feeblemindedness. (3) Manic depression gives manic depression or dementia praecox. (4) Involution psychosis gives dementia praecox. (5) Senile psychosis gives any form of psychosis, imbecility or epilepsy.—Thus all roads seem to lead to dementia praecox and thence to feeble-mindedness.—His results further indicate that insanity among siblings tends to be similar, and that it is more often associated

with low-grade mentality than superior. This is at variance with the popular notion of the close relationship between genius and insanity.—The high incidence of tuberculosis with insanity often leads to mistaken inferences.—The extreme frequency of tuberculosis in the total population must be remembered as well as the fact that the insane, by reason of their deterioration, tend to live in conditions predisposing to the disease.—Two other students, Koller and Diem, discovered that insane aunts and uncles occur as frequently in families of sane as of insane and that, therefore, collateral insanity is relatively unimportant unless associated with parental insanity.—These studies demonstrate that our knowledge is inadequate to warrant theories of neuropathic heredity and how imperative such research is.—*Miriam C. Gould.*

427. Nachtsheim, H. **Der Mechanismus der Vererbung.** [**The mechanism of heredity.**] Naturw. Wochenschr. 18: 105–114. 1919.

428. Nachtsheim, H. **Berichtigung.** [**A correction.**] Zeitschr. indukt. Abstamm. Vererb. 20: 295. 1919.

429. Nakahara, Waro. **A study on the chromosomes in the spermatogenesis of the stone-fly, Perla immarginata Say, with special reference to the question of synapsis.** Jour. Morphol. 32: 509–529. *Pl. 1–3.* 1919.—Ten chromosomes appear in the spermatogonia division. The chromosome group consists of two pairs of V's, a pair of rods, two spherules (m-chromosomes), and two unpaired rods, one of which is much longer than the other. These last are interpreted as the X- and Y-chromosomes, respectively. Preparatory to the first spermatocytic division a double spireme forms out of the resting nucleus, and this process the author interprets as a precocious split for the second spermatocytic division, which follows the first without a resting stage. Homologous chromosomes are connected to each other telosynaptically in the spireme; later, the members of each pair bend toward each other at the synaptic point and become reunited parasynaptically before the metaphase, thus forming rings and tetrads.—*Bertram G. Smith.*

430. Nelson, J. C. **Monomorphism in Equisetum Telmateia Ehrh.** Amer. Fern Jour. 9: 93–94. 1919.

431. Nicolas, G. **Variations de l'androcée du Stellaria media L. en Algérie.** [**Variations of the androecium of Stellaria media L. in Algeria.**] Bull. Soc. Hist. Nat. Afr. Nord. 9: 135–137. 1918.

432. [Norstedt, C. T. O.] [Rev. of: Harms, U. **Über die Geschlectsvertheilung bei Drya octopetala L. nach Beobachtungen in Kgl. Botanischen Garten Berlin-Dahlem.** (**Concerning sex ratios in Drya octopetala in the Kgl. Botanical Garden Berlin-Dahlem.**) Ber. Deutsch. Bot. Ges. 36: 292–300. *Fig. 5–10.* 1918.] Bot. Notiser 1918: 247. 1918.

433. Northrop, J. H. **Concerning the hereditary adaptation of organisms to higher temperature.** Jour. Gen. Physiol. 2: 313–318. 1920.—The experiments described were performed with races of *Drosophila* raised on sterile yeast cultures and handled with bacteriological care to prevent the entrance of bacteria into the breeding flasks. The incubators employed to maintain the higher temperatures were controlled within 0.2° to 0.3°C. of the desired temperatures by means of an original device regulating the flow of water through the jackets. *Drosophila* will develop at 32.5°C.; the rate of development increases from 10° up to 27.5°, but from 27.5° the rate falls. If the higher temperature in which a fly is raised occasions a lasting adaptation, it would be expected that eggs from such a fly would show increased resistance to high temperature. It was found that flies raised at 20°C. produce eggs that are capable of full development when raised in temperatures 29° and 32°C., but when raised in a temperature of 33° they will not go beyond the pupal stage. Flies raised in incubators at 32° produce eggs that will develop into adults when raised at 29°, but at 32° and 33° they will not even form larvae. The difference in these two sets of results is not due to deleterious effects

of increased temperature upon the eggs before they are laid, because the flies raised at 20° did not tend to produce eggs any less resistant after they had been laying in the high temperature for a week or 10 days. Cultures of flies could not be held at 30° for successive generations; but if the adults of each generation were removed from high temperature for 24 hours or more within a week after they hatched, the culture could be continued for the rest of the time at this temperature. One culture was continued in 30° by means of this intermittent cooling for ten generations and another culture was raised for 15 generations uninterruptedly at 28°; in neither case did there appear any sign of adaptation. The flies were still unable to produce more than one generation at a continuous temperature of 29° or over. "There is no evidence of any hereditary adaptation to higher temperature."—*E. C. MacDowell.*

434. OBERSTEIN, O. **Über das Vorkommen echter Knospenvariationen bei pommerschen und anderen Kartoffelsorten.** [Occurrence of true bud variation in Pommeranian and other varieties of potato.] Deutsch. Landw. Presse 1919: 560–561. *1 pl.* 1919.—See Bot. Absts. 5, Entry 296.

435. OHLY. **Züchterische Beobachtungen in einer Merinofleischschafherde.** [Breeding observations in a Merino sheep herd.] Mitteil. Deutsch. Landw. Ges. 1918: 235. 1918.

336. PASCHER, A. **Oedogonium, ein geeignetes Objekt für Kreuzungsversuche an einkernigen haploiden Organismen.** [Oedogonium, a suitable object for the study of crossing in uninucleate haploid organisms.] Ber. Deutsch. Bot. Ges. 36: 168–172. 1918.—Importance of study of results of crossing haploid organisms is emphasized, as illustrated by the work of Burgeff with *Phycomyces* and of Pascher with *Chlamydomonas.* Author reports successful crosses between two species of *Chara* and between two species of *Spirogyra.* After a discussion of the advantages, disadvantages, and difficulties offered by various groups of algae for work of this nature, it is reported that species of *Oedogonium* have shown themselves very favorable for hybridization experiments. Most species of this genus are easily cultivated; the isolation of single filaments and the bringing them together in desired combinations within a confined space, such as a small tube, offer no difficulties; the filaments with maturing oospores can be transferred to agar, where they readily complete their development; the zoospores of different species are marked by characteristic differences in such respects as the shape of the cell as a whole and the form of the anterior end; and the oospore, on germinating, gives rise to four zoospores, whose nuclei result from the reduction divisions, and which resemble, except in size, the zoospores produced by vegetative cells of the same species. In making a cross, the female at least must belong to a dioecious species. Probably dioecious forms with dwarf males are especially suitable. In cultures containing several species, the author has found forms which, especially in the characters of the oospores, betrayed a hybrid nature. It is probable that some forms which have been described as species were really hybrids. A list of species of *Oedogonium* is given which are recommended for experiments in hybridization.—*C. E. Allen.*

437. PEARL, RAYMOND. [Rev. of: EAST, EDWARD M., AND DONALD F. JONES. **Inbreeding and outbreeding: their genetic and sociological significance.** *14 x 21 cm., 285 p., 46 fig.* J. B. Lippincott: Philadelphia, 1919.] Science 51: 415–417. April 23, 1920.—See Bot. Absts. 4, Entry 571.

438. [PEARSON, KARL.] **Quadrature coefficients.** Biometrika 12: 000. Nov., 1919.—Formulae from Biometrika I, p. 276, are reprinted as preface to a table by P. F. EVERITT to facilitate the calculation of areas within a curve.—*John W. Gowen.*

439. PETRÉN, A., AND OTHERS. **Angaende skrivelse till Konungen med begäran om utredning och förslag i fråga om upprättandet av ett svenskt rasbiologiskt institut.—Motion n:o 7 i Första Kammaren.** [Concerning a writing to the Swedish government proposing an extrication of and a project to establish a Swedish eugenical institute. Motion n:o 7. in the first Chamber of the parliament. Bihang till riksdagens protokoll 1920. *190 x 225 mm., 27 p.* Stock-

holm, 1920.—Mentions reasons for and importance of establishing a race-biological institute. Parliament is asked to demand a special proposal for the organization of such an institute.—*K. V. Ossian Dahlgren.*

440. Piltz, J. **Über homologe Heredität bei Zwangsvorstellungen. [On homologous heredity in hallucination.]** Zeitschr. ges. Neur. u. Psych. 43. 1918.

441. Plunkett, C. R. **Genetics and evolution in Leptinotarsa.** Amer. Nat. 53: 561–566. Nov.–Dec., 1919.—Tower's work is almost entirely in agreement with the modern Mendelian theory of heredity. Where there is apparent disagreement, critical evidence is lacking because of Tower's failure to subject the individuals he worked with to a rigorous genetic analysis.—*Alexander Weinstein.*

442. Ragionieri, Attilio. **Un bel problema per i biologi: Sulla comparsa dell' odore nei fiore delle "roselline di Firenze" (Ranunculus asiaticus var.). [A good problem for biologists: On the appearance of odor in the flowers of the "Florentine roselline" (Ranunculus asiaticus).]** Bull. R. Soc. Toscana Orticult. 44: 87–94. 1919.—See Bot. Absts. 4, Entry 1832.

443. Rasmuson, Hans. **Genetische Untersuchungen in der Gattung Godetia. [Genetical investigation within the genus Godetia.]** Ber. Deutsch. Bot. Ges. 37: 399–403. 1919.—A very condensed preliminary note about author's experiments with *Godetia Whitneyi* and *G. amoena.* Branching habit, leaf-characters, color, size, form and doubleness of the flowers, are analyzed.—*K. V. Ossian Dahlgren.*

444. Raum, J. **Ein weiterer Versuch über die Vererbung die Samenfarbe bei Rotklee. [A further study on the inheritance of seed color in red clover.]** Zeitschr. Pflanzenzücht. 7: 148–155. Dec., 1919.

445. Rebel, H. **Ein neuer Tagfalterhybrid. [A new butterfly hybrid.]** Verhandl. K. u. K. Zool. Bot. Ges. Wien 68: 273–276. 1918.

446. Richet, C., and H. Cardot. **Mutations brusques dans la formation d'une nouvelle race microbienne. [Sudden mutations in the formation of a new race of microbes.]** Compt. Rend. Acad. Sci. Paris 168: 657–663. 1919.

447. Roberts, Herbert F. **A practical method for demonstrating the error of mean square.** School Sci. Math. 19: 677–692. Nov., 1919.—This paper treats of the mean, the standard deviation and coefficient of variation with especial reference to practical methods of illustrating the error of the mean square to students of little training in mathematics.—*John W. Gowen.*

448. Roemer, Th. **Über Lupinenzüchtung. [On Lupine breeding.]** Deutsch. Landw. Presse 1919: 174–175. 1919.—See Bot. Absts. 5, Entry 299.

449. Rother, W. **Phyllokakteen Kreuzungen. [Phyllocactus crosses.]** Monatsschr. Kakteenkunde 29: 32–33. 1919.—Reciprocal crosses of *P. Wrayi* and *P. Vogelii* are described and differentiated.—*A. S. Hitchcock.*

450. Ružička, Vladislav. **Restitution und Vererbung. Experimenteller, kritischer und synthetischer Beitrag zur Frage des Determinationsproblems. [Restitution and heredity. Experimental critical and synthetic contribution to the problem of determination.]** Julius Springer: Berlin, 1920.

451. St. John, Harold. **Two color forms of Lobelia cardinalis L.** Rhodora 21: 217–218. 1919.—Describes variation in color of flowers of *Lobelia cardinalis.* A form with rose-colored flowers, found in New Hampshire, is named *f. rosea.* One with white flowers was named *alba* by A. Eaton in 1836.—*T. D. A. Cockerell.*

452. Schindler, F. **Bedeutung der Landrassen unserer Kulturpflanzen. [Significance of local varieties of our cultivated plants.]** Deutsch. Landw. Presse 1918: 155. 1918.

453. Schmidt, Johs. **La valeur de l'individu a titre de générateur, appréciée suivant la methode du croisement diallèle. [Individual potency appraised by the method of diallel crossing.]** Compt. Rend. Trav. Lab. Carlsberg 14: 1–33. 1919.—See Bot. Absts. 5, Entry 302.

454. Schroeder. **Entstehung und Vererbung von Missbildungen an der Hand eines Hypodaktylie-Stammbaumes. [Origin and inheritance of deformities in a hypodactylous pedigree.]** Monatsschr. Geburtshilfe Gynäkologie 48: 210–222. *3 pl. 7 fig.* 1918.

455. Shamel, A. D. **Performance records of avocados based on citrus experiments.** California Citrograph 5: 68, 86–88. *1 fig.* Jan., 1920.—Description of methods recommended for obtaining records of yield and quality of fruit, hardiness, and other horticulturally important characteristics of avocado trees, as basis for selection of desirable types for propagation. Organization suggested similar to the "bud selection department" of the California Fruit Growers' Exchange, which last season sold 230,000 citrus buds taken from superior trees.—*H. B. Frost.*

456. Siegel, W. **Das Recht des Gemüsezüchters. [The right of the vegetable breeder.]** *8vo.* Frick: Wien. 1919.—See Bot. Absts. 5, Entry 304.

457. Siemens, H. W. **Erbliche und nichterbliche Disposition. [Hereditary and nonhereditary disposition.]** Berlin. Klin. Wochenschr. 56: 313–316. 1919.

458. Siemens, H. W. **Über die Grundbegriffe der modernen Vererbungslehre. [On the fundamental concepts of modern genetics.]** Münchener Med. Wochenschr. 65: 1402–1405. 1918.

459. Siemens, H. W. **Was ist Rassenhygiene? [What is race hygiene?]** Deutschlands Erneuerung 2: 280–282. 1918.

460. Smith, L. H. **The life history and biology of the pink and green aphid (Macrosiphum solanifolii Ashmead).** Virginia Truck Sta. Bull. 27: 27–79. *12 fig.* 1919.—Much variation among individuals is found with respect to size of parts, color and reticulation within well-known pink and green varieties. No inheritance of size variations has been noted. Strains that differ from one another have been obtained. Sexual forms are not usually produced in Virginia. Spring migrants are chiefly of green variety. Nineteen first-born and eight last-born generations were reared from May to November, and 34 first-born generations in a twelve-month period. Four molts occur. Average age at beginning of reproduction is eleven days, average number of young produced by viviparous female is 45 during lifetime averaging 31 days.—*A. Franklin Shull.*

461. Snell, K. **Farbenänderung der Kartoffelblüte und Saatenanerkennung. [Color changes of the potato blossom and the recognition of varieties.]** Der Kartoffelbau 1919: 1–3. 1919.—See Bot. Absts. 5, Entry 306.

462. Sommer, K. **Über Kartoffelzüchtung und vergleichende anbauversuche mit Neuzüchtungen auf der Domäne Ellischau. [Potato breeding and comparative cultural tests of new varieties on the Ellischau estate.]** Nachr. Deutsch. Landw. Ges. Österr. 1919: 190–193. 1919.—See Bot. Absts. 5, Entry 307.

463. Stahel, G. **Eerste verslag over de werkzaamheden ten behoeve van de selectie van Koffie en Cacao. [First report on the effectiveness of selection in coffee and cacao.]** Dept. Landbouw in Suriname (Paramaribo) Bull. 36. *23 p.* 1919.—See Bot. Absts. 5, Entry 308.

464. Stieve, H. **Über experimentell, durch veränderte äussere Bedingungen hervorgerufene Rückbildungsvorgänge am Eierstock des Haushuhnes (Gallus domesticus). [On degenerative processes in the ovary of domestic fowl produced experimentally by changed external conditions.]** Arch. Entwicklungsmech. Organ. **44**: 530–588. *10 fig.* Sept., 1918.—Laying fowls were removed from their normal quarters and placed in close confinement. After various intervals the birds were killed and the ovaries examined. In all cases egg production ceased. If the birds were well fed, production was not resumed. The large ova were not resorbed for several months, though degenerative changes took place in the nucleus, which extended to smaller and smaller ova, the longer the birds were kept. If, however, the birds were starved or kept on limited diet for a time, and then fed suitably, the large ova were quickly resorbed, the degenerative changes did not extend to the small ova, and production was resumed after a comparatively brief interval.—*H. D. Goodale.*

465. Stout, A. B. **Further experimental studies on self-incompatibility in hermaphrodite plants.** Jour. Genetics **9**: 85–129. *Pl. 3–4.* Jan., 1920.—Two self-sterile plants of *Verbascum phoeniceum* were crossed. In F_1, 58 plants were self-sterile, 9 bore some seeds, and 2 were highly self-fertile. From a highly self-fertile plant of this species there were raised (in addition to 27 plants with contabescent anthers) 5 self-sterile plants, 2 plants with some seeds, and 5 highly self-fertile plants.—Sowings made from open-fertilized or commercial seeds of *Eschscholtzia californica*, *Nicotiana Forgetiana*, *Brassica pekinensis*, and *Raphanus sativus*, showed a majority of self-sterile, and a minority bearing few or many seeds. The descendants of each of two self-fertile plants of *Nicotiana Forgetiana* showed a majority of more or less self-fertile plants.—In *Cichorium intybus*, 10 plants were uniform as to self-fertility or self-sterility throughout the blooming period. Of the descendants of 3 self-fertile plants, 244 were self-sterile, and 107 bore some seeds. In the next selfed generation, 205 plants were self-sterile, and 266 self-fertile in various degrees.—It is concluded that self-sterility in some species is highly variable.—*John Belling.*

466. Sturtevant, A. H. **Contributions to the genetics of Drosophila melanogaster. III. Inherited linkage variations in the second chromosome.** Carnegie Inst. Washington Publ. **278**: 305–341. Washington, D. C. 1919.—The data presented demonstrate two genes in second chromosome of *Drosophila melanogaster*, each of which, in females heterozygous for it, greatly decreases crossing-over in region in which it lies. Both genes were found in same female, in stock from Nova Scotia. C_{IIl}, located to left of black, makes star black = 0, and black purple very small. C_{IIr}, located between purple and plexus, greatly reduces purple speck region. Homozygous C_{II} shows no effect on crossing-over; homozygous C_{IIl} not tested. No crossing-over in males, as always.—C_{III}, located in right end of third chromosome, greatly decreased crossing-over between spineless and rough when heterozygous, but increases it when homozygous. $C_{III,II}$, in third chromosome, when heterozygous decreases crossing-over in third chromosome, but increases purple curved region of second.—Mechanism of these effects is still unknown. Other linkage variations are caused by sex, age, temperature, and genetic factors. In all cases, linear order of genes is unchanged, and flies of same constitution, under like conditions, give consistent results. The methods and results are striking confirmation of chromosome view of heredity.—*C. R. Plunkett.*

467. Sturtevant, A. H. **A new species closely resembling Drosophila melanogaster.** Psyche **26**: 153–155. *1 fig.* Dec., 1919.—Describes *Drosophila simulans*, new species that has hitherto been confused with *D. melanogaster*. New form is common and widely distributed. Specimens can be separated easily only by means of male genitalia. Female *melanogaster* × male *simulans* produces only daughters, unless the mother carries a Y-chromosome. The hybrids are all sterile.—*A. H. Sturtevant.*

468. Sturtivant, Grace. **Registration of new varieties.** Gard. Chron. **67**: 73. Feb. 14, 1920.—Plant patents seem impossible in the United States; but the registration of new varieties is important. It is suggested that higher awards should be given for plants in gar-

dens than for those at exhibitions. The custom of bracketing the breeder's name after the name of the variety is spreading among *Iris* specialists. Parentage should be put on record.—*John Belling.*

469. Sumner, F. B. **Continuous and discontinuous variations and their inheritance in Peromyscus.** Amer. Nat. 52: 177–208. *12 fig.* April–May, 1918.—Discusses in this first paper structural and pigmental differences in the western deer mouse, *Peromyscus maniculatus* (Wagner) based on collections from four climatically different localities in California,—Eureka, Berkeley, LaJolla, and Victorville. Humidity and rainfall are in a descending, and mean annual temperature in an ascending, order for localities as given. Considers hair color including microscopical structure, skin color, length of body, tail, foot, and ear, and number of tail vertebrae, illustrating by histograms and ordinary graphs.—Finds for pigmentation, intensive and extensive, series is Eureka>Berkeley>LaJolla>Victorville. For tail length Eureka>LaJolla>Berkeley and Victorville. For number of caudal vertebrae, Eureka>LaJolla>Victorville. For foot length, Eureka>LaJolla, Berkeley and Victorville. Ear length LaJolla>Eureka and Victorville>Berkeley. General conclusions reserved for final paper.—*L. B. Walton.*

470. Tammes, T. **Die Flachsblüte.** [**The flower of flax.**] Recuéil· Trav. Bot. Neérland. 15: 185–227. *22 fig.* 1918.—See Bot. Absts. 5, Entry 310.

471. Taylor, H. V. **The popularity and deterioration of potatoes.** Gard. Chron. 67: 108. Feb. 28, 1920.—New potato varieties are usually lower in quality than old standard varieties but at the same time are more resistant to diseases and adverse conditions. With cultivation and propagation the qualities improve, but vigor and disease resistance decreases. These simultaneous changes are held responsible for the appearance of six varieties which have attained popularity and each after ten to fifteen years have been succeeded in turn by another new variety.—*J. L. Collins.*

472. Thellung, A. **Neure Wege and Ziele der botanischen Systematik erläutert am Beispiele unserer Getreidearten.** [**New methods and purposes of botanical taxonomy illustrated by examples of our cereal species.**] Naturw. Wochenschr. 17: 449–458. 465–474. *3 fig.* 1918.

473. Thellung, A. **Über geschlechtsbegrenzte Speziesmerkmale (zu dem Aufsatz von Brehm).** [**On sex-limited species characters (in response to von Brehm).**] Naturw. Wochenschr. 18: 144. 1919.

474. Thomas, Roger. **The improvement of "Tinnevellies" cotton.** Agric. Jour. India 14: 315–330. 1919.

475. Turesson, Göte. **The cause of plagiotropy in maritime shore plants.** Contributions from the plant ecology station, Hallands Väderö, No. 1. Lunds Universitets Årsskrift. N. F. 16[2]: 1–33. *15 tables, 4 fig., 2 pl.* 1919.—The prostrate form of some shore plants is demonstrated to depend upon geotropism induced by brilliant sunlight ("photocliny"). In obscure light the geonegative reaction becomes predominant.—From one hereditary point of view it is interesting to find that the prostrate vegetation can be made up of two genetically different elements, viz., modificatory prostrate forms, and hereditary prostrate variations. Both forms are sometimes found within the same systematic species. *Atriplex latifolium, A. ratulum* and *Chenopodium album* have each a forma "*prostratum*," which is constantly plagiotropic; the main species are only plagiotropic in intense light and erect in ordinary light. When growing together on exposed beach it may be difficult to separate the two types, and cultivating of them becomes necessary. By self-fertilization the *prostratum* form of both the *Atriplex*-species is found to breed true to plagiotropy.—"The hereditary prostrate variations differ physiologically from the prostrate modifications in being more sensitive to light; they respond to conditions of illumination which leave the latter unaffected and in a vertical position." Author supposes that the prostrate races have come into existence by dropping out of "height"-determining factors.—*K. V. Ossian Dahlgren.*

476. Urban, J. **Hochpolarisierende Rübe und ihre Nachkommenschaft. [High-polarizing beets and their progeny.]** Zeitschr. Zucker Industr. Böhmen 42: 387–391. 1919.—See Bot. Absts. 5, Entry 312.

477. Vaerting, M. **Die verschiedene Intensität der pathologischen Erblichkeit in ihrer Bedeutung für die Kriegsdegeneration. [Different intensity of pathological inheritance and its significance for war degenerations.]** Der Frauenarzt. 1918.

478. van der Wolk, P. C. [German rev. of: van der Wolk, P. C. **Onderzoekingen over blijvende modificaties en hun betrekking tot mutaties. (Researches on permanent modifications and their relations to mutations.)** Cultura 31: 82–105. *1 pl.* 1919. (See Bot. Absts. 3, Entry 296.)] Zeitschr. Pflanzenzücht. 7: 142–144. Dec., 1919.

479. Vernet, G. **Biométrie et homogénéité. [Biometry and homogeneity.]** Bull. Agric. Inst. Sci. Saigon 2: 15–26. 1920.

480. Vieillard, P. **Note sur la sélection des riz par la constitution de lignées pures et sur les hybridations des riz. [Note on the selection of rice by establishment of pure lines and on the hybridization of rice.]** Bull. Agric. Inst. Sci. Saigon 2: 11–15. 1920.

481. Vogt, A. **Vererbung in der Augenheilkunde. [Heredity in ophthalmology.]** Münchener Med. Wochenschr. 66: 1–5. 1919.

482. Volkart, A. **40. und 41. Jahresbericht. Schweizerische Samenuntersuchungs- und Versuchsanstalt in Oelikon-Zürich. [40th and 41st annual report. Swiss seed-control and experiment station in Oerlikon-Zürich.]** Landw. Jahrb. Schweiz. 1919: 1–40. 1919.—See Bot. Absts. 5, Entry 313.]

483. von Bubnoff, Serge. **Über einige grundlegende Prinzipien der paläontologischen Systematik. [Some fundamental principles of paleontological taxonomy.]** Zeitschr. indukt. Abstamm. Vererb. 21: 158–168. Sept., 1919.—Wedekind was followed in his application of the statistical rules of variation to paleontological material. Two very common Triassic ammonites from one locality were studied in hundreds of specimens. A form had been separated from each and named as a species on account of a single and doubtful difference. When the variates were seriated, the supposed separate forms gave in each case a single typical variation curve along with the species. This shows that the difference in question was not sufficient to distinguish species, or even varieties; and races, or "elementary species," cannot be dealt with in paleontology.—A correlation between two or more characteristics was obtained by comparing different stages of growth, or by comparing closely allied species. Characteristics which are correlated in this fashion should vary together if the variation is genetic. They did not vary together in a trial of individuals of the same species. Hence this correlation is a test of specific difference.—*John Belling.*

484. von Caron-Eldingen. **Physiologische Spaltungen ohne Mendelismus. [Physiological segregation without Mendelism.]** Deutsch. Landw. Presse 1919: 515–516. 1919.—See Bot. Absts. 5, Entry 314.]

485. von Caron-Eldingen. **Mutationen und Doppelkörner. [Mutations and double grains.]** Deutsch. Landw. Presse 45: 618. *3 fig.* 1918.

486. von Caron-Eldingen. **Physiologische Spaltungen oder vegetative Mutation (Meinungsaustausch). [Physiological splitting or vegetative mutations.]** Deutsch. Landw. Presse 46: 56. 1919.

487. von Graevenitz, Luise. **Ein merkwürdiges Resultat bei Inzuchtsversuchen. [A remarkable result in an inbreeding experiment.]** Zeitschr. indukt. Abstamm. Vererb. 21: 169–173. Sept., 1919.—Effects of four different types of pollination compared on the off-

spring of three plants, *Petunia*, *Digitalis* and *Oenothera*. Flowers of individual plants treated with pollen from following sources: (1) from the same flower, (2) from other flowers on the same plant, (3) from a sister plant, (4) from a plant of a different strain. In all but the first the flowers were castrated. For (1) and (2) the same lot of pollen was used and applied at the same time. Fifty-two plants of *Petunia* were pollinated in this way and the progenies of each, numbering at least 50 individuals in each class, were weighed. The results show that in 37 cases the (2)-pollinated plants were heavier than (1) while in 15 cases the reverse holds. The other two types of pollination resulted in still heavier plants on the average according to the dissimilarity of the parents. Four plants of *Digitalis* treated in like manner show the same result, the cross-pollination between different flowers of the same plant give heavier offspring than self-pollination within the individual flower. *Oenothera* gave no differences. *Antirrhinum*, although not fully investigated, shows a difference between the pollinations. Author is unable to find any circumstances which might account for these effects and considers them to be biologically not understandable.—*D. F. Jones.*

488. von Oettingen. **Die Vererbung erworbener Eigenschaften (aus dem Werke der Pferdenzucht von Oberlandstallmeister von Oettingen). [The inheritance of acquired characters (from the work in horse-breeding by von Oettingen).]** Deutsch. Landw. Tierzucht. **23**: 7. 1919.

489. von Ryx, G. **Ein neues Beispiel einer Knospenmutation bei den Kartoffeln. [A new example of bud mutation in potatoes.]** Deutsch. Landw. Presse 2. *1 fig.* 1918.

490. von Tschermak, A. **Der gegenwärtige Stand des Mendelismus und die Lehre von der Schwächung der Erbanlagen durch Bastardierung. [The present status of Mendelism and the doctrine of the weakening of hereditary units through hybridization.]** Naturw. Wochenschr. **17**: 509–611. 1918.

491. von Tschermak, Erich. **Über Züchtung landwirtschaftlich und gärtnerisch wichtiger Hülsenfrüchter. [Breeding of agriculturally and horticulturally important legumes.]** Arb. Deutsch. Landw. Ges. **1919**: 80–106. 1919.

492. von Tschermak, Erich. **Bastardierungsversuche mit der grünsamigen Chevrier-Bohne. [Hybridization studies with the green-seeded Chevrier bean.]** Zeitschr. Pflanzenzücht. **7**: 57–61. June, 1919.

493. von Tschermak, E. **Beobachtungen bei Bastardierung zwischen Kulturhafer und Wildhafer. [Observations on hybridizations between cultivated oats and wild oats.]** Zeitschr. Pflanzenzücht. **6**: 207–209. 1918.

494. von Tschermak, E. **Beobachtungen über anscheinende vegetative Spaltungen an Bastarden und über anscheinende Spätspaltungen von Bastardnachkommen speziell Auftreten von Pigmentierungen an sonst pigmentlosen Deszendenten. [Observations on apparent vegetative splitting in hybrid offspring, especially the occurrence of pigmentation on otherwise pigmentless descendants.]** Zeitschr. indukt. Abstamm. Vererb. **21**: 216–232. *1 fig.* Nov., 1920.

495. von Ubisch, G. **Gerstenkreuzungen. [Barley crosses.]** Landw. Jahrb. **53**: 191–244. *3 pl., 23 fig.* 1919.—See Bot. Absts. 5, Entry 315.

496. Waller, A. E. **Xenia.** School Sci. Math. **19**: 150–157. Feb., 1919.—Popular account of xenia to which nothing new is added.—See also Bot. Absts. 5, Entry 115.—*J. H. Kempton.*

497. Walter, F. K. **Über "familiäre Idiotie." [On familial idiocy.]** Zeitschr. ges. Neur. u. Psych. **40**. 1918.

498. WEBBER, HERBERT JOHN. **Selection of stocks in citrus propagation.** California Agric. Exp. Sta. [Berkeley] Bull. 317: 267–301. *4 tables, 14 fig.* Jan., 1920.—The individual trees in citrus orchards are always markedly variable in yield, doubtless partly because of variation in the stocks used in budding. Sweet orange and sour orange are principal citrus stocks in California. Seeds of each species have usually been collected indiscriminately; seedlings are always highly variable, yet few are usually discarded in nursery.—Tests at Citrus Experiment Station showed that large, intermediate and small nursery trees of three standard Citrus varieties retained their original size rank after two years in orchard, though selected in nursery budded from "performance-record" trees, where many of smaller stocks had been discarded at transplanting and some also at budding. Sweet-orange and sour-orange seedlings selected in nursery rows for variation in leaf form, habit, etc., and budded on sour-orange stocks in duplicate, indicate presence of numerous genetic types, some undesirable, among ordinary nursery stocks. Measurements in nursery of sour-orange stocks sorted at transplanting showed great variation, with much greater average size from the seedlings originally larger.—Possible factors in stock variation discussed. Probably seedlings small because of small embryos in polyembryonic seeds, crowding in seed bed, etc., as well as those genetically weak, are undesirable as stocks. Recommendations include: (1) planting of seeds from trees budded to selected good stock varieties, (2) rigorous elimination of small seedlings at transplanting and budding, and of small budded trees when ready for orchard planting.—*H. B. Frost.*

499. WEIBULL, C. G. **Weibullsholm 1870–1920, en återblick. [Weibullsholm 1870–1920, a retrospective review.]** *18 p., 11 fig.* W. WEIBULLS Illustrerade Årsbok (Landskrona) 15 (1920). 1919.—Account of the evolution and working methods of Weibull's station for plant improvement.—*K. V. Ossian Dahlgren.*

500. WEINGART, W. **Künstliche Befruchtung von Kakteen. [Artificial fertilization of cacti.]** Monatsschr. Kakteenkunde 29: 106–107. 1919.—The author gives the results of self and cross pollination of several cactuses, mostly species of *Cereus.*—*A. S. Hitchcock.*

501. WOLFF, FRIEDRICH. **Ein Fall dominanter Vererbung von Syndaktylie. [A case of dominant inheritance of syndactyly.]** Arch. Rassen u. Gesellschaftsbiol. 13: 74–75. 1918.—One man in a family of five was syndactyl. Both of his parents, his sister and his three brothers were normal, and there seems to have been no previous history of syndactyly in this family. Married to a normal woman, he had seven children, all syndactyl. Each of these has married a normal individual and the combined number of grandchildren is now eighteen, of whom eight are syndactyl. In this family the syndactyly is somewhat more marked in males.—*C. H. Danforth.*

502. YAMPOLSKY, CECIL. **The occurrence and inheritance of sex intergradation in plants.** Amer. Jour. Bot. 7: 21–38. Jan., 1920.—A general discussion of sex intergrades based on the author's studies of *Mercurialis annua*, on various other studies of sex-intergrades and sex polymorphism in plants and in animals, and on a survey of data on sex forms in orders of seed plants as given in ENGLER and GILG's "Syllabus der Pflanzenfamilien."—In the monocots, 10 out of 11 orders representing 22 families have hermaphroditic, monoecious, dioecious and polygamous individuals, and in dicots 31 of the 40 orders including 90 families have certain representatives with two or more of the various types of sex. This distribution, shown in tables for orders and families (not for species) reveals that "practically every order has families which contain forms that show more than one kind of distribution of sex elements." The various terms used in describing sex conditions in plants are defined and species illustrating them are cited. It is pointed out that the obvious facts of sex distribution in plants, together with the results of experimental studies of heredity in polygamous or intersexual forms support the doctrine of varying sex potencies in germ cells rather than a sex-determination based on segregation of fixed unit factors.—*A. B. Stout.*

503. YLPPO. **Über das familiäre Vorkommen von Icterus neonatorum gravis.** [On familial occurrence of Icterus neonatorum gravis.] Münchener Med. Wochenschr. 65: 98. 1918.

504. ZANDER, L. **Der Einfluss der Bastardierung auf die Honigbildung.** [The influence of hybridization on honey formation.] Zeitschr. Angew. Entomol. 5: 88–93. 1918.

505. ZIEGLER, H. E. **Zuchtwahlversuche an Ratten.** [Selection experiments on rats.] Festschr. 100-jähr. Best. Kgl. Württ. Landw. Hochschule Hohenheim 1919: 385–399. 1919.

HORTICULTURE

J. H. GOURLEY, *Editor*

FRUITS AND GENERAL HORTICULTURE

506. CONDIT, I. J. **The Kaki or oriental persimmon.** California Agric. Exp. Sta. Bull. 316: 231–266. *20 fig.* 1919.—A discussion is given of the history of the persimmon, *Diospyros, Spp.*, its introduction into the United States and the botany of the reproductive parts. Different varieties of the Oriental species of persimmon, *Diospyros kaki*, are discussed at length from the standpoint of their morphology, astringency, soil requirements, methods of propagation and care of the trees, and methods of harvesting, processing and marketing the fruit. A table of analysis of different varieties of persimmons is given and a brief discussion of the insect enemies and diseases.—*W. P. Kelley.*

507. DETJEN, L. R. **The limits in hybridization of Vitis rotundifolia with related species and genera.** North Carolina Agric. Exp. Sta. Tech. Bull. 17. *25 p.* 1919.—See Bot. Absts. 4, Entry 562.

508. GARDNER, V. R. **Pruning the apple.** Missouri Agric. Exp. Sta. Circ. 90. *20 p. 11 fig.* 1920.

509. HENDRICKSON, A. H. **Plum pollination.** California Agric. Exp. Sta. Bull. 310. *28 p. 5 fig.* 1919.—A considerable number of varieties of two different species of plums are grown commercially in California, namely, the Japanese, *Prunus triflora*, and the European, *P. domestica*. Of the seventeen varieties studied all except four are self-sterile. No evidence of inter-sterility between different varieties was noted, but certain varieties are more effective pollinators than others. Comparative study of different orchards indicated that the common honey bee is an effective agent in promoting cross-fertilisation between the different varieties of plums.—*W. P. Kelley.*

510. SHAW, P. J. **Fourteenth Annual Report of the Nova Scotia Agricultural College and Farm. Part 5.—Report of the Professor of Horticulture.** Prov. of Nova Scotia Ann. Rept. Sec. for Agric. 1918: 75–100. 1919.

511. SHEWARD, T. **Fruit trees in pots for winter forcing.** Gard. Chron. Amer. 23: 360. *1 fig.* 1919.

512. SMITH, ARTHUR. **A lesson on fall preparation of the ground for spring planting.** Gard. Chron. Amer. 23: 341–343. 1919.

513. TRUELLE, A. **La situation des terrains a-t-elle de l'influence sur la richesse saccharine des pommes a cidre? [Has the location of the soil an influence on the sugar content of cider apples?]** Ann. Sci. Agron. Francaise et Etrangère 36: 107–116. 1919.—Pomologists have always held that the soil and exposure are among the most important factors affecting the chemical composition of cider fruits. Some data are published in which are given the density at 15° and total sugar expressed as grams of fermentable glucose. Twelve varieties of apples were studied but only the most commercially important six are reported on. The data are

grouped and considered under the headings of (1) those for trees grown on slopes and plateaus and (2) those for trees grown in valleys, a comparison being made for each variety grown in the two situations. The results show considerable variation in the sugar content, there being greater variation among those grown in the valleys. According to the author the following points are indicated by the results at hand: (1) The topographic position exercises an influence upon the production of sugar in certain varieties of cider apples. (2) The effect of the location on the sugar content is not uniform. In some varieties it is greater when grown on the higher elevations and with others it is greater when they are grown in valleys. (3) The differences in the weights of sugar in the juice from the apples grown on the uplands and in the valleys vary from 1 to 10.88 grams per liter. (4) The effect of topography on the sugar content of cider apples is generally feeble. The effect of topography is less than that of variety, which depends mainly on the composition of the soil.—*A. B. Beaumont.*

514. Tufts, Warren P. **Pollination of the Bartlett pear.** California Agric. Exp. Sta. Bull. 307: 369–390. *8 fig.* 1919.—The majority of the varieties of pears grown in California bloom for comparatively brief periods only, but all of them produce an abundance of pollen. Artificial pollination experiments showed that Bartlett pears are partially self-sterile when grown in certain localities and wholly so in others. All the other commercial varieties are capable of cross fertilizing the Bartlett variety. It was noted that the fruit resulting from cross-fertilization with pollen from a different variety tended to drop less freely in June than was the case with self-fertilized fruits. It is recommended that other varieties of pears be planted intermittently throughout an orchard of Bartlett pears as a means of promoting cross-fertilization. [See Bot. Absts. 4, Entry 798.]—*W. P. Kelley.*

515. Tufts, Warren P. **Almond pollination.** California Agric. Exp. Sta. Bull. 306: 337–366. *15 fig.* 1919.—It is shown that all the common varieties of almonds grown in California are self-sterile to a large extent and certain of them are inter-sterile. The different varieties may be roughly divided into two classes on the basis of the time of blooming, and considerable differences were noted in the amounts of pollen produced by the different varieties. Experiments demonstrated that cross-pollination can be effected between certain varieties very readily whereas other varieties are inter-sterile. It is shown that mixed planting of inter-fertile varieties in the same orchard results in increased yields of fruit. The inter-pollinating relationships of the different varieties are shown tabularly. The effects of meteorological conditions and insects on pollination are briefly discussed. The common honey bee is though to be the best pollinating agent. [See Bot. Absts. 4, Entry 797.]—*W. P. Kelley.*

FLORICULTURE AND ORNAMENTAL HORTICULTURE

516. Acosta, Celsa. **Sobre el cayeput. [The cajuput.]** Revist. Agric. Com. y Trab. 2: 535–537. *3 fig.* 1919.—Description of cajuput tree (*Melaleuca leucadendron* Linn.) and its uses.—*F. M. Blodgett.*

517. Arango, Rodolfo. **La palma real, su belleza ornamental y utilidad práctica. [The royal palm as an ornamental and useful plant.]** Revist. Agric. Com. y Trab. 2: 557–559. *2 fig.* 1919.

518. Baxter, Samuel Newman. **How nurserymen may best compete for the Christmas tree market.** Florists' Exchange 49: 133. 1920.—Ordinary nursery ground is too valuable for growing large Christmas trees; but small trees are gaining in favor with dwellers in small apartments and can be profitably grown. Nursery-grown, bushy stock is more attractive than the wild, the supply of which may soon become exhausted or unavailable. The 1- to 2-foot size could be offered in 6- or 8-inch pots, and the 2- to 4-foot size in larger pots or tubs, both at reasonable prices. Frequent transplanting is unnecessary; thinning of plants in the nursery row and shearing will assist in making bushy specimens. Figures are given of expected yield per acre over a ten-year period.—*L. A. Minns.*

519. Esler, John G. **A rhododendron king.** Florists' Exchange 49: 169. 1920.—Mr. W. K. Labar for the past fifteen years has collected native rhododendrons all over the Blue Ridge from Pennsylvania to North Carolina, selling them, as well as azaleas, kalmias and leucothoës, to parks, cemeteries and nurserymen. He has secured about 100 acres of wooded hillside with northern exposure, and will specialize in the above mentioned plants and others of similar nature. He is planting some of these by the thousand, using small collected plants and seedlings.—*L. A. Minns.*

520. Gibson, Addison H. **The poinsettia.** Gard. Chron. Amer. 23: 366. 1919.

521. Gibson, H. **Hardy shrubs that can be forced.** Gard. Chron. Amer. 23: 335, 336. 1919.

522. Gibson, Henry. **Forcing herbaceous plants and bulbs for winter flowers.** Gard. Chron. Amer. 23: 359. 1919.

523. Griffiths, David. **Producing domestic Easter lilies.** Florists' Exchange 49: 134. 1920.—Notes on growing Easter lilies up to the present are added to Griffith's article in Florists' Exchange 48: 775. 1919. Nine batches of bulbs now in the greenhouses of the Bureau of Plant Industry, Department of Agriculture, Washington, D. C., are mentioned, all of which promise interesting data in this investigational work. It is suggested that each grower of Easter lilies might advantageously do a little experimental work for himself.—*L. A. Minns.*

524. Hammond, Bertha B. **Forcing hyacinths for winter bloom.** Gard. Chron. Amer. 23: 337, 338. *Fig. 1-6.* 1919.

525. Holzhausen, Axel. **Laeliocattleya suecica nov. hybr.** Svensk. Bot. Tidskr. [Stockholm] 13: 97-99. 1919.

526. Matthews, Edwin. **Transplanting a mammoth yew tree in winter.** Florists' Exchange 49: 83. *1 fig.* 1920.—An English yew, 25 feet in height, 30 feet in circumference, about 80 years old, and weighing, with the ball of soil attached, approximately 5 tons was moved about one-half mile at Beverly, New Jersey, in January, 1918, and reset on the grounds of the owner. It was raised out of its former situation by means of rollers and windlass, raised onto a strong dray wagon by means of jack-screws, and drawn to its destination by six horses. Adverse conditions made the task formidable, but subsequent good care makes the removal appear to be successful up to the present.—*L. A. Minns.*

527. Moore, Henry I. **Descriptive list of hardy and semi-hardy primulas.** Gard. Chron. Amer. 24: 401, 402. 1920.

528. Moore, Henry I. **The city rose garden.** Gard. Chron. Amer. 23: 361. 1919.

529. Pleas, Sarah A. **A plea for seedling peonies.** Flower Grower 6: 123, 124. *1 fig.* 1919.

530. Rothe, Richard. **Landscape possibilities with brook and natural stream.** Gard. Chron. Amer. 23: 393, 394. *4 fig.* 1920.

531. Sakamoto, Kiyoshi. **The Japanese garden and how to construct it.** Florists' Exchange 49: 61, 63, 138. *9 fig.* 1920.—A Japanese garden must be made to appear as if it were a piece of natural scenery. The noblest sentiment evoked comes from the correct placing of each object—cottage, tree, herb or stone. Only large gardens can be successfully arranged to present different aspects according to season. An ordinary garden may better be made to appear much the same the year round. Evergreens are the foundation planting, set off by deciduous trees. The main types of garden are described: (1) the plain-garden, reproducing a plain, usually of considerable extent, good examples of which are the Tokiwa Garden and

the gardens of the Imperial Shrines of Ise; (2) the cypress garden, which may be small, only a section cut apart from a larger garden and representing a forest scene in miniature; and (3) the thicket garden, small, seeming to lead one to a dense wood beyond.—*L. A. Minns.*

532. Saunders, A. P. **American Iris Society.** Florists' Exchange 49: 285. 1920.—The meeting for the formation of the American Iris Society was held at the Museum Building of the New York Botanical Garden, Bronx Park, New York, on January 29, 1920. Sixty persons were present, among whom were many of the trade, and amateur Iris specialists. Dr. N. L. Britton, Director of the New York Botanical Garden, delivered the opening address. He told of the Iris garden begun in the New York Botanical Garden, and invited members of the newly-formed Iris Society to make free use of the library of the Botanical Garden.—The work of the Iris Society has been carefully planned. There will be test and exhibition gardens established, Iris shows with suitable prizes, and investigations made in history, classification of garden varieties, culture and pests of the Iris. A constitution was approved and officers elected of whom John C. Wister of Philadelphia is president, and R. S. Sturtevant of Wellesley, Massachusetts, is secretary.—*L. A. Minns.*

533. Smith, Arthur. **The care and culture of house plants.** Gard. Chron. Amer. 23: 372–375. 1919.

534. Smith, Arthur. **Putting the garden to bed for the winter.** Gard. Chron. Amer. 23: 368–371. 1919.

535. White, E. A. **Hubbard Gold Medal awarded to rose "Columbia."** Florists' Exchange 49: 171. 1920.—The Executive Committee of the American Rose Society has recently voted to award to the hybrid tea rose Columbia, registered in 1917 by E. G. Hill, of Richmond, Indiana, the Gertrude M. Hubbard Gold Medal for the best rose of American origin introduced during the last five years. This medal, the highest honor the American Rose Society can confer on a hybridizer, has been bestowed but once; in 1914 it was given to M. H. Walsh of Woods Hole, Massachusetts, for the introduction of the climbing rose "Excelsa."—*L. A. Minns.*

VEGETABLE CULTURE

536. Olmstead, W. H. **Availability of carbohydrate in certain vegetables.** Jour. Biol. Chem. 41: 45–58. 1920.—The amount of carbohydrate available to the body from certain vegetables, usually used in low carbohydrate diets for diabetic patients, was determined (1) by the use of diastase and copper reduction, (2) by feeding to phloridinized dogs. The results by these two methods were—cabbage (1) 4.4 per cent, (2) 5.0 per cent, cauliflower (1) 2.8 per cent, (2) 3.4 per cent, spinach (2) 1.2 per cent, lettuce (1) 1.0 per cent. The amount in cabbage was reduced about 90 per cent by thrice cooking.—*G. B. Rigg.*

537. Tracy, W. W. **Growing tomato seed.** Seed World 7[9]: 18–19. 1920.

538. Work, P. **Vegetable gardening on eastern muck soil.** Jour. Amer. Peat Soc. 13: 27–36. 1920.—Muck soils have proved to be preeminently adapted for the production of onions, celery and summer lettuce and they are well suited for several other crops.—*G. B. Rigg.*

539. Zimmerman, H. E. **Tomato grafted on potato.** Amer. Bot. 25: 144. *1 fig.* 1919.

HORTICULTURE PRODUCTS

540. Baughman, Walter F., and George S. Jamieson. **The composition of Hubbard squash seed oil.** Jour. Amer. Chem. Soc. 42: 152–157. 1920.

541. Haynes, Dorothy, and Hilda Mary Judd. **The effect of methods of extraction on the composition of expressed apple juice, and a determination of the sampling error of such**

juices. Biochem. Jour. 13: 272–277. 1919.—The following points were taken up: (1) does rapid freezing by liquid air produce any alteration in character of the sample, (2) does freezing render tissues freely permeable to all those constituents of the cell sap present in expressed juice, (3) the probable error due to individual variability in apples used. Comparisons were made of PH values, freezing points, time and fall of viscometer, conductivity, acidity, and determination of sugars. No real difference was found between liquid air and freezing mixture method. Tissues were freely permeable to acids and sugars but colloids were held back as indicated by changing viscosity. Samples varied greatly resulting in a large probable error. Authors conclude that neglect of sampling errors in previous work of this nature vitiates much data.—*A. R. Davis.*

542. Jamieson, George S., and Walter F. Baughman. **Okra seed oil.** Jour. Amer. Chem. Soc. 42: 166–170. 1920.

MORPHOLOGY, ANATOMY AND HISTOLOGY OF VASCULAR PLANTS

E. W. Sinnott, *Editor*

543. Baccarini, P. **Notule teratologiche. [Teratological notes.]** Nuovo Gior. Bot. Ital. 25: 225–247. 1918.—Abnormalities in flower development and morphology were noticed among members of diverse plant groups: *Delphinium Ajacis, Brassica Rapa, Isatis tinctoria, Viburnum Sandankwa, Dahlia variabilis, Cypripedium* sp., *Carlina vulgaris* and *Anchusa italica.* The abnormalities consist in depression, entire disappearance or malformation of floral parts, notably the essential parts of the flower. In some cases, for example in *Delphinium Ajacis,* the reduction in the number of carpels suggests the reappearance of characters found at present in the Staphysagria group.—*Ernst Artschwager.*

544. Bassler, Harvey. **A sporangiophoric lepidophyte from the Carboniferous.** Bot. Gaz. 68: 73–108. Aug., 1919.—See Bot. Absts. 3, Entry 1597.

545. Bexon, Dorothy. **Observations on the anatomy of teratological seedlings. II. On the anatomy of some polycotylous seedlings of Centranthus ruber.** Ann. Botany 34: 81–94. *9 fig.* 1920.—The vascular anatomy of seedlings of *Centranthus ruber* showing all degrees of polycotyly from very incomplete tricotyly to complete tetracotyly is described. The hemitricotylous material is divisible into three groups: (a) Two bundles, one from each half of the incompletely split cotyledon, approach and fuse at various levels to form one pole of the diarch root, the other pole being formed by the bundle from the other cotyledon. (b) The two bundles remain distinct for a distance in the hypocotyl forming with the bundle from the other cotyledon a triarch condition, which eventually becomes reduced to diarchy by the fusion of the two bundles from the same cotyledon. (c) One bundle from the split cotyledon fails to rotate, retains its collateral structure and finally disappears. In the tricotyls a triarch condition is usually established, and becomes reduced to the diarch condition either by the disappearance of one arm or by the fusion of the two. The hemitetracotyls and tetracotyls for the most part show conditions like those described under (a) and (b) above with the modifications resulting from the splitting of both original cotyledons instead of one. One hemitetracotyl showed double structure throughout and evidently represented a twinned condition. It is suggested that the twinning may be due either to the fusion of distinct embryos or to the partial separation of the daughter cells resulting from the division of the embryo initial.—*W. P. Thompson.*

546. Bobilioff, W. **De inwendige bouw der schorselementen van Hevea Brasiliensis. [The structure of cell elements in the bark of Hevea Brasiliensis.]** Arch. Rubbercult. Nederlandsch-Indië 3: 222–231. 1919.—Paper deals principally with the structure of the laticiferous vessels of *Hevea* and their cytology in connection with the physiological significance of latex.

The author points out that protoplasm and nuclei occur in the laticiferous vessels, but that the nuclei are larger than those of other cortex cells. Both nuclei and vacuoles occur in the protoplasm where they can be seen after the caoutchouc has been dissolved out. Therefore the latex of *Hevea* is probably cell-sap, which generally occurs in the vacuoles of the laticiferous vessels. Sometimes many of the nuclei of laticiferous vessels unite in one place, hence it seems that the nuclei have the faculty of moving. Author also observes that the nuclei pass from one vessel into another through the wall openings.—*W. E. Cake.*

547. Bürgerstein, A. **Beiträge zur Naturgeschichte der Scoglien und Kleineren Inseln Süddalmatiens. 8. Anatomische Beschreibung des Holzes einiger Sträucher und Halbsträucher. [The natural history of the smaller islands of southern Dalmatia. 8. Anatomical description of the wood of some shrubs and undershrubs.]** Denkschr. K. Akad. Wiss. Wien. (Math.-Nat. Kl.) 92: 329–334. 1916.

548. Chirtoiü, Marie. **Remarques sur le Symplocos Klotzschii et les affinités des Symplocacées. [Remarks on Symplocos Klotzschii and the affinities of the Symplocaceae.]** Bull. Soc. Bot. Genève 10: 350–361. *5 fig.* 1918.

549. Chirtoiü, Marie. **Observations sur les Lacistéme et la situation systematique de ce genre. [Observations on the species of Lacistema and the systematic position of this genus.]** Bull. Soc. Bot. Genève 10: 317–349. *18 fig.* 1918.

550. Clute, Willard N. **Peloria.** Amer. Bot. 25: 148. 1919.

551. Coulter, J. M. **Perennating fruit of Cactaceae.** [Rev. of: Johnson, Duncan S. The fruit of Opuntia fulgida. A study of perennation and proliferation in the fruits of certain Cactaceae. Carnegie Inst. Publ. 269. *62 p., 12 pl.* 1918.] Bot. Gaz. 68: 151. 1919.

552. Coulter, J. M. **Root-nodules.** [Rev. of: Spratt, Ethel R. A comparative account of the root-nodules of the Leguminosae. Ann. Botany 33: 189–199. *5 fig.* 1919. (See Bot. Absts. 3, Entry 1139.)] Bot. Gaz. 68: 311. 1919.

553. Coulter, J. M. **Suspensor of trapa.** [Rev. of: Tison, A. Sur le suspenseur du Trapa natans L. Rev. Gén. Bot. 31: 219–228. *5 fig.* 1919. (See Bot. Absts. 3, Entry 2451.)] Bot. Gaz. 68: 312. 1919.

554. Cremata, Merlino. **Un fenomeno curioso. [A curiosity.]** Revist. Agric. Com. y Trab. 2: 509. *2 fig.* 1919.—Several cases are cited where the royal palm has become branched. —*F. M. Blodgett.*

555. Dixon, Henry H. **Mahogany and the recognition of some of the different kinds by their microscopic characters.** Sci. Proc. Roy. Soc. London 15: 431–486. *22 pl.* 1918.

556. Eberstaller, Robert. **Beiträge zur Vergleichenden Anatomie der Narcisseae. [Comparative anatomy of Narcissus.]** Denkschr. K. Akad. Wiss. Wien. (Math.-Nat. Kl.) 92: 87–105. *3 pl.* 1916.

557. Esmarch. **Über den Wundverschluss bei geschnittenen Saatkartoffeln. [Wound healing in cut seed potatoes.]** Fühl. Landw. Zeit. 67: 253–256. 1918.—True periderm formation on the exposed surfaces of cut seed potatoes takes place only, and most rapidly, when the tubers are kept in a fairly moist place. The practice of leaving the cut tubers to dry in the air results only in the drying in of the upper cell layers which may be accompanied by a suberization of the walls. It is questionable whether a crust formed in such a way affords real protection against parasitic bacteria and fungi.—*Ernst Artschwager.*

558. GERTZ, OTTO. **Proliferation av Honhänge hos Alnus glutinosa (L.) I. Gaertn. [Proliferation of the female catkins of Alnus glutinosa.]** (Résumé and legends of illustrations in German.) Svensk. Bot. Tidskr. [Stockholm] 13: 71–74. 1919.—Author describes and illustrates a case of proliferated female catkins in *Alnus glutinosa* not heretofore reported in literature.—*W. W. Gilbert.*

559 GRIER, N. M. **Note on proliferative power of Pinus sp.** Ohio Jour. Sci. 20: 21–23. 1919.

560. GROVES, JAMES. **Sex-terms for plants.** Jour. Botany 58: 55–56. 1920.—A brief note continuing the discussion of the terminology of plants begun in Jour. Botany 57. The codification of botanical terminology is very necessary. Authors are now constantly inventing new terms and piling up a mass of terminology which cannot but retard and embarrass future workers. This problem should be dealt with in future meetings of the International Botanical Congress.—*K. M. Wiegand.*

561. HAWTREY, S. H. C. **Notes on a few useful plants and home industries of Paraguay.** South African Jour. Indust. 3: 35–41. 1920.

562. HILL, J. BEN. **Anatomy of Lycopodium reflexum.** Bot. Gaz. 68: 226–231. *5 fig.* 1919.—The chief points of interest are the presence of typical cortical roots and the various "types" of stele in the stem. The development and differentiation of the tissues in the steles of the cortical roots parallel those in the stele of the stem. The xylem arrangement may be radial, parallel-banded, or radial so modified as to consist of an inner cylinder of xylem inclosing a small strand of phloem, the last being most frequent. The author's previous suggestion that in *Lycopodium* all xylem arrangements may occur in the same stem is confirmed.—*H. C. Cowles.*

563. HIRSCHT, KARL. **Verschlossenblütige Pflanzen im Zimmergarten. [Cleistogamous flowers in a window garden.]** Monatsschr. Kakteenkunde 29: 103–104. 1919.—The cleistogamous flowers of *Anacampseros filamentosa* Sims. are described.—*A. S. Hitchcock.*

564. JAUCH, BERTHE. **Quelques points de l'anatomie et de la biologie des Polygalacées. [Certain details of the anatomy and biology of Polygalaceae.]** Bull. Soc. Bot. Genève 10: 47–84. *15 fig.* 1918.

565. JAUFFRET, AIME. **La détermination des bois de deux Dalbergia de Madagascar, d'après les caractères de leurs matières colorantes. [Identification of wood of Dalbergia by staining reactions.]** Compt. Rend. Acad. Sci. Paris 168: 693–694. 1919.—The wood of two species of *Dalbergia* from Madagascar showed very characteristic specific reactions when treated with alcohol, sulphuric acid, caustic soda, ammonia, iron perchloride, bisulphite of soda, ether, chloroform, and benzene. The alcoholic solution of the powdered wood of each species also gave a characteristic spectrum. Such characters offer a basis for the identification of species in the absence of other parts of the plant.—*F. B. Wann.*

566. JOHANSSON, K. **Fyllomorfi och diafys hos Geranium pyrenaicum L. [Phyllomorphy and diaphysis of Geranium pyrenaicum L.]** Svensk. Bot. Tidskr. [Stockholm] 13: 99. 1919. —A brief description of cases of phyllomorphy and diaphysis in *Geranium pyrenaicum* is given and references made to cases of teratology in other species of *Geranium.*—*W. W. Gilbert.*

567. KONDO, M. **Ueber die in der Landwirtschaft Japans gebrauchten Samen. [Seeds used in Japanese agriculture.]** Ber. Ohara Inst. Landw. Forsch. 1: 261–324. *17 fig.* 1918.— See Bot. Absts. 5, Entry 37.

568. McMURRAY, NELL. **The day flower.** Amer. Bot. 25: 150. 1919.—The flower of *Commelina communis* is described.—*W. N. Clute.*

569. MILLER, E. C. **Development of the pistillate spikelet and fertilization in Zea mays L.** Jour. Agric. Res. 18: 255–265. *Pl. 19–32.* 1919.—Study made on three varieties of maize: Pride of Saline, Freed White Dent, and Sherrod White Dent. The development of the pistillate spikelet is briefly described.—In the development of the embryo sac there is no degeneration of megaspores; the megasporocyte nucleus by three divisions gives rise to the eight nuclei of the sac, as in *Lilium*. The antipodals multiply and form a tissue of from 24 to 36 cells in the base of the sac.—The silk is receptive to pollen not only at the stigmatic surface, but also along the greater portion of its length. The pollen tube may penetrate the silk at once or grow along the surface for some distance and penetrate later. Around the two vascular bundles of the silk are sheaths of cells with rich contents; it is between these cells that the tube grows. The tube penetrates into the embryo sac and liberates the two male nuclei, which are formed before the shedding of the pollen grain. One of them fuses with the egg nucleus, while the other unites with the two polar nuclei, which do not fuse until this time. About 26 to 28 hours elapse between pollination and fertilization.—The endosperm develops rapidly, filling the sac with tissue in 36 hours; the embryo by this time has 14 to 16 cells. [See Bot. Absts. 4, Entry 679.]—*L. W. Sharp.*

570. MILLER, WARD L. **Polyxylic stem of Cycas media.** Bot. Gaz. 68: 208–221. *11 fig.* 1919.—The normal cylinder begins its differentiation as high up as the meristem, the others beginning theirs successively lower, and each one in the cortex outside the next inner cylinder. Protoxylem and protophloem are developed during the early activities of the normal cylinder, the protoxylem elements usually being scalariform, as in the primary xylem. The secondary xylem is characteristically pitted. In the first cortical cylinder most of the xylem elements are pitted, neither protoxylem nor protophloem being observed. In both cylinders there is a relatively large number of suberized bast fibers. All cortical cylinders are similar in origin and development, and probably are related in appearance to alternating periods of rest and activity.—*H. C. Cowles.*

571. MORVILLEZ, F. **L'appareil conducteur foliaire des Legumineuses: Papilionacées et Mimosées. [Leaf traces in the Leguminosae: Papilionatae and Mimosoideae.]** Compt. Rend. Acad. Sci. Paris 168: 787–790. *9 fig.* 1919.—Ten types of vascular supply in the petioles of members of the sub-families Papilionatae and Mimosoideae are described and figured. In a previous paper (Compt. Rend. 167: 205. 1918) the leaf traces in the Caesalpinioideae were described.—The three sub-families of the Leguminosae present types of leaf traces with medullary strands similar to those of the Chrysobalanoideae of the Rosaceae; this character is encountered even in such widely separated genera as *Swartzia*, *Affonsea* and *Bocoa*. The most highly specialized forms possess the simpler trace.—Subdivisions of the Papilionatae agree in leaf trace anatomy with the exception of the Astragaleae, in which are encountered the various types characteristic of the other tribes. This may represent a stock from which the others have been derived. Moreover, the Astragaleae, through the Sophoreae, seem to be related to types possessing medullary strands, thus constituting an assemblage of closely related forms, to which are attached the different sub-families of the Leguminosae.—*F. B. Wann.*

572. NELSON, J. C. **Monomorphism in Equisetum Telmateia Ehrh.** Amer. Fern Jour. 9: 93–94. 1919.

573. NELSON, J. C. **Another "freak" Equisetum.** Amer. Fern Jour. 9: 103–106. *Pl. 6.* 1919.—Linn County, Oregon, is a new locality for *Equisetum fluviatile* L. Among the specimens collected was one, *E. fluviatile* var. *polystachyum*, which had 31 branches of the two upper whorls bearing strobiles at the tip.—*F. C. Anderson.*

574. SAHNI, B. **On certain archaic features in the seed of Taxus baccata, with remarks on the antiquity of the Taxineae.** Ann. Botany 34: 117–134. *7 fig.* 1920.—It is suggested that the Palaeozoic seeds *Cardiocarpus*, *Cycadinocarpus*, *Mitrospermum*, and *Taxospermum*, all

of which probably belonged to the Cordaitales, form a series illustrating a general tendency, a continuation of which has resulted in the production of the type of seed found in *Taxus*, as well as in *Torreya* and *Cephalotaxus*. This tendency may be summarised as follows: The point of origin of the "outer" system of vascular strands shifts nearer and nearer the subnucellar pad of tracheids which gives rise to the "inner" system. During this process the bundles of the outer system cut through the "stone." The canals through the stone then move forward toward the micropyle so that for an increasing distance the bundles come to lie inside the stone. At the culmination of the process when they lie entirely within the stone the condition found in *Taxus* is reached. In this genus the "inner" system of bundles has disappeared. The seeds of *Torreya* and *Cephalotaxus* are derived from the same source by a modification of the same tendency. On the basis of this theory these three genera are the nearest existing relatives—apart from *Ginkgo*—of the Cordaitales and like *Ginkgo* have been derived directly from the Cordaitales. It is proposed to place them in a separate group the Taxales, distinct from Coniferales and nearer *Ginkgo*.—*W. P. Thompson.*

575. St. John, Harold. **The genus Elodea in New England.** Rhodora 22: 17–29. 1920. —See Bot. Absts. 5, Entry 451.

576. Salisbury, E. J. **Variation in Anemone apennina, L., and Clematis vitalba, L., with special reference to trimery and abortion.** Ann. Botany 34: 107–116. *9 fig.* 1920.—This paper furnishes additional data supporting the author's previously published views concerning the essential trimery of the Ranunculaceous flower and the causes of variation in the numbers of the constituent parts. Curves are given showing the variation in the number of stamens, carpels, and perianth parts in a large number of flowers of the species studied. The curves show marked periodicity, the crests occurring at multiples of three. In more than half the flowers of *Anemone apennina* the stamens and carpels are in multiples of three. Evidence is given to show that congenital fission is the chief cause of variation in number, though transformation of stamens into perianth parts was also observed.—*W. P. Thompson.*

577. Schaffner, John H. **Dieciousness in Thalictrum dasycarpum.** Ohio Jour. Sci. 20: 25–34. 1919.—Intermediate forms between extremes of staminateness and carpellateness are described. Great diversity of sexual expression is found on different branches of the same inflorescence. It is concluded that maleness or femaleness is determined by the physiological state at the inception of the sporophylls; or that if sex has been determined earlier, it is later reversed. A general survey of the origin and nature of dieciousness in sporophytes is given, showing evolutionary gradations from the bisporangiate to the monosporangiate condition in various groups.—*H. D. Hooker, Jr.*

578. Seward, A. C. [Rev. of: Chamberlain, C. J. **The living cycads.** Univ. Chicago Science Ser. *172 p. 91 fig.* Univ. Chicago Press: Chicago, 1919.] New Phytol. 18: 262. 1919.

579. Small, James. **The origin and development of the Compositae. Miscellaneous topics.** New Phytol. 18: 129–176. *Fig. 64–78.* 1919.—See Bot. Absts. 5, Entry 720.

580. Small, James. **The origin and development of the Compositae. General conclusions.** New Phytol. 18: 201–234. *Fig. 79.* 1919.

581. Souèges, R. **Embryogenie des Polygonacées. Developpement de l'embryon chez le Polygonum Persicaria L. [Development of the embryo of Polygonum persicaria L.]** Compt. Rend. Acad. Sci. Paris 168: 791–793. *8 fig.* 1919.—The two-celled proembryo of *Polygonum persicaria L.* gives rise, by a series of transverse divisions, to six layers of cells, the upper two being derived from the apical cell and the lower four from the basal cell. The two layers produced from the apical cell give rise respectively to the cotyledons, as in the Ranunculaceae and Cruciferae, and to the upper portion of the hypocotyl. In the Ranunculaceae and Cruciferae the corresponding layer gives rise to the complete hypocotyl. The four layers derived from the basal cell of the two-celled proembryo give rise respectively to (1) the lower portion of the hypocotyl; (2) the root cap; and (3) and (4) a rudimentary suspensor.—*F. B. Wann.*

582. Spratt, Amy Vera. **Some anomalies in monocotyledonous roots.** Ann. Botany 34: 99–105. *Pl. 3, 1 fig.* 1920.—Members of several monocotyledonous natural orders show an anomalous root condition consisting in the filling in of a large pith with scattered vascular strands. These may be formed by secondary growth (*Dracaena*) or differentiated at the growing point (*Pandanus*, *Yucca*) and at later stages may form a solid stele in some cases. The secondary thickening in *Dracaena* may occur in the pericycle or in cortical layers.—*W. P. Thompson.*

583. Sprecher, A. **Étude sur la semence et la germination du Garcinia mangostana L.** [**A study of the seed and germination in Garcinia mangostana L.**] Rev. Gén. Bot. 31: 513–531, 609–634. *Pl. 5–7, 34 fig.* 1919.—In the East Indian "mangosteen," a member of the Guttiferae, the ovule is anatropous and has two integuments. During the development of the embryo sac the nucellar cells are absorbed, the sac coming to life directly against the inner integument. The cells of the latter bud into the sac and form an embryo, which becomes detached from the integument and is completely surrounded for a time by the endosperm cytoplasm with its free nuclei; these soon disappear. When fully developed the embryo has the form of a swollen tubercle which represents the hypocotyl; there is no trace of root, stem, or cotyledons. Two or three such embryos are occasionally developed in one embryo sac, forming a compound tubercle. The central cylinder which differentiates in the tubercle usually lies along the longitudinal axis of the latter, but in many cases it develops in an oblique or transverse position. Normal fertilization and embryogeny also occur.—The course of the vascular bundles in the flower and fruit is followed, and it is shown that the white pulp in which the seed lies (usually only one seed matures) represents the endocarp; this separates at an early stage from the red mesocarp, becomes divided into sections, and grows fast to the integument. The histological changes occurring during the development of the fruit are described.—At germination a root and a stem grow out from the embryonal tubercle and develop very slowly. If the stem and a portion of the tubercle be removed a new stem is regenerated. In polyembryonic seeds more plantlets develop from the tubercle. The primary root has no root hairs, but the walls of certain epidermal and hypodermal cells remain thin; water enters at these points.—The arrangement of vascular bundles in the seedling and the histology of its various parts are briefly described. In root, stem, leaf and fruit there is a system of secretory canals which arise schizogenously.—*L. W. Sharp.*

584. Stout, A. B. **Intersexes in Plantago lanceolata.** Bot. Gaz. 68: 109–133. *2 pl.* Aug., 1919.—See Bot. Absts. 3, Entry 1518.

585. Styger, Jos. **Beiträge zur Anatomie des Umbelliferen-früchte.** [**Contribution on the Anatomy of Umbelliferous Fruits.**] Schweiz. Apotheker Zeitg. 57: 199–205, 228–235. *7 fig.* 1919.—See Bot. Absts. 5, Entry 831.

586. Turrill, W. B. **Observations on the perianth in Ranunculus auricomus and Anemone coronaria.** New Phytol. 18: 253–256. *3 fig.* 1919.—The author describes transition stages between stamens and petals, petals and sepals, and sepals and bracts in *Ranunculus*; and notes a sepal occurring in the whorl of bracts in *Anemone.*—*I. F. Lewis.*

587. Weatherwax, Paul. **Paraffin solvents in histological work.** Bot. Gaz. 68: 305–306. Oct., 1919.—The sinking of paraffin in the replacement of xylol may be avoided by running a current of cold air through the melted paraffin, thus causing it to harden as a frothy mass of lessened specific gravity. Before allowing it to harden, the mass is kneaded to secure finer grain and a more even distribution of the air bubbles. The author, however, does not find any special disadvantages in the old method, and sees no valid reason for the rather general abandonment of the use of chloroform as a medium for the introduction of paraffin.—*H. C. Cowles.*

588. Weingart, Wilh. **Vom Reif des Cereus trigonus Haw. var. guatemalensis Eichl.** [**The bloom on Cereus trigonus var. guatemalensis.**] Monatsschr. Kakteenkunde 29: 80–84. 1919.—The author shows that the bloom contains resin as well as wax.—*A. S. Hitchcock.*

589. Weingart, Wilh. **Sphärite im Hypoderm von Cereen.** [Sphere crystals in the hypoderm of Cereus.] Monatsschr. Kakteenkunde 29: 45-48. 1919.—An account is given of the sphere crystals in *Cereus Hirschtianus* and *C. Lauterbachii*, and of the effect upon them of various reagents. The spherites contain no proteids and are allied to inulin. They constitute reserve material.—*A. S. Hitchcock.*

590. Woodward, R. W. **Further notes on Philotria.** Rhodora 21: 218-219. 1919.—In a recent issue (Rhodora 21: 114. 1919.), writer reported what appeared to be *Philotria angustifolia* growing in brackish water at Old Lyme, Connecticut. On revisiting the station in August 1919 both flowers and fruit were examined while fresh, and from this examination detailed descriptions of the staminate and pistillate flowers and the fruit are given. Writer has not had an opportunity to verify his identification by comparison with authentic material. but believes that it is *P. angustifolia* or some species closely related to it.—*James P. Poole.*

MORPHOLOGY AND TAXONOMY OF ALGAE

E. N. Transeau, *Editor*

591. Børgesen, F. **The marine algae of the Danish West Indies. Vol. 3. Rhodophyceae.** Dansk Bot. Ark. 3: 145-240. *Fig. 149-230.* 1917.—This part completes the family Squamariaceae, from p. 144, 1915, of the same volume. (This family contributed by Mme. A. Weber-van Bosse), and includes the families Hildenbrandiaceae, Corallinaceae (the subfamily Melobesieae by Mme. Paul Lemoine, text in French) and part of the Ceramiaceae. New are *Amphiroa rigida* Lamour. var. *antillana* Børgesen; *Mesothamnion caribaeum*, nov. gen. & sp. Børgesen; *Antithamnion antillarum* Børgesen; *Spyridia aculeata* var. *disticha*, and its forma *inermis* Børgesen. New combinations: *Lithophyllum accretum* (Fosl. & Howe) Lemoine; *Lithophyllum* (?) *propinquum* (Fosl.) Lemoine; *Melobesia* (*Lithoporella*) *atlantica* (Fosl.) Lemoine; *Melobesia* (*Litholepis*) *affinis* (Fosl.) Lemoine; *Porolithon mamillare* (Harv.) var. *occidentale* (Fosl.) Lemoine; *Porolithon Boergesenii* (Fosl.) Lemoine. Mme. Lemoine's treatment of the Melobesieae has a key to the 20 species, and list of the other species known from the Antilles as a whole; also a comparison with the species of other regions, showing a strong resemblance to those of the Mediterranean, and a somewhat less marked though still distinct resemblance to those of the Indo-Pacific. 19 of the 20 species are figured, either in section or in habit or both; most of them for the first time. Two species are recorded for the first time in America: *Jania adhaerens* Lamour., of the Red Sea, Indian Ocean and Japan; and *J. decussato-dichotoma* Yendo, of Japan. *Jania* sp., *Griffithsia* sp., *Callithamnion* sp., *Antithamnion* sp., are described and the last two figured; probably new, but sterile.—*Frank S. Collins.*

592. Børgesen, F. **The marine algae of the Danish West Indies. Vol. 3. Rhodophyceae.** Dansk Bot. Ark. 3: 241-304. *Fig. 231-307.* 1918.—Completes the family Ceramiaceae and begins the family Rhodomelaceae. New are *Laurencia chondrioides* Børgesen; *Polysiphonia sphaerocarpa* Børgesen.—*Frank S. Collins.*

593. Børgesen, F. **The marine algae of the Danish West Indies. Vol. 3. Rhodophyceae.** Dansk. Bot. Ark. 3: 305-368. *Fig. 308-360.* 1919.—Completes the family Rhodomelaceae and covers the families Delesseriaceae, Bonnemaisoniaceae, Gigartinaceae, and Rhodophyllidaceae. New are *Dasya caraibica* Børgesen; *Cottoniella arcuata* Børgesen, nov. gen. et sp. *Dasya* sp. is described and figured, probably new species but sterile. For *Lophocladia trichoclados* are described and figured the cystocarps and antheridia, hitherto unknown. As in previous parts of this paper, full descriptions are given of all species, and many details are described and figured for the first time.—*Frank S. Collins.*

594. Boyer, Charles S. **Rare species of North American Diatomaceae.** Bull. Torrey Bot. Club 47: 67-72. *Pl. 2.* 1920.—The following new species of diatoms are described: *Auliscus floridanus*, *A. hyalinus*, *Dimerogramma intermedium*, *Glyphodesmis tumida*, *G. campechi-*

ana, *Synedra anguinea*, *S. incisa*, *Eunotia Stevensonii*, *Pinnularia Hagelsteinii*, *Nitzchia semicostata*, and *Surirella Palmeri*. *Navicula Attwooddii* M. Perag. and an abnormal form of *Aulodiscus oregonus* Harv. & Bail. are discussed.—*P. A. Munz.*

595. Bristol, B. Muriel. **On the alga-flora of some desiccated English soils: an important factor in soil biology.** Ann. Botany 34: 35–80. *Pl. 11.* *18 fig.* 1920.—By means of water cultures it is show that there is a widely distributed plant association in cultivated soils consisting of moss protonema and algae. Sixty-four species and varieties of algae were identified. All these algae can withstand from four to twenty-six weeks desiccation. Descriptions of the algae including six new species are given.—*E. N. Transeau.*

596. Bullock-Webster, G. R. **A new nitella.** Irish Nat. 28: 1–3. *Pl. 1.* 1919.—*Nitella spanioclema*, a new species collected at Lough Shannach, County Donegal, Ireland.—*W. E. Praeger.*

597. Carter, Nellie. **On the cytology of two species of Characiopsis.** New Phytol. 18: 177–186. *3 fig.* 1919.—See Bot. Absts. 5, Entry 118.

598. Church, A. H. **Historical Review of the Florideae II.** Jour. Botany 57: 329–334. 1919 (continued from *Ibid.* 57: 304).—The Florideae represent an independent line of evolution in the sea from the ancestral stage of encysted plankton-flagellates, attaining somatic and reproductive specialisation along their own lines. Nuclear migrations and haustorial connections of the carpospore are but an extension of secondary pit-connections and migration in the somatic organization. Cenocytic decadence of the trophocyte is paralleled by the secondary coenocytic organisation in the vegetative soma of distinct generic types.—Progressive differentiation of the sex mechanism leads through inevitable stages to oogamy and fertilisation *in situ*, following the failure of the oospore to be discharged, thus giving rise to many separate phyla of algae. Though efficient in economy of materials, this method leaves dispersal unprovided for.—Most important in the Floridae, however, is not the fertilization *in situ* with a parasite zygote and a sporophyte generation producing spores, but the presence of three successive generations as follows: (I.) Gametophyte, (II.) Carposporophyte (diploid) and (III.) Tetrasporophyte (haploid). In (I.) there is the most complete economy in the sexual process. The gametes are expressed as mere nuclei, a condition otherwise attained only in the Angiosperms. There is also post-sexual nutrition. This is made possible by the mechanism of the pit-connections left open at the base of the young carpogonium. In (II.) the generation is asexual. Whether it be haploid or diploid does not matter, but there has been no inducement to a haploid condition. It is a very much reduced stage. In (III.) the spores are immediately dispersed and take the small chance of immediate germination. They grow to a free autotropic soma, but there is a reduction to the haploid condition at the formation of tetraspores.—The haploid spores on germination give a haploid soma which is normally free and autotropic, and which may be sexual and repeat the sequence, though it may as well be asexual. Of special interest are cases where the tetraspore formation is wanting and reduction is otherwise provided for, but the locus of the process is wholly subsidiary and secondary. The condition in *Scinaia* and *Nemalion* in this respect is discussed. —The clue to the peculiar behavior of the zygote and young carposporophyte in its relation to the auxiliary cells is seen in its practically holoparasitic habit. The passage of food material quickly is rendered possible by the mechanism of secondary pit-connections dependent on the soft penetrable wall-membrane.—The clearest view of the Florideae is that they consist of a multitude of distinct phyla as the survivors of a specialized and circumscribed ancient race of Marine Algae. All of the living representatives are on a closely comparable physiological plane, but the phyla diverge as to somatic construction and organisation and as to internal economy, becoming more specialised in relation to the parasitic carposporophyte.—The phases of haustorial connection, progressively more intricate and devastating in relation to the parental thallus they drain, constitute but one aspect of the question. The production of the cystocarpic wall after fertilisation passing to the initiation of these structures before fertilisation represents a specialization of great significance. A true phytogenetic classi-

fication should thus combine: (1) the auxiliary cell standpoint of Schmitz, with (2) the special feature of thallus-organisation, and (3) adult cystocarpic-differentiation, more clearly recognised as significant by the older algologists (Harvey). [See also Bot. Absts. 4, Entry 1014.]—*K. M. Wiegand.*

599. COULTER, J. M. **Alaria.** [Rev. of: YENDO, KICHISABURO. **A monograph of the genus Alaria.** Jour. Coll. Sci. Imp. Univ. Tokyo, 43[1]: *145 p. 19 pl.* 1919. (See Bot. Absts. 5, Entry 612.)] Bot. Gaz. 68: 151–152, 1919.

600. DUCELLIER, F. **Contribution à l'Etude de la flore desmidiologique de la Suisse.** [**A contribution to the study of the Desmid flora of Switzerland.**] Bull. Soc. Bot. Genève 10: 85–154. *3 pl., 134 fig.* 1918.—The many species of *Desmids* found at five different stations are enumerated with detailed notes and figures. Many of the species are new to Switzerland.—*W. H. Emig.*

601. DUCELLIER, F. **Etude critique sur Euastrum ansatum Ralfs et quelques-unes de ses variétés Helvétiques.** [**A critical study of Euastrum ansatum Ralfs and some of the Swiss varieties.**] Bull. Soc. Bot. Genève 10: 35–46. *29 fig.* 1918.—*Euastrum ansatum* Ralfs var. *simplex* n. var., also the new varieties *commune*, *dideltiforme*, *robustum*, and *rhomboidale*, are figured and described in detail.—*W. H. Emig.*

602. DUCELLIER, F. **Trois Cosmarium nouveaux.** [**Three new forms of Cosmarium.**] Bull. Soc. Bot. Genève 10: 12–16. *3 fig.* 1918.—The three new forms of *Cosmarium* described and figured include: (1) *C. obliquum* Nordst. form *minutissima* n. form., the smallest known form of this species; *C. crassangulatum* Borge, var. *Champesianum* n. var. differs from the species in size and the papillose nature of the cell wall; and *C. Hornavanense* (Schmidle) Gutwinski form *Helvetica* n. form.—*W. H. Emig.*

603. GHOSE, S. L. **A new species of Uronema from India.** Ann. Botany 34: 95–98. *15 fig.* 1920.—*Uronema indicum* from Lahore, India, is described, bringing the number of species in this genus up to four.—*E. N. Transeau.*

604. GROVES, JAMES. **Sex-terms for plants.** Jour. Botany 58: 55–56. 1920.—See Bot. Absts. 5, Entry 560.

605. HODGETTS, WILLIAM J. **Roya anglica G. S. West. A new Desmid; with an emended description of the genus Roya.** Jour. Botany 58: 65–69. 1920.—The author has compiled the account of this new species from descriptions, notes and drawings by G. S. WEST. The form of the vegetative cell is cylindrical or subcylindrical, unconstricted and very slightly tapering toward each end where it is subtruncate. The zygospore is globose with a hyaline smooth wall. The chief distinctions on which *Roya* can be retained as a genus are: (1) the simple structureless nature of the wall; and (2) the delay in the division of the chloroplast until the cell has reached maturity.—*K. M. Wiegand.*

606. HOWE, MARSHALL A. **Observations on monosporangial discs in the genus Liagora.** Bull. Torrey Bot. Club. 47: 1–8. *Pl. 1, fig. 25–29.* 1920.—The genus *Liagora* of the marine red algae of the family Nemalionaceae is made up of species mostly dioecious, some being monoecious. Some species often have small flat orbicular discs of a deep red color and bearing on their distal surface a few sporangia the contents of which remain undivided. The lack of any obvious genetic connection between these and the *Liagora* makes them appear to be independent of it. They probably arise from gonidia, gemmae, or aplanospores which come from terminal or subterminal cells of the assimilatory filaments of the *Liagora*.—*P. A. Munz.*

607. LUCAS, A. H. S. **Notes on Australian Marine Algae, II. Description of four new species.** Proc. Linnean Soc. New South Wales 44: 174–179. *Pl. 6.* 1919.—*Laurencia infestans* is described and figured and *Falkenbergia olens*, *Polysiphonia zostericola* and *Trichodesmium scoboideum* are discussed in detail.—*Eloise Gerry.*

608. Meister, Fr. **Zur Pflanzengeographie der schweizerischen Bacillariaceen.** [On the plant geography of the Swiss Bacillariaceae.] Bot. Jahrb. 55 (Beiheft): 125–159. *2 fig.* 1919. —Brun in 1880 described 32 genera and 182 species from Switzerland; Meister in 1912 listed 45 genera and 376 species, or including varieties 621 forms. The greater proportion of the Swiss diatoms occurred also in the Tertiary. About one-half of the Tertiary diatomaceous flora has persisted down to the present: thus when compared with the Phanerogams the conservative nature of the diatoms is apparent. The number that have appeared since Tertiary times is less than those that have perished so that the diatoms seem to be a waning group. Tertiary species common to central France and Hungary must have arisen in pretertiary times, therefore in the Cretaceous or Jurassic. The oldest known genera of fresh water Bacillariaceae are Epithemia, Rhopalodia, and Melosira. The Swiss diatom flora shows a much closer relation to the west European than to the east European tertiary flora. Several old tertiary forms are now found living only in Switzerland. The oldest forms from the Oligocene or those of the Miocene of west or east Europe now inhabit the bottoms of the Swiss lakes. There are many diatoms in the Alps and in the colder lakes of the lowlands that occur elsewhere only in the far north or in central Asia. Meister believes that Diatoms have migrated from central Asia to the Alps and the Arctic region rather than the reverse. Why are there so many endemic species in Switzerland when diatoms are generally so ubiquitous? There is no good explanation at present, but the author assumes as a working hypothesis that the relics have descended from preglacial times and that conditions during or before the glacial period were different from what is generally believed. Meister shows that many diatoms inhabit both highland and lowland lakes and are therefore not sensitive to variations in warmth; more than three-fourths were found to be indifferent. Extensive lists are given in various portions of the paper. *Navicula acuta* n. sp. is described.—*K. M. Wiegand.*

609. Pilger, R. **Ueber Corallinaceae von Annobon.** [On the Corallinaceae of Annobon.] Bot. Jahrb. 55: 401–435. *55 fig.* 1919.—This paper is the first report on the algae collected in 1911 by Dr. J. Mildbraed on Annobon, the smallest of the Guinea Islands, where the black calcareous rocks support a rich flora of marine algae. These lime-loving algae inhabit a zone between low and high tides which is wider on the west coast where the waves are high, and narrowest on the north where there is simply the swell of the ocean. The Corallinaceae are often very delicately colored. The decalcified material was imbedded in paraffin, sectioned and stained with Ruthenian red, Bismark brown, chlor-iodide of zinc, or haematoxylin. A brief account of the genus Goniolithon Foslie is given together with Foslie's diagnosis of the genus, and Harvey's description of *G. mamillare* (Harvey) Fosl., the only species found by Mildbraed. This species ranges from Brazil and Terra del Fuego to Cape Verde and Algoa Bay. Foslie suggests that *G. mamillare* may be a juvenile form of *G. brassica-florida.* Pilger gives extended descriptions of his material under four headings: (1) female material, (2) tetraspore material, (3) inner structure of the conceptacle projections and the branches, and (4) structure of the cortex. The female material formed thin crusts on the rocks with a smooth or uneven surface, the crusts sometimes being proliferous. The tetrasporic material produced crusts on stones or mussels. The cell structure, cell division, plasma membrane, and chromatophores are described, and illustrated. The reaction of the different cells to chlor-iodide of zinc is discussed. Elongation of the cell-rows takes place always by the division of the uppermost cell of the row, and the cells are connected in the direction of growth by peculiar double-faced pits. Heterocysts are formed in *Goniolithon* in 2's or 3's on the surface of the "Vorsprung." The whole floor of the tetrasporic conceptacle is covered with 4-parted tetrasporangia. The female conceptacles are in most cases empty or contain merely the remains of carpospores. The cortex is differentiated into a hypothallus and a perithallus. The cells of the perithallus are rich in starch. *Lithophyllum africanum* Foslie occurs on Annobon. This species forms cornice-like projections from the rock 15 cm. or even 30 cm. broad. The little fan-shaped ends of the inconspicuous branches are arranged story-like on the surface of the mass. *L. Kotschyanum* Unger is next described. This species is often attached to the larger species, *L. africanum.* It has a thin crust with a different type of branching. The crust does not show a real hypothallus with cells running at an angle to those of the peri-

thallus, and there are no pit connections between the cells in the lower layers of the crust. *Lithophyllum leptothalloideum* and *L. Mildbraedii* are described as new. These are thin crustaceous species, the former growing on *L. africanum*, the latter on stones and rocks. The anatomy and reproductive bodies of each are described. *Amphiroa annobonensis* also is described as new. Extended observations are made on the cellwall and tetrad cohesion of the cells in the Corallinaceae. The author finds that a middle lamella is present and therefore the whole gelatinous mass seeming to lie between the cells is really cell wall, and the calcium carbonate is actually deposited in the wall. The contributions of Yendo and Mme. Lemoine are cited. The author differentiates between pores due to the breaking down of the wall and true pits; and considers this to be of systematic value.—*K. M. Wiegand.*

610. Reverdin, L. **Le Stephanodiscus minor nov. spec. et revision du genre Stephanodiscus. [Stephanodiscus minor n. sp. and a revision of the genus Stephanodiscus.]** Bull. Soc. Bot. Genève 10: 17–20. *22 fig.* 1918.—A new species of *Stephanodiscus* (Diatom) with three to six silicious appendages is described and compared with the other two species of the same genus.—*W. H. Emig.*

611. Smith, Catharine, W. **Variation in the number of ribs in Costaria costata.** Publ. Puget Sound Biol. Sta. 2: 207–312. 1919.—While the number of ribs reported in literature is 3–5, the author finds that the number may be as high as 11. The number is not necessarily constant throughout the length of the same frond.—*T. C. Frye.*

612. Yendo, K. **A monograph on the Genus Alaria.** Jour. Coll. Sci. Imp. Univ. Tokyo 43: 1. 1919.—Deals mainly with the taxonomy of this genus but considers briefly various morphological details, the economic uses, and the distribution of these large marine algae. A considerable bibliography is included.—*G. J. Peirce.*

613. Zimmermann, Ch. **Quelques diatomées nouvelles ou curieuses. [Some new or peculiar diatoms.]** Broteria Ser. Bot. 17: 97–100. *Pl. 3 (5 fig.).* 1919.—Eight of the nine forms mentioned are proposed as new, viz.: *Navicula cardinaliculus* var. *margaritacea*, *N. Jequitinhonhae*, *N. Torrendii* with var. *capitata* and forms *typica* and *nana*, *N. mutica* var. *rhombica*, *Achnanthes lanceolata* var. *brasiliensis*, *Fragilaria undulata* var. *brasiliensis*. All forms mentioned are figured, and all are from Brazil.—*Edward B. Chamberlain.*

MORPHOLOGY AND TAXONOMY OF BRYOPHYTES

Alexander W. Evans, *Editor*

614. Amann, Jules. **Additions à la flore des mousses de la Suisse. [Additions to the moss flora of Switzerland.]** Bull. de la Murithienne 1916–18: 42–66. 1919.—The author first gives a large number of localities extending the range of species listed in the Flore des mousses de la Suisse [see Bot. Absts. 4, Entry 1032], and mentions a number of forms now first reported for the region, usually accompanying these reports with short descriptive notes. He proposes the following species as new: *Amblystegium ursorum*, *Bryum perlimbatum*, *Ceratodon mollis*, *Desmatodon spelaeus*, *Didymodon riparius* (not Kindb.), *Lesquereuxia glacialis*, *Mnium adnivense*, *Pseudoleskeella ambigua*, *Ptychodium abbreviatum*, *P. albidum*, and *P. pallescens*. Under the genus *Ptychodium* he gives a key to the Swiss species and states that between *Lesquereuxia saxicola* and *Ptychodium plicatum* a long series of transitional forms occurs, so that sterile material can not be definitely determined. At the conclusion of the paper he lists, as an example of the general trend of moss associations in the high-alpine region, twenty-one mosses and one hepatic found growing on the mica-schist of the Combin de Corbassière (Pennine Alps) at an altitude of 3600–3700 m.—*Edward B. Chamberlain.*

615. Bristol, B. Muriel. **On the gemmae of Tortula mutica, Lindb.** Ann. Botany 34: 137–138. *5 fig.* 1920.—This is a note recording the discovery of a specimen of *Tortula mutica* which bore numerous gemmae scattered over the surface of the leaf. The gemmae are borne on one-celled stalks and generally consist of two or four cells.—*W. P. Thompson.*

616. Britton, E. G. **Mosses of Bermuda.** Bryologist 22: 87. 1919.—This list of twenty-two species is an enumeration without comment of the forms mentioned in the recently issued Flora of Bermuda by Britton and others.—*Edward B. Chamberlain.*

617. Brotherus, V. F., and W. W. Watts. **The mosses of North Queensland.** Proc. Linnaean Soc. New South Wales 43: 544–567. 1918.—In a foreword the second author gives a brief description of the region where most of his collections were made and refers to the important work on the mosses of Queensland done by F. M. Bailey. He calls attention to the fact that the species of North Queensland are Malasian rather than Australian in their affinities. He notes further that, as a result of his explorations, one new genus and fourteen new species have been brought to light, that 17 other genera and 30 other species have been added to the flora of Australia, and that numerous species, heretofore known only from other parts of Australia, can now be definitely recorded from Queensland. In the main body of the work a list of species is given, with localities and occasional notes on distribution, and the new genus and new species are described. The new genus, *Pterobryidium* Broth. & Watts., is related to *Pterobyropsis* Fleisch. and is based on a single species. The new species are the following, Brotherus and Watts being the authorities except where otherwise noted: *Brachymenium Wattsii* Broth., *Bryum kurandae*, *Campylopus Wattsii* Broth., *Chaetomitrium entodontoides*, *Dicranoloma Wattsii* Broth., *Ectropothecium serriofolium*, *Floribundaria robustella*, *Fissidens cairnensis*, *F. kurandae*, *Pterobryidium australe*, *Pterobryopsis filigera*, *Syrrhopodon cairnensis*, *Taxithelium Wattsii* Broth. and *Trichosteleum elegantulum*. To these should be added *Mniodendron comatulum* Geheeb, a manuscript species here described for the first time. —*A. W. Evans.*

618. Chamberlain, Edward B. [Rev. of: Amann, J., and C. Meylan. **Flore des mousses de la Suisse.** [Flora of the mosses of Switzerland.] Genève, 1918. (See also Bot. Absts. 4, Entry 1032.)] Bryologist 22: 41–43. 1919.—The reviewer criticises the method employed by the authors in the citation of authorities for binomials. In all cases the original authority for the species is given, but when this name appears in parentheses the authority for the combination is not indicated in any way; the reader, therefore, unless thoroughly conversant with the literature, is in doubt as to "whether the combination be 'new' or not." In other respects the reviewer speaks in high terms of the work.—*A. W. Evans.*

619. Corbière, L. **Deux mousses africaines également françaises.** [Two African mosses occurring likewise in France.] Rev. Bryologique 41: 84–85. 1914. [Issued in 1919.]—In this paper (which is to be continued) the discovery of *Grimmia Pitardi* Corb. in the department of Var in southern France is announced. The species was described in 1909 from specimens collected in Tunis and has since been recorded from Tripoli. A full description is included. —*A. W. Evans.*

620. Dixon, H. N. **Rhaphidostegium caespitosum (Sw.) and its affinities.** Jour. Botany 58: 81–89. 1920.—The author's first impression of *Rhaphidostegium sphaerotheca* (C. M.) Jaeg., obtained from material collected on Table Mountain, Cape Colony, led him to believe that it was a well-marked species. Further study, however, showed that this was not the case but that the Table Mountain specimens, which were exceptionally large and fine, belonged to an extensive "Formenkreis," the usual material of which was small and commonplace. In this "Formenkreis" the author was able to include a number of specimens from South and Central Africa and from the Mascarene Islands, some of which had been referred to other species or even to other genera. Previous experience suggested that when a plastic species had a wide African distribution it was well to look further and see if it might not be identical with some South American or Australian species. Acting on this hypothesis the author was able to demonstrate that *R. sphaerotheca* was really a synonym of the American *R. caespitosum* (Sw.) Jaeg., a species originally described by Swartz from West Indian material under the name *Hypnum caespitosum*. He was able to show further, by the study of numerous type-specimens, that *R. caespitosum* had many synonyms and that it had a cosmopolitan range in the tropical and subtemperate portions of the Southern Hemisphere, even extending into

the temperate zone. The best specific characters for this polymorphic species are derived from the perichaetial leaves and capsules, and it seems to be impossible to divide it into definite groups, since the variations run in different directions and are little correlated. Some of the species referred in this paper to *R. caespitosum*, as synonyms, are the following: *Hypnum lithophilum* Hornsch., *Hypnum loxense* Hook., *Leucomium Robillardii* (Duby) Jaeg., *Pterogoniella Stuhlmanni* Broth., *R. agnatum* (Hampe) Jaeg., *R. caespitans* Schimp., *R. Catillum* (C. M.) Jaeg., *R. cucullatifolium* (Hampe) Jaeg., *R. Dicnemonella* (C. M.) Broth., *R. Duisaboanum* (Mont.) Jaeg., *R. fluminale* (C. M.) Broth., *R. inconspicuum* (Hornsch.) Jaeg., *R. Kegelianum* (C. M.) Jaeg., *R. perlaxum* (C. M.) Par., *R. Sauloma* (C. M.) Broth., *R. sphaerotheca* (C. M.) Jaeg., *R. subsphaericarpum* (Hampe & C. M.) Jaeg., *Sematophyllum subnervatum* Mitt., and *Stereodon tristiculus* Mitt. In all 58 synonyms are given.—*K. M. Wiegand.*

621. DOUIN, CH., AND L. TRABUT. **Deux hepatiques peu connues.** [Two little known hepatics.] Rev. Gén. Bot. 31: 321–328. *Pl. 9, 1 fig.* 1919.—Two liverworts from Algeria, *Corbierella algeriensis* Douin & Trabut and *Riccinia perennis* (Steph.) Trabut, are described. The first is probably the same as *Exormotheca Holstii* Steph. but, on account of certain peculiar characters, is made the type of the new genus *Corbierella* Douin & Trabut. The second species, which was originally described by Stephani under the name *Riccia* (*Ricciella*) *perennis*, is the only member of the genus *Riccinia* Trabut, proposed in 1916. The writers regard it as intermediate between the Ricciaceae, with which it is classed, and the Marchantiaceae.—*L. W. Sharp.*

622. EVANS, ALEXANDER W. **Notes on New England Hepaticae.—XV.** Rhodora 21: 149–169. *Pl. 126, 14 fig.* 1919.—Under the name *Nardia obscura* the writer describes and discusses a new species, closely related to *N. hyalina* (Lyell) Carringt. and *N. obovata* (Nees) Lindb. It has been observed in several mountainous localities, especially in the White Mountains, and seems to retain its distinctive features even while exhibiting a considerable range of variability. For the sake of comparison the features of *N. hyalina* and *N. obovata* are likewise discussed. Another species, *Cephalozia Loitlesbergeri* Schiffn., is reported for the first time from New England, the first American record having been based on material from Nova Scotia. Additions to local state floras include the following: *Jungermannia sphaerocarpa*, *Calypogeia sphagnicola* and *Notothylas orbicularis* from Maine; *Riccardia pinguis*, *Pellia Neesiana*, and *Lophocolea alata* from Massachusetts. According to the census given at the close of the paper 191 Hepaticae are now known from New England, including 142 from Maine, 151 from New Hampshire, 129 from Vermont, 79 from Rhode Island, 145 from Connecticut, and 62 from all six states.—*James P. Poole.*

623. HARRIS, G. T. **On Schistostega osmundacea Mohr.** Jour. Quekett Microsc. Club II, 13: 361–374. *2 pl., 2 fig.* 1917.—This moss thrives in crevices, caves, burrows, etc., facing north to northeast, where it is moist but not wet. The capsules are deciduous. Gemmae are formed on the protonema and are disseminated by animals. The flask-shaped cells found on protonema are separation-cells, remaining after the liberation of gemmae. The protonema is made up largely of obconic light-cells, whose structure is discussed, and is almost completely used up in mature plants.—*L. B. Walker.*

624. KASHYAP, S. R. **The androecium in Plagiochasma appendiculatum L. et L. and P. articulatum Kashyap.** New Phytol. 18: 235–238. *2 fig.* 1919.—At the end of the paper the author summarizes his conclusions as follows: "Three different considerations, therefore, show that the androecium of *P. appendiculatum* and *P. articulatum* is really homologous with that of the higher Marchantiales in being a branch-system. . . . . (1). The arrangement of the scales at the tip of the lobes which is very similar to that at the tip of the vegetative lobe. (2). The repeated branching of the receptacle, two or three times in some specimens. (3). The invariably acropetal succession of the antheridia in all lobes exactly as in *Marchantia*, the highest genus of the Marchantiales."—*I. F. Lewis.*

625. Luisier, A. **Les mousses de Madère.** [Mosses of Madeira.] Broteria Ser. Bot. 17: 112–142. 1919.—This article is the sixth of a series covering a complete discussion of Madeiran mosses and includes the genera *Bryum* (in part) to *Thamnium.* No new forms are described, but short notes, references, and reprints of original descriptions are appended to many of the species mentioned. More extended and critical discussions occur for *Bryum serrulatum* Card., *Pogonatum subaloides* (C. M.) Jaeg., *Neckera intermedia* Brid., and the species of *Leucodon* and of *Echinodium.*—*Edward B. Chamberlain.*

626. Melin, Elias. **Sphagnum angermanicum n. sp.** Svensk. Bot. Tidskr. [Stockholm] 13: 21–25. *3 fig.* 1919.—Under the above name the author describes and illustrates a new species of *Sphagnum* which he found in Ångermanland, Sweden, in 1915, 1916 and 1917. It is related to *S. molle* Sulliv.—*W. W. Gilbert.*

627. Paul, H. **Einige für den Bayerischen Wald neue Pflanzen.** [Plants new to the Bavarian Forest.] Mitteil. Bayer. Bot. Ges. Erforsch. Heim. Flora 3: 467–468. 1918.—The author lists the following four plants as additions to the known flora of the Bavarian Forest: *Scutellaria minor* L., *Sphagnum subbicolor* Hpe., *Bryum cyclophyllum* Br. Eur., and *Catharinaea Hausknechtii* Broth. He describes the exact localities where these plants were found and enumerates various other species, both spermatophytes and bryophytes, which grew in association with them.—*A. W. Evans.*

628. Potier de la Varde, R. **Observations sur quelques espèces du genre Fissidens.** [Observations on certain species of the genus Fissidens.] Rev. Bryologique 41: 85–92. *Pl. 1.* 1914. [Issued in 1919.]—In the first part of this paper (which is to be continued) the value of apparent dioecism as a specific character in the genus *Fissidens* is discussed. This condition is brought about when the same protonema gives rise to distinct male and female shoots. It thus represents a special form of monoecism, and the student of mosses is cautioned against attributing positive dioecism to a species until the protonemal relations have been established. In the second part of the paper the status of *F. tamarindifolius* Wils. is considered, and the conclusion is reached that it represents a form of *F. inconstans* Schimp. This conclusion is based on the study of a long series of European specimens ranging from England to Italy and the Tirol. The figures were drawn from material collected in the department of the Manche in France.—*A. W. Evans.*

629. Seymour, M. E. **Mosses of the Cascade Mountains, Washington, collected by J. A. Allen.** Bryologist 22: 85–86. 1919.—This is a list, without comment, of the specimens issued in the somewhat uncommon set of exsiccati mentioned in the title.—*Edward B. Chamberlain.*

MORPHOLOGY AND TAXONOMY OF FUNGI, LICHENS, BACTERIA AND MYXOMYCETES

H. M. Fitzpatrick, *Editor*

630. Adams, J. F. **The alternate stage of Pucciniastrum Hydrangeae.** Mycologia 12: 33–35. 1920.—Along a path about which hydrangeas and hemlocks were numerous, *Hydrangea arborescens* was found to be heavily infected with *Pucciniastrum Hydrangeae* while the hemlocks were infected with a *Peridermium* stage resembling *P. Peckii.* Inoculation with aeciospores on *Hydrangea arborescens grandiflora* proved successful, mature uredinia appearing in about 12 days. Three other species of cultivated hydrangeas and a species of *Vaccinium* failed to show infections. Because of differences in hosts and in morphology, the *Peridermium* is considered distinct from *P. Peckii* and is technically described, the name *P. Hydrangeae* (Berk. & Curt.) comb. nov. being proposed.—*H. R. Rosen.*

631. Arnaud, G. **Les Astérinées.** [The "Astérinées."] Ann. École Nation. Agric. Montpellier 16: 1-288. *Pl. 1-53, 22 fig., maps 1-3.* July 1917-August, 1918.—The name Astérinées is given to a group of black, saprophytic fungi which, although taxonomically heterogeneous, are homogenous from the standpoint of biology and climatology. Nearly all Pyrenomycetes belong to this group. The work is divided into three parts: (1) Comparative morphology. (2) Special taxonomy and morphology. The two groups of Pyrenomycetes, viz., Microthyriales and Dothideales are studied and described in detail. (3) Climatology and geographical distribution. A bibliographical index of the most important publications pertaining to these fungi is given.—*F. F. Halma.*

632. Arthur, J. C. **Errors in double nomenclature.** Bot. Gaz. 68: 147-148. Aug., 1919.—Attention is called to the difficulties which confront taxonomists working with parasitic fungi, because of the necessity of having accurate taxonomic knowledge of hosts as well as of parasites. Occasion is taken to correct an error in a previous paper by the author (Bot. Gaz. 65: 470-471. 1918. See Bot. Absts. 1, Entry 385). Two species there described as new, *Puccinia Nicotianae* and *P. Acnisti*, growing respectively on *Nicotiana tomentosa* and *Acnistus arborescens*, turn out to be one species, and the host of the two also is identical. The common host now appearing to be *A. aggregatus*, the correct name of the parasite is *Puccinia Acnisti.*—*H. C. Cowles.*

633. Bachmann, E. **Neue Flechtengebilde.** [New lichen structures.] Ber. Deutsch. Bot. Ges. 36: 150-156. *Pl. 3.* 1918.—Studies of microtome sections of the thalli of limestone-inhabiting lichens containing *Chroolepus* or *Scytonema* as gonidia show three new points: (1) Spheroidal cell-clusters made up of groups of pseudoparenchymatous cells and storing oils; (2) "Hyphal knots," similar in structure, but without the oil and believed to serve for water-storage; and (3) Wandering gonidia, which are free from connection with the hyphae, occur more deeply situated than the usual gonidia, and are yellow-red instead of green.—*L. W. Riddle.*

634. Bokura, U. **A bacterial disease of lily.** Ann. Phytopath. Soc. Japan 1[2]: 36-90. *Pl. 1-2.* 1919.—See Bot. Absts. 4, Entry 1243.

635. Breed, R. S., and H. J. Conn. **The nomenclature of the Actinomycetaceae.** Jour. Bact. 4: 583-602. 1919.—A review of the literature relative to the proper generic names to be used in the family Actinomycetaceae is given, followed by a discussion in which the conclusion is reached that the generic name *Actinomyces* Harz should be used rather than *Streptothrix* Corda, *Streptothrix* Cohn, *Discomyces* Rivolta, or *Actinocladothrix* Afanasiev and Schultz. *Nocardia* Trevisan may be used as a subdivision of the genus *Actinomyces*. *A. bovis* Harz may be considered as the type species.—*Chester A. Darling.*

636. Burger, Owen F. **Sexuality in Cunninghamella.** Bot. Gaz. 68: 134-146. Aug., 1919.—See Bot. Absts. 3, Entry 2096.

637. Chou, Chung Ling. **Notes on fungous diseases in China.** [Text in Chinese.] Khu Shou [Science-Publication of the Chinese Science Society] 4: 1223-1229. *Fig. 1-16.* 1919.—See Bot. Absts. 5, Entry 732.

638. Clark, Paul F., and W. H. Ruehl. **Morpholgical changes during the growth of bacteria.** Jour. Bact. 4: 615-625. 1919.—Seventy strains of bacteria representing 37 species, many of which were pathogenic forms, were studied as to the variation in size, shape, characteristic groupings, and staining when grown for different periods of time on ordinary culture media. The conclusions were that in all strains examined excepting those of the diphtheria group and possibly *B. mallei* the organisms found in cultures four to nine hours old are much larger than in older cultures. The period when the largest organisms are found corresponds closely to the period when the cells are dividing rapidly. In the diphtheria group the organisms in cultures of from 4 to 9 hours old are definitely smaller and more solid staining than in older cultures.—*Chester A. Darling.*

639. Colosi, G. **Contributo alla conoscenza dei Licheni della Sardegna. [Some Lichens of Sardinia.]** Malpighia 28: 458–471. 1919.—Lists 115 species and varieties, two of the varieties being new.—*L. W. Riddle.*

640. Dittrich, G. **Über Vergiftungen durch Pilze der Gattungen Inocybe und Tricholoma. [Poisoning caused by species of Inocybe and Tricholoma.]** Ber. Deutsch Bot. Ges. 36: 456–459. 1918.—*Inocybe sambucina* is reported to have caused severe poisoning. However, since the species has not been identified with certainty and no specimens could be obtained the following season, it is doubtful whether the fungus in question or some other organism was responsible. *Tricholoma tigrinum* and related species are considered harmless by some and poisonous by others.—*Ernst Artschwager.*

641. [Dodge, B. O.] Anonymous. **Index to American mycological literature.** Mycologia 12: 55–58. 1920.

642. Doidge, E. M. **An interesting group of leaf fungi.** South African Jour. Nat. Hist 1: 164–171. *4 fig.* 1919.—An account of the *Perisporiaceae* and *Microthyriaceae*, is given in a non-technical style, including methods of collecting, preserving and making microscopic preparations.—*E. M. Doidge.*

643. Dufrenoy, Jean. **Sur les maladies parasitaires des Chenilles processionaires des Pins d'Arachon. [Concerning parasitic maladies of caterpillars.]** Compt. Rend. Acad. Sci. Paris 168: 1345–1346. 1919.—Brief descriptions of various bacteria and higher fungi found growing as parasites on certain larvae. Those mentioned are: *Bacterium pityocampae*, *Streptococcus* sp., *Beauveria globulifera* (J. Beauverie) and *Penicillium* sp.—*V. H. Young.*

644. Emile-Weil, P., and L. Gaudin. **Contribution à l'étude des onychomycoses—Onychomycoses à Penicillium, à Scopulariopsis, à Sterigmatocystis, à Spicaria. [Contribution to the study of onychomycoses due to Penicillium, etc.]** Arch. Méd. Exp. et Anat. Path. Paris 28: 452–467. *Pl. 12, 4 fig.* 1919.—Mycoses of the nails are not uncommon, but have been given little study. They are mostly confined to the toe-nails, particularly those of the big toes. The pathological aspects are discussed briefly. Reports are given of *Penicillium brevicaule* var. *hominis* (=*Scopulariopsis Koningi* Vuill.), *Scopulariopsis cinerea* n. sp., *Spicaria unguis* n. sp., *Sterigmatocystis unguis* n. sp. The cultural characteristics of these fungi are described, as well as their morphology. Their mode of infection is probably through lesions. They frequently follow frost injury.—*E. A. Bessey.*

645. Eriksson, Jakob. **Sur l'hétéroecie et la specialisation du Puccinia caricis, Reb. [On heteroecism and specialization in Puccinia caricis Reb.]** Rev. Gén. Bot. 32: 15–18. 1920.—After a large number of collections and inoculation experiments author divides *Puccinia caricis* into 3 species: *P. Caricis diffusa*, with aecidia on Urtica and Ribes; *P. caricis-Urticae* (*P. Urticiae-caricis*, Kleb.), with aecidia on Urtica; and *P. caricis-Ribis* (*P. Ribesii, Caricis*, Kleb.), with aecidia on Ribes. Under the last named are 3 sub-species: *P. Caricis-Ribis, diffusa*; *P. caricis-Grossulariae*; and *P. caricis-Ribis-nigri*.—*L. W. Sharp.*

646. Fragoso, Romualdo Gonzalez. **Notes and communications at the session of Oct. 1, 1919.** Bol. R. Soc. Española Hist. Nat. 19: 429–430. Oct., 1919.—The President of the Society commented on certain species distributed as new in the last fascicle of Maire's "Mycotheca Boreali-Africana," particularly with reference to their relation to the mycological flora of Spain. *Puccinia Scirpi-littoralis* (Pat.) Maire, II, III. The species encountered on *Scirpus* in regions where there is no *Limnanthemum nymphoides* may be this new species; *P. Laguri-Chamaemoly* Maire, O, I–II, III, probably occurs in southern Spain where *Allium Chamaemoly* occurs; *P. madritensis* Maire, O, I–II, III, is probably the species reported as an aecidium on *Clematis cirrhosa* from the Balaeric islands and referred to *P. Agropyri* Ell. & Ev.; *Uromyces Cuenodii* Maire, II, III; *Entyloma Eryngii-tricuspidati* Maire, probably occurs in southern Spain; *Physoderma Ornithogali* Maire attacks *Ornithogalum narbonnense* which occurs in all parts of Spain.—*O. E. Jennings.*

647. Fries, Thore C. E. **Onygena equina (Willd.) Pers. funnen i Halland. [Onygena equina (Willd.) Pers. found in Holland.]** Svensk. Bot. Tidskr. [Stockholm] 13: 107. *Fig. 1.* 1919.

648. Gunn, W. F. **Some Irish Mycetozoa.** Irish Nat. 28: 45–48. 1919.—The number of Irish Mycetozoa should approximate those recorded from Great Britain and further exploration is desired. A list of thirty-eight species and varieties from new localities and one, *Hemitrichia vesparium*, new to Ireland is given.—*W. E. Praeger.*

649. Hadden, N. G. **The Uredineae of West Somerset.** Jour. Botany 58: 37–39. 1920. —This paper is a list of the rusts found within a few miles of Porlock in Somerset, England. The list is said to contain an unusually large number of species, a number of which are rare and interesting. The nomenclature is that of Ramsbottom's list of British Uredinales.—*K. M. Wiegand.*

650. Harris, J. E. G. **Contributions to the biochemistry of pathogenic anaerobes. VIII. The biochemical comparison of microorganisms by quantitative methods.** Jour. Path. and Bact. 23: 30–49. *Fig. 1–2.* 1919.—See Bot. Absts. 5, Entry 936.

651. Herre, A. W. C. T. **A list of lichens from southeastern Alaska.** Publ. Puget Sound Biol. Sta. 2: 279–285. 1919.—A taxonomic report on the lichens collected by the members of the U. S. Bureau of Soils Kelp Exploration Expedition to Southeastern Alaska in 1913. 86 species and varieties were collected, of which 19 were not previously known from Alaska. The range of others is extended.—*T. C. Frye.*

652. Herrmann, E. **Behandlung und Untersuchung der Trockenpilze. [Treatment and examination of mushrooms for drying purposes.]** Pharm. Zentralhalle Deutschland 60: 5–7, 21–25. *Pl. 1, fig. 16.* 1919.—A description of edible fungi and tests for distinguishing these from poisonous fungi.—*H. Engelhardt.*

653. Hollande, A. Ch. **Formes levures pathogènes observées dans le sang d'Acridium (Caloptenus italicus L.). [Pathogenic yeast forms observed in the blood of Acridium.]** Compt. Rend. Acad. Sci. Paris 168: 1341–1344. *1 fig.* 1919.—Marchal has reported yeast forms in the blood of the caterpillars of *Cochylis*. It is probable that these are parasitic. Author in 1918 discovered a form of yeast in the blood of crickets (*Caloptenus italicus*). The normal limpid blood of the insect assumes a milky appearance when the insect is affected and death ensues. Disease was produced by injection of blood of diseased individuals into normal ones. It was possible to produce the disease in *Psophus stridulus L.* but in other forms experimented upon the yeast cells were rapidly destroyed by the leucocytes. Organism is described and figured. The organism was obtained in pure culture on various media and in certain cases filamentous forms have appeared. Author suggests two possibilities, viz., that a yeast and another fungus are present and secondly that the yeast form is merely one stage in the life cycle of a filamentous form. It is proposed to inoculate insects with both forms to settle this point.—*V. H. Young.*

654. Keissler, K. v. **Beiträge zur Naturgeschichte der Scoglien und Kleineren Inseln Suddalmatiens. 4. Fungi. [The natural history of the smaller islands of southern Dalmatia. 4. Fungi.]** Denkschr. K. Akad. Wiss. Wien. (Math.Nat. Kl.) 92: 299–300 1916.—Only six species are listed. These are *Diplodina Sandstedei*, *Didymosphaeria* sp., *Hysterium angustatum*, *Stictis radiata*, *Scutula Aspiciliae*, and *Leciographa centrifuga*.—*H. M. Fitzpatrick.*

655. Kempton, F. E. **Origin and development of the pycnidium.** Bot. Gaz. 68: 233–261. *6 pl.* Oct., 1919.—Pycnidia originate and develop by two main methods, meristogenous and symphyogenous, the meristogenous method resolving itself into two modes, simple and compound. Variations of the meristogenous method are found in *Coniothyrium pyriana* and *Sphaeronaemella fragariae*. The symphyogenous method is less often found and is variable. Acervuli arise as do pycnidia, simple acervuli by the simple meristogenous mode, and complex ones usually by the compound meristogenous or symphyogenous method. Complex subicles

usually arise symphyogenously, although they may arise by the compound meristogenous mode. Simple sporodochia usually originate by the simple meristogenous method. Complex sporodochia, with a large base or subicle, usually arise either by the compound meristogenous mode or symphyogenously. The pseudo-acervulus of the species of *Pestalozzia* studied arises and develops as a pycnidium which breaks open and appears like an acervulus. The simple meristogenous development is the one more often found in the Sphaeropsidales, while the compound meristogenous and symphyogenous modes are the more usual in the Melanconiales and Tuberculariaceae.—*F. E. Kempton.*

656. Lagerberg, Torsten. **Onygena equina (Willd.) Pers. från Dalarna. [Onygena equina (Willd.) Pers. found in Dalarna.]** Svensk. Bot. Tidskr. [Stockholm] 13: 108. *Fig. 1.* 1919.

657. Lendner, A. **Les mucorinées géophiles recoltées a Bourg-Saint-Pierre. [The soil mucors collected at the village of St. Peter, Switzerland.]** Bull. Soc. Bot. Genève 10: 362-376. *3 fig.* 1918.—Six cultures of *Mucor* which were obtained from the soil near St. Peter contained two new varieties and two new species. *Mucor Ramannianus* Moeller was frequently found in coniferous forests, sometimes on *Sphagnum* of peat bogs. *M. plumbens* Bonorden was common in the air and soil. *M. hiemalis* Wehmer (−) var. *albus* n. var produced occasional zygospores with the + strain of the species. *M. hiemalis* (+) var. *toundrae* n. var. differs from the species in its habit of growth in cultures. *M. Jauchae* n. sp. was isolated from the soil of a fir forest. *M. vallesiocus* n. sp. was obtained from the soil. of a meadow.—*W. H. Emig.*

658. Lettau, G. **Schweizer Flechten. [Some lichens of Switzerland.] Part I.** Hedwigia 60: 84-128. **Part II.** Hedwigia 60: 267-312. 1918.—An enumeration of the lichens and of some fungus-parasites of lichens, arranged by localities with critical notes. No new species are described.—*L. W. Riddle.*

659. Licent, Eug. **La forme ascophore du Clasterosporium fungorum (Fr.) Sacc. (Amphisphaeria fungorum n. sp. Eug. Licent.) [The ascogenous form of C. fungorum (Fr.) Sacc.]** Compt. Rend. Acad. Sci. Paris 170: 60-62. *1 fig.* 1920.—*C. fungorum* is transferred from the Mucedineae to the genus *Amphisphaeria* of the Ascomycetes. The author has discovered and describes the asci-containing perithecia which appear in November beneath the dark-colored conidiferous filaments of this fungus when growing upon the white fructifications of *Corticium calceum* Persoon or *C. lacteum* Fries as a host. These perithecia develop until they project almost entirely free from the conidial layer, attaining a diameter of 0.2 to 0.5 mm.—*C. H. and W. K. Farr.*

660. Magnusson, A. H. **Material till Västkustens Lavflora. [Material for the Lichen Flora of the West Coast.]** Svensk. Bot. Tidskr. [Stockholm] 13: 75-92. 1919.—The author gives a list of several hundred species of lichens found by him on the west coast and adjacent islands of Sweden together with brief notes on their habitat and abundance.—*W. W. Gilbert.*

661. Malme, Gust. O. A. **Lichenes suecici novi. [New Swedish lichens.]** Svensk. Bot. Tidskr. [Stockholm] 13: 26-31. 1919.—Author gives Latin descriptions of six new species of lichens of the genera *Lecidea*, *Catillaria*, and *Rhizocarpon.*—*W. W. Gilbert.*

662. McCulloch, Lucia. **Basal glume rot of wheat.** Jour. Agric. Res. 18: 543-551. *Pl. 62-63.* 1920.—See Bot. Absts. 5, Entry 749.

663. Merewschkowsky, C. **Note sur une nouvelle forme de Parmelia vivant à l'état libre. [A new form of Parmelia living in an unattached condition.]** Bull. Soc. Bot. Genève 10: 26-34. *1 fig.* 1918.—*Parmelia conspersa* (Ehrh.) Ach. forma *vaga* n. form occurs in abundance on a certain plateau with all the vegetative characteristics of a steppe. On the steppe, conditions for growth are unfavorable to the production of new lichen plants by the development of fungous spores and algal cells. As a consequence of the arid conditions this lichen does not have fruiting bodies but multiplies by the fragmentation of the thallus.—*W. H. Emig.*

664. Murrill, William A. **Corrections and additions to the polypores of temperate North America.** Mycologia 12: 6–24. 1920.—Since the publication of the polypores in the *North American Flora* much additional information has been obtained on this group involving clearer identity of some of the forms previously described and adding a number of species not previously listed. Various changes are accordingly made or suggested involving the reclassification of numerous forms.—*H. R. Rosen.*

665. Murrill, W. A. **Fungi from Hedgcock.** Mycologia 12: 41–42. 1920.—Twelve species of polypores collected by Hedgcock and others are listed.—*H. R. Rosen.*

666. Murrill, W. A. **Collecting fungi at Yama farms.** Mycologia 12: 42–43. 1920.—Describes an interesting collecting tour in a large tract of virgin land near Poughkeepsie, New York. Nearly 100 species of fungi were collected and several of the more interesting polypores and agarics are mentioned.—*H. R. Rosen.*

667. Murrill, W. A. **Trametes serpens.** Mycologia 12: 46–47. 1920.—American specimens referred to *Trametes serpens* Fr. are found to have smaller pores and are "otherwise distinct" from those of Europe. The distribution of the American plant is given and it is compared with a Philippine specimen, *Elmeriana setulosa*, which it seems to match.—*H. R. Rosen.*

668. Murrill, W. A. **The genus Poria.** Mycologia 12: 47–51. 1920.—Historical sketch of the genus *Poria* as used by mycologists before the time of Persoon together with Persoon's interpretation of the genus is presented. *P. medullapanis* (Jacq.) Pers., one of the species upon which Persoon based the genus, is thoroughly described, its variations noted and a large number of American collections of this species which were examined by the writer are listed.—*H. R. Rosen.*

669. Murrill, W. A. **Collecting fungi near Washington.** Mycologia 12: 51–52. 1920.—Brief notes of mycologists and of a few fleshy fungi seen during a collecting trip around Washington, D. C.—*H. R. Rosen.*

670. Northrup, J. H., Lauren H. Ashe, and R. R. Morgan. **A fermentation process for the production of acetone and ethyl alcohol.** Jour. Indust. Eng. Chem. 11: 723–727. *2 fig.* 1919.—The general characteristics of a new organism, *Bacillus acetoethylicum*, are given according to the descriptive chart of the Society of American Bacteriologists, but a formal diagnosis is postponed for a later paper. [See Bot. Absts. 4, Entry 1515.]—*B. M. Duggar.*

671. Olivier, H. **Les lichens pyrénocarpés de la flore d'Europe.** [The pyrenocarpic lichens of Europe.] Bull. Geog. Bot. 28: 146–152, 168–183. 1918.—First two installments of a compilation of all the described genera, species, and varieties, of pyrenocarpic lichens of Europe, with keys and diagnoses. The two parts cited include the genera *Normandina*, and *Endocarpon* (taken in a broad sense), and the key to *Polyblastia.*—*L. W. Riddle.*

672. Paine, Sydney G., and W. F. Bewley. **Studies in bacteriosis. IV.—"Stripe" disease of tomato.** Ann. Appl. Biol. 6: 183–202. *Pl. 8–9, 5 fig.* 1919.—See Bot Absts. 5, Entry 756.

673. Paine, Sydney G., and H. Stansfield. **Studies in bacteriosis. III.—A bacterial leaf-spot disease of Protea cynaroides, exhibiting a host reaction of possibly bacteriolytic nature.]** Ann. Appl. Biol. 6: 27–29. *Pl. 2, fig. 3–6.* 1919.—See Bot. Absts. 5, Entry 757.

674. Pethybridge, G. H., and H. A. Lafferty. **A disease of tomato and other plants caused by a new species of Phytophthora.** Sci. Proc. Roy. Dublin Soc. 15: 487–503. *3 pl.* 1919.—See Bot. Absts. 4, Entry 1335.

675. PFEILER, W., AND F. ENGELHARDT. **Zeigt der Ferkeltyphus-Bacillus (Bac. Voldagsen Dammann und Stedefeder) ein Labiles biochemisches und agglutinatorisches Verhalten? [Does the Ferkel typhus bacillus (Bac. Voldagsen Dammann und Stedefeder) show a labile biochemical and agglutination relation?]** Zeitschr. Immunitätsforsch. u. exp. Therapie 28: 434–445. 1919.—The authors show that *Bac. Voldagsen* Damm. & Stedef. is distinct from the other members of the paratyphosus group and that it does not approach the characters of the group on long continued cultivation on artificial media.—*C. W. Dodge.*

676. PLITT, CHARLES C. **A short history of lichenology.** Bryologist 22: 77–85. 1919.

677. PUYHAUBERT, A., AND R. JOLLY. **Note sur un cas de mycètome à grains noirs provoqué par un champignon du genre Madurella. [Notes on a case of mycetoma with black granules, caused by a fungus of the genus Madurella.]** Arch. Méd. Exp. et Anat. Path Paris 28: 441–445. *5 fig.* 1919.—A skin disease of a native of the Ivory Coast was shown to be due to infection with a fungus probably identical with *Madurella mycetomi* (Laveran) Brumpt. The fungus grown in pure culture on carrot gave abundant mycelium with numerous small black sclerotia, but no spores.—*E. A. Bessey.*

678. RICK, J. **Contributio II ad monographiam agaricinorum brasiliensium. [Second contribution to a monograph of Brazilian agarics.]** Broteria Ser. Bot. 17: 101–111. 1919.—The article is a sequel to one published in Broteria for 1905. After briefly outlining difficulties of study, the author lists 92 species or varieties mostly collected in the vicinity of Parecy Novo. Notes upon spore measurements, color, appearance, and habitat, based upon fresh material, accompany some of the species. *Tricholoma sulphurellem, Clitocybe nauseosa, Collybia sericea, Mycena sulphureo-conspersa, Leptonia rosea, L. straminea, L. albo-serrulata, L. olivacea, L. fuligineo-straminea, Pholiota pusilla, P. rosea, Inocybe megalospora, Psalliota haemorrhoidaria var. straminea,* and *Schizophyllum album* are proposed as new. Note is made that *Lactarius Russula* as previosly reported by the author is probably *L. trivialis.*—*Edward B. Chamberlain.*

679. ROBERTSON, W. F. **A starch-splitting bacterium found in cases of diabetes mellitus.** Jour. Path. and Bact. 23: 122–123. 1919.

680. ROMELL, L. **Svamplitteratur, särskilt för studium av hymenomyceter [hattsvampar). [Mycological literature, especially for the study of the hymenomycetes (hat fungi).]** Svensk. Bot. Tidskr. [Stockholm] 13: 110–112. 1919.—A list of European literature on the hymenomycetes and related fungi is given comprising thirty titles.—*W. W. Gilbert.*

681. ROSEN, H. R. **Ergot on Paspalum.** Mycologia 12: 40–41. 1920.—*Paspalum floridanum* is recorded as a new host for *Claviceps* spp. It is noted that attacked spikelets fall with pedicels attached to them in contrast to the fall of normal spikelets in which the pedicels remain attached to the rachis.—*H. R. Rosen.*

682. SEAVER, FRED J. **Photographs and descriptions of cup-fungi—VIII. Elvela infula and Gyromitra esculenta.** Mycologia 12: 1–5. *Pl. 1.* 1920.—Comparison between descriptions and illustrations of *Elvela* (*Helvella*) *infula* and *Gyromitra esculenta* leads the writer to believe that these names are referable to the same fungus. Differences noted by various authors are explainable as variations. The name *Elvela infula* Schaeff. is adopted because of its priority; 11 synonyms are listed and the plant is redescribed and illustrated.—*H. R. Rosen.*

683. SERNANDER, R. **Subfossile Flechten.** Flora 112: 703–724. *7 fig.* 1918.

684. SPEARE, A. T. **Further studies of Sorosporella uvella, a fungous parasite of noctuid larvae.** Jour. Agric. Res. 18: 399–439. *Pl. 51–56.* 1920.—*Sorosporella uvella* is recorded for America. It is found to be related to the verticillaceous hyphomycetes rather than to the Entomophthorales. It produces chlamydospores and thin-walled conidia. Yeast-like vege-

tative cells, occurring in the blood of infected insects, are ontogenetically related to other phases in the development of the organism. Fruiting structures of the Isaria type have been observed in culture and in moist chamber. An ascigerous stage has not been observed. An emended description is presented.—The organism produces a disease of noctuid larvae (cut worms) and in infection experiments a mortality of from 60 to 90 per cent was obtained.—Ingestion of vegetative cells by phagocytes was observed, the process being followed apparently by the destruction of the phagocytes. Phagocytosis is discussed at some length, also certain phases of insect control by means of fungous parasites.—A bibliography of 24 titles is appended.—*D. Reddick.*

685. Stevens, F. L. **Three new fungi from Porto Rico.** Mycologia 12: 52-53. 1920.—The following fungi collected by the writer and briefly described by Mr. Lamkey are presented: *Microstoma ingaicola* Lamkey sp. nov. producing witches' brooms on *Inga laurina*, *M. pithecolobii* Lamkey sp. nov. producing spots on *Pithecolobium saman*, and *Perenoplasmopara portoricensis* Lamkey sp. nov. producing spots on *Melia azedarach.*—*H. R. Rosen.*

686. Stevens, F. L., and Nora Dalbey. **A parasite of the tree fern (Cyathea).** Bot. Gaz. 68: 222-225. *2 pl.* Sept., 1919.—A fungus collected on *Cyathea arborea* in Porto Rico has characters suggesting relationship with Microthyriaceae, Perisporiaceae, Dothidiaceae, and Phacidiaceae; the authors incline to place it in the last-named group, proposing for it a new generic name, *Griggsia*. The type species is described as *Griggsia cyathea.*—*H. C. Cowles.*

687. Strasser, P. Pius. **Siebenter Nachtrag zur Pilzflora des Sonntagberges (N.-Ö.) 1917. [7th addition to the fungus flora of Sonntagberg.]** Verhandl. Zool.-Bot. Gesell. Wien. 68: 97-123. 1918.—A list of species is given accompanied by the data of collection and critical notes. The material was in most cases examined by von Höhnel and a considerable number of species and a few genera are listed as new and are attributed to him. Some of these have been published elsewhere by von Höhnel but others are designated here as unpublished, the binomial being followed by the citation "v. H. nov. spec. in litt." These fall in many groups of the fungi but since technical descriptions of these will be given elsewhere by von Höhnel they need not be listed here.—*H. M. Fitzpatrick.*

688. Takahashi, R. **On the fungous flora of the soil.** Ann. Phytopath. Soc. Japan 1[2]: 17-22. 1919.—The author isolated several fungi from the soil of the test garden of the Tokyo Imperial Agricultural College by using soil extract gelatin-agar (+60, Fuller's scale). The isolation is made at two different periods, the one in September, 1915, and the other in February, 1916. The result of the experiments is listed as follows: In 1915 (a) In the soil obtained from 2 cm. below the surface: *Mucor racemosus*, *Aspergillus oryzae*, *A. fumigatus*, *Penicillium roseum*, *P. candidum*, *P.* sp. No. 1., *Chaetomium crispatum*, *Stemphylium verruculosum*, and *Penicillium* sp. No. 2.; (b) 5 cm. below the surface: *Aspergillus fumigatus*, *A. niger*, *Penicillium humicola*, *P. candidum*, *Allescheriella nigra*, *Acrostalagmus* sp., and *Helminthosporium subulatum*; (c) 8 cm. below the surface: *Aspergillus fumigatus* and *Trichoderma Koningi*; (d) 12 cm. below the surface: *Penicillium Duclauxi*, *Penicillium* sp. No. 2, *Chaetomium olivaceum* and *Alternaria tenuis*. In 1916 (f) 2 cm. below the surface: *Rhizopus nigricans*, *Aspergillus oryzae*, *A. niger*, *A. glaucus* and *A. nidulans*; (g) 8 cm. below the surface: *Mucor adventitus*, *M. circinelloides*, *Zygorhynchus Mölleri*, *Rhizopus nigricans* and *Botrytis cinerea.*—*T. Matsumoto.*

689. Tanaka, Tyôzaburô. **New Japanese fungi-notes and translations—VIII.** Mycologia 12: 25-32. 1920.—The following fungi are described: *Phytophthora Carica* (Hora) Hori causing a fruit rot of *Ficus Carica*, *Capnodium Tanakae* Shirai and Hora sp. nov. saprophytic on fruits of *Citrus grandis*, *Gloeosporium foliicolum* Nishida sp. nov. causing a spotting of fruits and leaves and a blighting of twigs of *Citrus* spp., *Dactylaria Panici-paludosi Sawada* sp. nov. on living leaves of *Panicum paludosum*, *Dactylaria Leersiae* Sawada sp. nov. on living leaves of *Leersia hexandra* and *Dactylaria Costi* Sawada sp. nov. on living leaves of *Costus speciosus.*—*H. R. Rosen.*

690. TSUJI, R. **On the morphology and the systematic position of Cercosporella persica Sacc. and Clasterosporium degenerans Syd.** (Japanese.) Ann. Phytopath. Soc. Japan 1[3]: 23–35. *Fig. 1–2.* 1919.—A fungus found on the leaves of a peach tree in Japan proved to be identical with *Cercosporella persica* Sacc. collected on a similar host and determined by W. G. FARLOW in the United States. This fungus is closely related to *Clasterosporium degenerans* Syd. on the leaves of *Prunus Mume* and *Armeniaca*, in that its conidiophores are produced on creeping hyphae emerging from stomatal openings, and also in color, shape, and mode of septation of their conidia, etc. He comes to the conclusion that these two species should be included under the same genus, and the name *Clasterosporium persicum* (Sacc.) Tsuji is proposed for the first-named species.—*T. Matsumoto.*

691. VUILLEMIN, PAUL. **Remarques sur les mycétomes. Hommage à la mémoire de R. Jolly.** [Remarks on mycetomas. Tribute to the memory of R. Jolly.] Arch. Méd. Exp. et Anat. Path. Paris 28: 446–451. 1919.—Gives a discussion of the different types of mycetomes and of the fungi producing them, in particular *Madurella mycetomi* (Laveran) Brumpt.—*E. A. Bessey.*

692. WAKSMAN, SELMAN A. **Cultural studies of species of Actinomyces.** Soil Sci. 8: 71–215. *Pl. 1–4.* 1919.—See Bot. Absts. 5, Entry 998.

693. WATSON, W. **The bryophytes and lichens of calcareous soil.** Jour. Ecol. 6: 189–198. 1918.—Gives lists of calciphile and calcifuge species, arranged by habitats as they occur in England; also a list of "indifferent" species. [See Bot. Absts. 4, Entry 309.]—*L. W. Riddle.*

694. WEIMER, J. L. **Variations in Pleurage curvicolla (Wint.) Kuntze.** Amer. Jour. Bot. 6: 406–409. 1919.—Variation in this species was studied to determine the taxonomic value of certain characters. The number of spores in the ascus is apparently 128, 256, or 512. The spore size in the strain studied is approximately the same as that recorded for other strains of the species, but the size of the perithecia is somewhat more variable. Secondary spore appendages, supposed to be a constant taxonomic character for the species, were not demonstrated.—*E. W. Sinnott.*

695. WEIMER, J. L. **Some observations on the spore discharge of Pleurage curvicolla (Wint.) Kuntze.** Amer. Jour. Bot. 7: 75–77. 1920.—Author reports that this species is able to discharge its spores to a height of 45 cm. above the fruiting surface of the culture, probably higher than can any other Ascomycete yet studied. This is due in part to the fact that the spore mass discharged is rather large and heavy, comprising some 500 spores and a quantity of gelatinous substance. Experiments show that the spore discharge is strongly and positively heliotropic, but that reflected light seems to exert a stronger stimulus than does direct light.—*E. W. Sinnott.*

696. WESTON, WILLIAM H. **Repeated zoospore emergence in Dictyuchus.** Bot. Gaz. 68: 287–296. *1 pl., 1 fig.* Oct., 1919.—The non-sexual reproduction of the fungus studied shows it to be a species of *Dictyuchus*, but exact determination was impossible, because sexual reproduction was not observed. *Dictyuchus* differs from all other Saprolegniaceae, save perhaps *Aplanes*, in that during spore formation the walls of adjacent spores unite with one another and with the enveloping sporangium membrane to form a polygonally chambered indehiscent structure. The zoospores which emerge from the sporangiospores come to rest and encyst as usual, but from these encysted spores ("cystospores") in turn laterally biciliate zoospores may emerge. This repeated emergence of laterally biciliate zoospores has not previously been reported in any of the Saprolegniaceae.—*H. C. Cowles.*

697. WHELDON, J. A. **Llanberis lichens.** Jour. Botany 58: 11–15. 1920.—A list of lichens compiled in the district around Llanberis in August, 1919. Many lichens known to occur in this district were not seen, while some rare species were observed. Few corticole species were collected as most of the time was spent above tree line. The arrangement is

that of A. Lorrain Smith's British Lichens. The list contains the names of about 125 species and a number of varieties. One species, *Bilimbia cambrica*, is described as new.—*K. M. Wiegand.*

698. Wilson, G. H. **A method for the simultaneous demonstration of gram-positive and gram-negative organisms in sections.** Jour. Path. and Bact. 23: 123–124. 1919.

699. Winslow, C. E. A., I. J. Kligler, and W. Rothberg. **Studies on the classification of the colon-typhoid group of bacteria with special reference to their fermentative reactions.** Jour. Bact. 4: 429–503. 1919.—The authors review rather completely the literature of the colon-typhoid group and arrange the whole series into six groups based mainly upon their fermentation of various carbohydrates. Several cultures are studied and classified. Seventeen species are included in the entire six groups and characteristics of each species given. The commonly called *B. paratyphosus A* is designated as *B. paratyphosus* and *B. paratyphosus B.* as *B. schottmulleri*, a new name; the name *B. morgani* is given to the formerly-called Morgan bacillus.—*Chester A. Darling.*

700. Yasuda, A. **Kinrui-Zakki 87.** [**Notes on fungi, 87.**] Bot. Mag. Tokyo 33: 112–114. 1919.—Three species of *Hymenomycetes* found in Japan *Stereum boninense*, *Hydnum violascens*, and *Tomentella fusca*, are reported. The first-named species was first described by the author under the name *Hymenochaete boninensis* Yasuda. [See Bot. Absts. 4, Entry 1196.]—*T. Matsumoto.*

701. Yasuda, A. **Kinrui-Zakki 88.** [**Notes on fungi, 88.**] Bot. Mag. Tokyo 33: 140–141. 1919.—Three species of *Hymenomycetes* found in Japan, *Polyporus Greenii*, *Stereum rimosum*, and *Clavaria amethystina*, are reported, of which the first-named species is new to science, its morphological characters being as follows: Pileus stipitate, corky, brown, covered with fine hairs, circular in outline 4 to 5 cm., slightly convex, triangular in section, azonate, context brown, thick, mouths grayish brown, angular, 1 to 2 mm.; spores light brown, ellipsoid, smooth, 8–9×5–5.5μ; stipe 2 to 3.5 cm. high, 1.1 to 1.5 cm. in diameter, slightly narrowed at the base, concolorous with the pileus, covered with fine hairs. Growing on the ground, Settsu, Japan. [See Bot. Absts. 4, Entry 1197.]—*T. Matsumoto.*

702. Yasuda, A. **Kinrui-Zakki 89.** [**Notes on fungi, 89.**] Bot. Mag. Tokyo 33: 167–169. 1919.—Three species of *Hymenomycetes* found in Japan, *Polystictus scopulosus*, *Coniophora arida*, and *Hypocrea citrina*, are reported. The first-named species is new to science; morphological characters as follows: Sporophore stipitate, coriaceous, 6.5 to 9 cm. high; pileus thin, fan-shaped, 4.5–6 cm. in length, 5 cm. in width, margin irregularly waved, chestnut brown, covered with depressed scales, context whitish; stipe short, lateral, smooth, yellowish; mouths grayish, tubes short, angular, 0.2 to 0.3 mm.; spores ellipsoid, smooth, light brown, 7 by 5 μ. Growing on the stem of Alnus sp. [See Bot. Absts. 4, Entry 1198.]—*T. Matsumoto.*

703. Yasuda, A. **Kinrui-Zakki 90.** [**Notes on fungi, 90.**] Bot. Mag. Tokyo 33: 189–191. 1919.—Three species, namely *Stereum japonicum*, *Chaetosphaeria tristis*, and *Lycoperdon spadiceum*, are reported. The first-named species is new to science; morphological characters as follows: Fructification ruspinate, coriaceous, 8 to 15 cm., hymenial layer light brown, velvety, upper part of context concolorous with the hymenium, lower part grayish brown, cystidia club-shaped, light brown, encrusted with crystals of calcium oxalate; spores spherical, hyaline, smooth, 4 μ. Growing on stems.—*T. Matsumoto.*

704. Zahlbruckner, A. **Beiträge zur Naturgeschichte der Scoglien und Kleineren Inseln Süddalmatiens. 5. Lichenes.** [**The natural history of the smaller islands of southern Dalmatia. 5. Lichens.**] Denkschr. K. Akad. Wiss. Wien. (Math.-Nat. Kl.) 92: 301–322. 1916.—New species are described and various nomenclatorial changes are made in the genera, *Verrucaria*, *Dermatocarpon*, *Arthonia*, *Arthothelium*, *Roccella*, *Lecanactis*, *Lecidea*, *Gyalecta*, *Pertusaria*, *Lecanora*, *Ramalina*, *Protoblastenia*, *Caloplaca*, *Xanthora*, *Buellia*, *Rinodina*. One hundred and twenty-six species are listed.—*H. M. Fitzpatrick.*

705. Zschacke, H. **Die mitteleuropaeischen Verrucariaceen.** [The Verrucariaceae of central Europe.] Hedwigia 60: 1-9. 1918.—Two earlier papers with the same title have been published. The present paper is based on collections made in Switzerland, while the author was interned. An enumeration of species is given with citations of localities and some critical notes. *Staurothele geoica* is described as a new species.—*L. W. Riddle.*

PALEOBOTANY AND EVOLUTIONARY HISTORY

E. W. Berry, *Editor*

706. Barett, A. **Contribution to the study of the "Siphoneae verticillatae" of the Calcare di Villanova-Mondovi.** [The verticillate Siphoneae of the Villanova-Mondovi limestone.] Atti Soc. Ital. Sci. Nat. e Mus. civ. St. Nat. Milano 58: 216-236. 1919.—The "Calcare di Mondovi" typically exposed—as the name indicates—in the region of Mondovi (Piedmont) and in particular in the massif of Villanova a few kilometers from the town, is rich in diminutive triassic algae, which, about 1865, Prof. Bruno recognised for the first time there. Different specimens of like fossils were studied by Gümbel and by Zittel who referred them to the Muschelkalk and the Wettersteinkalk horizons. For this work Barett examined some specimens of the Calcare di Villanova at the Museo Geologico di Torina, sent by Prof. Bruno, and especially the abundant material of his own collecting not only from Villanova, but also from other localities of the surrounding calcareous zone: M. Calvario, Gravagna, Moline and Torre, Peveragno. Material of different appearance according to the source or origin, but always crystalline, so that the fossils, although superficially seemingly well preserved are profoundly metamorphosed in the interior, rendering their preparation and study most difficult. Barett recognised the presence of the following Diploporidi in the Calcare di Mondovi: *Kantia debilis* Gümbel, *K. philosophi* Pia, *K. dolomitica* Pia, *Tentlosporella gigantea* Pia, *T. hercules* Sapp., *T. vicetina* Tornquist, and in addition the following, which he proposes as new: *Kantia philosophi* var. *gracilis* n. var., *K. monregalensis* sp. n., and *K.* (?) *Brunnoi* sp. n. He describes and figures them all.—Despite the great number of specimens examined, their different origins or sources, and the extraordinary abundance of the individuals contained in them, this study, because of the above-mentioned difficulty, has not yielded as great results as might have been hoped; nevertheless from this it stands proved that in the Calcare di Mondovi there are also encountered the *Kantia philosophi* and *dolomitica* typical of the Muschalkalk, and the *Teutlosporella gigantea* and *T. vicentina*, hitherto not noted; and there results then the confirmation that the horizon is to be referred to the lower Neotriassic.—*R. Pampanini.*

707. Benson, M. **Cantheliophorus, Bassler: New records of Sigillariostrobus (Mazocarpon).** Ann. Botany 34: 135-137. 1920.—Evidence is given to show that specimens described by Bassler as proving the existence of a sporangiophoric lepidophyte and referred to a new genus *Cantheliophorus*, as well as similar specimens previously described by Nathorst, are really examples of Sigillarian microsporophylls.—*W. P. Thompson.*

708. Berry, E. W. **The evolution of flowering plants and warmblooded animals.** Amer. Jour. Sci. 49: 207-211. Mar., 1920.—Discusses the correlation between the two and the dependence of the latter on the former.—*E. W. Berry.*

709. Bertrand, Paul. **Les zones vegetales du terrain houiller du Nord de la France.** [Plant zones of the coal regions of Northern France.] Compt. Rend. Acad. Sci. Paris 168: 780-782. 1919.—A table of the location and vertical extent of the plant zones in the coal deposits of Northern France.—*F. B. Wann.*

710. Cockerell, T. D. A. **Carpolithes macrophyllus a Philadelphus.** Torreya 19: 244. 1919.—*Carpolithes macrophyllus* Ckll., described in Torreya 11: 235, is transferred to *Philadelphus*, but very likely belongs to *P. palaeophilus* Ckll. (1908).—*J. C. Nelson.*

711. Conklin, E. G. **The mechanism of evolution.** [1] Sci. Monthly 10: 170–181. 1920.—This is a discussion of Mendelism in which the author concludes that the law, especially as regards the segregation of inheritance factors, is of universal occurrence—that there is no other type of inheritance. Alternative inheritance with dominant and recessive characters, purity of germ cells, monohybrids, dihybrids, etc., factorial theory of heredity, blending inheritance, species hybrids, and unequal reciprocal hybrids are discussed in relation to the above conclusion. [See also next following Entry, 712.]—*L. Pace.*

712. Conklin, E. G. **The mechanism of evolution.** [2] Sci. Monthly 10: 269–291. *Fig. 11–21.* 1920.—This paper takes up the cellular basis of ontogeny and phylogeny. There is no fundamental difference between germ cells and somatic cells. Nucleus and cytoplasm are fundamentally different chemically, morphologically and physiologically.—Mitosis furnishes the necessary mechanism for the accurate division of the cell, and the persistent identity of the chromosome is accepted. The suggestion is made that chromomeres are probably much more constant than chromosomes.—The mechanism of heredity is to be found in the germ cells. Genetics and cytology must cooperate in correlating features of the germ cell with the phenomena of heredity. The similarity of chromosomes of the spermatozoon and of the egg, the reduction division, the doubling of chromosomes in fertilization, the sex-chromosomes, sex-linked characters, linkage of characters, chromosomal localization and cross-overs are all presented as favoring the localization of the genes in the chromosomes. [See also next preceding Entry, 711.]—*L. Pace.*

713. Coulter, J. M. **Cones of Williamsonia.** [Rev. of: Arber, E. A. Newell. **Remarks on the organization of the cones of Williamsonia gigas.** Ann. Botany 33: 173–179. *5 fig.* 1919. (See Bot. Absts. 3, Entry 1143).] Bot. Gaz. 68: 152. 1919.

714. Grandori, Luigia. **Su di un seme mesozoico di pteridosperma e sulle sue affinità con forme paleozoiche e forme viventi.** [On a Mesozoic pteridosperm seed and its affinities with Paleozoic and recent forms.] Atti d'Accad. Veneto-Trentino-Istriana. 8: 107–116. *3 fig., 1 pl.* 1915.

715. Grandori, Luigia. **Sulle affinità delle Pteropsida fossili, studio critico.** [On the affinities of the fossil Pteropsida.] Atti d'Acad. Veneto-Trentino-Istriana 8: 163–195. *7 fig.* 1915.

716. Knowlton, F. H. **A dicotyledonous flora in the type section of the Morrison formation.** Amer. Jour. Sci. 49: 189–194. Mar., 1920.—Records the presence of an Upper Cretaceous flora similar to that of the Dakota sandstone from the type locality of the Morrison formation near Golden, Colorado. The Morrison formation has yielded a varied dinosaur fauna and there has been much controversy as to whether it was of Jurassic or Lower Cretaceous age.—*E. W. Berry.*

717. Principi, Paolo. **Le Dicotiledoni fossili del giacemento oligocenico di Santa Giustina e Sassello in Liguria.** [The fossil dicotyledons of the Oligocene of Santa Giustina and Sassello in Liguria.] Mem. Desc. Carta Geol. d'Italia 6: 1–294. *Pl. 1–86.* 1916 (1919). —Liguria is one of the classic regions of Tertiary geology. The Oligocene of Sta. Giustina and Sassello record the transition from continental to delta and then lagoonal or estuary to littoral conditions of deposition followed by a recurrence of lagoonal conditions at the base of the middle Oligocene and littoral again at the top of the middle Oligocene. The fossil plants which are the subject of the memoir come from the basal beds or Sannoisian stage. Previous accounts of this flora have been published by Sismonda in 1859 and 1865, and Squinabol in the period from 1889 to 1892 described the Cryptogams, Gymnosperms and Monocotyledons.—The dicotyledons recorded number 339 forms, the most varied genera being Quercus Juglans, Myrica, Ficus, Laurus, Cinnamomum, and Rhamnus. Eighty-six new species are described in the following genera: Castanea, Dryophyllum, Quercus, Juglans, Juglandophyllum, Myrica, Comptonia, Populus, Protoficus, Ficus, Artocarpidium, Artocarpus, Cocculites,

Cocculus, Laurus, Persea, Cinnamomum, Magnolia, Anona, Sterculia, Dombeyopsis, Pterospermites, Bombax, Sapindus, Malpighiastrum, Celastrus, Rhamnus, Aralia, Dewalquea, Cornus, Terminalia, Lomatia, Amelanchier, Prunus, Machaerium, Aristolochia, Chrysophyllum, Diospyros, Apocynophyllum, Alstonia, Viburnum and Carpites.—The flora shows a curious mingling of temperate and tropical types and contains very many more of the former than does the known North American floras of corresponding age.—*E. W. Berry.*

718. SAHNI, B. **On certain archaic features in the seed of Taxus baccata, with remarks on the antiquity of the Taxineae.** Ann. Botany 34: 117-134. *7 fig.* 1920.—See Bot. Absts. 5, Entry 574.

719. SCHLAGINTWEIT, O. **Weichselia Mantelli im nordöstlichen Venezuela. [Weichselia Mantelli in northeast Venezuela.]** Centralb. Min. Geol. Paläont. 1919: 315-319. 1919.—Records this ubiquitous Mesozoic fern from Santa Maria, Venezuela, in a shale thought to be Neocomian in age.—*E. W. Berry.*

720. SMALL, JAMES. **The origin and development of the Compositae. Miscellaneous topics.** New Phytol. 18: 129-176. *Fig. 64-78.* 1919.—This is chapter 12, in which miscellaneous topics are presented. A table of known fossil remains of Compositae and their localities is accompanied by critical notes and comments. The composites are believed to have arisen in late Cretaceous or early Eocene. From the point of origin in the northern Andean region of South America, migration occurred chiefly along mountain ranges. By the end of the Eocene the differentiation of types and wide dispersal was accomplished.—A summary of cytology, with original figures based on *Senecio*, follows. Spermatogenesis, oogenesis, and the history of the embryo sac are discussed, with a special account of the antipodals. The chromosomes are treated from the standpoint of phylogeny. A table is given of the number in all composites so far as known.—The nature and distribution of the latex system in the tribes are discussed.—Last are brief accounts of seedling structure, pericarp, anatomy, phytochemistry, and pappus in the Compositae.—The bibliography contains 173 titles.—*I. F. Lewis.*

721. SMALL, JAMES. **The origin and development of the Compositae. General conclusions.** New Phytol. 18: 201-234. *Fig. 79.* 1919.

722. STOPES, MARIE C. **New Bennettitean cones from the British Cretaceous.** Phil. Trans. Roy. Soc. London B, 208: 389-440. *5 pl.* 1918.—Bennettites albianus, sp. nov., is described from a cone found in the Gault (or Albian) of Folkestone Warren. The fruit is an ovulate cone, not less than 70 mm. in diameter and probably much more. The innumerable seeds, 600 or more revealed in a single transverse section, are five-ribbed, much elongated, torpedo-shaped, 5-6 mm. long and about 1.2 mm. in greatest diameter. The seed with its many layered integument is inclosed in a cupule-like extension of tubular cells of the stalk. The micropyles are blocked by plugs of nucellar tissue. Around the apex of the seed, interseminal scales are completely mutually fused not only with each other but with the seed tissues. The embryos contain two cotyledons and both the radicle and the hypocotyl are relatively massive. The scales are externally covered by a well marked "plastid layer" which runs around the collar of the micropyle.—The complete fusion of the stony scales must have meant that there was great stability and strength in the hard, uniform shell which surrounded the fruit. This solid shell firmly enclosed the ripe seeds, which did not rattle about in it loose, for the ribbed apices of the seeds were wedged into the solid mass. It is not impossible that the hard fruit had considerable drought to withstand. It certainly seems fitted to do so.—The extraordinarily great size of Bennettites albianus raises a point of general interest. In many families of animals giant forms appear shortly before the extinction of the group. This new Bennettites possesses the largest cone of the family and was taken from the highest and latest geological horizon in which the group is known. May it then be considered in the same light as the animal giants—namely a burst of glory before extinction? Any conclusion on this point, no matter how tempting, must not be accepted too readily. A giant fruit in many

of the cycadales may be borne on small plants. The giantism of the animals approaching extinction was not in their reproductive organs but in their general bodies. The comparison with animals is, therefore, insecure and rests on too many assumptions. Paleobotanical evidence is made up of too few isolated cases to point a general law of evolution.—Bennettites maximus Carruthers is also figured and described in detail for the first time. The only specimen of this is from the lower Greensand in the Isle of Wight. This specimen consists of a large trunk containing a number of cones. Sections made of this trunk show a number of cones. These are bisporangiate. The male organs were developing at the time the plant was petrified. The female receptacle was at that time undifferentiated, meristematic tissue. Sections have been made, however, of one cone showing the ovule rudiment and the surrounding tissue.—*A. E. Waller.*

723. Stopes, Marie C. **On the four visible ingredients in banded bituminous coal: Studies in the composition of coal, No. 1.** Proc. Roy. Soc. London B, 90: 470–487. *Pl. 11–12.* 1919.—Proposes names fusain, durain, clarain, and vitrain for four recognizably distinct ingredients of banded bituminous coal. These types are distinctive (a) in effect on sensitive plates (b) chemical and physical behaviour (c) in microscopic details.—*Paul B. Sears.*

724. Wilson, W. J. **Notes on some fossil plants from New Brunswick.** Geol. Surv. Canada, Summary Rept. 1917 F: 15–17. 1918.—Publication of identifications and notes on specimens and photographs submitted to Robert Kidston. The material came from the Carboniferous of Rothwell, New Brunswick.—*E. W. Berry.*

PATHOLOGY

G. H. Coons, *Editor*

C. W. Bennett, *Assistant Editor*

725. Anonymous. **Celery leaf-spot disease or blight.** Jour. Dept. Agric. Ireland 20: 86–89. *3 fig.* 1920.

726. Anonymous. **A new disease of pears, new to the continent of America.** Agric. Gaz. Canada 6: 951–952. *4 fig.* Oct., 1919.—Specimens of pears received by the Division of Botany, Dominion Department of Agriculture, from Kentville, Nova Scotia, showed an unusual rot. Nearly full grown pears showed one or more large, circular, dark-brown spots which were quite firm in texture. *Phytophthora cactorum* was obtained in culture from the spots. Only the fruit upon the low hanging branches showed the disease, which suggests that the infection may originate from surrounding infected vegetation. Control measures, chiefly prophylactic are suggested.—*O. W. Dynes.*

727. Appel, Otto, and Johanna Westerdijk. **Die Gruppierung der durch Pilze hervorgerufenen Pflanzenkrankheiten.** [The classification of plant diseases due to fungi.] Zeitschr. Pflanzenkrankh. 29: 176–186. 1919.—The authors point out the advantages of a classification based upon symptomology, particularly to students of phytopathology. They suggest five main groups, viz.: rots, spots, fungus coverings, increase of tissues, and vascular diseases. Each main group is divided into auxiliary groups, thus: "Rots," for instance, is subdivided into rot of seeds, of seedlings, of roots, of tubers, of bulbs, of rhizomes; basal stem rots; general stem rots; rots of buds and flowers, of fruits, of wood, of bark; and dry rots. The group "Increase of tissues" covers witches' brooms, galls, and flower and fruit transformations (ergot, smuts, etc.). Each group is discussed, reviewing examples.—*H. T. Güssow.*

728. Baker, C. F. **A contribution to Philippine and Malayan technical bibliography. Work fundamental to plant pathology and economic entomology.** Philippine Agric. 8: 32–37. 1919. See Bot. Absts. 5, Entry 1238.

729. Barre, H. W. **Report of the division of botany.** South Carolina Agric. Exp. Sta. Ann. Rept. 32: 29–34. 1919.—A summary of the work on the following projects is given: Cotton anthracnose, angular leaf spot of cotton, bacterial content of milk, plant disease survey, coöperative research.—*G. H. Coons.*

730. Blin, H. **La pourriture des griffes d'asperges.** [Asparagus root-rot.] Rev. Hortic. 91: 325–326. *1 fig.* Aug., 1919.—This disease is due to *Rhizoctonia violacea* which attacks many other types of plants. All portions of asparagus plants which are attacked should be carefully dug up and burned. The soil should then be disinfected with carbon-bisulfide (about 250 grams per square meter) or preferably formaldehyde (about 60 grams per square meter). Either of these should be forced into the soil at several places with a syringe. Sulfo-carbonate of potassium (300 grams in 100 liters of water) has also been used successfully. The soil is first removed from the hills which have been attacked and these are then sprayed lightly with the mixture. The following year, before hilling-up a second treatment is given. Before replanting infected areas they should be thoroughly disinfected during the winter and the clumps dipped in the disinfecting solution. Following any of these treatments the soil should be well fertilized, since the disinfection destroys the soil organisms present. Care should be taken to avoid such disinfectants as may leave harmful residues in the soil treated. As a matter of precaution, it is better not to replant infected areas for 2 or 3 years.—*E. J. Kraus.*

731. Boas, Friedrich. **Beiträge zur Kenntnis des Kartoffelabbaues.** [Contribution to the knowledge of deterioration in potatoes.] Zeitschr. Pflanzenkrankh. 29: 171–176. 1919.—The author states that minute differences in the hydrogen-ion concentrations may have marked effects upon metabolic processes.—This caused him to inquire whether, in plant diseases, especially in leaf roll or curly disease of potatoes, there could be determined any differences in the hydrogen-ion concentrations existing in sound and diseased plants.—He ascertained from his experiments (describing technique employed) that, without exception, the cell sap of sound plants showed appreciably more acidity than that of diseased plants. The acid metabolism of diseased plants is plainly disturbed. In determining the albumen metabolism that might be expected under the circumstances, author determines that the diseased potato stems are flooded with amino acids, while the sound tissues are free, or only show traces of these acids. Examining then into the catalase contents of diseased and sound plants, he finds obvious differences in his experimental varieties, inasmuch as the diseased portions show an increase in catalase contents over the sound ones; but not all experiments gave identical results. (Bibliography.)—*H. T. Güssow.*

732. Chou, Chung Ling. **Notes on fungous diseases in China.** (Text in Chinese.) Khu Shou [Science-Publication of The Chinese Science Society] 4: 1223–1229. *46 fig.* 1919. —The author gives a detailed description of symptoms and morphology of fifteen fungous diseases found in the locality of Nanking: *Peronospora parasitica* on *Brassica juncea*, *Peronospora effusa* on spinach, *P. vicae* on peas, *P. schleideniana* on onion leaves, *Alternaria brassicae* on *Brassica pekinensis*, *Cercospora cruenta* on beans, *Ustilago crameri* on wheat, *U. avenae* on oats, *Urycystis tritici* on wheat, *Ustilago shiriana* on bamboo, *Erysiphe graminis* on barley, *Pleospora gramineum* on barley, *Exoascus deformans* on peach leaves, *Aecidium mori* on mulberry stems, and *Sclerotinia cinerea* on cherries.—*Chunjen C. Chen.*

733. Cook, Mel. T. **Philippine plant diseases.** [Rev. of: Reinking, Otto A. **Philippine economic-plant diseases.** Philippine Jour. Sci. A, 13: 165–274. *43 fig., 22 pl*, 1918. (See Bot. Absts. 2, Entry 1308.)] Bot. Gaz. 68: 310–311. 1919.

734. Cook, Melville T. **Report of the department of plant pathology.** Ann. Rept. New Jersey Agric. Exp. Sta. 1918: 299–302. 1919.

735. Cook, Mel. T. **Potato diseases in New Jersey.** New Jersey Agric. Exp. Sta. Circ. 105. *38 p.* 1919.—Along with descriptions and illustrations of the common potato diseases

the results of the spraying tests for a period of six years and the rules governing seed certification in several States are given.—*Mel. T. Cook.*

736. Cook, Mel. T. **Seed and soil treatment for vegetable diseases.** New Jersey Agric. Exp. Sta. Circ. 106. *4 p.* 1919.

737. Cook, Mel. T., and J. P. Helyar. **Diseases of grain and forage crops.** New Jersey Agric. Exp. Sta. Circ. 102. *16 p.* 1918.

738. Crain, C. C. **Warm bath for wheat.** Sci. Amer. 121: 579. *1 fig.* 1919.—Popular account is given of treatment for smut.—*Chas. H. Otis.*

739. Darnell-Smith, G. P. **Dry rot in timber.** Australian Forest. Jour. 2: 314–316. 1919.—See Bot. Absts. 5, Entry 175.

740. Edson, H. A., and M. Shapovalov. **Temperature relations of certain potato-rot and wilt-producing fungi.** Jour. Agric. Res. 18: 511–524. *9 fig.* 1920.—Single strains of *Fusarium coeruleum*, *F. discolor* var. *sulphureum*, *F. eumartii*, *F. radicicola*, *F. tricothecioides*, and a northern and a southern strain of *Verticillium albo-atrum* were grown on 2 per cent potato agar without sugar at temperatures ranging from 1° to 40°. Minimum temperature for all forms is around 5°; maximum for *F. coeruleum*, *F. tricothecioides* and *V. albo-atrum*, ("northern") 30° or slightly less, for *F. oxysporum*, about 37°, for *F. radicicola* about 39°, and for the remaining, slightly under 35°; optimum for *F. oxysporum* and *F. radicicola* about 30°; for the remaining about 25°.—A certain degree of correlation exists between the temperature relations of these organisms in pure cultures and their geographical distribution and seasonal occurrence. This is particularly striking in the case of the 2 wilt-producing fungi, *F. oxysporum* and *V. albo-atrum*.—A temperature of about 4° should hold Fusarium tuber rots in check during storage. The susceptibility of *V. albo-atrum* to high temperatures suggests the possibility of a heat treatment for seed tubers harboring the fungus.—Temperature tests in certain cases may serve as a useful supplementary method for the identification of fungi exhibiting contrasting thermal relationships.—*D. Reddick.*

741. Ellis, J. H. **The stage of maturity of cutting wheat when affected with black stem rust.** Agric. Gaz. Canada 6: 971. 1919.—See Bot. Absts. 5, Entry 20.

742. Fragoso, Romualdo Gonzalez. **Notes and communications at the session of Oct. 1, 1919.** Bol. R. Soc. Española Hist. Nat. 19: 429–430. 1919.—See Bot. Absts. 5, Entry 646.

743. Fragoso, R. G. **Enfermedades del almendro. [Diseases of the almond.]** Bol. R. Soc. Española Hist. Nat. 19: 458. Oct., 1919. [Review of an article by A. Ballester, published as a leaflet by Dir. Gen. Agric. Spain, in April, 1919.] The reviewer presents critical discussion of the publication and takes issue with several statements. *Clasterosporium carpophilum* is reported as a serious parasite, especially in its conidial stage (*Coryneum beijerinckii*). The following disease producing species omitted by Fragoso are cited: *Puccinia pruni* and *P. cerasi*, *Gloeosporium amygdalinum*, *Fusicoccum amygdali*, and *Cercospora circumscissa*.—*O. E. Jennings.*

744. Gauba, Th. **Das Hopfenmissjahr 1918. [An off-year for hops.]** Der Bierbrauer 46: 161–162. 1918.—Very grave losses (30 to 50 per cent) in Austria, Hungary and Germany occasioned by early attack of hop aphis followed by sooty mold and mildew. [Through abstr. of Matouschek in Zeitschr. Pflanzenkr. 29: 193. 1919.]—*D. Reddick.*

745. Geschwind, A. **Die der Omorikafichte (Picea omarica Panc.) schädlichen Tiere und parasitischen Pilze. [Insect enemies and diseases affecting P. o.]** Naturw. Zeitschr. Forst.- und Landw. 16: 387–395. 1918.—Diseases mentioned are caused by *Herpotrichia nigra*, *Lophodermium macrosporum* and *Trametes pini*.

746. Hecke, Ludwig. **Die Frage der Bekämpfung des Getreiderostes. [The problem of controlling cereal rusts.]** Nachrichten Deutsch. Landw. Gesell. Österreich. n. s. 2: 140-142. 1918.—In wheat regions of Austria the rusts cause damage to cereals in the following order: to wheat, yellow rust, brown rust (*P. dispersa, P. triticina*), black rust (*P. graminis*); to rye the same; to oats, black, crown rust (*P. coronifera*); to barley, dwarf rust (*P. simplex*), black rust, yellow rust. The yellow is the most destructive in rust years; brown rust attacks late. Black rust is injurious chiefly in hilly sections. In southern part, *P. maydis* is general and injurious. [Through abstr. by Matouschek in Zeitschr. Pflanzenkr. 29: 210. 1919.]—*D. Reddick.*

747. Jehle, R. A., and others. **I. Control of cotton wilt. II. Control of cotton anthracnose and improvement of cotton.** Bull. North Carolina Dept. Agric. 41[1] (Supplem.) 5-28. *Fig. 1-6, and 1-5.* 1920.—The first part of this report contains the results of field demonstrations in several counties of the Coastal Plain section of North Carolina, in the control of cotton wilt. Dixie Wilt Resistant cotton was successfully grown on infested lands in this section. The report, furthermore, includes data on the known distribution of wilt in North Carolina and factors favoring its prevalence and spread.—The second part deals with demonstrations of the value in cotton anthracnose control, of the selection of disease free seed and improvement through breeding of these selected strains. Cleveland Big Boll and Dixie Wilt Resistant cotton were employed.—*R. A. Jehle.*

748. Krout, Webster S. **Common diseases of celery.** New Jersey Agric. Exp. Sta. Circ. 112. *12 p.* 1919.

749. McCulloch, Lucia. **Basal glumerot of wheat.** Jour. Agric. Res. 18: 543-551. *Pl. 62-63.* 1920.—This disease is widespread in U. S. A. and occurs on leaf, head and grain of wheat (*Triticum*). A dull, brownish black area appears at the base of the glumes, involving usually only the lower third but at times extends over nearly the whole surface. Sometimes the discoloration is on the inner surface of the glume. Dissection of affected spikelets shows more evidence of disease on the inner surfaces than on the outer. The grains inclosed in diseased glumes vary from nearly perfect to ones in which the germ end varies in color from a slight brown to charcoal black.—The disease is caused by *Bacterium atrofaciens* n. sp., for which a technical description is presented. The parasite is a white, polar-flagellated rod producing green fluorescence in ordinary culture media. Group number, 221.2322123.—Artificial infections were secured on leaf and head, the incubation period being about four days.—*D. Reddick.*

750. Merino, G. **Bud-rot.** Philippine Agric. Rev. 12[2]: 92-96. *4 pl.* 1919.—A brief compilation of data on the budrot of the coconut palm.—*E. D. Merrill.*

751. Moore, J. C. **Experiments with parasitic fungus on the cacao thrips.** Report on the Agricultural Department, Grenada, 1917-18. Imperial Department of Agriculture, Barbados. 1918.—Spraying experiments on thrips infesting cacao trees with cultures of the fungus *Sporotrichum globuliferum*, parasitic on *Heliothrips rubrocinctus*, Giard., are here noted. Although carried out under difficulties, the following points have been demonstrated: (1) The fungus was readily distributed amongst thrips in the field; (2) Under favourable conditions of atmospheric humidity the fungus caused death of large numbers of both young and adult thrips on the inoculated trees; (3) The fungus spread by natural agencies to trees outside the inoculated area. Several considerations of practical importance remain to be determined.—*J. S. Dash.*

752. Morgenthaler, Otto. **Über die Mikroflora des normalen und muffigen Getreides. [Microflora of normal and of musty grain.]** Landw. Jahrb. Schweiz. 32: 551-571. 1918.—Healthy grain sown in plates shows chiefly bacteria, especially *Bact. herbicola*, and no fungi. Musty grain yields many fungous thalli and few bacteria. *Penicillia* are abundant but are not responsible for the odor. What organism does impart the characteristic odor was not

determined.—Musty grain intended for human consumption should be washed thoroughly and the light grains skinned off. [Through abst. by MATOUSCHECK in Zeitschr. Pflanzenkr. 29: 203–204. 1919.]—*D. Reddick.*

753. MÜLLER, K. **Die Bekämpfung der Rebenperonospora nach der Inkubations-Kalendermethode. [Control of grape downy mildew by the incubation-period method.]** Jahresber. Vereinig. Angew. Bot. 16:21–28. 1918.—Based on the investigations of ISTVÁNFFI and coworkers regarding the relation between incubation period and outbreaks of *Plasmopara*, and telluric conditions. Experimental trials made in Baden show that dates of outbreaks can be forecast with sufficient certainty to give growers warning in time to make protective treatments. [Through abst. by SEELIGER in Zeitschr. Pflanzenkr. 29: 205. 1919.]—*D. Reddick.*

754. NOWELL, W. **Bracket fungi of lime trees and the critical period in the development of young lime trees.** Report on the Agricultural Department, Dominica, 1917–18. 11–14. Imperial Department of Agriculture, Barbados. 1919.—The author, as Mycologist to the Imperial Department of Agriculture, reports on the prevalence of smaller fungi, of which *Nectria* and *Stilbum* spp. are the most noticeable, on the dead branches of lime trees. While functioning mostly as saprophytes, these fungi may, in certain types of cases, become weak parasites affecting principally the wood. Interesting suggestions are given on the treatment of lime trees during the critical period of their development.—*J. S. Dash.*

755. PAINE, S. G., AND W. F. BEWLEY. **"Stripe" disease of tomatoes.** Jour. Ministry Agric. Great Britain 26: 998–1000. 1920.—A brief popular account is given of "stripe" disease of tomatoes occurring chiefly in greenhouses, caused by a bacillus closely related to, if not identical with, *Bacillus lathyri*. The disease affects the tissues of the stems, leaves, and fruits which become stained a dark brown color. Suggested preventive and remedial measures consist in avoiding seed from fruit grown in an infected area, in disinfection of the soil where an attack has occurred, in using a balanced fertilizer, in using care in pruning the plants, and in altering the temperature and humidity to favor a more hardy development of the plants.—*M. B. McKay.*

756. PAINE, SYDNEY G., AND W. F. BEWLEY. **Studies in bacteriosis. IV.—"Stripe" disease of tomato.** Ann. Appl. Biol. 6: 183–202. *Pl. 8–9, 5 fig.* 1919.—The symptoms appear as brown to black sunken areas or stripes on the stem, as yellow to brown blotches on the leaves, as brown sunken patches on the fruit, and as brown discoloration of the root cortex. Infection appears usually to take place underground, but the disease may be spread from plant to plant above ground. A soft rapid growth of the plants renders them more susceptible to attack.—*Macrosporium solani* may occur as a saprophyte on the lesions.—Lesions occur also in the pith and cortex. The disease is assigned to a bacterial growth which advances from the root up the stem in the pith, and works outward, causing swelling and browning of the cell walls as it passes to the exterior, then spreads upward in the outer cortical layers and epidermis.—Bacteriolysis apparently may occur in the plant tissue, since some diseased spots seemed to be sterile.—The organism is described, and appears to be identical with *Bacillus lathyri*, differing only in a slightly higher resistance to heat and apparently greater ability to reduce nitrates.—An organism apparently identical with *Aplanobacter michiganense* was also isolated from affected plants, but did not reproduce the disease. [See also next following Entry, 757.]—*G. R. Bisby.*

757. PAINE, SYDNEY G., AND H. STANSFIELD. **Studies in Bacteriosis III.—A bacterial leaf-spot disease of Protea cynaroides, exhibiting a host reaction of possibly bacteriolytic nature.** Ann. Appl. Biol. 6: 27–29. *Pl. 2, fig. 3–6.* 1919.—The disease is characterised by dome-shaped reddish-brown blisters or by sunken spots on the leaves.—The host cells are thought to be able to kill and perhaps dissolve the bacteria. There is production of a resin-like substance in which the bacteria become imbedded. The host cells become disorganized. A red pigment allied to phloro-tannin red was produced in the spots.—The parasite was isolated from but

few of the spots. Infection experiments proved the pathogenicity of the organism isolated. —*Pseudomonas proteamaculans* n. sp. is given as the cause of the disease. [Group number is 221.1313023.) [See also next preceding Entry, 756.]—*G. R. Bisby.*

758. Petch, T. **Rubber diseases.** Tropic. Agriculturist 52: 27-34. 1919.—The red root disease (*Poria hypobrunnea*) occurs in Ceylon and Java, in limited areas. The identification of the disease is somewhat difficult but is unmistakable in young trees, where the top root bears external mycelium which forms stout, red strands which sometimes unite into a continuous red sheet. Internally the strands are white. The mycelium turns brown and finally black with age. The diseased wood of young trees is somewhat soft and friable and permeated with red sheets which often follow the annual rings. In older trees the entire mycelium may be black.—The disease spreads largely from decaying stumps and logs of trees killed by the fungus. It is held somewhat in check by the careful removal of all felled trees and old stumps including all diseased lateral roots.—White stem blight and top canker are described briefly.—*R. G. Wiggans.*

759. Ramsbottom, J. K. **Experiments on the control of narcissus eelworm in the field.** Jour. Roy. Hortic. Soc. 44: 68-72. *Fig. 18, 19.* 1919.—Three series of experiments for the control of *Tylenchus devastatrix* are reported. Applications of sulphate of potash alone and in combination with sulphate of ammonia, superphosphate and bone meal did not decrease the attacks. The same was true when various chemicals were applied to the soil. Following an affected crop of narcissus, rye, oats, clover, lucerne peas, broad beans, rye grass, onions, wheat, chives, buckwheat, and potatoes were planted, of which only onions became infested. —*J. K. Shaw.*

760. Reinking, O. A. **Host index of diseases of economic plants in the Philippines.** Philippine Agric. 8: 38-54. 1919.—A host index is presented, showing diseases of about one hundred economic plants in the Philippines. The hosts are arranged alphabetically, and under each host are given the organisms (fungi and bacteria) associated with it and the names of the diseases. In addition to known parasitic forms, saprophytic organisms are included. —*S. F. Trelease.*

761. Robson, R. **Root-knot disease of tomatoes.** Jour. Roy. Hortic. Soc. 44: 31-67. *Fig. 14-17.* 1919.—Root-knot of tomatoes (*Heterodera radicicola*) was controlled by applying 1,000 pounds cyanide of sodium (or of potassium) per acre to the subsoil. The application of 300 pounds of mercuric chloride also controlled the nematode. The cost of treatment in any of the above methods was approximately £50 per acre. No deleterious effect upon the growing crops was noted as a result of applying the above compounds at the rates per acre given. Mercuric chloride applied at the rate of 775 pounds per acre had a decided stunting effect.—*H. A. Jones.*

762. Rosenbaum, J., and Charles E. Sando. **Correlation between size of the fruit and the resistance of the tomato skin to puncture and its relation to infection with Macrosporium tomato Cooke.** Amer. Jour. Bot. 7: 78-82. 1920.—As tomatoes grow larger, their resistance to infection by *Macrosporium tomato* greatly increases. This difference in immunity is apparently not due to chemical differences between young and old fruit. Infection may be obtained with fruits of all degrees of maturity when the skin is injured or removed previous to infection. Stomata or other natural openings in the skin are absent. As the fruit develops, the cuticle increases markedly in thickness. Authors show that coincidently with this, the skin of the fruit becomes more resistant to mechanical puncture with a needle. They suggest that ability to resist infection may be due to the ability of the skin to resist puncture by the fungous filament.—*E. W. Sinnott.*

763. Rumbold, Caroline. **The injection of chemicals into chestnut trees.** Amer. Jour. Bot. 7: 1-20. *7 fig.* 1920.—See Bot. Absts. 5, Entry 964.

764. Rumbold, Caroline. **Effect on chestnuts of substances injected into their trunks.** Amer. Jour. Bot. 7: 45–56. *2 pl.* 1920.—See Bot. Absts. 5, Entry 965.

765. Schander, and Fritz Krause. **Die Krankheiten und Schädlinge der Erbse. [Diseases and insect pests of peas.]** Flugbl. Abt. Pflanzenkr. Kaiser Wilhelms-Inst. Landw. Bromberg 29–30.—July, 1918.

766. Schröder, P. **Ein flacher Hexenbesen. [A flat witches' broom.]** Mitt. Deutsch. Dendrol. Gesell. 1918: 290. *1 pl.* 1918.—On a spruce tree, 35 years old, growing at Hohen-Luckow (Mecklenburg) there is a broom 1.45 m. across and flat in form. [Through absts. by Matouschek in Zeitschr. Pflanzenkr. 29: 200. 1919.]—*D. Reddick.*

767. Speare, A. T. **Further studies of Sorosporella uvella, a fungous parasite of noctuid larvae.** Jour. Agric. Res. 18: 399–439. *Pl. 51–56.* 1920.—See Bot. Absts. 5, Entry 684.

768. Spieckermann. **Schädigung der Kulturpflanzen durch zu hohen Säuregehalt des Bodens. [Injury to cultivated plants through too high acidity of soil.]** Landw. Zeitg. Westfalen u. Lippe 1918: 255–256. 1918.—Superphosphate and sulfate of ammonia had to be used for fertiliser instead of the customary Thomas slag and nitrate of soda. Rye, oats and potatoes showed injury. The soil was found high in acidity and the lime content greatly reduced. [Through abstr. by Matouschek in Zeitschr. Pflanzenkr. 29: 198. 1919.]—*D. Reddick.*

769. Stevens, H. E. **Citrus scab.** Florida Grower 21[1]: 9. 1920.—Description and etiology of the disease with recommendations for control by spraying.—*H. R. Fulton.*

770. Uzel, H. **Rotfäule der Zuckerrübe. [Red rot of sugar beet.]** Zeitschr. Zuckerind. Böhmen 43: 138–139. 1918.—Red rot (*Rhizoctonia violacea*) occurs mostly in wet fields. Diseased plants should be removed and destroyed. Land should be drained and quick lime worked in. It should not be planted to sugar beet, fodder beet, alfalfa, red clover, serradella, potato, asparagus, or fennochio as these plants are attacked by the fungus. Mycelium from rotten beets passes with wash water to compost and back to land. Mycelium also may winter in the wash tanks. Rotten beets can not be used for feeding as the fungus persists in manure. [Through abstr. by Matouschek in Zeitschr. Pflanzenkr. 29: 213. 1919.]—*D. Reddick.*

771. Vincens, F. **Maladies de l'Hévéa dues au Diplodia. [Diseases of Hevea due to Diplodia.]** Bull. Agric. Inst. Sci. Saigon 1: 321–329. 1919.—A general discussion of diseases of *Hevea* caused by *Diplodia*, with preventive treatment and remedies.—*E. D. Merrill.*

772. Winston, J. R., and W. W. Yothers. **Bordeaux-oil emulsion.** Florida Grower 23[3]: 9. Jan. 18, 1920.—Directions are given for combining Bordeaux mixture and oil emulsions. Experimental results are reported of the successful use of this combination spray against certain insects and fungous diseases of citrus.—*H. R. Fulton.*

773. Wormald, H. **A phytophthora rot of pears and apples.** Ann. Appl. Biol. 6: 89–100. *Pl. 3, 2 fig.* 1919.—*Phytophthora cactorum* was obtained from pears and apples in England. The fruit often fell prematurely. Inoculation experiments demonstrated the pathogenicity of the fungus. In one case after inoculation the hyphae were found to invade the seeds of pear. One experiment suggested that zoospores might cause infection through the uninjured skin of the pear.—The sporangia germinated either by germ tubes directly, by zoospores which escaped rapidly with the hyaline plug of the sporangium forming a vesicle around them at first, or by production of germ tubes by the zoospores within the sporangium.—The zoospores appeared to utilize the anterior cilium as the organ of locomotion.—Oospores were found. Measurements are given of the various spores and sporangia.—The fungus obtained from either apple or pear would rot both fruits.—Sanitation and spraying are suggested as control measures.—*G. R. Bisby.*

774. Zweigelt, Fritz. **Biologische Studien an Blattläusen und ihren Wirtspflanzen. [Biological studies of aphides and their host plants.]** Verhandl. Zool.-Bot. Gesell. Wien 68: 124–142. *4 fig.* 1918.—Part 1. Mechanics of sap extraction by aphides. Part 2. Anatomy and etiology of aphis galls and the rôle of the plant in formation of roll galls. Part 3. Rôle of the insect in formation of galls. [Through abstr. by Matouschek in Zeitschr. Pflanzenkr. 29: 217–219. 1919.]—*D. Reddick.*

PHARMACOGNOSY AND PHARMACEUTICAL BOTANY

Heber W. Youngken, *Editor*

775. Albertus, Halvar. **Bidrag till kännedom om hesperidinliknande kropparsförekomst inom familjen Labiatae. [Contribution to the knowledge of the occurrence of Hesperidin-like bodes in the family Labiatae.]** Svensk. Farm. Tidskr. 23: 609. 1919.—A microscopic study was made of the stems, leaves, and in some cases the flowers of over 100 members of the family Labiatae for the presence of hesperidin-like bodies. When found, their solubility in caustic soda solution, concentrated sulphuric acid, concentrated ammonia and chloral hydrate was determined.—*A. M. Hjort.*

776. Anonymous. **Production of Pyrethrum flowers in Japan.** Sci. Amer. Supplem. 88: 305. 1919. [From Commerce Reports.]—A short article on the growing of *Chrysanthemum parthenium*, with cost and production statistics.—*Chas. H. Otis.*

777. Anonymous. **Why the castor-oil plant is called Palma Christi.** Sci. Amer. Supplem. 88: 376. 1919.

778. Babe, E., and Teodoro Cabrera. **Clitorina, nuevo reactivo indicador de acidos y alcalis. [Clitorina, a new chemical indicator.]** Revist. Agric. Com. y Trab. 2: 537–539. *1 fig.* 1919.—The name "Clitorina" is given to an indicator made by extracting with 95 per cent alcohol the coloring matter from the flowers of a double blue variety of butterfly pea, *Clitoria ternatea* L. This was found to be superior to phenolphthalein for detecting minute adulterations of milk with potash solutions. It was also found to be superior to phenolphthalein and tincture of cochineal as an indicator in some other reactions.—*F. M. Blodgett.*

779. Ballard, C. W. **The identification of gums by the phenylhydrazine reaction.** Jour. Amer. Pharm. Assoc. 9: 31–38. *Fig. 1–15.* 1920.—Author has made a study of the character of the osazones prepared from different drugs as althaea, peach kernels, sassafras pith, brown mustard, yellow mustard, elm bark, apricot kernels, tragacanth, acacia, quince seed, linseed, indian gum, and bitter almond kernels. Method of application of test is given with sketches and description of the osazones from the various drugs.—*Anton Hogstad, Jr.*

780. Beal, George D., and Thomas S. Hamilton. **The "Shaking-out" method for the quantitative estimation of alkaloids. II.** Jour. Amer. Pharm. Assoc. 9: 9–15. 1920.—Lead acetate when used as a clarifier for alkaloidal extracts has no harmful effect upon the extraction of the alkaloid by immiscible solvents, and that the addition of sodium chloride after clarification increases the quantity of alkaloid removed at a single extraction. Employing the use of amyl alcohol for morphine determinations a residue of anhydrous morphine could be obtained.—*Anton Hogstad, Jr.*

781. Beath, O. A. **The chemical examination of three species of larkspur.** Wyoming Agric. Exp. Sta. Bull. 120: 55–88. *Pl. 1–11, 4 fig.* 1919.—A bulletin in four parts, dealing with the poisonous properties of the three species, *Delphinium barbeyi*, *D. glaucescens*, and *D. geyeri*. Part 1 is general in its scope, dealing with the distribution, a review of the literature, losses to stock, toxicity as effected by age, acidity, seasonal variations of the poisons, characteristic symptoms. Part 2 deals with the experimental methods employed including the determination of the crude alkaloids, preparation and properties of water extracts, and the

extractive value of the solvents. Part 3 deals with the chemical analysis of the three species at different growth stages and of the principal organs of the plant at each stage. Part 4 deals with the method of treatment for Larkspur poisoning. A bibliography of the works cited is given at the end of the article.—*James P. Poole.*

782. Beythien, A. **Gewürze und Gewürz-Ersatz im Kriege. [Spices and spice substitutes in war.]** Zeitschr. Untersuch. Nahrungs- u. Genussmittel 38: 24–33. 1919.—Current prices of spices and substitutes and composition of latter which include cauliflower-, cabbage-, celery- and mushroom-extracts, cinnamon, lemon, almond, and caraway oils, and synthetic benzaldehyde and vanilla. Many substitutes found fraudulent.—*H. G. Barbour.*

783. Buc, H. E. **Delicate test for strychnine.** Jour. Assoc. Official Agric. Chem. 3: 193. 1919.—Method of making the test is given.—*F. M. Schertz.*

✓ 784. Burque, L'abbé F.-X. **L'Identité du Poglus. [The identity of Poglus.]** Le Naturaliste Canadien 46: 145–148. Jan., 1920.—The author criticizes the determination of the species by Frère Marie-Victorin in the previous monthly issue. He closes an interesting discussion by the presentation of evidence that the Indians of the region (the Hurons of Lorette) have actually been calling no less than three species of the *Umbelliferae* by the same name, "Poglus,"—namely, *Archangelica atropurpurea*, *Ligusticum* sp.? and *Heracleum* sp.? He thinks that *Archangelica* is most likely the beneficial species for influenza. [See also Bot. Absts. 5, Entry 811.]—*A. H. MacKay.*

785. Chalmers, D. F. **Report on the operations of the Department of Agriculture. Burma.** 1919: 1–15. 1919.—See Bot. Absts. 5, Entry 11.

786. Clair, H. W. **Scottish Chamomiles.** Chem. and Druggist 91: 1512. 1919.—A comparison between the dried flowers of the "single-flowered" variety of *Anthemis nobilis*, known as Scottish chamomile, and the "double-flowered" variety of the same plant, known as English chamomile. The Scottish Chamomile, formerly cultivated to a considerable extent in the Deeside district of Scotland is more bitter and aromatic than the "double-flowered" variety and of greater value as an internal tonic medicine. The "double-flowered" variety was not obtained by ordinary cultivation from the "single-flowered" type, but by collecting seed from "sport" plants, and by a careful process of selection from these deviating forms a strain which retained the habit of producing "double flowers" was obtained. The Scottish chamomile is used but slightly outside of Scotland.—*E. N. Gathercoal.*

787. Clevenger, Joseph F., and Clare Olin Ewing. **Partial analyses of 330 American crude drugs.** Jour. Amer. Pharm. Assoc. 8: 1010–1029. 1919.—The examinations of these 330 crude drugs include scientific and trade names, part employed, color of powder, total and acid-insoluble ash; total, and volatile ether extracts (with colors and odors); and general remarks as to cleanliness of sample. [See also next following Entry, 788.]—*Anton Hogstad, Jr.*

788. Clevenger, Joseph F., and Clare Olin Ewing. **Partial analyses of 330 American crude drugs.** Jour. Amer. Pharm. Assoc. 9: 15–30. 1920.—Conclusion of article from *Ibid.* 8: 1029. 1919. [See also next preceding Entry, 787.]—*Anton Hogstad, Jr.*

789. Cushny, Arthur R. **The properties of optical isomers from the biological side.** Pharm. Jour. 103: 483. 1919.—See Bot. Absts. 5, Entry 879.

790. Dussel, G. B. **Kort overzicht over den Landbouw op Curaçao. [A short survey of the agriculture on Curaçao.]** Pharm. Weekblad 56: 1512–1514. 1919.—Most of the Curaçao Aloes comes from the Island of Aruba, but large areas are cultivated on Curaçao and Bonaire. The cultivation and propagation is very easy and inexpensive. The cuttings of old plants are set in rows about 0.5 m. apart, when in due time a short stem and rosettes of leaves will be produced, and, after the rains, a flowering stem, which divides into two or more branches,

develops. In the dry season the leaves are cut off and placed in a V-shaped container slanting on one side in order to allow the juice to drain. This is collected in empty coal-oil cans and the contents of the cans is then transferred to large copper kettles in which the juice is concentrated to the desired consistence; it is then run into paper lined petroleum boxes or into gourds. The plant, which prefers a dry, chalky soil, yields aloes for about 12 years; after this time it has to be dug up and the soil is properly manured and replanted.—*H. Engelhardt.*

791. Escobar, Romulo. **La Cicuta. [Cicuta.]** Agricultor Mexicano 36: 6–8. 1920. Description of the plant of water hemlock (*Cicuta* sp.), symptoms of the poisoning induced in sheep, and methods of eradicating the plant.—*John A. Stevenson.*

√ 792. Ewe, George E. **Chinese cantharides. [Mylabris Cichorii.] A worthy candidate for admission to the U. S. P.** Jour. Amer. Pharm. Assoc. 9: 257–263. 1920.—Upon experimentation, employing a series of physiological tests on horses, it was found that *Mylabris Cichorii* has a vesicating and rubefacient power equal to the U. S. P. varieties. The cantharidin content on the average was found to be 50 per cent greater than the U. S. P. varieties. Author also states that the material is cheaper and more available at the present time.—*Anton Hogstad, Jr.*

793. Ewe, George E. **The assay of calabar beans and its preparations.** Jour. Amer. Pharm. Assoc. 8: 1006–1009. 1919.—Author was unable to obtain satisfactory results with the present U. S. P. method for the assay of calabar beans and its preparations. He believes the loss to be due partly to incomplete extraction and partly to decomposition of alkaloids by numerous manipulations and vigorous heating treatments and by long exposure to light required in carrying out the process. Methods of the writer are given for the assay of the drug and its preparations.—*Anton Hogstad, Jr.*

794. Ewing, C. O. **White pine bark adulterated with elm bark.** Jour. Amer. Pharm. Assoc. 9: 253. 1920.—Upon examination a shipment of white pine bark collected in Michigan was found to contain elm bark. The outer part of the bale, to the depth of about 1 foot, consisted almost entirely of the rossed outer bark of *Ulmus fulva.*—*Anton Hogstad, Jr.*

795. Ewing, Clare Olin, and Arno Viehoever. **Acid-insoluble ash standards for crude drugs.** Jour. Amer. Pharm. Assoc. 8: 725–730. 1919.—Upon reviewing the analyses of a considerable number of domestic and imported crude drugs with regard to their content of ash and acid-insoluble ash, authors noted in a number of instances where a striking discrepancy occurred between the general run of analyses and the U. S. P. and N. F. standards. The authors suggest, as an expression of their personal opinion that an extension of ash standards including limits for acid-insoluble ash would be very much preferable to present standards and that it should not only be included in the U. S. P. but should be extended. The authors then discuss the question of ash contents of several drugs, namely asafoetida, hydrastis, hyoscyamus, mustard, rhubarb and sassafras, emphasizing the need of acid-insoluble ash standards. Simple method of writers included for determining acid-insoluble ash content.—*Anton Hogstad, Jr.*

√ 796. Farwell, Oliver A. **Cramp bark, highbush cranberry.** The Druggist 2: 13. 1920.—It has been known since 1913 that the commercial Cramp Bark is the product of *Acer spicatum* Lam. and not the true *Viburnum Opulus* var. *americanum*, Mill, as required in the National Formulary and as stated in the text books. Farwell now produces evidence to show that as long ago as 1870 the Acer bark had displaced the true Viburnum bark.—*Wm. B. Day.*

797. Fishlock, W. C. **Bay leaves (Pimenta acris).** Report on the Agricultural Department, Tortola, 1917–18, 6. Imperial Department of Agriculture, Barbados. 1919.—A reference is made here to the existence of "false" or bad varieties of the bay tree whose leaves yield an oil of inferior quality for making bay rum.—*J. S. Dash.*

798. French, Harry B. **Review of the drug market.** Jour. Amer. Pharm. Assoc. 8: 843–844. 1919.—A general discussion of the effect of the signing of the Armistice on the drug market. Writer states that the general tendency of American crude drugs has been to greatly advance in price since the signing of the Armistice and that this tendency will continue for the next several months. Chemicals have a tendency to decline and European crude drugs will be obtainable at lower prices as soon as they can finance shipments and transportation can be arranged.—*Anton Hogstad, Jr.*

799. Fuller, H. C. **Report on alkaloids.** Jour. Assoc. Official Agric. Chem. 3: 188–193. 1919.—It is recommended that in conducting assays for strychnine, reliance be placed on a gravimetric determination and not on a determination obtained by volumetric means.—*F. M. Schertz.*

800. Gathercoal, E. N. **The permanency and deterioration of some vegetable drugs twenty-five years of age.** Jour. Amer. Pharm. Assoc. 8: 711–716. 1919.—Examination of some 144 crude drugs which were prepared some twenty-five years ago and which had been kept in glass-stoppered bottles, showed that most of the drugs were very well preserved and which compared with the present U. S. P. and N. F. requirements. Among the drugs much depreciated were Orange and Lemon peels, Labiatae herbs and a number of leaf drugs (Buchu, Boneset, Coltsfoot, Witchhazel, Matico, Gaultheria, and probably Pilocarpus).—*Anton Hogstad, Jr.*

801. Greig-Smith, R. **The germicidal activity of the Eucalyptus oils. Part I.** Proc. Linnean Soc. New South Wales 44: 72–92. *Fig. 1.* 1919.—Eucalyptus oil as listed in Materia Medica is from *E. globulus.* Many oils of other origin are sold under this name. The Baker and Smith classification of oils is followed in these tests to determine the toxic effect of 40 to 50 specimens of crude and refined oils. *E. polybractea* (Blue Mallee), *E. cinerea* (Argyle apple), *E. australiana* (Narrow-leaf peppermint), and *E. dives* (Broad-leaf peppermint) are at present the chief sources of commercial oils in New South Wales. The test-organisms employed to determine the toxicity of the oils were *Micrococcus aureus* and *Bac. coli communis* from serum suspensions. The activity and quality of the oil was found to vary strikingly even within the same tree and also with different specimens of a species. It was affected by altitude and growth conditions in general. On the whole these oils had lower toxicity than phenol. The results of the tests are given in nine tables. The main constituents seemed relatively insignificant with reference to toxic action. Bactericidal power was proportional to the acidity of the oil and assisted by although not caused by it alone. The iodide reaction was no criterion as to the germicidal value of the oils. The vapors of the oils had decided bacterial action.—*Eloise Gerry.*

802. Griebel, C., and A. Schäfer. **Thymus Serpyllum L. als Majoranpulververfälschung. [Wild thyme as imitation marjoram powder.]** Zeitschr. Untersuch. Nahrungs- u. Genussmittel 38: 141–145. 1919.—The chief morphological characteristics of marjoram and of wild and common thyme are compared.—*H. G. Barbour.*

803. Grimme, C. **Altes and Neues ueber Capsella Bursa pastoris. Mittheilung aus dem Institut fuer angeswandte Botanik. [Old and new facts about capsella bursa pastoris. Communication from the institute for applied botany.]** Pharm. Zentralhalle Deutschland 60: 237–242, 248–251. 1919.—Shepherd's purse has been used since times immemorial as a home remedy, as diuretic and antipyretic. Recent investigations have shown that the drug possesses strongly hemostyptic properties and can be used as a substitute for the high-priced and in Germany unobtainable golden seal. The chemistry of the drug is still to be investigated but the medicinal value seems to be partly due to mustard oil which is present in all parts of the plant, but especially in the seeds.—*H. Engelhardt.*

804. Guérin, P. [Rev. of: Étienne, P. **Étude anatomique de la famille des Épacridees. (Anatomic study of the family Epacrideae.)** Thése Doct. Univ. Pharm. Toulouse. *222 p. 116 fig.* 1919.] Bull. Sci. Pharm. 26: 533. 1919.—The author describes the anatomic structure of

the branches and leaves of 26 members of the family *Epacrideae*. The *Epacrideae* appear to take in Oceania the place which the *Ericaceae*, to which they are nearly related, take in South Africa.—*H. Engelhardt.*

805. Guérin, P. [Rev. of: Berger, Marie-Gaston. **Étude organographique, anatomique et pharmacologique de la famille des Turnéracées. (Organographic, anatomic and pharmacologic study of the family Turneraceae.)** *870 p., 53 pl.* Bigot Frères: Paris, 1919.] Bull. Sci. Pharm. 26: 533. 1919.—The six genera of the family Turneraceae can easily be distinguished from each other by their anatomic structure. The author believes that the Turneraceae must be considered as a special family, but if taken away from the Bixaceae, they should be counted to the family Passiflorae. The author further deals with the medicinal use of the members of this family, and especially of that of damiana (Turnera aphrodisiaca) and with the various substitutes offered for this drug.—*H. Engelhardt.*

806. Hart, Fanchon. **A microscopical method for the quantitative determination of vegetable adulterants.** Jour. Amer. Pharm. Assoc. 8: 1032–1034. 1919.—The areas of the various tissues present are totalled by the aid of an ocular micrometer used in conjunction with a stage micrometer and from these figures the author calculates the percentage of impurities. The author gives method of examination for black pepper adulterated with pepper shells and checks results obtained by measuring the shells and powdered kernel portions in a 10 minim graduate.—*Anton Hogstad, Jr.*

807. Hatcher, Robert A. **Standardization of digitalis. A preliminary report.** Jour. Amer. Pharm. Assoc. 8: 913–914. 1919.—The author reports the results obtained by separating the principles of Digitalis into two groups, namely, the chloroform-soluble fraction and the water-soluble fraction. The chloroform-soluble fraction was found to be more readily absorbed and more lasting in its effects while the water-soluble fraction being more actively emetic. Author believes that Digitalis should be assayed in reference to the chloroform-soluble fraction and that this fraction may be made available for intravenous use, since it mixes perfectly with water.—*Anton Hogstad, Jr.*

808. Jones, J. **Bay oil.** Report on the Agricultural Department, Dominica, 1918–19: 5. Imperial Department of Agriculture, Barbados. 1919.—Two samples of oil from varieties of *Pimenta acris*, namely Bois d'Inde and Bois d'Inde Citronelle, grown in Dominica, are reported on. The latter variety contains a smaller percentage of Phenols, and has a strong odour of citral, and the suggestion is made that it may have some commercial value in the manufacture of toilet preparations.—*J. S. Dash.*

809. Jones, J. **Camphor.** Report on the Agricultural Department, Dominica, 1918–19: 5–7. Imperial Department of Agriculture, Barbados. 1919.—Results of distillations of leaves, twigs and prunings from three plots showed that two of them were of true camphor trees, yielding both camphor and oil, while the other was not, the material from it producing oil only.—*J. S. Dash.*

810. Keenan, G. L. **The microscopical identification of mowrah meal (Bassia) in insecticides.** Jour. Amer. Pharm. Assoc. 9: 144–147. *Fig. 1–3.* 1920.—In the examination of products designated as ant and worm eradicators, author detected the presence of mowrah meal, which he states resembles cocoa powder in general appearance. The powder consists largely of the powdered cotyledons and occasional fragments of seed coat. Chloral hydrate reveals the presence of yellowish-brown masses occurring separately as isolated fragments and also in characteristic group arrangement. The uses of mowrah meal and a morphological description of *Bassia latifolia* are also included. With bibliography.—*Anton Hogstad, Jr.*

811. Marie-Victorin, Fr. des E. C. **L'identité du Poglus (Heracleum lantatum, Michx). [The identity of Poglus of the Hurons of Lorette.]** Le Naturaliste Canadien 46: 121–124. Dec., 1919.—The Indians (Hurons) of Lorette, Province of Quebec, have been using the root of

Poglus with wonderful success against epidemic influenza. M. l'Abbe F.-X. Burque. (*Ibid.* 45: 67–70. 1918) had identified it with *Angelica atropurpurea* L. (*Archangelica atropurpurea* (L.) Hoffm.).—The author accompanied by M. Edouard Laurin visited Bastien, the local Indian chief, who pointed out a young specimen of *Poglus* which had not yet its radical leaves. The abundant pubescence showed it could not be *Angelica.* Further examination convinced him it was *Heracleum lanatum* Michx. (la Berce laineuse). Chief Bastien insisted on the powerful febrifuge properties of the plant, and cited extraordinary cases of cures. It was believed to be the cause of the protection of the tribe from the epidemic. The Hurons collect the root in autumn, and use the infusion.—The author then quotes authorities on the properties of *Heracleum*, notes its distribution, and describes its appearance and habitat. [See also Bot. Absts. 5, Entry 784.]—*A. H. MacKay.*

812. Merrill, E. C. **Preliminary study of some of the physical and chemical constants of balsam Peru.** Jour. Assoc. Official Agric. Chem. 3: 194–197. 1919.—The method for the determination of the iodine value of cinnamein by Hanus, as at present employed, is unsatisfactory and furthermore may be entirely inadequate as an index of the character of pure Peru balsam. The employment of such physical constants as viscosity, surface tension, optical rotation and refractometer observation may prove of value in the final interpretation of the character of Peru balsam.—*F. M. Schertz.*

813. Nelson, E. K. **The constitution of capsaicin, the pungent principle of capsicum. II.** Jour. Amer. Chem. Soc. 42: 597–599. 1920.

814. O'Brien, J. F., and J. P. Snyder. **Deterioration of high-test American grown Digitalis.** Jour. Amer. Pharm. Assoc. 8: 914–919. 1919.—Assays of the tincture and fluidextract of Digitalis made from American-grown Digitalis from the state of Washington, after being kept for a period of two and one half years, under conditions which closely paralleled those of the average drug store, showed that these preparations did deteriorate and that the deterioration was practically the same in both preparations. By the guinea pig method the loss in activity was from 330 to 175 per cent, a loss of 47 per cent; the one hour frog method the loss in activity was from 264 to 120 per cent, a loss of 55 per cent; by the cat method the loss in activity was from 250 to 175 per cent, a loss of 30 per cent. However, all the preparations after standing this length of time still retained sufficient activity for them both to be considered standard preparations.—*Anton Hogstad, Jr.*

815. Passerini, N. **Sul potere insetticida del Pyrethrum cinerariaefolium Trev. coltivato a Firenze in confronto con quello di alcune altre Asteracee. [A comparison of the insecticidal value of Pyrethrum cinerariaefolium Trev. grown at Florence with other members of the Asteraceae.]** Nuovo Gior. Bot. Italiano 26: 30–45. 1919.—Both as regards rapidity of action and effectiveness *Pyrethrum cinerariaefolium* Trev. is superior as an insecticide to other members of the *Asteraceae.* If ground into a fine powder, the heads, foliage, stems and roots of the plant are equally effective; however, the most rapid action is obtained from the heads of the plant.—*Ernst Artschwager.*

816. Petrie, J. M. **The occurrence of methyl laevo-inositol in an Australian poisonous plant.** Proc. Linnean Soc. New South Wales 43: 850–867. *2 fig.* 1918.—*Heterodendron oleaefolium* Desf. (Sapindaceae) a large, drought-resistant shrub, endemic to Australia, which has been described as a valuable forage plant was suspected of causing fatalities to cattle and horses. It was found to be strongly cyanogenetic. It contains the methyl ester of laevo-rotary inositol and the method of extraction and characteristics and properties of the compound are given in detail. The amount isolated was equivalent to 0.65 per cent of the dried (at 100°C.) leaves. It is not optically isomeric with pinite of Maquenne, which is the methyl dextro-inositol, possessing a different melting point and optical rotation. It is apparently identical with Tanret's quebrachite and has been previously recorded for three plants only—*Aspidosperma querbracho* (Apocyanceae), *Hevea brasiliensis* (Euphorbiaceae)

and *Grevillea robusta* (Proteaceae). The occurrence of this compound is exceedingly rare, in contrast to the inactive inositol which exists as a plastic substance in most plants. *Heterodendron* also contains a cyanogenetic glucoside.—*Eloise Gerry.*

817. Pittenger, Paul S. **Preliminary note on a new pharmaco-dynamic assay method.** Jour. Amer. Pharm. Assoc. 8: 893–900. 1919.—Writer states that the goldfish method is unquestionably the simplest so far proposed and can be easily carried out by those not specially skilled in the pharmacodynamic art. A tincture of Digitalis should have a minimum lethal dose of 2.85 when assayed by this method. Results of the authors experiments are recorded as well as details of methods employed including a list of apparatus necessary for the experiments.—*Anton Hogstad, Jr.*

818. Pittenger, Paul S., and George E. Ewe. **The standardization of Piscidia Erythrina (Jamaica dogwood).** Amer. Pharm. Jour. 91: 575–583. *Fig. 1–3.* 1919.—The similarity between the action of Jamaica dogwood and that of Cannabis, suggested the possibility of employing similar methods of standardization. The following tentative standard was adopted: Fluidextract of Jamaica dogwood should be of such strength that it will produce incoordination in dogs in doses of 0.55 mils per kilo weight of animal and should not produce incoordination in doses less than 0.5 mils per kilo, the drug being administered by capsule after fasting the animal for 12 hours. A series of experiments were conducted to assay Jamaica dogwood preparations according to the piscidin content, but on account of the contamination with resinous matter it was difficult to obtain the piscidin in a pure state, therefore as the authors state we are without a reliable chemical means of accurate standardizing Jamaica dogwood preparations, but that they can be accurately standardized by the physiological assay method.—*Anton Hogstad, Jr.*

819. Reens, Emma. **The Coca de Java.** [Javanese coca.] Bull. Sci. Pharm. 26: 497–505. 1919.—A detailed study of the cultivation and propagation of the coca tree is given together with data on collecting the leaves, the extraction and purification of the alkaloid. The author states that while in South America the leaves of *E. bolivianum* and *E. peruvianum* are altogether used, in the East Indies and especially in Java *E. spruceanum* or *E. novogranatense* is cultivated.—*H. Engelhardt.*

820. Robson, W. **Bay trees (Pimenta acris).** Report on the Agricultural Department, Montserrat, 1917–18: 17. Imperial Department of Agriculture, Barbados. 1919.—A record is given of the yield of Bay leaves and oil from a plot for seven consecutive years. The results of 41 distillations during 1917 are given. From these it was found that the average Phenol content was 55 per cent, being 5 per cent higher than the average for 1914–16.—*J. S. Dash.*

821. Robson, W. **Ajowan Plant (Carum copticum).** Report on the Agricultural Department, Montserrat, 1917:18: 19–22. Imperial Department of Agriculture, Barbados. 1919.—Interesting cultural and distillation trials are recorded with this plant. The percentage of oil in the seed was found to be 3, while the per cent Thymol in the oil is given as 40 to 45.—*J. S. Dash.*

822. Robson, W. **American horsemint (Monarda punctata).** Report on the Agricultural Department, Montserrat, 1917–18: 22–23. Imperial Department of Agriculture, Barbados. 1919.—Small trials with this plant gave satisfactory results, the oil obtained from distillation containing about 44 per cent by weight of Thymol.—*J. S. Dash.*

823. Rowe, L. W. **Maintaining frogs for test purposes.** Jour. Amer. Pharm. Assoc. 8: 928–930. *1 fig.* 1919.—A description with sketch of a tank for maintaining frogs for test purposes.—*Anton Hogstad, Jr.*

824. Rowe, L. W. **Digitalis standardization. A consideration of certain methods of biological assay.** Jour. Amer. Pharm. Assoc. 8: 900–912. 1919.—Experiments were performed, first to determine whether any relationship exists between the results of assays by the cat

and frog methods; second, to determine the accuracy of the cat method and third to suggest certain modifications of the method, in order to make it more practical for commercial assay work. Sufficient data was not obtained with cats to absolutely prove that they are as unsatisfactory as dogs but from data reported indicates that there is no real consistency between the results obtained when using the cat and those obtained with the frog. Author states that it seems most logical to conclude that no relationship exists between the minimum lethal doses of heart tonic preparations to cats, dogs, and frogs, but that the frog method is the most accurate of the three. With bibliography.—*Anton Hogstad, Jr.*

825. Sayre, L. E., and G. N. Watson. **Final report on the alkaloids of Gelsemium.** Jour. Amer. Pharm. Assoc. 8: 708–711. 1919.—Investigations by the authors seem to show that there does not exist in the drug any such alkaloid as Gelseminine, but that this constituent (so-called) is a compound body consisting of several alkaloids having different properties. Methods are given for the separation of these various substances including Sempervirene, Gelsemic acid, Gelsemine and another substance named by the authors "Gelsemidine"—not "Gelseminine"—since gelseminine, the name formerly given to the amorphous alkaloids of gelsemium, has been proved conclusively to be not a single alkaloid but a mixture of three alkaloids. Another substance was also obtained which was strongly alkaloidal in appearance and behavior and very much like Lloyd's Emetoidine, which the authors state might be called "Gelsemoidine." Physical descriptions of these substances follow.—*Anton Hogstad, Jr.*

826. Sprinkmeyer, H., and O. Gruenert. **Über Vanillinerzeugnisse.** [Vanilla products.] Zeitschr. Untersuch. Nahrungs- u. Genussmittel 38: 153–155. 1919.—Deterioration of vanilla and related substances in mixtures.—*H. G. Barbour.*

√ 827. Stanford, Ernest E., and Clare Olin Ewing. **The resin of man-root (Ipomoea pandurata (L.) Meyer) with notes on two other Convolvulaceous resins.** Jour. Amer. Pharm. Assoc. 8: 789–795. *Fig. 1.* 1919.—Alcoholic extracts of three Convolvulaceous roots gave the following results: *Ipomoea pandurata* (Man-root) 4.65 per cent of resin; *I. batata* (Sweet potato) 0.56 per cent of resin; *I. discoidesperma* Donn. Sm. (Yellow morning glory) 6.5 per cent of resin. The resin of man-root possessed mild cathartic properties, that of the sweet potato failed to demonstrate any cathartic action. The material on hand of the yellow morning glory was insufficient for adequate tests. Examination of the extracts of man-root and sweet potato with various organic solvents showed them like other Convolvulaceous resins to be of complex composition and partly of glucosidal nature. No chemical examination was made of the resin of yellow morning glory. Descriptions of the roots are included. With bibliography.—*Anton Hogstad, Jr.*

828. Steel, I. **Plantago in medicine.** Australian Nat. 4: 105–107. 1919.—Its uses as a native home remedy. Some references to its properties in English literature.—*T. C. Frye.*

829. Stockberger, W. W. **Commercial drug growing in the United States in 1918.** Jour. Amer. Pharm. Assoc. 8: 807–811. 1919.—A report on the progress of the cultivation of a number of drugs as Belladonna, Cannabis, Digitalis, Calendula, Sage and Henbane. Summarizing the total production the author states that in the case of Belladonna approximately 83 tons of herb (including leaves and stems), and 11 tons of root; 60 tons of Cannabis; 9000 to 10,000 pounds of Sage have been produced in the United States during 1918. No figures were given for the production of Calendula. Very little success has been attained in the commercial cultivation of Henbane. Digitalis has not been placed on an established commercial basis as yet.—*Anton Hogstad, Jr.*

830. Stroup, Freeman P. **A chemical test to distinguish between caffeine and theobromine.** Amer. Jour. Pharm. 91: 598–599. 1919.—Employing the use of potassium bichromate and sulphuric acid, the author states that it is a simple matter to distinguish between caffein and theobromin, according to the colors produced.—*Anton Hogstad, Jr.*

831. Styger, Jos. **Beiträge zur Anatomie des Umbelliferenfruchte. [Contribution on the anatomy of umbelliferous fruits.]** Schweiz. Apotheker Zeitg. 57: 199–205, 228–235. *7 fig.* 1919.—A description of the macroscopic and microscopic characteristics of the fruits of *Angelica Archangelica*, *F. Narthex*, *F. galbaniflua*, *F. angulata*, *Pastinaca sativa*, *Heracleum Spondylium*, *Laserpitium Siler*, *L. marginatum*, *Opopanax chironeum*, and *Daucus Carota*. *Angelica Archangelica* is winged and its mesocarp is composed for the most part of loosely arranged, porous and reticulately thickened parenchyma with large intercellular-air-spaces; its vittae are distributed above the inner epidermis and in the ribs. *Ferula Narthex* shows a band of thick-walled, punctated cells in the inner mesocarp and giant vittae in the mesocarp. *F. galbaniflua* is distinguished from *F. Narthex* by having vittae in the ribs as well as the mesocarp. *F. angulata* possesses vittae in mesocarp and ribs, its outer epidermis and the cell layers lying directly beneath are strongly thickened but not woody, and hesperidin crystals exist in all the epidermal cell glands. *Pastinaca sativa* shows vittae alongside vascular bundles, a sclerenchyma band in the inner mesocarp and finely punctated parenchyma in its winged ribs. *Herecleum Spondylium* has a sclerenchyma band in the inner mesocarp and finely punctated thick-walled parenchyma in the wings outside of the bundles. *Laserpitium marginatum* has elliptical vittae while those of *L. Siler* are triangular, as viewed in cross section. *Opopanax chironeum* shows cells of epidermis, wings and within vascular bundles with elliptical punctations; *Daucus Carota* has delicate spines growing from secondary ribs, and bristle-hairs only on primary ribs. [See also next following Entry, 832.]—*H. W. Youngken.*

832. Styger, Jos. **Beiträge zur Anatomie des Umbelliferenfruchte. [Contribution on the anatomy of Umbelliferous fruits.]** Schweiz. Apotheker Zeitg. 57: 243–250. 1919.—An analytical key, based upon a pharmacognic system, to the 50 Umbelliferous fruits described by the author in preceding pages of this serial. These are placed in 3 main groups, viz.: I. Without oil containing elements. II. With secretion sacs. III. With oil reservoirs (vittae). The first two of these captions have but one representative each, viz.: *Conium maculatum* and *Hydrocotyle vulgaris* respectively. The third group includes two subdivisions: 1. With commissurral vittae only. 2. With dorsal and commissural vittae. Further grouping of these subdivisions is based upon presence of one or more vittae in mesocarp, sclerenchyma plates, hairs, strongly thickened and lignified parenchyma elements in mesocarp, secondary vittae, and distribution of the vittae in inter-rib and rib regions. [See also next preceding Entry, 831.]—*H. W. Youngken.*

833. Sutton, Richard L. **Ragweed dermatitis.** Jour. Amer. Med. Assoc. 73: 1433–1435. 1919.—The important part played by anaphylaxis in the causation of various eruptions has long been recognized. Anaphylaxis has been defined as "a state of hypersusceptibility of the organism to foreign substances, which is brought about by the introduction of certain foreign substances and their cleavage products." C. Walker has pointed out that certain proteins, including those of ragweed pollen may cause dermatitis in predisposed persons. The author describes four cases of ragweed dermatitis. In two of them the common ragweed. *Ambrosia elatior*, was the chief offender. The giant ragweed, *A. trifolia*, the mugwort, *A. psilostachya*, and the bur marsh-elder, *Iva xanthifolia*, probably occupy lesser rôles. All have been shown to cause hay fever. Pollen vaccine treatment gave beneficial results.—*Wm. B. Day.*

834. Thurston, Azor. **Oil of sandalwood and its adulteration.** Jour. Amer. Pharm. Assoc. 9: 36–37. 1920.—A compilation of the refractive indices and optical rotations of some 42 samples of commercial sandalwood oils with a few additional notes. With bibliography. —*Anton Hogstad, Jr.*

835. Viehoever, Arno. **The pharmacognosy laboratory, its activities and aims.** Jour. Amer. Pharm. Assoc. 8: 717–725. 1919.—A detailed account of the activities and aims of the Pharmacognosy Laboratory, Bureau of Chemistry, U. S. Department of Agriculture, prepared in the hope that other workers engaged in pharmaceutical and related research, may be induced

to prepare similar statements, sufficiently detailed to indicate the nature of their studies, though the work may still be in progress.—Part I is devoted to a discussion of Crude Drug Control in which the author discusses various phases of the work, as domestic trade; import trade; elimination of inert and objectionable material in crude drugs and spices; extension of standardization of purity for drugs; value of volume weight determinations; pharmacopoeial work; prevention of waste and utilization of waste crude drug products.—Part II is devoted to the investigations of the pharmacognosy laboratory which cover a wide range of pharmaceutical and chemical research. Author also discusses the coöperative work of the laboratory with various institutions, laboratories, etc. With bibliography.—*Anton Hogstad, Jr.*

836. Vierhout, P. **Het Winnen van Curacao-Aloe. [The production of Curaçao aloes.]** Pharm. Weekblad. 56: 1510–1512. *Pl. 1, fig. 3.* 1919.—A description of methods of collecting aloes in Curaçao.—*Abstractor.*

837. Wirth, E. H. **A study of Chenopodium ambrosioides var. anthelminticum and its volatile oil.** Jour. Amer. Pharm. Assoc. 9: 127–141. *22 fig.* 1920.—The author has made a study of the oil of Chenopodium which falls under the heading of the "western oils" in order to compare same with the Maryland variety, the latter according to general opinion has been claimed to be superior to the former.—A detailed discussion as to the composition of the oil is given, the western oil agreeing with the Maryland oils, save in the amount of ascaridol which is present in the latter from 60 to 80 per cent and in the former the average was 42 to 45 per cent. Specific gravity of western oil 0.934 compared to a specific gravity of 0.955–0.980 as stated in the U. S. P. Upon subjecting an oil with a specific gravity of 0.934 to steam distillation, one fraction, 70 to 75 per cent had a specific gravity of 0.900 and 25 to 30 per cent had a specific gravity of 1.000, thereby showing that the western oil might be fractionated on a commercial basis. Experiments found this to be impracticable owing to the waste involved. —An exhaustive pharmacognostic study of *Chenopodium ambrosioides anthelminticum* is given, in which the author, by microchemical tests, employing 5 per cent KOH in 95 per cent alcohol, shows that the oil is not contained in the seeds but occurs only in the glandular hairs and here only in the large thin-walled terminal hairs. The hairs upon the leaves were found to contain oil but no glandular hairs were noted on the stems, which thus eliminates using stem portions for the production of the oil. Flowers also contain oil, which sets forth the value of subjecting the plant to distillation at the time of flowering.—*Anton Hogstad, Jr.*

838. Wong, Ying C. **Opium in China.** Amer. Jour. Pharm. 91: 776–784. 1919.—An interesting account of this gigantic evil which has cost China billions of dollars and, more important than that, has led millions and millions of her strong citizens into wreck and misery. Author discusses in detail the history and cultivation of the poppy; interesting synonyms and their application to the different grades of opium; opium smoking; suppression of the poison.—*Anton Hogstad, Jr.*

839. Wunschendorff, M. E. **La racine d'Atractylis gummifera. [The root of Atractylis gummifera.]** Jour. Pharm. et Chim. 20: 318–321. 1919.—The writer gives an account of the earlier investigations of the root by Lefranc. He succeeded in isolating about 4 per cent of a petroleum-ether soluble resin, which was insoluble in water and alcohol, but gave pseudo-solutions with chloroform, benzene, carbon tetrachloride, carbon disulphide, etc. It had all the characteristics of caoutchouc and could easily be vulcanised. He further isolated tannic acid, several sugars and a substance which probably was identical with Lefranc's potassium atractylate. The ash, 14.8 per cent, was rich in silica and iron.—*H. Engelhardt.*

840. Yamamoto, R. **On the insecticidal principle of Chrysanthemum cinerariifolium.** Ber. Ohara Inst. Landw. Forsch. 1: 389–398. 1918.—Pyrethron, the insecticidal principle, is a yellow, transparent, neutral syrup, having a saponification value of 216 and iodine value of 116. It is easily saponified with alcoholic potash and loses its insecticidal power after saponification. The power of this pyrethron is reduced either by heating or exposure to the air for a long time. Pyrethron has germicidal as well as insecticidal powers.—*H. S. Reed.*

841. Youngken, Heber W. **Observations on Digitalis Sibirica.** Jour. Amer. Pharm. Assoc. 8: 923–928. *14 fig.* 1919.—A botanical investigation of *Digitalis Sibirica* Lindley, including a description of the plant, histology of leaf, stem and root. Author found that a tincture prepared from the dried leaves according to the U. S. P. method for tincture of digitalis and when assayed by the one-hour frog method, showed the tincture to be three quarters over the strength required for the U. S. P. tincture of digitalis.—*Anton Hogstad, Jr.*

PHYSIOLOGY

B. M. Duggar, *Editor*

Carroll W. Dodge, *Assistant Editor*

GENERAL

842. Bechhold, H. **Colloids in biology and medicine.** [Translated from the second German edition, with notes and emendations by Jesse G. M. Bullowa.] *XV + 464 p., 54 fig.* Van Nostrand Co.: New York, 1919.—Proof sheets of the original were received in 1915 and 1916, but the translation has been brought practically up to date by numerous insertions and notes. The work is divided into four parts as follows: I. Introduction to the study of colloids, 127 p. II. Biocolloids, 83 p. III. The organism as a colloid system, 144 p. IV. Toxicology and pharmacology, microscopical technic, 77 p.—The strictly biological (physiological) aspects deal in a larger measure with animal structures and behavior, due largely to the greater specialization in such organisms, but the plant material is in nowise neglected.—*B. M. Duggar.*

843. Haldane, J. S. **The new physiology and other addresses.** *VII + 156 p.* Charles Griffin & Co., Ltd.: London, 1919.—This small volume embodies six addresses under the following titles: (1) the relation of physiology to physics and chemistry; (2) the place of biology in human knowledge and endeavour; (3) the new physiology; (4) the relation of physiology to medicine; (5) the theory of development by natural selection; and (6) are physical, biological, and psychological categories irreducible? Each topic includes some discussion more or less directly relating to the field, problems, or development of physiology and physiological concepts. Special emphasis is placed upon arguments designed to strengthen the claims of biology as an independent science, and with these the distinctive field of physiology as a fundamental branch of this science. Despite the accumulation of facts relating to the "physical and chemical sources and the ultimate destiny of the material and energy passing through the body" there is "an equally rapidly accumulating knowledge of an apparent teleological ordering of this material and energy." The old "vital force" could never become a working hypothesis; on the other hand, physico-chemical explanations of the mechanism of such processes as respiration are difficult and disappointing, while such assumptions applied to heredity "make the physico-chemical theory of life unthinkable." Nevertheless "we need not sit down in despair, for we can look for other working conceptions."—*B. M. Duggar.*

844. McLean, F. T. **Opportunities for research in plant physiology in the Philippines.** Philippine Agric. 8: 27–31. 1919.—A short article pointing out some of the advantages of the Philippine Islands as a place for research in plant physiology.—*S. F. Trelease.*

845. Willows, R. S., and E. Hatschek. **Surface tension and surface energy and their influence on chemical phenomena.** *2nd ed. VIII + 115 p., 21 fig.* Text-books of chemical research and engineering. Blakiston's Son & Co.: Philadelphia, 1919.—The new edition does not depart from the first in presenting for both biologist and chemist a concise discussion of the fundamental laws of surface tension and surface energy without necessarily applying these to specific phenomena. An additional chapter deals with complex phenomena including such topics as stable emulsions, the theory of dyeing, also tanning.—*B. M. Duggar.*

PROTOPLASM, MOTILITY

846. Chambers, Robert. **Changes in protoplasmic consistency and their relation to cell division.** Jour. Gen. Physiol. 2: 49-68. 1919.—See Bot. Absts. 5, Entry 119.

DIFFUSION, PERMEABILITY

847. De Vries, O. **Verband tusschen het soortelijk gewicht van latex en serum en het rubbergehalt van de latex. [The relation between the specific gravity of latex and serum and the rubber content of latex.]** Arch. Rubbercult. Nederlandsch-Indië 3: 183-206. 1919.—See Bot. Absts. 5, Entry 183.

848. Dixon, H. H., and W. R. G. Atkins. **Osmotic pressures in plants. VI. On the composition of the sap in the conducting tracts of trees at different levels and at different seasons of the year.** Sci. Proc. Roy. Dublin Soc. 15: 51-62. 1918.—The aim of this paper is the study of sap composition at different levels in the same tree and the examination of similar trees during the various seasons of the year. Three trees of *Acer macrophyllum*, two each of *Ilex aquifolium* and *Cotoneaster frigida*, and one each of *Arbutus unedo* and *Ulmus campestris* were employed.—The sap was centrifuged from the fresh conducting wood of the trees. It was found to vary greatly in color and in content of both sugars and salts. During the late autumn and winter while the trees are dormant the osmotic pressure is small and approximately constant throughout the wood sap. The upper portions of the stem and the roots have slightly greater pressure than the central portions. In the early spring large quantities of sugars from the storage cells of the wood parenchyma and the medullary rays are added to the sap. This is followed by a marked increase in osmotic pressure from root to crown, the greater increase occurring in the upper part of the tree. During late spring the concentration of salts is very much greater than in early spring. At this time the concentration of sugars is still high, being about half the maximum concentration.—In *Acer macrophyllum*, sucrose is present in quantity. In the root this amounts to 0.6 per cent in October and 1 per cent in February. In the stem at 10 m. level, where the highest concentrations are recorded, 0.5 per cent sucrose is found in October and 5.5 per cent in February. The reducing sugars are not found at all or only in minute traces. In the other trees both reducing sugars and sucrose were found, the latter usually predominating. In the spring the reducing sugars consisted of the hexoses and maltose, at other times the latter is absent.—In the evergreens, *Arbutus unedo* and *Ilex aquifolium*, and in the sub-evergreen, *Cotoneaster frigida*, neither great seasonal changes nor gradients from roots to crown were observed. At certain seasons the roots may have slightly higher concentrations than the stems.—*A. E. Waller.*

849. Loeb, Jacques. **Electrification of water and osmotic pressure.** Jour. Gen. Physiol. 2: 87-106. 1919.—Experimenting with the amphoteric electrolytes $Al(OH)_3$ and gelatin the author finds that water diffuses through collodion membranes into solutions of metal gelatinates or aluminates as if the water were positively charged, and into their acid salts as if it were negatively charged. The turning point for the sign of electrification of water seems to be near, or to coincide with, the isoelectric points, which is a hydrogen ion concentration about 2 times 10^{-5} N for gelatin and about 10^{-7} N for $Al(OH)_3$. When diffusing into solutions of metal gelatinates the rate is determined by the charge of the cation, the rate being approximately 2 to 3 times as great into solutions containing the monovalent cations of Li, Na, K, NH_4 as into those of the divalent cations of Ca or Ba at the same concentrations of gelatin and hydrogen ions. When diffusing into acid salts of gelatin, water—apparently negatively charged—diffuses less rapidly into a solution of gelatin sulfate than into a solution of gelatin chloride or nitrate of the same gelatin and hydrogen ion concentrations. "If we define osmotic pressure as that additional pressure upon the solution required to cause as many molecules of water to diffuse from solution to the pure water as diffuses simultaneously in the opposite direction through the membrane, it follows that the osmotic pressure cannot depend only on the concentration of the solute but must depend also on the electro-

static effects of the ions present and that the influence of ions on the osmotic pressure must be the same as that on the initial velocity of diffusion. This assumption was put to a test in experiments with gelatin salts for which a collodion membrane is strictly semipermeable and the tests confirmed the expectation."—*O. F. Curtis.*

850. Shull, C. A. **Permeability.** [Rev. of: Williams, Maud. **The influence of immersion in certain electrolytic solutions upon permeability of plant cells.** Ann. Botany 32: 591–599. 1918. (See Bot. Absts. 2, Entry 304.)] Bot. Gaz. 68: 232. 1919.

851. Stiles, Walter, and Franklin Kidd. **The comparative rate of absorption of various salts by plant tissue.** Proc. Roy Soc. London 90 B: 487–504. *Tables 1–10, fig. 1–7.* 1919.—Rate of absorption of various chlorides, sulphates, nitrates, and potassium salts from solutions 0.02N was measured by the electrical conductivity method, using discs of carrot and potato.—Initial absorption was rapid, possibly in proportion to ionic mobility. This is followed by a long period of almost logarithmic approach to equilibrium. The final quantity absorbed is independent of the initial rate in the case of any given salt.—Initial absorption rates are in the following order: Kations, K (Ca, Na), Li (Mg, Zn), Al; anions, SO_4, NO_3 Cl.—The final absorption order is, kations, K, Na, Li (Ca, Mg); anions NO_3, Cl, SO_4. Monovalent ions are at equilibrium in much greater quantity than divalent ions in the cases studied. The rate and extent of intake of one ion of a salt may be affected by the nature of the other ion. From aluminium sulphate aluminium is rapidly absorbed, and the sulphate ion slowly.—It is pointed out that there is essential agreement with other workers.—*Paul B. Sears.*

852. Thoday, D. **The "osmotic hypothesis:" a rejoinder.** New Phytol. 18: 257–259. 1919.—This is an answer to certain criticisms brought forward by Stiles and Jørgensen.—*I. F. Lewis.*

WATER RELATIONS

853. Cribbs, James E. **Ecology of Tilia americana. I. Comparative studies of the foliar transpiring power.** Bot. Gaz. 68: 262–286. *13 fig.* 1919.

854. Dosdall, Louise. **Water requirement and adaptation in Equisetum.** Plant World 22: 1–13, 29–34. *5 fig.* 1919.—See Bot. Absts. 4, Entry 217.

855. Flood, Margaret G. **Exudation of water by Colocasia antiquorum.** Sci. Proc. Roy. Dublin Soc. 15: 505–512. *2 pl.* 1919.—An inquiry into the question of whether the water exuded from the leaf-tips of *Colocasia* was conduction water, or whether it was secreted from a special gland led to the following considerations: 1. It had been related to transpiration and called a nocturnal "liquid transpiration" supplanting the diurnal vaporous one. 2. The drops were sometimes seen to be ejected for short distances, coming through small pores. 3. It had been stated that the water was secreted by a hydathode and that the secretion was simple filtration. 4. Modern observations had shown that the freezing point of the exudate differed little from distilled water, and that its electrical conductivity was less than that of tap-water.—A colloid (India ink mixed with gelatine) was successfully passed through the end pore and up into the canals, after some preliminary experimentation. When the leaf-tip was attached to a water reservoir, after severance from the leaf, drops of water continued to be exuded. This amounted to 6 cc. in 20 hours. These experiments (and the last-mentioned repeated, substituting a 0.3 per cent starch solution) prove that there was no continuous membrane between the depression and the water channels. Anaesthetizing the tip did not slow up the rate of dropping, showing that the water must be urged forward from below in the plant and not exuded by the action of the tip alone. Cutting the leaf-blade anywhere results in copious exudation from the veins at every cut. The same occurs when the petioles are cut. The conclusions from these observations and experiments, made when the soil about the plant was damp and the air saturated, is that no gland or epithem functioning in secretion is present in the leaf-tip. The phenomenon must depend upon the normal transfer of water through the plant. [See also Bot. Absts. 4, Entry 1406.]—*A. E. Waller.*

856. Shull, C. A. **Curing timber.** [Rev. of: Stone, Herbert. **The ascent of the sap and the drying of timber.** Quart. Jour. Forest. 12: 261–266. 1918.] Bot. Gaz. 68: 310. 1919. —See Bot. Absts. 5, Entry 233.

MINERAL NUTRIENTS

857. Espino, Raf. B. **Methods in nutrition experiments.** [Rev. of: Schreiner, Oswald, and J. J. Skinner. **The triangle system for fertilizer experiments.** Jour. Amer. Soc. Agron. 10: 225–246. 1918.] Plant World 22: 53–54. 1919.

858. Girard, Pierre. **Schème physique pour servir a l'étude de la nutrition minérale de la cellule.** [**Physical scheme to serve for a study of the mineral nutrition of the cell.**] Compt. Rend. Acad. Sci. Paris 168: 1335–1338. 1919.—The object of this work is to demonstrate *in vitro* the principles which bear on the differential permeability of the plasma membrane. By the use of barium chloride combined with various other chemicals, using a gold beater's skin, the author finds that differential permeability can be demonstrated. The phenomenon is explained on an electrical basis and is attributed to the ionization of the substances in solution.—*V. H. Young.*

859. Hoagland, D. R. **Relation of nutrient solution to composition and reaction of cell sap of barley.** Bot. Gaz. 68: 297–304. 1919.—The osmotic pressures in the sand and water cultures of barley are reflected in the cell sap of the tops and roots. The electrical conductivity of the nutrient solution has a marked influence on the conductivity of the sap, both in tops and in roots; the conductivity of the sap is from 4 to 50 times greater than that of the nutrient solution. The sap from the tops of plants in all cultures had almost the same P_H value, approximately 6.0. Plants were grown in 6 different soils and in every case the sap concentration was much greater than that of the soil solution. Emphasis is placed on the dynamic nature of the relation between the soil solution and the plant.—*H. C. Cowles.*

860. Le Clerc, J. A., and J. F. Breazeale. **Effect of lime upon the sodium-chlorid tolerance of wheat seedlings.** Jour. Agric. Res. 18: 347–356. *Pl. 38–47.* 1920.—The work was done with reference to "alkali" soils. Soil, sand, and solution cultures were used, since inert material might affect the toxic limits of dissolved salts. It is found that plants in soil and sand show higher tolerance to alkali salts than solution cultures. This is not due entirely to the physical effect of the presence of solid particles of different degrees of fineness, but also to certain soluble substances which are present in very small quantities.—Very small amounts of calcium oxide and calcium sulfate overcome the toxic effects of sodium chlorid and sodium sulfate. Magnesium sulfate and barium chlorid are slightly antagonistic to sodium chlorid, while potassium chlorid, sodium nitrate, sodium phosphate, ferric chlorid, and alum had no effect on its toxicity.—The presence of lime did not prevent the entrance of sodium chlorid or sodium sulfate into the plant. The antagonistic effect of lime seems to be due to some other cause than its effect on permeability.—*D. Reddick.*

861. Livingston, B. E. [Under Notes and Comment, no special title.] Plant World 22: 26–27. 1919.—A discussion of work by F. W. Gericke on a preliminary test of the influence of temperature upon the physiological balance of the nutrient solution as related to germination in wheat. Stress is laid on the need of quantitative definition of all effective conditions in experimental work.—*Chas. A. Shull.*

862. Shive, John W. **Relation of moisture in solid substrata to physiological salt balance for plants and to the relative plant-producing value of various salt proportions.** Jour. Agric. Res. 18: 357–378. 1920.—Three different degrees of moisture were maintained in sand cultures, 40, 60 and 80 per cent of the water-retaining capacity of the sand. Tests were made with 36 different sets of salt proportions of the three salts, monopotassium phosphate, calcium nitrate, and magnesium sulfate in solutions with each of the moisture percentages noted. The solutions, all having an initial total osmotic concentration of 1.75 atmospheres, were

supplied to the sand cultures in such quantities as to produce the different standards of moisture. Culture solutions were renewed every third day. Daily water loss was restored daily. Wheat was grown for 28 days.—The physiological balance of the nutrient solutions producing the best yields of tops and roots was not altered by variations in the moisture content of the sand. A slight shifting of the balance, as affecting growth, is indicated for the growth of 9 high-yielding cultures, as a whole, out of the series of 36, with each increase in the moisture content of the cultures, from a position in the series characterized by lower partial concentration of potassium phosphate to one of higher partial concentration of this salt, and correspondingly lower ones of calcium nitrate and magnesium sulfate.—Good physiological balance and optimum total concentration of a nutrient solution for plants is not alone sufficient to produce the best growth of which the solution is capable when it is diffused as a film on the particles of a solid substratum. An optimum degree of moisture is essential to impart to the soil (sand) solution its maximum physiological value. The plant-producing value of any fertilizer treatment is thus determined largely by the moisture conditions of the substratum.—The lowest percentage of moisture employed corresponds with low yields of tops and roots, lowest transpiration rates, and with lowest water requirement ratios. The highest moisture content is associated with low yield of tops and roots, with high transpiration rates, and with the highest water requirement ratios. The medium degree of moisture is correlated with the highest yields of tops and of roots, high transpiration rates, and medium water requirement ratios.—*D. Reddick.*

863. Steinkoenig, L. A. **Relation of fluorine in soils, plants and animals.** Jour. Indust. Eng. Chem. 11: 463–465. 1919.—After reviewing the literature the author reports fluorine determinations of 9 soils, using Merwin's determination with modifications, which is given in detail. Fluorine occurs in amounts averaging 0.03 per cent. Three soils contained but 0.01 per cent, and in one case it was not found. Soils carrying stones made up of mica schist contain relatively higher amounts,—Hagerstown loam 0.11–0.15 per cent, York silt loam 0.05 per cent. Fluorine is in the soil in such minerals as biotite, tourmaline, muscovite, apatite, fluorite and phlogopite. Plants absorb fluorine and thus it is available for animals, which latter may also obtain it from spring water. [See also Bot. Absts. 4, Entry 1636.]—*C. R. Hursh.*

864. Stiles, Walter, and Franklin Kidd. **The influence of external concentration on the position of the equilibrium attained in the intake of salts by plant cells.** Proc. Roy. Soc. London 90 B: 448–470. *Tables 1–13, 6 fig.* 1919.—Salt intake by discs of carrot and potato tissue was measured by changes in electrical conductivity of the external solution. The initial concentrations used varied from N/10 to N/5000. Carrot is considered more suitable than potato because of less exosmosis into distilled water. Toxic salts, e. g., copper sulfate, produce greater exosmosis in both distilled water and in solutions.—The ratio between final internal and final external concentration is called the absorption ratio. The initial rate of absorption is roughly proportional to the concentration of the external solution; but the final absorption ratio, at equilibrium, diminishes as concentration of the external solution increases. The equation of the absorption ratio is given as $y = KC^m$, where y is the final interval and C the final external concentration. This happens to be the adsorption equation, but no basis was found for postulating the mechanism of salt intake.—*Paul B. Sears.*

PHOTOSYNTHESIS

865. Pulling, H. E. **Physiological problems of photosynthesis.** [Rev. of: Henrici, Marguerite. **Chlorophyllgehalt und Kohlensäure-Assimilation bei Alpen- und Ebenenpflanzen.** Verhandl. Naturforsch. Ges. Basel 30: 43–136. 1918.] Plant World 22: 123–126. 1919.

METABOLISM (GENERAL)

866. Armstrong, E. Frankland. **The simple carbohydrates and the glucosides.** 3rd ed. *IX+239 p.* Monographs on Biochemistry. Longmans, Green & Co.: London, 1919.—No new chapters have been added since the second edition of this work, but much new material

has been incorporated. Important among the special advances necessitating the revision are (1) the discovery of a third isomeric form of glucose differing from the pentaphane ring forms in structure serving to throw new light on the constitution of sucrose, and (2) definite data for the characterization of carbohydrates as regards the relationship of optical rotatory power to structure.—*B. M. Duggar.*

867. Ayers, S. Henry, and Philip Rupp. **Simultaneous acid and alkaline bacterial fermentations from dextrose and the salts of organic acids respectively.** Jour. Infect. Diseases 23: 188-216. 1918.—The quantitative fermentation of dextrose by *Bacillus coli* and *B. aerogenes* into formic, acetic, lactic, and succinic acids is shown, with the accompanying changes in H-ion concentration. The reversion of reaction is explained as the formation of carbonates or bicarbonates from the formic acid salts, as the changes in P_H agree quite closely with the disappearance of the formic acid. Simultaneous fermentations of acid from dextrose and of alkali from citrate are shown with the alkali-forming group of bacteria.—*W. H. Chambers.*

868. Behrend, Robert, and George Heyer. **Über die Oxydation der Muconsäure. Synthese der Schleimsäure. [Concerning the oxidation of muconic acid. Synthesis of mucic acid.]** Ann. Chem. **418**: 294-316. 1919.—As an average of 12 tests under controlled conditions the action of potassium permanganate upon muconic acid yielded, per 100 molecules of the acid, 21 molecules of oxalic acid, 11 molecules of tartaric acid, a trace of mucic acid, and unidentified products. Oxidation by sodium chlorate and osmic acid yielded, per 100 molecules, 32 molecules of mucic acid and small amounts of other products.—*W. E. Tottingham.*

869. Besson, A., A. Ranque, and C. Senez. **Action biochimique des microbes sur les sucres et les alcools. [Biochemical action of bacteria on sugars and alcohols.]** Compt. Rend. Soc. Biol. **81**: 930-933. 1918.—Fermentation of the common sugars and alcohols by bacteria of the colon-typhoid-dysentery group and other organisms is tabulated, with emphasis on the constancy of the property of gas production.—*W. H. Chambers.*

870. Besson, A., A. Ranque, and C. Senez. **Sur la vie du coli-bacille en milieu liquide glucosé. [On the life of B. coli in liquid glucose-containing media.]** Compt. Rend. Soc. Biol. **82**: 76-78. 1919.—The time relation between growth and fermentation is shown. Gas and acid production commenced when multiplication ceased. More than one-half of the acid was produced in the first hour.—*W. H. Chambers.*

871. Besson, A., A. Ranque, and C. Senez. **Sur la vie des microbes dans les milieux liquides sucrés. [On the life of bacteria in liquid sugar-containing media.]** Compt. Rend. Soc. Biol. **82**: 107-109. 1919.—The action of different bacteria on glucose is shown to be similar to that of *Bacillus coli*, the cultures becoming sterile in 6 days. The acid and gas production of *B. coli* from different sugars and alcohols is reported.—*W. H. Chambers.*

872. Besson, A., A. Ranque, and C. Senez. **Sur la vie du coli-bacille en milieu liquide glucosé. Importance des doses de glucose. [On the life of Bacillus coli in liquid glucose-containing media. Importance of amounts of glucose.]** Compt. Rend. Soc. Biol. **82**: 164-166. 1919.—The relation of amounts of glucose to titratable acid, death of the culture, disappearance of the sugar, and time of gas fermentation is reported. They found reversion of reaction with 0.2 per cent or less of glucose, and death of the culture in 6 days with 0.4 per cent or more. —*W. H. Chambers.*

873. Bourquelot, E., and Bridel. **Application de la methode biochemique a l'étude de plusieurs d'Orchidées indigènes. Découverte d'un glucoside nouveau, la "loroglossine." [Discovery of a new glucoside, "loroglossine," in one of the indigenous orchids.]** Compt. Rend. Acad. Sci. Paris **168**: 701-703. 1919.—Preparation and properties of the glucoside "loroglossine" from *Loroglossum hircinum* Rich. are described.—*F. B. Wann.*

874. Bunker, J. W. M. **The determination of hydrogen ion concentration.** Jour. Biol. Chem. 41: 11–14. 1920.—An electrode and a vessel are described which have been in use a long time, meeting the requirements of quick, accurate determinations in large numbers.—*G. B. Rigg.*

875. Church, A. H. **The ionic phase of the sea.** New Phytol. 18: 239–247. 1919.—This is a discussion of sea water as the "primary source of 'life'" from the standpoint of the modern physico-chemist. The ionization of the salt content of sea water is discussed, particularly in relation to the ions of carbonic acid. Far-reaching analogies are pointed out between living substance and sea water; the latter is even considered to be "the primordial material of which protoplasmic units are but individualized particles or segregated centres of actions, still more complex, but of the same category."—*I. F. Lewis.*

876. Clevenger, Clinton B. **Hydrogen-ion concentration of plant juices. I. The accurate determination of the hydrogen-ion concentration of plant juices by means of the hydrogen electrode.** Soil Sci. 8: 217–226. 1919.—The apparatus is essentially that described by Clark and Lubs with modifications to prevent foaming of the plant juice and to simplify both the shaking apparatus and the temperature. To prevent contact between the electrodes and plant juice during saturation with hydrogen the juice is placed in dropping funnels attached to the electrode vessels. To reduce contact potential, contact between the plant juice and the saturated potassium chloride solution is made by means of a scratch around the cock connecting the two. Duplicate measurements agree within 0.1 millivolt.—*William J. Robbins.*

877. Clevenger, Clinton B. **Hydrogen-ion concentration of plant juices. II. Factors affecting the acidity or hydrogen-ion concentration of plant juices.** Soil Sci. 8: 227–242. 1919. —Determinations of acidity should be made as quickly as possible after cutting the plant and extracting the juice, as the acidity of plant juice may decrease or increase on standing. The roots of cow pea are generally more acid than the leaves and the leaves more acid than the stems. The acidity in the roots of cow pea during a 24 hour period is rather constant, being higher during the day. In the leaves and stems the acidity drops during the afternoon, rising during the night and reaching a maximum in the morning. The acidity of the roots of plants appears to be correlated with the reaction of the soil, but the acidity of the tops of the plants studied was greater on limed than on unlimed soil.—*William J. Robbins.*

878. Colin, H. **Utilization du glucose et du levulose par les plantes supérieures. [Utilization of glucose and levulose by higher plants.]** Compt. Rend. Acad. Sci. Paris 168: 697–699. 1919.—The proportion of glucose to levulose in green leaves of beet is often less than 1, but increases down the midrib and in the petiole. Etiolated leaves of beet, artichoke, and chicory showed a larger proportion of dextrose than of levulose, whereas in the storage organs of these plants the reverse is true. It is assumed that these two sugars must either be transported at unequal rates or that they are utilized in unequal amounts. The author concludes that it is more probable that the glucose is oxidized in the cell in preference to levulose, the latter playing an essential rôle in tissue formation. Thus respiration is less intense in the petiole than in the blade, and less in etiolated leaves than in green leaves.—*F. B. Wann.*

879. Cushny, Arthur R. **The properties of optical isomers from the biological side.** Pharm. Jour. 103: 483. 1919.—The living plant discriminates between laevo and dextrorotatory bodies because it is itself optically active, but no optically active substances have as yet been synthetically produced by man. Because of this phenomenon of discrimination by the living plant and the fact that an optically active alkaloid, such as cinchonine, can be used to separate a mixture of laevo and dextro tartrates, and the further fact that vegetable and animal organisms that act upon asymmetric bodies generally destroy the substance that occurs in nature but will not destroy the non-natural isomer, the author declares that "until life appeared no optically active body existed, and without life and its products there would be none today." Further, this optical activity is the most persistent evidence of life, since an

optically active alkaloid or acid, centuries after the plant that produced it is destroyed, will still retain its activity, and the occurrence of any optically active substance, such as petroleum, proves that it must have been derived from living tissues.—Experiments with hyoscyamine, a laevorotatory substance and its isomer atropine, optically inactive, consisting of equal parts of laevo and dextro hyoscyamine, demonstrated that *l*-hyoscyamine had the same physiological effect on peripheral nerve-endings as twice the quantity of *dl*-hyoscyamine (atropine). Again, a comparison, by the effect on blood pressure, indicates that natural adrenaline (*l*-adrenaline) was twice as powerful as synthetic adrenaline (*dl*-adrenaline) and that *d*-adrenaline (obtained from *dl*-adrenaline) was without activity.—*E. N. Gathercoal.*

880. Haas, A. R. C. **The electrometric titration of plant juices.** Soil Sci. 7: 487–491. *1 fig.* 1919.—An electrometric apparatus is described for determining the buffer action, acid and alkali reserve, and the total and actual acidities of plant juices. Rhubarb juice has a greater actual acidity and greater buffer action than that of soy bean tops.—*William J. Robbins.*

881. Haynes, Dorothy, and Hilda Mary Judd. **The effect of methods of extraction on the composition of expressed apple juice, and a determination of the sampling error of such juices.** Biochem. Jour. 13: 272–277. 1919.—See Bot. Absts. 5, Entry 541.

882. Jones, Harry. **Some factors influencing the final hydrogen-ion concentration in bacterial cultures with special references to streptococci.** Jour. Infect. Diseases 26: 160–164. 1920.—The composition of the medium, the initial reaction and any other conditions which favor or hinder abundant growth of a given organism should be considered in order to obtain accurate information regarding its final hydrogen ion concentration.—*Selman A. Waksman.*

883. Knudson, L., and E. W. Lindstrom. **Influence of sugars on the growth of albino plants.** Amer. Jour. Bot. 6: 401–405. 1919.—Albino corn seedlings grown both on agar and in water culture were supplied with sugar (sucrose and glucose). On agar, they all lost weight, but those supplied with sugar lost considerably less than controls which had no sugar. Results with plants grown in the dark were essentially the same as with those grown in the light. In water culture the albino seedlings made an appreciable gain when provided with sugar, and lived much longer than the controls, but ultimately died. The better growth in water culture is explained as probably due to higher concentration of sugar and higher temperatures at which the plants were grown. Roots of plants supplied with sugar often continued to live for some time after the shoots died. The substitution of asparagin for nitrates in the culture solutions caused practically no difference in growth. The authors explain the failure of albino plants to thrive when sugar is supplied as due to the inability of the plant to absorb sugar rapidly, and to the relatively slow rate of its conduction.—*E. W. Sinnott.*

884. Kremers, R. E., and J. A. Hall. **On the identification of citric acid in the tomato.** Jour. Biol. Chem. 41: 15–17. 1920.—The presence of citric acid in the tomato has been shown by means of its triphenacyl ester.—*G. B. Rigg.*

885. Meinicke, E. **Die Lipoidbindungsreaktion.** [The lipoid-fixation reaction.] Zeitschr. Immunitätsforsch. u. Exp. Therapie 27: 350–363. 1918; 28: 280–326. 1919.—Antibodies are probably globulins, or at least inseparable from them by any known method. In the reaction between serum and the extract, the colloids of the latter force the NaCl equivalent of the serum globulins from solution, probably by removing NaCl. This reaction is stronger in positive sera. An immunized organism reacts more quickly and more intensively following a recent addition of antigen than the control. The possibility of a specific, more intensive reaction resides not only in the cell but also in the serum itself. The intensity of the reaction seems due to the fact that the NaCl equivalent of the most labile substance in the system is forced out of solution by the most stable substance present. The various forms of immunity reactions are only the expression of the different reagents acting in various combinations in such a system, hence it is possible to combine various forms of reactions. In the so-called

inactivation of sera the reaction possibilities of the serum globulins are changed in two ways: it is separated from the NaCl equivalent with more difficulty; and the salt removal acts on the protein molecule itself, as is seen in the closer binding of the salt on warming the sera.—*C. W. Dodge.*

886. Morishima, Kanichiro. **Phenol red-china blue as indicator in fermentation tests of bacterial cultures.** Jour. Infect. Diseases 26: 43–44. 1920.—An indicator is proposed consisting of phenol red and decolorized china blue for fermentation tests of bacterial cultures. The production of acid causes first a bright green color changing to a deep blue, when too much acid is formed. The production of alkali is indicated by a pink color.—*Selman A. Waksman.*

887. Posternak, M. S. **Sur la constitution du principe phospho-organique de réserve des plantes vertes. [On the constitution of the phospho-organic principle in the reserve of green plants.]** Compt. Rend. Acad. Sci. Paris 169: 37–42. 1919.—An attempt is made to determine whether or not the phospho-organic reserve of plants is or is not a hexa-phosphate of inosite. The author plans experiments on the synthesis of this compound to determine whether or not 3 molecules of water are held as water of crystallization or are an essential constituent of the molecule.—*V. H. Young.*

888. Sherman, H. C. **Protein requirement of maintenance in man and the nutritive efficiency of bread protein.** Jour. Biol. Chem. 41: 97–109. 1920.—The proteins of wheat, corn, and oats appear to be about equally efficient in human nutrition, and need only be supplemented by small amounts of milk in order to be fully as efficient as the proteins of ordinary mixed diets.—*G. B. Rigg.*

889. Steenbock, H., and P. W. Boutwell. **Fat-soluble vitamine. III. The comparative value of white and yellow maizes.** Jour. Biol. Chem. 41: 81–96. *pl. 2.* 1920.—The occurrence of yellow pigment and the growth-promoting property attributed to the presence of the fat-soluble vitamine seem to be intimately associated in the maize kernel.—*G. B. Rigg.*

890. Willaman, J. J. **The function of vitamines in the metabolism of Sclerotinia cinerea.** Jour. Amer. Chem. Soc. 42: 549–585. 1920.—The basal medium for these tests was Currie's mineral solution plus asparagin (as a source of nitrogen), plus sucrose. Growth was completed in 10 days; sporulation began the third or fourth day. "The amount of vegetation is not proportional to the concentration of the juice, the fungus being unable to utilize the greater amounts of nutrients in the same degree that it does the lesser." "Reproduction is more abundant on the peach juice than on the others," i.e., prune juice and apricot juice. "The higher concentrations are not necessarily the optimum for reproduction." The fungus can make excellent growth on either asparagine or glycine, providing the growth-promoting material of the 2 cc. of prune juice is also present. It was also shown that diammonium hydrogen phosphate alone would not produce growth; that wort alone will support growth fairly well; and the two together make an excellent medium for growth. The vitamine preparation served to make the ammonia nitrogen more useful to the fungus. Vitamine B was obtained by means of Lloyd's alkaloidal reagent. It adsorbs the vitamine from an acid solution and releases it in an alkaline one Pectin interferes with the adsorption of the vitamine. 65 per cent alcohol, to which a few drops of 1 per cent H_2SO_4 was added, was used for the preparation of the vitamine. When pectin has been removed, the vitamine will pass through a colloidin sac.—The author presents evidence that two vitamines are concerned in the life cycle of *Sclerotinia cinerea.* One enables vegetative growth to take place and is more readily adsorbed by Fuller's earth on an alcohol medium; the other enables the fungus to sporulate well and is more readily adsorbed in an aqueous medium. Evidence given shows that the shuffling of the nitrogen and sugar constituents of the medium will not of itself determine the occurrence or non-occurrence of reproduction in *Sclerotinia.* Both vitamines must be present if reproduction is to occur. Other evidence presented would indicate the presence of but one vitamine.—*J. M. Brennan.*

891. Willaman, J. J. **Colorimeter and indicator method.** [Rev. of: Duggar, B. M., and C. W. Dodge. **The use of the colorimeter in the indicator method of H-ion determination with biological fluids.** Ann. Missouri Bot. Gard. 6: 61-70. 1919. (See Bot. Absts. 4, Entry 1449.)] Bot. Gaz. 68: 232. 1919.

892. Zellner, Julius. **Zur Chemie der höheren Pilze. XIII. Über Scleroderma vulgare Fr. und Polysaccum crassipes DC. [Chemistry of the higher fungi.]** Akad. Wiss. Wien (Monatshefte für Chemie) 39: 603-615. 1918.—Following the general plan of his earlier studies the author reports the presence of mannit, cholin, and viscosin among the substances investigated in *Scleroderma*. In *Polysaccum* it is noteworthy that no mannit occurs. With this species special attention was devoted to a tannoid pigment. In neither fungus could the author demonstrate either invertase, maltase, or diastase.—*B. M. Duggar.*

893. Zoller, H. F. **Quantitative estimation of indole in biological media.** Jour. Biol. Chem. 41: 25-36. 1920.—Indole is an important product of the metabolism of certain microorganisms. A simple, rapid, reliable method for its determination has been evolved, requiring only the reagents and apparatus common to most laboratories.—*G. B. Rigg.*

894. Zollér, H. F. **Influence of hydrogen ion concentration upon the volatility of indole from aqueous solution.** Jour. Biol. Chem. 41: 37-44. 1920.—The range of most rapid volatilization of indole from the aqueous solutions studied is from P_H 8.0 to 10.5. Results suggest that the practice of steam distillation can be supplanted by direct distillation when the reaction of the solution is taken into account.—*G. B. Rigg.*

METABOLISM (NITROGEN RELATIONS)

895. Anonymous. [Rev. of: Lakon, Georg. **Der Eiweissgehalt panachierter Blätter, geprüft mittels des makroskopischen Verfahrens von Molisch. (The protein content of mottled leaves tested by the macroscopical method of Molisch.)** Biochem. Zeitschr. 78: 145-154. 1917.] Biedermann's Zentralbl. Agrikulturchem. 47: 251. 1918.—White-variegated leaves of many species of plants, especially, *Acer negundo*, furnish suitable material for the macroscopical demonstration of the protein reaction according to Molisch. The protein-rich green places in the leaves give a very strong color while the protein-poor albescent places are only slightly colored. Protein-rich and protein-poor places are directly related to the presence and absence of chromatophores, in the leaf. In the case of yellow 'panachierten,' chromatophores are found in the leaf tissues and so one finds them rich in protein. The investigation supports the views of Molisch in that the principal masses of proteins of the leaves occur in the chromatophores. When submitted to the xanthoproteic reaction leaves which contain anthocyanin first take on a red color when placed in nitric acid solution, because, in spite of the decolorization, they contain anthocyanin in the colorless isomeric form.—*F. M. Schertz.*

896. Berman, N., and L. F. Rettger. **Bacterial nutrition: further studies on the utilization of protein and non-protein nitrogen.** Jour. Bact. 3: 367-388. 1918.—The utilisation of different brands of commercial peptones by proteolytic and non-proteolytic bacteria is probably related to the simpler nitrogen-containing substances. The liquefaction of gelatin was not a necessary indication of the proteolytic property of an organism. The availability of casein for bacterial use is shown before and after digestion with trypsin.—*W. H. Chambers.*

897. Bonazzi, Augusto. **On nitrification. III. The isolation and description of the nitrite ferment.** Bot. Gaz. 68: 194-207. *pl. 14.* 1919.—This paper presents the results of the study of an organism, capable of forming nitrates from ammonia, isolated in a pure state from Wooster [Ohio] soil after many unsuccessful attempts. A review is given of the pertinent literature, and the methods are described by which the organism was isolated and its cultural characteristics determined. The cultural solution used throughout was the one recommended by Omelianski, of the following composition: H_2O, 1000 cc.; $FeSO_4$, 0.4 gram; $MgSO_4$, 0.5 gram; K_2HPO_4, 1 gram; NaCl, 2 grams; and $(NH_4)_2SO_4$, 2 grams. Solid media

used were gypsum block, magnesium carbonate block, magnesium carbonate and ammonium-magnesium-phosphate block, ammonium sulphate washed agar, and silicic acid jelly. The best results were obtained with Winogradsky's silicic acid jelly. Incubation of all cultures was made at 28 to 30°C. At this temperature cultures were obtained which nitrified as much as 8.04 mgm. of ammoniacal nitrogen in 26 days of incubation. The organism is not motile. Its thermal death point was found to lie between 50° and 55°C., when the vitality of the organism, after heating 5.5 minutes at the required temperature, was tested at rest in Omeliansky's solution containing basic magnesium carbonate. The organism occurs in a large form ±1.25μ in diameter and in a small coccus form which the author names β. He concludes that the megalococcus isolated by these methods is very similar to that described by Winogradsky from South American soils and should be classed as a species of the genus *Nitrosococcus*.—*D. H. Rose.*

898. Brackett, R. N., and H. F. Haskins. **Report on nitrogen.** Jour. Assoc. Official Agric. Chem. 3: 207-217. 1919.—See Bot. Absts. 5, Entry 1003.

899. Conn, H. J., and J. W. Bright. **Ammonification of manure in soil.** Jour. Agric. Res. 16: 313-350. 1919.—A foreword by Conn refers largely to previous studies of spore-formers and non-spore-formers. Under the title "What soil organisms take part in ammonification of manure?" Bright shows the predominance of *Pseudomonas fluorescens* and *Pseudomonas caudatus* in manured soil and gives the results of an investigation of their function in Dunkirk silt clay loam.—Fresh horse or cow manure was added to the soil in the ratio of 1:20. In addition to plate counts direct microscopic examinations were made. Not only was the unsterilised material used but also the sterilised to which was added the pure cultures. The latter was used both separately and in combination.—In unsterilized soil which was kept in pots the data show a rapid increase in non-spore-formers. After 7 days they were never less than 92.5 per cent, while in certain cases they were as high as 97 per cent. The results from experiments conducted in flasks are not so striking, yet the same relation holds. Isolations showed only 2.8 per cent which form spores.—The growth of *Ps. fluorescens* and *Ps. caudatus* in sterilised manured soil compared with that of a spore-former, *Bacillus cereus*, shows that the spore-former had increased in 7 days only 8.3 times while the two former organisms had increased respectively 110 and 132 times over the original inoculation. When these three organisms were in association *Ps. fluorescens* and *Ps. caudatus* rapidly gained the ascendancy over *B. cereus*, the latter soon sporulating and remaining in this condition.—A test of the ammonia production and cell count in soil of the above three organisms in pure culture shows *B. cereus* to be the most powerful ammonifier. The two non-spore-forming organisms gave many times more cells per gram of manured soil. However, when the three organisms were grown in association there was no increase in total ammonia formed and in cell counts the two non-spore-formers had gained the ascendancy. *B. cereus* was not found although 2.3 million per gram were present at the beginning.—The taxonomic study by Conn includes a description of *Ps. fluorescens*, *Ps. aeruginosa*, *Bacterium termo* and *Ps. putida* with a brief summary of characters of typical *Ps. fluorescens* and *Ps. caudatus*.—*J. K. Wilson.*

900. Dakin, H. D. **On amino acids.** Biochem. Jour. 12: 290-317. 1918.—Some new methods are presented for the extraction of amino acids by partially miscible solvents. A new amino acid, hydroxyglutanic acid, and a new peptide from caseinogen, isoleucylvaline, have been isolated and studied.—*W. H. Chambers.*

901. Frear, William, Walter Thomas, and H. D. Edmiston. **Notes on the use of potassium permanganate in determining nitrogen by the Kjeldahl method.** Jour. Assoc. Official Agric. Chem. 3: 220-224. 1919.—See Bot. Absts. 5, Entry 1005.

902. Hendrick, Ellwood. **Micro-organisms in plant chemistry and nitrogen fixation. An account of the development and application of micro-organisms useful to plant growth—fixation of nitrogen in the soil.** Chem. and Mettallurg. Eng. 19: 574-576. *6 fig.* 1918.—This is

a popular account of the utilization of a muck swamp, and among the products described is that designated "inoculant"—a material in which 28 strains of legume bacteria and 5 strains of *Azotobacter* are grown.—*G. M. Armstrong.*

903. Hirsch, Paul. **Die Einwirkung von Mikroorganismen auf die Eiweisskörper.** [The action of micro-organisms on proteins.] *IX+355 p., 7 fig.* Die Biochemie in Einzeldarstellungen IV [Edited by A. Kanitz]. Gebrüder Borntraeger: Berlin, 1918.—This number in the above biochemical series is essentially an amino acid reference book and follows naturally No. III, by M. Siegfried, on partial protein hydrolysis ("Über partielle Eiweisshydrolyse"). This monograph takes up the secondary cleavages of the proteins, the decomposition of the amino acids. The first part discusses the chemistry of the amino acids and of their proteolysis by bacteria and fungi, with one section on ergot. Part 2 gives chemical and biological methods for isolating and determining the amino acid cleavage products. Part 3 gives the physical and chemical properties of the products and their derivatives, and part 4, the synthesis of some of them.—*W. H. Chambers.*

904. Holm, George E. **A modification of the apparatus for the determination of arginine nitrogen by Van Slyke's method.** Jour. Amer. Chem. Soc. 42: 611-612. 1920.

905. Levene, P. A. **The structure of yeast nucleic acid. V. Ammonia hydrolysis.** Jour. Biol. Chem. 41: 19-23. 1920.—On mild hydrolysis with 5-per cent ammonia at a temperature of 100°C. yeast nucleic acid is broken up into four nucleotides. Three have already been reported. A fourth, crystalline cytidinphosphoric acid, has now been isolated.—*G. B. Rigg.*

906. Long, Esmond R. **A study in fundamentals of the nutrition of the tubercle bacillus: the utilization of some amino acids and ammonium salts.** Amer. Rev. of Tuberculosis 3: 86-108. *2 fig.* 1919.—The experiments performed are concerned primarily with the growth of human tubercle bacilli on media of known chemical composition. The hydrolysis of proteoses and peptones, as also the deaminization of some of the constituent amino acids, is reported. Good growth was afforded by glycerol media with urethane, glycocoll, and alanine as sources of nitrogen; likewise ammonia, methyl amine, and ethyl amine, as also the acid amids, were utilized. Ammonium salts of the dibasic acids oxalic, malonic, succinic, malic, and tartaric afforded excellent growth, but the ammonium salts of fatty, ketonic, and hydroxy acids did not permit growth. Between P_H 6.4 and P_H 7.8 the reaction of a glycerol peptone culture medium is unimportant in the growth of this organism. Regarding the course of catabolism, it is suggested that "the amino acids (that is, those studied—glycocoll and alanine) break up into ammonia and alcohols, perhaps with amines as intermediate stages, that hydroxy malonic acid (tartaric acid) is formed in the medium through the oxidation of glycerol, and that ammonium malonate and malonic ester, or closely allied compounds, are of great importance in the synthesis of the bacillus's organic substance."—*B. M. Duggar.*

907. Phelps, I. K., and H. W. Daudt. **Investigations of the Kjeldahl method for the determination of nitrogen.** Jour. Assoc. Official Agric. Chem. 3: 218-220. 1919.—See Bot. Absts. 5, Entry 1006.

908. Trowbridge, P. F. **Symposium on the determination of nitrogen in fertilizers.** Jour. Assoc. Official Agric. Chem. 3: 217-218. 1919.—See Bot. Absts. 5, Entry 1007.

METABOLISM (ENZYMES, FERMENTATION)

909. Anonymous. **Glycerin manufacture by the fermentation of sugar.** Sci. Amer. Supplem. 88: 315. 1919.—[From *Engineering*, Sept. 5, 1919.]—A method employing yeasts.—*Chas. H. Otis.*

910. Anonymous. [Rev. of: Biedermann, W. **Fermentstudien. 1. Mitteilung. Das Speichelferment. (Salivary ferments.)** Fermentforschung 1: 385-436. 1916.] Biedermann's Zentralbl. Agrikulturchem. 47: 279-280. 1918.—The reviewer credits the author with finding

that the time required for the hydrolysis of starch to dextrine under the action of salivary ferments is conversely proportional to the quantity of ferment. Saccharification is in no way parallel to dextrin formation but remains behind if the quantity of ferment is decreased. It is believed that the diastase enzyme consists of two components; an amylase which splits the starch molecule to dextrine, and a dextrinase which can attack only the dextrin group.—*F. M. Schertz.*

911. Anonymous. [Rev. of: Biedermann, W. **Fermentstudien. II. Mitteilung. Die Autolyse der Stärke.** (The autolysis of starch.) Wochenschr. Brauerei 34: 183–186. 1917.] Biedermann's Zentralbl. Agrikulturchem. 47: 280–281. 1918.—The reviewer indicates that previous work of Biedermann shows the rapid hydrolysis of boiled starch solution by saliva ash, which effect is due to a ferment liberated from the starch. It is now shown, according to the reviewer, that a similar hydrolysis occurs without adding any ash, if the starch solution is made at 70–90°C. Boiled solutions are hydrolyzed after a longer period, while extracts prepared by grinding starch in water hydrolyze rapidly. The diastatic power of the latter extract is similar to that of a very dilute solution of saliva, and completely transforms starch into sugar. Of the salivary salts calcium chloride promotes maximum diastatic action. The action of salivary ash in promoting the decomposition of starch solutions which have been subjected to boiling suggests that this mixture of salts promotes the formation of amylase from starch.—*F. M. Schertz.*

912. Anonymous. [Rev. of: Jacoby, Martin. **Über Fermentbildung.** (Formation of enzymes.) Biochem. Zeitschr. 79: 35–50. 1917.] Biedermann's Zentralbl. Agrikulturchem. 47: 281–282. 1918.—Traces of grape sugar were found to greatly increase the activity of enzymes on urea. Search was then made to see what building stones the enzymes used. According to the reviewer there were then tested a number of materials in relation to their action on the decomposition of urea. The formation of urease was greatly stimulated by d-glucose, d-galactose, glycerol, dl-glyceric aldehyde, dihydroxy acetone, pyroracemic acid, and lactic acid. A stimulatory action of less intensity was shown by d-fructose, d- and l-arabinose. Maltose, ethylene glycol, and propylene glycol produced little action, while d-mannose, d-sorbose, rhamnose, heptose, the polysaccharides, glucosides, and sugar alcohols had no action.—*F. M. Schertz.*

913. Anonymous. [Rev. of: Lombbroso, Ugo. **Über die Reversibilität der Enzymwirkungen. 1. Mitteilung. Spaltung und Synthesis der Fette durch eine Lipase.** (Cleavage and synthesis of fats by the action of one and the same lipase.) Arch. Pharmacol. Sperim. 14: 429–459. 1912.] Biedermann's Zentralbl. Agrikulturchem. 47: 287. 1918.—According to the reviewer it is shown that fat hydrolysis begins immediately at 37°C. and can proceed to 80 per cent of completion. Synthesis does not begin till after 30–40 hours and then does not proceed to a very great extent. The presence of bile neither increases nor retards the synthesis of fat but increases the hydrolysis. Warming at 40°C. for several hours destroys the lipolytic properties but the synthetic activities are not affected. The presence of glycerin lessens the harmful action of heat while oleic acid has no influence. The synthetic power of pancreatic juice is not increased if either glycerin or oleic acid remains in contact with it for a long time. Pancreatic juice which possesses synthetic properties has only small lipolytic capacities. Addition of fat slows down the synthetic activities but does not inhibit them. No synthesis could be demonstrated with the secretion of the small intestine in spite of a well developed lipolytic property.—*F. M. Schertz.*

914. Anonymous. [Rev. of: Schweizer, Karl. **Zur Kenntnis der Desaminierung.** (Deamination.) Biochem. Zeitschr. 78: 37–45. 1917.] Biedermann's Zentralbl. Agrikulturchem. 47: 282. 1918.—The setting free of ammonia (deamination) in the final stages of protein decomposition has been ascribed to the action of deaminases which, however, have not been isolated. A hydrolytic action was ascribed to the deaminase. Chodat and Schweizer in 1913 showed that tyrosinase possessed deaminizing properties and that deamination may

be due to the oxidising function of this enzyme. The author isolated tyrosinase from the potato and studied its action upon the amino acids. He detected formaldehyde, ammonia, and small quantities of carbon dioxide as decomposition products. He found that the presence of chlorophyll favored the action of tyrosinase. No deamination occurred when the oxygen was displaced by hydrogen or carbon dioxide. The author shows that the oxidising ferment tyrosinase has the ascribed properties of the deaminase and so makes the existence of a deaminate doubtful.—*F. M. Schertz.*

915. Barton, Arthur Willis. **The lipolytic activity of the castor and soy bean.** Jour. Amer. Chem. Soc. **42**: 620–632. 1920.—The author finds that the lipase from the castor bean splits the esters of fatty acids to a greater degree than does the soy bean lipase. Both seeds contain the same lipases. When lard or olive oil is used as substrate, ether and alcohol must be added before titration. Lipases from both sources act in the same ranges of acidity.—*J. M. Brannon.*

916. Carnot, P., and P. Gerard. **Mécanisme de l'action toxique de l'urease. [Mechanism of the toxic action of urease.]** Compt. Rend. Acad. Sci. Paris **169**: 88–90. 1919.—There are reported experiments *in vitro* and *in vivo* using the urease of soy beans, and an explanation is given of the toxic action of soy beans on the basis of the action of the urease contained in them.—*V. H. Young.*

917. Colin, H., and A. Chaudun. **Sur la loi d'action de la sucrase. Influence de la viscosité sur la vitesse d'hydrolyse. [On the law of action of sucrase: influence of viscosity on the rate of hydrolysis.]** Compt. Rend. Acad. Sci. Paris **168**: 1274–1276. 1919.—If saccharose is in excess with relation to the enzyme sucrase, the rate of hydrolysis is proportional to the viscosity of the solution.—*V. H. Young.*

918. Hérissey, M. H. **Sur la conservation du ferment oxydant des champignons. [The preservation of the oxidizing ferment (oxydase) of fungi.]** Jour. de Pharm. et Chim. **20**: 241–245. 1919.—The oxydases of fungi, especially of *Russula delica*, can easily be preserved in macerations with glycerin (1 part of the sliced fungus and 2 parts of glycerin). They may also be obtained by adding ether to the sliced fungus, allowing the mixture to stand for some time and then drawing off the lower aqueous liquid and keeping this together with an equal volume of ether, water, or glycerin in sealed tubes. The oxydases thus remain intact for more than 20 years and form a very important reagent for biologic tests.—*H. Engelhardt.*

919. Jacoby, M. **Über den vermeintlichen Abbau der Stärke durch Formaldehyde. [The supposed decomposition of starch by formaldehyde.]** Ber. Deutsch. Chem. Ges. **52B**: 558–562. 1919.—Formaldehyde action on starch has no relation to diastatic action; that is, formaldehyde is not a "diastase-model." The author disagrees with Woker and agrees with von Kauffman and Sallinger on this point.—*G. M. Armstrong.*

920. Kopeloff, Nicholas, and S. Byall. **Invertase activity of mold spores as affected by concentration and amount of inoculum.** Jour. Agric. Res. **18**: 537–542. 1920.—Spores of *Aspergillus Sydowi*, *A. niger*, and *Penicillium expansum* exhibit invertase activity in sugar solutions of concentrations varying from 10 to 70 per cent. Maximum activity occurs in concentrations between 50 and 60 per cent. An increase in the number of spores results in an increased invertase activity in a saturated sugar solution. About 5000 spores of *A. Sydowi* per cubic centimeter of saturated sugar solution cause inversion; but from 50,000 to 110,000 spores per cc. of the other two organisms are required.—*D. Reddick.*

921. McHargue, J. S. **The significance of the peroxidase reaction with reference to the viability of seeds.** Jour. Amer. Chem. Soc. **42**: 612–615. 1920.—The author thinks that the peroxidase reaction can be made use of in seed-testing laboratories for detecting non-viable seeds and for distinguishing between seed of high, medium, and low viability. Lettuce, alfalfa, and soy-bean seeds contain both oxidases and peroxidases. The peroxidase can be used to determine the rate at which seeds lose their viability.—*J. M. Brannon.*

922. Myers, R. C., and L. C. Scott. **Salivary amylase. I. A preliminary experimental study of its stability in saliva.** Jour. Amer. Chem. Soc. 40: 1713–1716. 1918.—Salivary amylase, sterilized by being passed through a Berkefeld filter, is relatively stable for one year with or without such preservatives as toluene, thymol, and chloroform; nevertheless, the preservatives mentioned are in a measure destructive, and in the order mentioned, beginning with the least destructive.—The causes which lower the stability of salivary amylase are not solely organisms and preservatives. The inherent chemical weakness of the enzyme molecule must be taken into account. Temperatures from 18 to 30°, light, and certain compounds in the saliva increase this weakness.—*C. R. Hursh.*

923. Northrup, John H. **Combination of enzyme and substrate. I. A method for the quantitative determination of pepsin. II. The effect of the hydrogen ion concentration.** Jour. Gen. Physiol. 2: 113–123. *fig. 1–3.* 1919.—The method described for the determination of pepsin depends on the change in conductivity of a digesting egg albumin solution. The author finds that the amount of pepsin removed from the solution by the substrate does not depend on the size of the particles of the substrate. The optimum H-ion concentration for the combination of enzyme and substrate corresponds to the optimum for digestion. The author suggests that the enzyme combines with the ionized protein.—*J. M. Brannon.*

924. Sabatier, Paul. **Ferments and catalyzers.** Sci. Amer. Supplem. 88: 274–275, 278–279. 1919. [Translated from *La Revue Scientifique* (Paris).]

925. Sallinger, Hermann. **Über die angeblichen diastatischen Eigenschaften des Formaldehyds.** [The alleged diastatic properties of formaldehyde.] Ber. Deutsch. Chem. Ges. 52B: 651–656. 1919.—The author thinks he has added proof to the view that starch is indifferent to the action of formaldehyde as an "enzyme."—*G. M. Armstrong.*

926. Shull, C. A. **Physiology of dormancy.** [Rev. of: (1) Crocker, William, and G. T. Harrington. **Catalase and oxidase content of seeds in relation to their dormancy, age, vitality, and respiration.** Jour. Agric. Res. 15: 137–174. *3 fig.* 1918 (See Bot. Absts. 2, Entry 173); (2) Harrington, G. T., and William Crocker. **Resistance of seeds to desiccation.** Jour. Agric. Res. 14: 525–532. 1918 (See Bot Absts. 1, Entry 1394).] Bot. Gaz. 68: 308–310. 1919.—A review of the data in these papers is introduced by the statement that this study "materially increases our knowledge of the physiology of dormancy and germination of seeds, throws much light on the problems of vitality and respiration, and is a general contribution of much significance to seed physiology.—*H. C. Cowles.*

927. Waksman, Selman A. **A method of testing the amylolytic action of the diastase of Aspergillus oryzae.** Jour. Amer. Chem. Soc. 42: 293–299. 1920.—The method used for obtaining pure starch was that developed by Sherman and associates. The author made a 2 per cent starch paste. This was divided into 10 cc. portions and brought to a temperature of 40°C. The proper amount of enzyme was added after this temperature had been reached. When the starch had all been hydrolysed, the solution lost its opaque color and became clear. In order to increase the accuracy of determining when hydrolysis was complete the dry starch was allowed to absorb a 0.5 per cent solution of neutral red. This evidently aided in determining when the solution passed from a colloidal to a clear state. The diastase from *Aspergillus oryzae* produces a good deal of glucose. It differs from malt and pancreatic diastase, as these produce chiefly maltose and but little glucose. The author finds that the Lintner method for measuring saccharogenic action of different enzymes upon starch should not be used for comparative studies of different enzymes, since the end-products are not the same in the case of the different enzymes.—*J. M. Brannon.*

928. Wood, Joseph T. **Note on trypsin and a new method of purifying enzymes.** Jour. Soc. Chem. Ind. 37: 313T–315T. 1918.—The author prepared a very pure enzyme solution by soaking Swedish filter paper in the impure trypsin solution, then drying quickly in a current of hot air. When such paper is soaked in water for 15 to 20 minutes, the enzyme is dis-

solved, but proteins are left behind. The pure solution gives no precipitate with safranin, contrary to the usual result with impure preparations. A polariscopic examination of the relatively pure solution shows no rotation. The solution thus obtained is 2½ times as strong as Grübler's trypsin. There is removed by the purification method mentioned about 35 per cent of extraneous matter.—*G. M. Armstrong.*

METABOLISM (RESPIRATION)

929. Bertrand, Gabriel. **Sur le mécanisme de la conservation des fruits dans l'eau froide.** [The mechanism of the preservation of fruits in cold water.] Compt. Rend. Acad. Sci. Paris **168**: 1285–1288. 1919.—The author has previously described (Compt. Rend. **168**: 1162) a method for preserving fruits for comparatively long periods in cold water. Later studies show that a considerable pressure is generated in sealed flasks containing fruit. It has been shown by Regnard that pressure may result in the death of minute animal forms. However, it has been shown that yeasts, etc., resisted greater pressures than were generated in the experiments performed. Cherries were preserved for eleven months under conditions where no pressure developed. It was found that the fruits absorb water and that salts, acids, sugars, and enzymes diffuse outward. Acidity incompatible with the growth of most bacteria was developed and numerous enzymatic changes resulted in the softening and transformation of the fruit. The author considers the most important factor in preservation is the exclusion of oxygen and the maintenance of a rigorous anaerobic condition such that even yeasts are unable to develop. Tests with guaiacum revealed an action similar to that of laccase. From the observations made, the author concludes that the chances of preserving fruit by this method depend: 1st, on the number and vitality of the organisms brought with the fruit; and 2nd, on the development of acidity and the initiation of biochemical processes resulting in the disappearance of O_2. Cut fruits were found to have poor keeping qualities due to their inability to resist the entrance of organisms.—*V. H. Young.*

930. Brooks, Matilda M. **Comparative studies on respiration. 8. The respiration of Bacillus subtilis in relation to antagonism.** Jour. Gen. Physiol. **2**: 5–15. 1919.—Suspensions of *Bacillus subtilis* in 0.75 per cent dextrose were subjected to various salt solutions and the rate of respiration, as indicated by the evolution of CO_2, was determined. NaCl and KCl, at concentrations of 0.15 M and 0.2 M respectively, increase the rate of respiration. At higher concentrations the rate is decreased, $CaCl_2$ increases the rate of respiration at a concentration of 0.05 M and decreases the rate at somewhat higher concentrations. A marked antagonism was observed between NaCl and $CaCl_2$ and between KCl and $CaCl_2$ in their effects on respiration. Antagonism between NaCl and KCl is slight and the antagonism curve shows two maxima.—*Otis F. Curtis.*

931. Gustafson, F. G. **Comparative studies on respiration. 9. The effects of antagonistic salts on the respiration of Aspergillus niger.** Jour. Gen. Physiol. **2**: 17–24. 1919.—Low concentrations of NaCl (0.125, 0.25, 0.5) and $CaCl_2$ (0.5 M) caused an increase in respiration of *Aspergillus niger* in the presence of 0.05 per cent dextrose as measured by the evolution of CO_2. Stronger concentrations of these salts (2 M NaCl and 1.25 M $CaCl_2$) decreased the respiration, probably through their osmotic effect in decreasing the water content of the mycelium. A mixture of 19 cc. of NaCl and 1 cc. of $CaCl_2$ (both 0.5 M) showed an antagonism, in that the respiration was normal, whereas each salt alone caused an increase. The effect of a substance on growth may differ from its effect on respiration, for, in the presence of 0.05 per cent dextrose, 0.5 M NaCl inhibited spore germination of *Aspergillus niger*, while 0.5 M $CaCl_2$ and various mixtures of the two salts did not inhibit spore germination.—*Otis F. Curtis.*

932. Osterhout, W. J. V. **Comparative studies on respiration. 7. Respiration and antagonism. Introductory note.** Jour. Gen. Physiol. **2**: 1–3. 1919.—The author briefly reviews the literature dealing with the effect of antagonistic salts on respiration and states that he has found pronounced antagonism between NaCl and $CaCl_2$ in their effects on this process. —*Otis F. Curtis.*

ORGANISM AS A WHOLE

933. Child, C. M. **A study of susceptibility in some Puget Sound algae.** Publ. Puget Sound Biol. Sta. 2: 249-267. 1919.—About 19 algae were used in the experimental work. These were tested from the standpoint of axial susceptibility, in respect to a few toxic agents. In all these the most actively growing regions were the most susceptibile to the poisons used. While differences in the permeability of the outer portions of cells may account for differences in susceptibility to certain poisons, they cannot account for all, since neutral red and certain other vital dyes probably kill from within the cell.—In *Ptilota pectinata* the differences in susceptibility of the different apical regions and axes enable one to picture the relative physiological conditions in the different parts, and make it possible to interpret to some extent the growth form in physiological terms. Apparently the inhibiting influence of a more actively growing tip is effective through a greater distance in the plant, than that of a less active tip. This is shown by the presence of alternate branching in the more active tips, and opposite branching in the less active ones. Thus activity and branch arrangement are correlated.—Experiments with a species of filamentous diatom, whose filaments are composed of bundle of gelatinous tubes in which are growing a Navicula type of diatom, show that this a pseudothallus is also most susceptible at the tips. Therefore either physiological correlation must exist between the tips and the other parts as in ordinary plants; or else growth and division are gradually inhibited by the gelatinous envelope, so that the individual diatoms at the tips of the pseudothallus are most active because they are in the most favorable situations. The pseudothallus reacts like a plant rather than like a colony.—*T. C. Frye.*

934. Gail, Floyd W. **Hydrogen ion concentration and other factors affecting the distribution of Fucus.** Publ. Puget Sound Biol. Sta. 2: 287-306. 1919.—The hydrogen ion concentration of the sea water is an important factor in distribution. The most favorable P_H is 8.0-8.2. At P_H 8.8 all growth ceases except the germination of oospores. Likewise in seawater of P_H 6.6 (and lower exponents) growth is insignificant or wanting, except in young plants, especially in temperatures above 17°C. Temperature is therefore another determining factor. Of the ranges tried, the lowest, 10.5° to 13°C. was the most favorable. When the temperature was permitted to rise to 30°C. for a part of the time, the growth was almost or wholly stopped. In the presence of much *Ulva* the P_H of the surrounding water is raised too high for *Fucus*. In tide pools the extremes of both temperature and P_H are too great. Both desiccation and light are also important factors.—*T. C. Frye.*

935. Garner, W. W., and H. A. Allard. **Effect of the relative length of day and night and other factors of the environment on growth and reproduction in plants.** Jour. Agric. Res. 18: 553-605. *Pl. 64-79. 35 fig.* 1920.—See Bot. Absts. 5, Entry 22.

936. Harris, J. E. G. **Contributions to the biochemistry of pathogenic anaerobes. VIII. The biochemical comparison of micro-organisms by quantitative methods.** Jour. Path. and Bact. 23: 30-49. *Fig. 1-2.* 1919.—A comparison was made from strictly quantitative data, (1) of the proteolytic and sugar-splitting properties of two anaerobes, *Bacillus sporogenes* and the Reading bacillus, and (2) the oxygen concentrations which permit or inhibit growth of these organisms. The two organisms are morphologically, and in cultural reactions, closely related.—Experimental methods are described for carrying out a comparison of the reactions of these organisms. Details are given of the apparatus used for fermentation experiments and of the methods for obtaining values for gas production, ammonia and amino-acid formation, production of volatile acids, and changes in hydrogen ion concentration and sugar content.—A simple method is described for determining the degree of oxygen toleration of organisms for routine purposes. It is suggested that results should be expressed in the form of the "aerobic index," which is defined.—The results are given in terms of fermentations of 5 different media and of determinations of the aerobic indices both of spores and young organisms on liquid and solid media.—From the results it is concluded that these two organisms are of the same race, but show small differences possibly acquired. In their biochemical behaviour

towards the five media used they are remarkably similar, but they show a somewhat striking difference in their powers of growing in the presence of oxygen.—The use of methods, such as these described, for investigations of the biochemical properties of bacteria in general is discussed, and a means is suggested for using these methods with aerobic organisms.—*W. W. Bonns.*

937. Hawkins, Lon A., and Rodney B. Harvey. **Physiological study of the parasitism of Pythium debaryanum Hesse on the potato tuber.** Jour. Agric. Res. 18: 275–297. *Pl. 35–37.* 1919.—See Bot. Absts. 4, Entry 1298.

938. Rosenheim, O. **Biochemical changes due to environment.** Biochem. Jour. 12: 283–289. 1918.—Only one-fourth the amount of chromogenic substance, probably flavone, was produced in the inflorescence of "Edelweiss" in London as in the native Alps. The difference is attributed to biochemical adaptation, possibly placing the flavones in a protective rôle against ultraviolet light.—*W. H. Chambers.*

939. Tevis, May. **Symbiotes or benevolent microbes and vitamines.** Sci. Amer. Supplem. 88: 282–283. 1919.—This paper is in the main a review of the theories and experiments of M. Paul Portier. According to these views, there are no simple organisms except bacteria, all higher organisms being in reality twofold—the organism itself and the microörganisms distributed throughout its tissues. The mitochondria, a definite number of which exist in each cell, are believed to be symbiotes, that is, polymorphic forms of bacteria. The cell apparently limits the number of symbiotes.—It is held that certain wasting diseases, such as scurvy, beri-beri, etc., are not due to the lack of vitamines, but are caused rather by a deficiency of symbiotes.—*Chas. H. Otis.*

GROWTH, DEVELOPMENT, REPRODUCTION

940. Anonymous. **Vertikales Wachstum der Bäume.** [Rev. of: Cambage, R. H. **The vertical growth of trees.** Jour. and Proc. Roy. Soc. New South Wales 52: 377–384. 1919. See Bot. Absts. 5, Entry 943.)] Naturwissenschaften 7: 354. 1919.

941. Buchanan, R. E. **Life phases in a bacterial culture.** Jour. Infect. Diseases 23: 109–125. 1918.—The growth of a culture of bacteria from initiation until death is divided into 7 phases, and mathematical formulae are presented to express the relation of the growth curve to time for each phase.—*W. H. Chambers.*

942. Budington, R. A. **Influence of certain ductless gland substances on the growth of plant tissues.** Biol. Bull. [Woods Hole] 37: 188–193. *Fig. 1.* 1919.—The growth of root-tips of *Allium* is retarded by the presence in their fluid nutrient environment of thyroid gland material, retradation being approximately proportional to the amount of thyroid substance present. The growth of the early leaves is not modified. Iodine, used as KI, in amounts equivalent to that in thyroid substances provoking marked modifications of growth, had no appreciable effect on growing root-tips. Pituitary substances up to two grains of the desiccated gland, and supra-renal substances up to one grain of the desiccated gland, in 120 cc. of nutritive solution had no effect on the growing root-tips. The experiments, which were limited to a single form, indicate that thyroid constituents may influence the rôle of protoplasmic action in cells other than those of animal tissues.—*J. E. Weaver.*

943. Cambage, R. H. **The vertical growth of trees.** Jour. and Proc. Roy. Soc. New South Wales 52: 377–384. 1919.—Vertical growth in the trees studied is practically limited to the terminal shoot, and it is very probable that when once definite branches are developed the portion of the axis below these increases in diameter but not in length.—*B. M. Duggar.*

944. Hibbard, R. P. **The condition of fruitfulness.** [Rev. of: Kraus, E. J., and H. R. Kraybill. **Vegetation and reproduction with special reference to the tomato.** Oregon Agric. Exp. Sta. Bull. 149. *90 p., 22 fig.* 1918.] Plant World 22: 23–24. 1919.

945. Stålfelt, M. G. **Über die Schwankungen in der Zellteilungsfrequenz bei den Wurzeln von Pisum sativum. [Variations in the frequency of cell division in the roots of Pisum sativum.]** Svensk. Bot. Tidskr. [Stockholm] 13: 61–70. 1919.—In experiments on the action of weak electric currents on roots of *Pisum sativum* the author observed a periodicity in cell divisions. The number of dividing cells was counted in 10 sections from each root. Since nuclear division is sensitive to external conditions these experiments were carried out in darkness at a constant temperature. The frequency of cell division in each root is periodic. The intensity of division shows distinct maxima and minima. The rhythm is independent of daily periodicity and therefore not synchronous in different roots. Periods of active division are succeeded by rest periods. The duration time of the phases of cell division is estimated in percentages of the total time required for division as follows: prophase, 32.78 per cent, metaphase, 36.96 per cent, anaphase 19.39 per cent, telophase, 10.95 per cent.—Pea roots of the same age and length were placed in a spiral of fine silver wire which carried 3 milliamperes at low potential. Roots were left in spirals 1 to 10 hours and examined for frequency of cell division. Roots so treated showed the maximum number of dividing cells. The maximum rate of division continues for several hours after stimulation. The author believes that the passage of the current acts as a stimulus which breaks the autonomous period of cell division. —*R. B. Harvey.*

946. Stout, A. B. **Intersexes in Plantago lanceolata.** Bot. Gaz. 68: 109–133. *2 pl.* 1919.—See Bot. Absts. 3, Entry 1517.

MOVEMENTS OF GROWTH AND TURGOR CHANGES

947. Cocks, E. **Making a plant tie itself into a knot.** Sci. Amer. 121: 579. *1 fig.* 1919. —A geotropic response.—*Chas. H. Otis.*

GERMINATION, RENEWAL OF ACTIVITY

948. Andronescu, Demetrius Ion. **Germination and further development of the embryo of Zea Mays separated from the endosperm.** Amer. Jour. Bot. 6: 443–452. *1 pl.* 1919.—Embryos of corn (with their scutella) were separated from their endosperms and germinated in water and in various culture media, of which 1 and 2 per cent sucrose solutions produced the best results. The young plants thus obtained were considerably smaller than those produced by whole kernels, but were otherwise identical with them. When the scutellum as well as the endosperm was removed, growth was very much reduced and the seedlings were unable to develop far.—Seedlings grown from embryos only and those grown from whole kernels were transplanted into soil and the plants obtained were essentially similar, except that the former were somewhat smaller than the latter. The author concludes that in germination and development the presence of endosperm is not essential, but is beneficial.—*E. W. Sinnott.*

949. Anthony, Stephen, and Harry V. Harlan. **Germination of barley pollen.** Jour. Agric. Res. 18: 525–536. *Pl. 60–61.* 1920.—The pollen of barley (*Hordeum*) germinates readily within a period of 5 minutes when proper moisture and temperature conditions are afforded. The moisture relation is extremely critical. In the experiments, moisture was supplied from a fragment of green leaf tissue placed in a dry mount of pollen in a Van Tieghem cell. Slight drying of pollen causes collapse of the cell wall and free moisture causes rapid swelling and bursting.—In field experiments the receptivity of the stigma was found to extend over several days. Pollen used in 8 successive stages of development (from immature to that obtained 2 days after dehiscence of the anther) gave satisfactory percentages of fertilization only when taken from anthers that were dehiscing or had only very recently opened.—No satisfactory means was found of storing barley pollen. A "study of the conditions governing fertilization in nature shows that conditions unfavorable to fertilizations are also unfavorable to progress in the development of pollen and vice versa. In this way natural fertilization is secured."—*D. Reddick.*

950. Kondo, M. **Ueber Nachreife und Keimung verschieden reifer Reiskörner (Oryza sativa). [After-ripening and germination of rice seeds in various stages of maturity.]** Ber. Ohara Inst. Landwirtsch. Forsch. 1: 361–387. 1918.—See Bot. Absts. 3, Entry 2805; 5, Entry 36.

951. Russell, E. J. **Report on the proposed electrolytic treatment of seeds (Wolfryn process) before sowing.** Jour. Ministry Agric. Great Britain 26: 971–981. 1920.—See Bot. Absts. 5, Entry 59.

952. Skårman, J. A. O. **Ett bidrag till frågan om temperaturens betydelse för frönas groning hos Geranium bohemicum L. [A report on the question of the importance of temperature for the growth of seed of Geranium bohemicum.]** Svensk. Bot. Tidskr. [Stockholm) 13: 93–97. 1919.—The author has observed that seeds of *Geranium bohemicum* are capable of withstanding very high temperatures and of remaining viable for many years. They also seem to require special conditions including exposure to considerable heat to bring about germination, as shown by their occurrence only on burned over land.—*W. W. Gilbert.*

953. Störmer. **Keimungshemmungen bei blauen Lupinen. [A case of arrested germination in blue lupines.]** Illustrierte Landw. Zeitg. 39: 12. 1919.—See Bot. Absts. 5, Entry 63.

RADIANT ENERGY RELATIONS

954. Daniel, Lucien. **Recherches sur le développement comparé de la laitue au soleil et a l'ombre. [Development of lettuce in sun and shade.]** Compt. Rend. Acad. Sci. Paris 168: 694–696. 1919.—The author reports the effect of shade on the development of lettuce plants and discusses in a general way the relation of illumination to the duration of species, giantism, and dimorphism.—*F. B. Wann.*

955. Schanz, F. **Effect of light on living organism.** Sci. Amer. Supplem. 88: 179. 1919. [Translated from *Meteorolog. Zeitschr.* (Braunschweig).]

956. Tsuji, T. **The action of ultra-violet rays on sugar-cane, pineapple and banana in Hawaii.** Sci. Amer. Supplem. 87: 327. 1919. [From *Louisiana Planter and Sugar Manufacturer.*]—Investigations on the connection between the action of ultra-violet rays and the formation of carbohydrates, acids, and other compounds.—*Chas. H. Otis.*

TEMPERATURE RELATIONS

957. Edson, H. A., and M. Shapovalov. **Temperature relations of certain potato-rot and wilt-producing fungi.** Jour. Agric. Res. 18: 511–524. *9 fig.* 1920.—See Bot. Absts. 5, Entry 740.

958. Potter, George F. **An apparatus for automatically changing the temperature of a chamber.** Amer. Jour. Bot. 7: 39–43. *3 pl.* 1920.—In order to obtain a uniform and known rate of temperature fall for experiments dealing with the injury of plant tissues by low temperatures, the author has devised a cooling chamber in which the rate of temperature change is automatically controlled by clockwork. This apparatus is described in detail.—*E. W. Sinnott.*

959. Shreve, Edith Bellamy. **A thermo-electrical method for the determination of leaf temperature.** Plant World 22: 100–104. *2 fig.* 1919.—A method of determining leaf temperatures without wounding the tissues is described. The apparatus consists of a pair of thermocouples and a portable galvanometer sensitive to 0.1°C., with damping key, arranged compactly on a board supported on a camera tripod. A reading can be made in a fraction of a second.—*Charles A. Shull.*

TOXIC AGENTS

960. Kidd, Franklin. **Laboratory experiments on the sprouting of potatoes in various gas mixtures. (Nitrogen, oxygen and carbon dioxido.)** New Phytol. 18: 248–252. 1919.—The following conclusions are reached: "1. Oxygen is harmful to the potato tuber in concentrations above 5–10 per cent. Oxygen 80 per cent kills in 4 to 5 weeks. Oxygen 5–10 per cent is the optimal concentration for sprouting. 2. The harmful action of oxygen is increased in the presence of carbon dioxide. 3. Carbon dioxide inhibits sprouting in a concentration of 20 per cent. This concentration is at the same time to some extent harmful. 4. Higher concentrations of carbon dioxide cause marked injury and death."—*I. F. Lewis.*

961. Kryz, Ferdinand. **Ueber den Einfluss von Ultramarin auf Pflanzen.** [On the effect of ultramarine on plants.] Zeitschr. Pflanzenkrankh. 29: 161–166. 1919.—Referring to his earlier experiments with soils containing graphite, the author recapitulates his results as follows. Seeds planted in soil containing a considerable amount of graphite are retarded in germination. Plant growth was retarded and arrested, while transpiration in sunflowers grown in graphite was increased. Since graphite is a chemically indifferent substance, the author raises the question as to whether the action of other indifferent substances would be similar in effect. He chooses ultramarine, describing it as a substance nearly indifferent chemically; stable in air, light, and alkalies; insoluble in water; and only slowly decomposed by acids and acid salts.—His observations are: germination of seeds does not occur very readily in soil containing ultramarine; growth is retarded; but there is no disturbance of transpiration; and neither a "poisonous" nor fatal effect is exerted by this substance. Intense spraying of leaves with ultramarine in water suspensions causes wilting and drying.—*H. T. Güssow.*

962. Richter. [Rev. of: Fallada, O. **Zur Rübensamenbeizung mit Schwefelsäure.** (Germination of beet seed after corrosion with sulphuric acid.) Österreich.-Ungar. Zeitschr. Zuckerindust. und Landw. 46: 22–34. 1917.] Biedermann's Zentralbl. Agrikulturchem. 47: 324–325. 1918.—A table is given showing the results of treatment of 100 beet seeds with sulphuric acid. Unsoaked seeds were treated as follows: with concentrated sulphuric acid; with sulphuric acid of 53°Bé. and for comparison some which were not treated with acid. Soaked (6 hours) seed were also treated as those above. The poorest germination was shown by the unsoaked seed treated with concentrated acid, and the best germination was shown by seed soaked for 6 hours and then treated with acid of 53°Bé. The seed and acid were heated for 20 to 25 minutes with steam and then the acid was allowed to act for two and one half hours. The number of seed germinated after 2, 3, 4, 6, and 14 days was recorded.—*F. M. Schertz.*

963. Richter. [Rev. of: Greisenegger, Ignaz K. **Versuch mit Samenrüben unter Verwendung von Mangansulfat als katalytischen Dünger.** (Experiments on seed beets using manganese sulfate as a catalytical manure.] Österreich.-Ungar. Zeitschr. Zuckerindust. und Landw. 46: 13–21. 1917.] Biedermann's Zentralbl. Agrikulturchem. 47: 320–323. 1918.—Pot experiments in sand and peat were conducted using Knop's nutrient solution for watering. Fifteen pots were used, placed in 3 groups. Group 1 had no manganese, group 2 had a small quantity of manganese (0.1773 grams or 25 kilograms per hectare), and group 3 had 4 times as much manganese as group 2. The yield of seed per pot was as follows: group 1, 56.3 grams; 2, 57.2 grams; 3, 69.8 grams. The stem yield was greatest in group 1 and least in group 2. In regard to the capacity for germination, 100 seed balls of group 1 produced 149 seedlings; 100 of group 2 produced 139 seedlings, and group 3 produced 131 seedlings. The seed of the above 3 groups were then planted in plots and fertilized (manganese lacking). The seed from the above group 1 produced 108.7 grams of sugar per beet; from group 2 the yield was 112.2 grams per beet; and from group 3, 94 grams. The yield per plot respectively was 4.54, 4.55 and 4.03 kgm. The average weight of each beet was respectively 578, 599, and 512 grams. Other data were worked out for the respective groups.—*F. M. Schertz.*

964. Rumbold, Caroline. **The injection of chemicals into chestnut trees.** Amer. Jour. Bot. 7: 1-20. *7 fig.* 1920.—Injection experiments were carried on in 1913 with 156 young Paragon chestnut trees grafted on native stock. Water, twenty-five inorganic substances (including three colloidal metals), twenty-five organic substances (including extracts of normal and of diseased bark), and five stains were injected. Various concentrations were used, and the amount entering the tree was measured in each case. In general, solutions were absorbed more readily than water, organic compounds more readily than inorganic ones and true solutions more readily than colloidal ones. The more concentrated the solution, the more rapidly it was absorbed. The rate of injection was most rapid in June and next in July, May, August, September, October, and April, respectively. The rate was more variable in the spring than in the summer or autumn, and was dependent to a considerable extent upon the rate of transpiration.—Previous literature on plant injection is reviewed at some length.—*E. W. Sinnott.*

965. Rumbold, Caroline. **Effect on chestnuts of substances injected into their trunks.** Amer. Jour. Bot. 7: 45-56. *2 pl.* 1920.—The author has injected a large number of substances into Paragon chestnut trees, as reported previously (see entry next preceding). The present paper describes the course of injected solutions in the tree, their effect on the tissues, and their influence on the parasitic fungus *Endothia parasitica*. Solutions travel usually in last annual ring of wood and were found to pass downward into the roots and upward into the leaves, and in one case even into the fruit. They are confined to a path but little wider than the diameter of the injection hole. The effect on the tree varied with the dilution of the solution and with the season at which injection was made. Certain substances, notably water, the alkali metals, colloidal metals, most organic compounds, certain dyes, and the water extract of normal bark, were without noticeable effect on the tree. A few, particularly weaker dilutions of alkali metals, apparently acted as slight stimulants. A third group, including the heavy metals, water extract of blight canker, and some others, were detrimental, causing the death of part or all of the tree. Particular solutions were often specific in their detrimental effects. Results as to the effect of injected solutions upon the blight fungus were very inconclusive. A little evidence is brought forward, however, which indicates that dilute solutions of lithium salts injected in the spring months may check somewhat the growth of the fungus canker.—*E. W. Sinnott.*

966. Stoklasa, J., in collaboration with J. Šebor, W. Zdobnický, F. Týmich, O. Horák, A. Němec, and J. Cwach. **Influence of aluminum ions on seed germination.** Sci. Amer. Supplem. 87: 318-320. 1919. [Translated from Biochem. Zeitschr. 91: 137-223. *fig. 1-16.* 1918.]

967. Wyeth, J. F. S. **The effect of acids on the growth of Bacillus coli.** Biochem. Jour. 12: 382-401. 1918.—Initial and final H-ion concentrations of *Bacillus coli* under varying conditions are determined, and it is found that the final reaction of the culture solutions depends on the initial H-ion concentration of the media, the buffer effect of the media, and the nature of the acid. There is a critical point in the H-ion concentration beyond which growth is completely inhibited.—*W. H. Chambers.*

ELECTRICITY AND MECHANICAL AGENTS

968. Baines, A. E. **Electrical conditions of the earth and atmosphere.** Sci. Amer. Supplem. 88: 290-291. 1919.—This article deals in part with plant life. The author believes that everything growing in the soil is charged or electrified by the earth,—the roots, stems, and veins being negative terminals, while the parts of the leaves between the veins act as aerolae, taking their charge from the positive air. An ordinary electrical current passes from air to earth and back again to air through the plant. If the soil is not moist to the root-depth, or if it does not contain electrolytes other than water, the plant is deprived of its supply of current and must suffer injury. It is claimed that if about 1 per cent of ferro sulphate

or other suitable electrolyte is mixed with the soil, or the ground is well watered with the mineral in solution, much of the water ordinarily required by plant life may be dispensed with. Potted plants so treated were kept alive in a warm greenhouse, exposed to the sun's rays, for three months without water. When vegetable life is said to be "resting" during the late autumn and winter months, it is probably due to lowered electrification.—*Chas. H. Otis.*

PHYSIOLOGY OF DISEASES

969. Anonymous. **Disease resistance in plants.** Gard. Chron. 65: 192. 1919.—This editorial is a popular consideration of the phenomenon of resistance in varieties of plants, suggesting briefly an explanation based on the presence and absence of certain chemical factors. The author suggests that the present status of the mechanism of immunity in animals may be a source of encouragement to plant pathologists.—*C. R. Hursh.*

970. Paine, Sydney G., and H. Stansfield. **Studies in Bacteriosis. III.—A bacterial leaf-spot disease of Protea cynaroides, exhibiting a host reaction of possibly bacteriolytic nature.** Ann. Appl. Biol. 6: 27–29. *Pl. 2, fig. 3–6.* 1919.—See Bot. Absts. 5, Entry 757.

971. Rose, D. H. **Infection as related to humidity and temperature.** [Rev. of: Lauritzen, J. T. **The relation of temperature and humidity to infection by certain fungi.** Phytopath. 9: 1–35. 1919.] Bot. Gaz. 68: 66–67. 1919.

MISCELLANEOUS

972. Anders, J. N. **Growing plants as health-giving agents.** Sci. Monthly 10: 63–69. 1920.—This is a popular presentation of the subject.—*L. Pace.*

973. Bobilioff, W. **De inwendige bouw der schorselementen ven Hevea brasiliensis. [The structure of cell elements in the bark of Hevea brasiliensis.]** Arch. Rubbercult. Nederlandsch-Indië 3: 222–231. 1919.—See Bot. Absts. 5, Entry 546.

974. Gagnespain, F. **Vegetable "plethora."** Sci. Amer. Supplem. 88: 220, 232. *1 fig.* 1919. [Translated from *La Rousse Mensuel* (Paris), April, 1919.]—Results of "over-feeding" of plants and differences in habitat between individuals of the same species.—*Chas. H. Otis.*

975. Glover, G. H., T. E. Newson, and W. W. Robbins. **A new poisonous plant, the whorled milkweed, Asclepias verticillata.** Colorado Agric. Exp. Sta. Bull. 246. *16 p. 13 fig.* 1918.—Serious losses of stock particularly sheep, are reported from southwestern Colorado due to *Asclepias verticillata*. The plant appears to be poisonous at all stages of growth and when dry. The symptoms of the affected animals are described. Death may result within 8 hours. The poisonous compound was not identified.—*C. R. Hursh.*

976. Harvey, R. B. **A thermo regulator with the characteristics of the Beckman thermometer.** Jour. Biol. Chem. 41: 9–10. *Pl. 1.* 1920.

977. Hibbard, R. P. **Preparation of seed potatoes.** [Rev. of: Appleman, C. O. **Physiological basis for the preparation of potatoes for seed.** Maryland Agric. Exp. Sta. Bull. 212: 79–102. *Fig. 1–11.* 1918.] Plant World 22: 91–92. 1919.

978. Nagel. **Kartoffellagerungsversuche. [Potato storage experiments.]** Illustrierte Landw. Zeitg. 39: 6. 1919.—See Bot. Absts. 5, Entry 46.

979. Weimer, J. L. **Some observations on the spore discharge of Pleurage curvicolla (Wint) Kuntze.** Amer. Jour. Bot. 7: 75–77. 1920.—See Bot. Absts. 5, Entry 695.

SOIL SCIENCE

J. J. Skinner, *Editor*

F. M. Schertz, *Assistant Editor*

GENERAL

980. Anonymous. **The value of lupins in the cultivation of poor, light land.** Sci. Amer. Supplem. 88: 265. 1919. [Abstract of paper read before Agricultural Section, British Assoc. Adv. Sci., by A. W. Oldershaw. (See Bot. Absts. 5, Entry 47.)] Reprinted in Sci. Amer. Supplem. 88: 321. 1919.

981. Bear, Firman E., and J. R. Royston. **Nitrogen losses in urine.** Jour. Amer. Soc. Agron. 2: 319–326. 1919.—The paper gives the results of losses of nitrogen from urine which has been stored under various conditions. Urine exposed to the air lost over 92 per cent of its nitrogen during 8 weeks when the average temperature was 38°C. When urine was not exposed to the air practically no losses took place. Litter allowed to dry out and remain dry lost 20 per cent of its nitrogen content while litter which was kept moist by daily additions of water lost over 97 per cent of its nitrogen. Samples protected with kerosene lost approximately 6 per cent of their nitrogen in 8 weeks.—*F. M. Schertz.*

982. Clevenger, Clinton B. **Hydrogen-ion concentration of plant juices. I. The accurate determination of the hydrogen-ion concentration of plant juices by means of the hydrogen electrode.** Soil Sci. 8: 217–226. 1919.—See Bot. Absts. 5, Entry 876.

983. Clevenger, Clinton B. **Hydrogen-ion concentration of plant juices. II. Factors affecting the acidity of hydrogen-ion concentration of plant juices.** Soil Sci. 8: 227–242. 1919. —See Bot. Absts. 5, Entry 877.

984. Conner, S. D. **The effect of zinc in soil tests with zinc and galvanized iron pots.** Jour. Amer. Soc. Agron. 12: 61–64. 1920.—The author found that acid soils when placed in zinc or galvanized pots, unless limed sufficiently, acted upon the zinc of the pots which were insufficiently protected by the granulated paraffine coating. The water-soluble Zn salts which were found in the soil caused the crops to fail the second season. The action of acid soils on Zn is evidence that soils contain true acids. No good protective coating for the pots was found.—*F. M. Schertz.*

985. Frear, William, and C. L. Goodling. **I. Cost of burning lime in the stack or heap. II. Supplementary report upon the limestone resources of Pennsylvania.** Pennsylvania Agric. Exp. Sta. Bull. 157. *23 p., 4 fig.* April, 1919.

986. Hepner, Frank E. **Wyoming forage plants and their chemical composition.** Wyoming Agric. Exp. Sta. Rept. 28: 117–128. 1917–18.—See Bot. Absts. 5, Entry 26.

987. Hoagland, D. R. **Relation of nutrient solution to composition and reaction of cell sap of barley.** Bot. Gaz. 68: 297–304. 1919.—See Bot. Absts. 5, Entry 859.

988. Kelley, W. P., and E. E. Thomas. **The effects of alkali on citrus trees.** California Agric. Exp. Sta. Bull. 318: 305–337. 1920.

989. Martin, J. C., and A. W. Christie. **Effect of variation in moisture content on the water-extractable matter of soils.** Jour. Agric. Res. 18: 139-143. 1919.—The water-soluble constituents of two soils of very different types have been studied at four moisture contents. The moisture contents approaching the air dry condition show a decided tendency to depress the nitrates and potash in both soils and the sulfates in the silty clay loam only. These depressions are reflected in the total dissolved material. The excess water in the sandy loam

soil causes a disappearance of nitrates and also decidedly depresses the potassium, calcium and magnesium, these losses also being reflected in the total solids extracted. Considerable variations in moisture contents of soils, provided the saturation point is not reached, do not appreciably modify the results obtained by the water-extraction method.—*F. M. Schertz.*

990. Russell, E. J. **Soil making.** Jour. Roy. Hortic. Soc. 44: 1-12. 1919.—This is a popular discussion of soils, soil changes and soil management, based largely on experiments at Rothamsted.—*J. K. Shaw.*

991. Shedd, O. M. **Effect of oxidation of sulphur in soils on the solubility of rock phosphate and on nitrification.** Jour. Agric. Res. 18: 329-345. 1919.—Compost experiments of rock phosphate, sulfur, soil and manure show after 24 months time, that about 17 and 84 per cent of the total phosphorus had been converted into a water-soluble and ammonium-citrate-soluble form, respectively. Sulphofication did not proceed as rapidly as when an inoculation was made with the sulphofying organism, and when this was done the time of the sulphofication may be considered to be reduced nearly one third. Composting under the same conditions but omitting the sulfur also showed favorable results in rendering the soil phosphate or that added in rock sulphate soluble, but not to the same extent as when sulphur was present. Nitrification was found to proceed to a certain extent regardless of the acid formed by the sulphur oxidation. The amounts of nitrogen found to be nitrified amounted to approximately 20 per cent of the total originally present. Sulphofication was found to take place in all of the soils examined but varied somewhat according to the type. When 25 and 50 mgm. of sulphur were added to 100 grams of soil, about the same percentage of the total was oxidized in a given time. Inoculation of mixtures of rock phosphate and sulphur was not sufficient to promote rapid sulphofication. It required in addition, soil or soil water. That the production of soluble phosphate was caused by the presence of sulphuric acid generated by the oxidation of the sulphur is demonstrated by the parallel rise in acidity and sulphate. The best conditions to promote the reaction are initial inoculation, high temperature, thorough aeration, and a fair moisture content. Other contributing factors are the proportions of the different ingredients and probably their mass. The acid phosphate made by this procedure has just as good a physical condition as the commercial product and would be cheaper if the time and labor involved in its manufacture are disregarded.—*F. M. Schertz.*

992. Shull, C. A. **Soil fertility.** [Rev. of: Van Alstine, E. The movement of plant food within the soil. Soil Sci. 6: 281-308. 1918. (See Bot. Absts. 2, Entry 1341.)] Bot. Gaz. 68: 312. 1919.

993. Takahashi, R. **On the fungous flora of the soil.** Ann. Phytopath Soc. Japan 1[2]: 17-22. 1919. See Bot. Absts. 5, Entry 688.

994. Watts, Francis. **The liming of soils.** West Indian Bull. 16: 332-341. 1918.—Compiled information.—*C. V. Piper.*

INFLUENCE OF BIOLOGICAL AGENTS

995. Barthel, Chr., and N. Bengtsson. **The influence of lime on the nitrification of barn-yard manure—nitrogen in arable soil.** Soil Sci. 8: 243-258. 1919. Manure or ammonium sulfate was added to limed and unlimed neutral and acid soils. Weekly determinations of the ammonia and nitrates were made. Lime stimulated the nitrification of the ammonium sulfate but exerted no favorable action on the nitrification of stable manure or in cases where the supply of lime was large impeded the nitrification.—*William J. Robbins.*

996. Hills, T. J. **Influence of nitrates on nitrogen assimilating bacteria.** Tropic. Agriculturist 52: 44-45. 1919.—Two lines of investigation, one on the influence of nitrate on azotobacter and the other on the influence of nitrate on B. radicicola in the soil, were briefly summarized without details of procedure. Full report given in Bull. Internat. Inst. of Agric., Sept., 1918.—*R. G. Wiggans.*

997. Jones, D. H., and F. G. Murdock. **Quantitative and qualitative bacterial analysis of soil samples taken in fall of 1918.** Soil Sci. 8: 259–267. 1919.—A surface and sub-surface sample of 46 soils representing 17 soil types in eastern Ontario were examined for total bacterial and mold counts on Brown's albumen agar, liquefier counts on a nutrient gelatine and Azotobacter, Ps. radicicola and Nocardia counts on a modified Ashby's agar. Only 3 samples had a very low total count. Azotobacter were found in 9 out of the 17 soil types and were absent in the light sandy soils and peat muck and shale types. Every soil type except yellow sand had fairly high numbers of Ps. radicicola and the sub-surface samples had a higher content than the surface samples. Molds were fairly uniform in numbers in all soils except a sandy clay loam and sandy clay shale in which they were absent. Nocardia were least numerous in sand but much alike in numbers in loams, peat mucks and shales.—*William J. Robbins.*

998. Waksman, Selman A. **Cultural studies of species of Actinomyces.** Soil Sci. 8: 71–215. *4 pl.* 1919.—The morphology, cultural characteristics and biochemical features of 41 species of Actinomyces are described and compared. A note is given on the habitat of each species. The cultural characteristics for each species include those on 13 or 14 different solid and liquid media. The utilization of different carbon or nitrogen compounds is also included in some cases. The biochemical features include nitrite formation, proteolytic action, change of reaction, inversion of sugar, diastatic action and growth on cellulose. Nearly all the Actinomyces studied reduce nitrates to nitrites and show diastatic and proteolytic activities. Most of the species studied grow on cellulose and half of them invert sugar. A key for the identification of the species based chiefly on biochemical characteristics is presented.—*William J. Robbins.*

FERTILIZATION

999. Calvino, M. **La fertilidad de la tierra y los abonos. III. El estiercol y los otros abonos organicos. [Manure and other organic fertilizers.]** Revist. Agric. Com. y Trab. 2: 540–543. *1 fig.* 1919.—Largely a translation of an article by Gino Beccari of the University of Pisa.—*F. M. Blodgett.*

1000. Calvino, Mario. **La fertilidad de la tierra y los abonos. [The fertility of the soil and fertilizers.]** Revist. Agric. Com. y Trab. 2: 501–503. 1919.

1001. Jones, Joseph. **Manurial experiments with cacao in Dominica.** West Indian Bull. 16: 342–353. 1918.—Reports results of plot experiments with various fertilizers.—*C. V. Piper.*

1002. Sampson, H. C. **Some factors which influence yield of paddy in comparative manurial experiments at the Manganallur Agricultural Station.** Agric. Jour. India 14: 739–746. 1919.—Experimental errors in field experiments are discussed, and the advantages and disadvantages of 1 year and long time fertilizer experiments given. No experimental data is given.—*J. J. Skinner.*

METHODS

1003. Brackett, R. N., and H. F. Haskins. **Report on nitrogen.** Jour. Assoc. Official Agric. Chem. 3: 207–217. 1919.—In the zinc-ferrous sulfate-soda method for nitrates the results of the different workers are too variable. The chief difficulty in the method lies in the distillation with the use of glass wool in the neck of the flask. Further work was recommended in the case of water-insoluble organic nitrogen. The Jones and Street method has been shown to be useful for distinguishing between good and bad organic ammoniates. Some difficulties in the method however are yet to be overcome. Results obtained with the Kjeldahl-Gunning-Arnold method using copper sulfate in lieu of oxide of mercury and with oxide of mercury alone, were very satisfactory, there being a good agreement and practically no difference in the averages. The oxide of mercury seems to be a little more effective and rapid in its catalytic action than copper sulfate and perhaps the digestion in the case of copper should be more prolonged than with mercury. The use of sodium sulfate in the place of potassium sulfate in the Gunning method and its modifications is to be studied.—*F. M. Schertz.*

1004. Fippin, Elmer O. **The truefast test for sour soil.** Jour. Amer. Soc. Agron. 12: 65–68. 1920.—The paper describes the chemical principles employed by the truefast test and points out the special features of the outfit. The manner of using the outfit is given.—*F. M. Schertz.*

1005. Frear, William, Walter Thomas, and H. D. Edmiston. **Notes on the use of potassium permanganate in determining nitrogen by the Kjeldahl method.** Jour. Assoc. Official Chem. 3: 220–224. 1919.—Results of the authors show that for the fertilizer mixtures represented the addition of permanganate caused a distinct loss of nitrogen. The loss depended somewhat upon the amount of permanganate but chiefly upon the time of the addition. If the addition was delayed for two minutes after removal from the flame no loss in nitrogen was observed.—*F. M. Schertz.*

1006. Phelps, I. K., and H. W. Daudt. **Investigations of the Kjeldahl method for the determination of nitrogen.** Jour. Assoc. Official Agric. Chem. 3: 218–220. 1919.—The hydrolysis of certain organic compounds of various constitutions was studied. In the presence of 0.7 gram of mercuric oxide, 10 grams of K_2SO_4 and 25 cc. of H_2SO_4, weights of the compound varying from 0.2 to 0.4 gram were hydrolyzed completely by 2.5 hours of boiling.—*F. M. Schertz.*

1007. Trowbridge, P. F. **Symposium on the determination of nitrogen in fertilizers.** Jour. Assoc. Official Agric. Chem. 3: 217–218. 1919.—The paper gives the answers of 38 station chemists and 17 commercial chemists, to a questionnaire on methods of determining nitrogen in fertilizers. Twenty-one chemists use a gram sample. Either mercury oxide or mercury is used by 41. Thirty-two do not use potassium permanganate at the close of the digestion. Sulphuric acid as standard is used by 31 chemists and 28 use sodium hydroxide to titrate the excess of acid. Cochineal is used as indicator by 42 chemists. Others use methyl red, methyl orange, congo red, sodium alizarin sulphonate, alizarin red and lacmoid. NH_4OH was compared with NaOH for titrating and out of 208 samples of fertilizer analyzed at different times 105 samples gave 0.01 per cent higher results with NaOH.—*F. M. Schertz.*

TAXONOMY OF VASCULAR PLANTS

J. M. Greenman, *Editor*

E. B. Payson, *Assistant Editor*

SPERMATOPHYTES

1008. Aellen, Paul. **Neue Bastardkombinationen im Genus Chenopodium.** [New Hybrid-combinations in the genus Chenopodium.] Rep. Sp. Nov. 15: 177–179. 1918. [Rep. Eu. & Med. 1: 257–259.]—The following new hybrid-combinations and new varieties are published: × *Chenopodium leptophylliforme* (*C. album* × *leptophyllum*), × *C. leptophylliforme* Aellen var. *glabrum*, × *C. pseudoleptophyllum* (*C. hircinum* × *leptophyllum*) and × *C. Binzianum* var. *obtusum*, × *C. basileense* [(*C. hircinum* × *striatum*) × *album* = *C. Haywardiae* × *album*].—*E. B. Payson.*

1009. Arthur, J. C. **New names for species of Phanerogams.** Torreya 19: 48–49. 1919. —In listing the hosts of Uredinales for the North American Flora, the author makes the following new combinations: *Senites Hartwegi* (*Zeugites Hartwegi* Fourn.), *Sanguinale pruriens* Trin. (*Panicum pruriens Trin.*), [Corrected (Torreya 19: 83. 1919) to read *Syntherisma* *pruriens* (Trin.) Arthur. nom. nov.], *Nymphoides Grayanum* (*Limnanthemum Grayanum* Griseb.), *Aureolaria virginica* (*Rhinanthus virginicus* L.), *Dasystephana spathacea* (*Gentiana spathacea* HBK.), and *D. Menzesii* (*Gentiana Menzesii* Griseb.).—*J. C. Nelson.*

1010. Balfour, Bayley. **Some late-flowering gentians.** Trans. Proc. Bot. Soc. Edinburgh 27: 246–272. 1918.—The author discusses several species of Asiatic gentians belonging to the section *Frigida* Kusnezow. Detailed descriptions of the species with synonymy are given and exsiccatae cited. The species treated are: *Gentiana Farreri* Balf. f., *G. Lawrencei* Burkill, *G. sino-ornata* Balf. f., *G. Veitchiorum* Hemsl., *G. ornata* Wall., and *G. prolata* Balf. f. —*J. M. Greenman.*

1011. Balfour, Bayley. **The genus Nomocharis.** Trans. Proc. Bot. Soc. Edinburgh 27: 273–300. 1918.—This article presents a consideration of the liliaceous genus *Nomocharis* of China and the Himalayas. The genus now embraces some 13 species of which the following are new: *Nomocharis Forrestii*, *N. saluenensis*, *N. tricolor*, and *N. Wardii.*—*J. M. Greenman.*

1012. Bitter, Georg. **Solanaceae quattuor austro-americanae adhuc generibus falsis adscriptae.** [**Four South American Solanaceae hitherto ascribed to the wrong genera.**] Rep. Sp. Nov. 15: 149–155. 1918.—**Solanocharis** is described as a new genus and to it assigned *S. albescens* (*Poecilochroma albescens* Britt.). The following new combinations are also made: *Jochroma Lehmannii* (*Poecilochroma Lehmanni* Damm.), *Vassobia dichotoma* (*Cyphomandra dichotoma* Rusby) and *Solanum Lauterbachii* (*Cyphomandra Lauterbachii* Hub. Winkl.).—*E. B. Payson.*

1013. Black, J. M. **Additions to the flora of South Australia. Nos. 13, 14.** Trans. Proc. Roy. Soc. South Australia 42: 38–61, *pl. 5–8*, 168–184, *pl. 15–18.* Dec. 24, 1918.—Important data are recorded concerning the flora of South Australia and the following plants are described as new: *Melaleuca quadrifaria* F. v. M., *Spyridium eriocephalum* Fensl. var. *adpressum*, *Limnanthemum stygium*, and *Dicrastylis verticillata*, *Stipa scabra* Lindl. var. *auriculata*, *Muehlenbeckia coccoloboides*, *Atriplex crassipes*, *A. campanulatum* Benth. var. *adnatum*, *Acacia rivalis*, *Frankenia foliosa*, *F. muscosa*, *F. cordata*, *F. serpyllifolia* Lindl. var. *eremophila*, and *Minuria rigida.*—*J. M. Greenman.*

1014. Bois, D. **Nothopanax Davadii.** Revue Horticole [Paris] 91: 212–213. *Fig. 67–68.* Jan., 1919.—See Bot. Absts. 3, Entry 1526.

1015. Brown, William H., and Arthur F. Fischer. **Philippine bamboos** Bur. Forestry, Dept. Agr. & Nat. Resources. [Manila.] Bull. 15. *32 p. Pl. 1–33.* 1918.—This paper deals primarily with the bamboos as a minor forest product of the Philippine Islands; nevertheless it is of interest to the taxonomist, since the authors include keys to the genera and recognize 30 or more species several of which are described and illustrated.—*J. M. Greenman.*

1016. Cardot, J. **Le cognassier de Delavay.** [**The quince of Delavay.**] Revue Horticole [Paris] 90: 131–133, *fig. 45–47.* 1918.—*Pirus Delavayi* Franchet (*Docynia Delavayi* Schneider) is transferred to the genus *Cydonia* as *C. Delavayi* Card.—*Adele Lewis Grant.*

1017. Challinor, R. W., Edwin Cheel, and A. R. Penfold. **On a new species of Leptospermum and its essential oil.** Jour. Proc. Roy. Soc. New South Wales 52: 175–180. Sept. 18, 1918.—*Leptospermum flavescens* var. *citratum* Bailey & White is raised to specific rank. Specimens on which this species is based were first collected at Copmanhurst, New South Wales, in 1911.—*J. M. Greenman.*

1018. Correvon, H. **Les Cyclamens sauvages.** [**The wild cyclamens.**] Revue Horticole [Paris] 90: 180–183, 196–198. 1918.—The author gives the results of several years of experience in growing various wild species of *Cyclamen*. A key by M. R. Buser to the cultivated species of this genus is included in which 24 species are listed.—*Adele Lewis Grant.*

1019. Dammer, U. **Zwei neue Solanaceen, Iochroma (Euiochroma) Weberbaueri und Cacabus multiflorus aus Peru. [Two new solanaceous plants, Iochroma (Euiochroma) Weberbaueri and Cacabus multiflorus from Peru.]** Rep. Sp. Nov. 15: 266–267. 1918.—The following species are described as new to science: *Iochroma Weberbaueri* and *Cacabus multiflorus*.—*E. B. Payson.*

1020. Dammer, U. **Eine neue Liliacee, Triocyrtis parviflora, aus Japan. [A new Liliaceous plant, Tricyrtis parviflora, from Japan.]** Rep. Sp. Nov. 15: 267–268. 1918.—*Tricyrtis parviflora* is described as a species new to science.—*E. B. Payson.*

1021. Dammer, U. **Neue Arten von Lachemilla aus Mittel- und Südamerika. [New species of Lachemilla from Central and South America.]** Rep. Sp. Nov. 15: 362–365. 1918.—The following species from Mexico, Costa Rica and Colombia are described as new to science: *Lachemilla Tonduzii*, *L. costaricensis*, *L. Purpusii*, *L. laxa*, *L. Uhdeana*, *L. Moritziana*, and *L. columbiana*.—*E. B. Payson.*

1022. Dinter, K. **Index der aus Deutsch-Südwestafrika bis zum Jahre 1917 bekannt gewordenen Pflanzenarten. II. [Index to the species of plants known from German Southwest Africa to the year 1917. II.]** Rep. Sp. Nov. 15: 340–355. 1918.—This alphabetical list, chiefly of flowering plants, includes a limited citation of synonyms and exsiccatae. The following new specific and varietal names or combinations are included: *Arctotis karasmontana*, *Asclepias filiformis* Benth. & Hook. var. *Buchenaviana*, *Atriplex sarcocarpus*, *Barbacenia minuta* (*Vellozia minuta* Baker), *Caralluma ausana* Dtr. & Brgr., *Cassia obovata* Collad. var. *pallidiflora*.—*E. B. Payson.*

1023. Engler, A. **Hieronymusia Engl., eine neue Gattung der Saxifragaceen. [Hieronymusia, a new genus of the Saxifragaceae.]** Notizblatt Königl. Bot. Gart. Mus. Berlin 7: 265–267. Oct. 1, 1918.—**Hieronymusia** is described and illustrated as a new genus of the Saxifragaceae. The genus is monotypic and is based on *Saxifraga alchemilloides* Griseb. (*Suksdorfia alchemilloides* (Griseb.) Engl.) a native of South America.—*J. M. Greenman.*

1024. Erikson, Johan. **Platanthera bifolia × montana i Blekinge** (one of the southern provinces of Sweden). (In Swedish.) Bot. Notiser 1918: 59–62. 1918.—*P. A. Rydberg.*

1025. Fraser, James. **A new grass, Koeleria advena Stapf.** Trans. Proc. Bot. Soc. Edinburgh 27: 302–303. 1918.—*Koeleria advena* Stapf is described as a new species of grass from specimens collected in the neighborhood of Edinburgh. The new grass appears to have been introduced into Scotland from eastern Spain or northwest Africa.—*J. M. Greenman.*

1026. Gamble, J. S. **Flora of the Presidency of Madras. Part III. Leguminosae-Caesalpinioideae to Caprifoliaceae.** 12½ × 18½ cm. *P. 391–577.* Adlard & Son & West Newman: London, 1919.—The present part begins with the subfamily Caesalpinioideae and continues through the Caprifoliaceae to the Rubiaceae in substantial accord with the Bentham and Hooker arrangement of families. The following new names and new combinations are included: *Delonix elata* (*Poinciana elata* L.), *Mimosa Prainiana*, *Rubus Wightii* (*R. rugosus* Wt., not Sm.), *Photinia Lindleyana* W. & A. var. *tomentosa*, *Jambosa Mundagam* (*Eugenia Mundagam* Bourd.), *J. Rama-Varma* (*Eugenia Rama-Varma* Bourd.), *J. occidentalis* (*Eugenia occidentalis* Bourd.), *J. Beddomei* (*Eugenia Beddomei* Duthie), *Syzygium Myhendrae* (*Eugenia Myhendrae* Bedd.), *S. Benthamianum* (*Eugenia Benthamiana* Wt.), *S. microphyllum* (*Eugenia microphylla* Bedd.), *S. montanum* (*Eugenia montana* Wt.), *S. Chavaran* (*Eugenia Chavaran* Bourd.), *S. malabaricum* (*Eugenia malabarica* Bedd.), *S. operculatum* (*Eugenia operculata* Roxb.), *S. Stocksii* (*Eugenia Stocksii* Duthie), *S. Jambolanum* DC. var. *axillare*, *Sonerila versicolor* Wt. var. *axillaris* (*S. axillaris* Wt.), *Trianthema triquetra* Rottl. var. *oblongifolia*, *Heracleum rigens* Wall. var. *multiradiatum*, *H. rigens* Wall. var. *elongatum*, *H. courtallense* (*H. rigens* Wall. var. *Candolleana* C. B. Clarke, in part), *H. Candolleanum* (*H. rigens* Wall. var. *Candolleana* C. B. Clarke, in part), *Schefflera micrantha* (*Heptapleurum rostratum* var.

micrantha C. B. Clarke), *S. Roxburghii* (*Aralia digitata* Roxb.), *S. venulosa* Harms var. *obliquinervia*, and *Alangium salvifolium* var. *hexapetalum* Wang. (*A. hexapetalum* Lamk.).—*J. M. Greenman.*

1027. Harms, H. **Araliaceae andinae.** Rep. Sp. Nov. 15: 245-254. 1918.—From the Andes of South America are described the following species as new to science or hitherto unpublished with a diagnosis: *Schefflera lasiogyne*, *S. Sodiroi*, *Oreopanax gnaphalocephalus*, *O. pariahuancae*, *O. Ruizii* Decne., *O. Sodiroi*, *O. brachystachyus* Decne, *O. brunneus* Decne., *O. ischnolobus*, *O. stenodactylus*, *O. Moritzii*, *O. mucronulatus*, *O. malacotrichus*, *O. palamophyllus*, *O. Trianae* Decne., *Aralia? Weberbaueri*.—*E. B. Payson.*

1028. Hassler, E. **Solanacea paraguariensia critica vel minus cognita.** Rep. Sp. Nov. 15: 113-121. 1918.—The first of two articles on solanaceous plants occurring in Paraguay gives critical notes on eight species of *Solanum* together with extensive citation of synonyms and exsiccatae. The following varieties new to science and new varietal combinations occur: *Solanum nudum* HBK. var. *pseudo-indigoferum*, *S. nudum* HBK. var. *micranthum* (*S. micranthum* W.), *S. verbascifolium* L. var. *typicum*, *S. Ipomoea* Sendt. var. *ipomoeoide* (*S. ipomoeoides* Chod. & Hassler), *S. Ipomoea* Sendt. var. *macrostachyum*, *S. malacoxylon* Sendt. var. *genuinum*, *S. malacoxylon* Sendt. var. *subvirescens*. Several new forms and subforms are also included.—*E. B. Payson.*

1029. Hassler, E. **Solanacea paraguariensia critica vel minus cognita. II.** Rep. Sp. Nov. 15: 217-245. 1918.—Critical notes, synonyms and citations of exsiccatae are given of 38 species, principally of the genus *Solanum*. The following new names and combinations in groups of specific and varietal rank as well as varieties new to science are published: *Solanum hirtellum* (*Atropa hirtella* Spreng.), *S. hirtellum* (Spreng.) Hassler var. *diminutum* Bitt., *S. verruculosum* (*Cyphomandra verrucuolsa* Hassler), *S. citrifolium* W. var. *typicum*, *S. citrifolium* W. var. *ochandrum* (*S. ochandrum* Dun.), *S. citrifolium* W. var. *leucodendron* (*S. leucocendron* Sendt.), *S. violifolium* Schott. var. *asarifolium* (*S. asarifolium* Kth. & Bouché), *S. pseudocapsicum* L. var. *typicum*, *S. pseudocapsicum* L. var. *Sendtnerianum*, *S. pseudocapsicum* L. var. *hygrophilum* (*S. hygrophilum* Schlechtd.), *S. pseudocapsicum* L. var. *ambiguum*, *S. torvum* Sw. var. *genuinum*, *S. torvum* Sw. var. *lanuginosum* (forma *lanuginosum* Sendt.), *S. bonariense* L. var. *paraguariense* (*S. paraguariense* Chod.), *S. lycocarpum* St. Hil. var. *genuinum*, *S. lycocarpum* St. Hil. var. *paraguariense*, *S. lycocarpum* St. Hil. var. *macrocarpum* (*S. grandiflorum* var. *macrocarpum* Hassler), *S. Balansae* (*S. Brownii* Chod.), *S. Balansae* Hassler var. *typicum*, *S. Balansae* Hassler var. *lyratifidum*, *S. Balansae* Hassler var. *ambiguum*, *S. Balansae* Hassler var. *genuinum*, *S. Balansae* Hassler var. *aureomicans*, *S. Balansae* Hassler var. *subinerme*, *S. robustum* Wendl. var. *laxepilosum*, *S. robustum* Wendl. var. *concepcionis*, *S. viridipes* Dun. var. *intermedium*, *Lycium chilense* Bert. var. *normale*, *L. chilense* Bert. var. *heterophyllum*, *L. Morongii* Britt. var. *typicum*, *L. Morongii* Britt. var. *indutum*, *Capsicum microcarpum* DC. var. *glabrescens*. Many new subspecies, forms and subforms are included or new combinations in these subspecific groups occur.—*E. B. Payson.*

1030. Herter, W. **Itinera Heteriana I.** Rep. Sp. Nov. 15: 373-381. 1918.—[Rep. Eu. & Med. 1: 309-317.]—I. *Cruciferae mediterraneae*. The author presents an alphabetical list of *Cruciferae* collected by himself in regions adjacent to the Mediterranean Sea with complete data for each collection. II. *Umbelliflorae mediterraneae*. A list, similar to the above, includes certain members of the families *Araliaceae*, *Umbelliferae* and *Cornaceae*.—*E. B. Payson.*

1031. Javorka, S. **Kisebb megjegyzések és újabb adatok. VI.** [**Minor observations and new data. VI.**] Bot. Közlemények. 17: 52-60. 1918.—Notes are recorded concerning several flowering plants of Hungary and one new form is characterised, namely *Draba Simonkaiana* Jav. f. *retyezátensis*.—*J. M. Greenman.*

1032. Jørgensen, E. **Ajuga pyramidalis × reptans.** Bergens Museum Aarbok 1917–1918. Naturvidenskabelig raekke 5: 1–4. 1918.—A hybrid between *Ajuga pyramidalis* and *A. reptans* L. is recorded and its important characters contrasted with those of the parent species.—*J. M. Greenman.*

1033. Knuth, R. **Geraniaceae Novae. I.** Rep. Sp. Nov. 15: 135–138. 1918.—The following species native to South Africa are published as new to science: *Pelargonium uniondalense, P. grandicalcaratum, P. rungvense, P. Patersonii, Monsonia stricta, M. alexandraensis*, and *M. Rudatisii.*—*E. B. Payson.*

1034. Koorders, S. H., and Th. Valeton. **Atlas der Baumarten von Java.** [Atlas of the species of trees of Java.] Roy. *8vo.* 1: *Pl. 1–200.* 1913; 2: *Pl. 201–400.* 1914; 3: *Pl. 401–600.* 1915; 4: *Pl. 601–800.* 1918. P. W. M. Trap. Leiden.—This work in four volumes of four numbers each, illustrates nearly 800 species of Javanese trees in detail. Many habit sketches and a few reproductions of photographs showing the general appearance of the trees are given. *Ormosia incerta* Krds. is described as new to science. Aside from this, descriptions are not given, but references are made to works in which descriptions do occur. The present atlas, although a complete work in itself, is intended to supplement previous publications of the same authors, especially the "Bijdragen tot de Kennis der Boomsoorten van Java" [Contributions to the knowledge of the tree species of Java]. [See also Bot. Absts. 4, Entry 1735]. —*E. B. Payson.*

1035. Léveillé, H., and A. Thellung. **Oenothera argentinae spec. nov.** Rep. Sp. Nov. 15: 133–134. 1918.—This species published previously with an insufficient diagnosis is here completely characterised. Its place of origin is perhaps the Argentine.—*E. B. Payson.*

1036. Maiden, J. H. **The tropical acacias of Queensland.** Proc. Roy. Soc. Queensland 30: 18–51. *Pl. 1–7.* 1918.—The author gives an annotated list of 60 recognised species of *Acacia* from Queensland, including the following which are described as new to science: *Acacia Bancrofti, A. curniveria, A. Whitei, A. argentea*, and *A. Armitii* (*A. deliberata* F. v. M., not A. Cunn.).—*J. M. Greenman.*

1037. Mez, Carl. **Sacciolepis, Mesosetum, Thrasya, Ichnanthus genera speciebus novis aucta.** Rep. Sp. Nov. 15: 122–133. 1918.—The following species are described as new to science: *Sacciolepis longissima, S. micrococcus, S. delicatula, S. Karsteniana, Mesosetum penicillatum, M agropyroides, Thrasya trinitensis, Ichnanthus lancifolius, I. Weberbaueri, I. lasiochlamys, I. verticillatus, I. montanus* (*Panicum inconstans* var. *montanum* Trin.), *I. peruvianus, I. trinitensis, I. polycladus, I. drepanophyllus, I. longiglumis, I. venezuelanus* and *I. Gardneri.*—*E. B. Payson.*

1038. Mottet, S. **Noveaux Trollius.** [New Trollius.] Revue Horticole [Paris] 90: 102–103. *1 pl.* 1918.—Two plants of this genus are described and illustrated. The first, *Trollius Ledebourii* Rchb. comes from Siberia while the second, *T. pumilus* var. *yunnanensis* Hort. is described as new and was grown from seed sent from Yunnan, China.—*Adele Lewis Grant.*

1039. Mottet, S. **Un Nouveau Columnea Hybride.** [A new Columnea hybrid.] Revue Horticole [Paris] 90: 168–170. *1 pl. and 1 text fig.* 1918.—The author describes and illustrates a new hybrid, *C. vedrariensis* Hort., resulting from a cross between *C. Schiedeana* Schlecht. and *C. magnifica* Oersted.—*Adele Lewis Grant.*

1040. Mottet, S. **Neillia, Physocarpus et Stephanandra.** Revue Horticole [Paris] 91: 236–238. *Fig. 77.* Feb., 1919.—See Bot. Absts. 3, Entry 1544.

1041. Mottet, S. **Nouveaux Viburnum de la Chine.** [New Viburnum of China.] Revue Horticole [Paris] 91: 262–264. *1 pl.* (*colored*). Apr., 1919.—See Bot. Absts. 3, Entry 1548.

1042. Porto, P. Campos. **O Cambuci (Paivaea Landsdorffii Berg.).** 21 × 27½ cm. *14 p. 9 fig.* Estabelecimento Graphico de Steele & C. Rio de Janeiro, 1920.—This pamphlet, published by the Botanical Garden in Rio de Janeiro, illustrates and gives a detailed account of *Paivaea Langsdorffii* Berg. with particular reference to its edible fruit.—*J. M. Greenman.*

1043. Pritzel, E. **Basedowia, eine neue Gattung der Compositen aus Zentral-Australien. [Basedowia, a new genus of the Compositae from central Australia.]** Ber. Deutsch. Bot. Gesellsch. 36: 332-337. *Pl. 12.* Oct. 18, 1918.—*Basedowia helichrysoides* is described and illustrated as a new genus and species of the Compositae. It is placed under the *Helichryseae* and is related to *Cassinia* and to *Helichrysum.*—*J. M. Greenman.*

1044. Pritzel, E. **Species novae ex Australia centrali.** Rep. Sp. Nov. 15: 356-361. 1918. —The following species and varieties new to science are characterised: *Triodia Basedowii, Crotalaria Strehlowii, Indigofera Basedowii, I. leucotricha, Swainsona phacoides* Benth. var. *erecta, Petalostyles spinescens, Heterodendron floribundum, Eremophila castelli Arminii, E. Leonhardiana, E. Strehlowii, Canthium lineare, Olearia arida, Rutidosis panniculata.*—*E. B. Payson.*

1045. Rogers, R. S. **Notes on Australian orchids, together with a description of some new species.** Trans. & Proc. Roy. Soc. South Australia 42: 24-37. *Pl. 2-4.* Dec. 24, 1918.—The author presents a synopsis with critical notes of several genera of orchids. The following species are new to science: *Calochilus cupreus, Pterostylis pusilla,* and *Prasophyllum regium.*—*J. M. Greenman.*

1046. Rubner, K. **Ein neues Epilobium (E. Graebneri) aus Westrussland. [A new Epilobium (E. Graebneri) from western Russia.]** Rep. Sp. Nov. 15: 179-180. 1918. [Rep. Eu. & Med. 1: 259-260.]—*Epilobium Graebneri* is characterised as a species new to science.—*E. B. Payson.*

1047. Schlechter, R. **Die Gattung Aganisia Ldl. und ihre Verwandten. [The genus Aganisia Ldl. and its relatives.]** Orchis 12: 24-42. *Pl. 2-5.* 1918.—The present article, which is continued from a previous number of this magazine (12: 6-16, *pl. 1.* 1917), includes a synoptical revision of *Koellensteinia* Rchb. f., *Paradisianthus* Rchb. f., *Warreella* Schltr., and **Otostylis** Schltr. n. gen. The following new species and new names are recorded: *Koellensteinia peruviana* from Peru, *K. eburnea* (*Cyrtopodium eburneum* Barb. Rodr. from Brasil, *K. Roraimae* from Guiana, *K. boliviensis* from Bolivia, *Paradisianthus neglectus* from Brasil, *P. micranthus* (*Zygopetalum micranthum* Barb. Rodr.) from Brasil, *Otostylis lepida* (*Aganisia nisia lepida* Lind. & Rchb. f.), *O. brachystalix* (*Zygopetalum brachystalix* Rchb. f.), and *O. venusta* (*Zygopetalum venustum* Ridl.).—*J. M. Greenman.*

1048. Schlechter, R. **Die Gattung Restrepia H. B. u. Kth. [The genus Restrepia HBK.]** Rep. Sp. Nov. 15: 255-270. 1918.—**Barbosella**, a new genus, is proposed for a number of Central and South American orchids formerly included under *Restrepia.* The following combinations result: *Barbosella australis* (*Restrepia australis* Cogn.), *B. Cogniauxiana* (*Restrepia Cogniauxiana* Speg. & Kränzl.), *B. crassifolia* (*Restrepia crassifolia* Edwall), *B. cucullata* (*Restrepia cucullata* Ldl.), *B. Dusenii* (*Restrepia Dusenii* Sampaio), *B. Gardneri* (*Pleurothallus Gardneri* Ldl.), *B. Kegelii* (*Restrepia Kegelii* Rchb. f.), *B. Löfgrenii* (*Restrepia Löfgrenii* Cogn.), *B. microphylla* (*Restrepia microphylla* Rodr.), *B. Miersii* (*Pleurothallus Miersii* Ldl.), *B. Porschii* (*Restrepia Porschii* Kränzl.), *B. prorepens* (*Restrepia prorepens* Rchb. f.), *B. rhyncantha* (*Restrepia rhyncantha* Rchb. f. & Warsc.), *B. varicosa* (*Restrepia varicosa* Ldl.). The author also gives a synopsis of the genus *Restrepia* with critical notes on each of the 21 recognised species. Three new sectional names are given as follows: *Pleurothallopsis, Eurestrepia* and *Achaetochilus.* The new name *Pleurothallus Edwallii* Dusen & Schltr. (*Restrepia pleurothalloides* Cogn.) is proposed.—*E. B. Payson.*

1049. Schlechter, R. **Die Gattung Sigmatostalix Rchb. f. [The Genus Sigmatostalix Rchb. f.]** Rep. Sp. Nov. 15: 139–148. 1918.—The species previously assigned to the genus *Sigmatostalix* are found to be very diverse structurally and from them have been segregated the new genera **Petalocentrum** and **Roezliella.** The three genera are compared critically and under each is given a key to the species properly assigned to them. Besides the new generic diagnoses the following species new to science and new combinations are included: *Petalocentrum pusillum* (*Sigmatostalix pusilla* Schltr.), *P angustifolium*, *Roezliella dilatata* (*Sigmatostalix dilatata* Richb. f.), *R. Wallisii* (*Sigmatostalix Wallisii* Rchb. f.), *R. reversa* (*Sigmatostalix reversa* Rchb. f.), *R. malleifera* (*Sigmatostalix malleifera* Rchb. f.), *R. Lehmanniana* (*Sigmatostalix Lehmanniana* Kränzl.), *Capanemia brachycion* (*Sigmatostalix brachycion* Griseb.), *C. Juergensiana* (*Rodriguezia Juergensiana* Kränzl.) and *C. pygmaea* (*Rodriguezia pygmaea* Kränzl.).—*E. B. Payson.*

1050. Schlechter, R. **Mitteilungen über einige europäische und mediterrane Orchideen. I. [Contributions concerning some European and Mediterranean Orchids. I.]** Rep. Sp. Nov. 15: 273–302. 1918. [Rep. Eu. & Med. 1: 274–302.]—I. The genera *Aceras*, *Himantoglossum* and *Anacamptis* are discussed in the light of their taxonomic history and present interpretation. Under each genus are listed the species and varieties belonging to it and critical notes are given. II. *Orchis persica*, a species new to science, is characterised. III. **Steveniella**, a new genus, is described and to it is referred 1 species, *S. satyrioides* (*Orchis satyrioides* Stev.). IV. *Gennaria* Parlat. is confirmed as worthy of generic rank. Specimens are cited for its single species, *G. diphylla* (Lk.) Parlat. V. *Platanthera parvula* is described as new to science.—*E. B. Payson.*

1051. Schlechter, R. **Odontioda × Fürstenbergiana Schltr., ein neuer bigenerischer Orchideenbastard. [Odontioda × Fürstenbergiana Schltr., a new bigeneric orchid-hybrid.]** Orchis 12: 19, 20. 1918.—This new bigeneric hybrid is the result obtained from crossing *Cochlioda vulcanica* Benth. with *Odontoglossum Eduardi* Rchb. f.—*J. M. Greenman.*

1052. Schlechter, R. **Orchidaceae novae et criticae. Decas LI–LIII.** Rep. Sp. Nov. 15: 193–209. 1918.—The following Guatemalan plants, chiefly from the collections of Bernoulli and Cario, are described as new to science: *Platanthera guatemalensis*, *Habenaria dipleura*, *H. latipetala*, *H. quinquefila*, *H. spithamaea*, *Pogonia debilis*, *Ponthieva pulchella*, *Pelexia guatemalensis*, *Spiranthes pulchra*, *Physurus humidicola*, *P. luniferus*, *P. trilobulatus*, *Microstylis acianthoides*, *M. lepanthiflora*, *Masdevallia guatemalensis*, *Stelis Bernoulii*, *S. Carioi*, *S. cleistogama*, *S. oxypetala*, *S. tenuissima*, *Pleurothallis Bernoullii*, *P. Carioi*, *P. lamprophylla*, *Epidendrum aberrans*, *E. lucidum*, *E. piestocaulos*, *E. verrucipes*, *Notylia guatemalensis*, *Leochilus major*, *Ornithocephalus tripterus*. Decas LIV. *ibid.* 210–217.—**Calogiossum**, a new genus of Madagascar orchids is described and to it are assigned the following new and transferred species: *C. flabellatum* (*Limodorum flabellatum* Thou.), *C. Humblotii* (*Cymbidium Humblotii* Rolfe), *C. magnificum*, *C. rhodochilum* (*Cymbidium rhodochilum* Rolfe). Additional new combinations in other genera are proposed as follows: *Platanthera Komarovii*, *Chloraea reticulata*, *Stelis ovatilabia*, *Dendrobium Casuarinae*, *Otostylis paludosa* (*Zygopetalum paludosum* Cogn.), *Oncidium Spegazzinianum* (*Leochilus Spegazzinianus* Kränzl.), *O. Waluewa* (*Waluewa pulchella* Regel), *Solenidium mattogrossense* (*Leochilus mattogrossensis* Cogn.), *Erycina diaphana* (*Oncidium diaphanum* Rchb. f.), *Pachyphyllum muscoides* (*Orchidotypus muscoides* Kränzl.), *P. cyrtophyllum* (*P. falcifolium* Schltr.).—*E. B. Payson.*

1053. Schlechter, R. **Orchidaceae novae et criticae. Decas LV–LVII.** Rep. Sp. Nov. 15: 324–340. 1918.—The following new species and varieties of Madagascan orchids are described: *Benthamia elata*, *Habenaria Perkoana*, *Cynosorchis diplorhyncha*, *C. Laggiarae*, *C. Laggiarae* var. *ecalcarata*, *Disperis Afzelii*, *Goodyeara Afzelii*, *Platylepis margaritifera*, *Bulbophyllum Afzelii*, *B. brachyphyton*, *B. Perkoanum*, *B. Laggiarae*, *B. melanopogon*, *B. mirificum*, *B. sarcorhachis*, *B. xanthobulbum*, *Lissochilus Laggiarae*, *Gussonea aurantiaca*,

Aerangis crassipes, *A. pumilio*, *A. venusta*, *Jumellea cyrtoceras*, *J. Ferkoana*, *Angraecum conchoglossum*, *A. Ferkoanum*, *A. dasycarpum*, *A. Laggiarae*, *A. melanostictum*, *A. mirabile*, *A. sarcodanthum*, *A. tenuispica*.—*E. B. Payson.*

1054. Schlechter, R. **Orchidaceae novae, in caldariis Horti Dahlemensis cultae.** [**New orchids cultivated in the Garden at Dahlem.**] Notizblatt Königl. Bot. Gart. Mus. Berlin 7: 268–280. Oct. 1, 1918.—The following new species of orchids are described: *Masdevallia paranaensis*, *Stelis diaphana*, *S. fragrans*, *S. Porschiana*, *S. robusta*, *S. thermophila*, *Pleurothallis lamproglossa*, *P. margaritifera*, *P. microblephara*, *P. mirabilis*, *P. paranaensis*, *P. Petersiana*, *P. rhabdosepala*, *Octomeria rhodoglossa*, *Encyclia laxa* native of Brasil, *Dendrobium dahlemense* from Sumatra, *Polystachya fulvilabia* from Kamerun, *Maxillaria phaeoglossa* and *M. xanthorhoda* native country unknown, and *Vanda Petersiana* from Burma.—*J. M. Greenman.*

1055. Schlechter, R. **Ueber einige neue Cymbidien.** [**On some new Cymbidiums.**] Orchis 12: 45–48. 1918.—The following new species and new hybrids are described: *Cymbidium Hennisianum* from *India*, *Cymbidium* × *Fürstenbergianum* (*C. Traceyanum* × *erythrostylum*), and *Cymbidium* × *magnificum* (*C. erythrostylum* × *Lowianum*).—*J. M. Greenman.*

1056. Schlechter, R. **Vanda × Herziana Schltr. n. hybr.** Orchis 12: 88, 89. 1918.—*Vanda* × *Herziana* is described as a new hybrid between *Vanda coerulea* and *V. suavis* Ldl. —*J. M. Greenman.*

1057. Schlechter, R. **Zwei neue Hybriden (Brassocattleya × Paulae Schltr. und Laeliocattleya × pulchella).** [**Two new hybrids.**] Orchis 12: 87. 1918.—*Brassocattleya* × *Paulae* was obtained by crossing *Cattleya aurea* with *Brassavola Perrinii* Rchb. f. and *Laeliocattleya* × *pulchella* was obtained by crossing the natural hybrid *Laelio* × *Crawshayana* with *Cattleya velutina* Rchb. f.—*J. M. Greenman.*

1058. Schneider, Camillo. **Weitere Beiträge zur Kenntnis der chinesischen Arten der Gattung Berberis (Euberberis).** [**Further contributions to the knowledge of the Chinese species of the genus Berberis (Euberberis).**] Oesterr. Bot. Zeitschr. 66: 313–326. 1916. *Ibid.* 67: 15–32, 135–146, 213–228, 284–300. 1918.—In this series of articles the author presents a revision of the Chinese species of *Berberis* recognizing 85 species and several varieties grouped in 10 sections. The following new species and new combinations are included: *Berberis phanera*, *B. Grodtmannia*, *B. Collettii*, *B. Willeana*, *B. Faberi*, *B. microtricha*, *B. Franchetiana*, *B. kansuensis*, *B. oritrepha*, *B. Wilsonae* Hensl. var. *subcaulialata* (*B. subcaulialata* Schn.), and *B. Wilsonae* Hemsl. var. *Stapfiana* (*B. vulgaris* var. *Stapfiana* Voss).—*J. M. Greenman.*

1059. Schulz, O. E. **Sisymbrium septulatum DC., eine bisher nicht genügend bekannte Art.** [**Sisymbrium septulatum DC., a species previously insufficiently known.**] Rep. Sp. Nov. 15: 369–372. 1918. [Rep. Eu. & Med. 1: 306–308.]—This species, described from incomplete material and confused by synonymy has been variously misinterpreted. A complete specific description is given and specimens are cited. The following new varieties are characterised: *S. septulatum* DC. var. *trichocarpum*, *S. septulatum* DC. var. *dasycarpum*, *S. septulatum* DC. var. *lasiocarpum*.—*E. B. Payson.*

1060. Small, James. **The origin and development of the Compositae.** *8vo. xi + 334 p., 6 pl., 79 text-fig.* William Wesley & Son: London, 1919. [Reprinted from the New Phytologist, Vols. xvi–xviii. 1917–1919.]—See Bot. Absts. 3, Entry 1142.

1061. Vierhapper, F. **Was ist Trifolium Pilczii Adamović?** [**What is Trifolium Pilczii Adamović?**] Oesterr. Bot. Zeitschr. 67: 252–264, 328–337. *Pl. 3.* 1918.—The author presents the results of a critical study of *Trifolium Pilczii* Adamovic, and discusses its relationship to *T. eximium* Steph. and *T. altaicum* Vierh.—*J. M. Greenman.*

1062. Viguier, R. **Les Araliacées cultivées.** [Cultivated Araliaceae.] Revue Horticole [Paris] 91: 228-229. Feb., 1919.

1063. Viguier, R. **Les Araliacées cultivées.** [Cultivated Araliaceae.] Revue Horticole [Paris] 91: 250-252. Mar., 1919.

1064. Von Wettstein, R. **Moltkea Dörfleri Wettstein und die Abgrenzung der Gattung Moltkea.** [Moltkea Dörfleri Wettstein and the demarcation of the genus Moltkea.] Oesterr. Bot. Zeitschr. 67: 361-368. *Pl. 3, 22 fig.* 1918.—The author describes in detail and illustrates *Moltkea Dörfleri* Wettst., discusses the relationship of the genus *Moltkea* to allied genera, and enumerates with the bibliography and synonymy eight species recognised under the above generic name.—*J. M. Greenman.*

1065. Wagner, Rudolf. **Erläuterungen zu Plumiers Abbildung der Anechites lappulacea (Lam.) Miers.** [Explanations to Plumier's illustration of Anechites lappulacea (Lam.) Miers.] Oesterr. Bot. Zeitschr. 67: 337-345. *3 fig.* 1918.

MISCELLANEOUS UNCLASSIFIED PUBLICATIONS

Burton E. Livingston, *Editor*

1066. Anonymous. **Palatability for sheep of certain New Zealand forest plants.** New Zealand Jour. Agric. 19: 293-294. 1919.

1067. Anonymous. **Lac cultivation in India.** Sci. Amer. Supplem. 88: 280. 1919. [From *Jour. Roy. Soc. of Arts.*]

1068. Anonymous. **Utilization of marine plants.** Sci. Amer. 121: 557. 1919.

1069. Anonymous. **Peat fuel for locomotives.** Sci. Amer. 121: 566. 1919.

1070. Baldwin, J. F. **Germination of grains.** Sci. Amer. 121: 626. 1919.—Reports of germination of grains of cereals found wrapped up with ancient Egyptian mummies are claimed to be fictitious.—*Chas. H. Otis.*

1071. Bussy, P. **Le latanier du Sud-Annam et sa fibre.** [The Bourbon palm of southern annam and its fiber.] Bull. Agric. Inst. Sci. Saigon 1: 377-380. 1919.—A discussion of the fibers produced by the palm *Corypha lecomtei* Becc.—*E. D. Merrill.*

1072. Caballero, A. **La Chara foetida A. Br., y las larvas de Stegomyia, Culex y Anopheles.** [Chara foetida A. Br. and the larvae of Stegomyia, Culex and Anopheles.] Bol. R. Soc. Española Hist. Nat. 19: 449-455. Oct., 1919.—In the botanical laboratory of the University of Barcelona it was noticed that an aquarium containing *Chara foetida* appeared not to breed mosquitoes as did other aquaria containing other aquatics (*Potamogeton fluitans*, *P. pectinatus*, *Elodea canadensis*, and *Apium nodiflorum*). Experiments were undertaken which indicated that a sufficient quantity of *Chara foetida*, probably not much more than one-eighth of the total volume of the container, caused the death of mosquito larvae by asphyxiation. The larvae of *Stegomyia* appeared somewhat more resistant than those of the other genera. The cultivation of *Chara foetida* is stated to be easy and economical and its use in tanks, ponds, etc., is recommended for preventing the development therein of mosquito larvae.—*O. E. Jennings.*

1073. Clarkson, Edward Hale. **The irresistible charm of the ferns.** Amer. Fern Jour. 9: 109-115. *Pl. 7-8.* 1919.

1074. Freund, Hans. **Ueber Kork-Ersatz.** [Substitutes for Cork.] Pharm. Zentralhalle Deutschland 60: 183-187. 1919.—The scarcity of cork in Germany necessitated the use of substitutes for this commodity. The author describes the various barks, piths, etc., used for this purpose.—*H. Engelhardt.*

1075. FUEHNER, H. **Goldregen Tabak. [Cytisus laburnum tobacco.]** Pharm. Zentralhalle Deutschland 60: 336-337. 1919.—The leaves of *Cytisus laburnum*, when subjected to a proper fermentation, furnish a product which can be used as a substitute for tobacco. The smoke does not smell disagreeable, does not irritate the mucous membranes and acts on the central nervous system in exactly the same way as tobacco.—*H. Engelhardt.*

1076. GRIEBEL, C. **Beiträge zum mikroskopischen Nachweis von pflanzlichen Streckungsmitteln und Ersatzstoffen bei der Untersuchung der Nahrungs- u. Genussmittel. [Microscopic demonstration of vegetable substitutes in food investigation.]** Zeitschr. Untersuch. Nahrungs- u. Genussmittel 38: 129-141. 1919.—Histological description of substitutes for bread and meal, preserves, spices, and coffee.—*H. G. Barbour.*

1077. HABERLANDT, G. **Food value of alfalfa used as a table vegetable.** Sci. Amer. Supplem. 88: 298, 312. 1919. [From *Die Naturwissenschaften* (Berlin).]

1078. HERTER, W. **Zur quantitativen Mikroanalyse der Nahrungs- und Futtermittel. [Quantitative micro-analysis of food.]** Zeitschr. Untersuch. Nahrungs- u. Genussmittel 38: 65-82. 1919.—Thorough theoretical discussion with numerous examples.—*H. G. Barbour.*

1079. HOWE, H. E. **Research and cotton.** Sci. Amer. 121: 606. 1919.—A brief résumé of what investigation has done in the past for this branch of the textile industries.—*Chas. H. Otis.*

1080. HOWE, H. E. **Using vegetable seeds.** Sci. Amer. 121: 554. 1919.

1081. KRAFFT, K. **Ergebnisse der Untersuchung von Ersatzmitteln im Jahre 1918 und Januar bis April 1919. [Investigation of food substitutes.]** Zeitschr. Untersuch. Nahrungs- u. Genussmittel 38: 213-221. 1919.—Substitutes for baking-powder and accessories, eggs, spices, extracts, flavorings, honey, preserves, fulminating powder, tea and coffee, tobacco, fruit juices, beer, and sausages.—*H. G. Barbour.*

1082. SMITH, E. PHILIP. **Pollinosis ("Hay-Fever").** Jour. Botany 58: 40-44. 1920.—A condensed account is given of the symptoms of hay fever. It is noted that the problem of treatment has heretofore been approached from the standpoint of the immunologist. The earliest work was that of DUNBAR and PRAUSNITZ, and their experiments were elaborate. A list is given of the plants found by these authors to cause hay fever. The present author adds various conifers to the list. He thinks the toxalbumen theory of Dunbar is scarcely tenable because the contents of the pollen grain are separated from the nasal membrane by the wall of the grain. Mechanical irritation or the production of substances on the surface of the pollen are the only alternatives left. The author finds mechanical irritation insufficient to account for the symptoms. On the other hand the grains are coated with tapetal debris often in the form of an oily substance. The oil was extracted by ether from the pollen of *Hibiscus* and was found to produce a blister when applied to the unbroken skin of the forearm. Very similar results were obtained with the pollen of *Plantago*. The cases of *Primula obconica* and *sinensis* are cited to show that such irritating oils are produced by plants. If this theory of the cause of hay fever is correct it will throw a new light on the whole problem and bring it into line with well-known cases of plant-dermatitis which cover quite a wide range of plant organisms.—*K. M. Wiegand.*

1083. STUART, G. A. D., AND E. J. BUTLER. **Report of the Director.** Sci. Rept. Agric. Inst. Pusa 1918-19: 1-10. 1919.—A summary of the more important scientific work for the year at the Pusa Institute (India).—*Winfield Dudgeon.*

1084. VERNET, G. **Sur les causes de la coagulation naturelle du latex d'Hevea brasiliensis. [On the causes of natural coagulation of the latex of Hevea brasiliensis.]** Bull. Agric. Inst. Sci. Saigon 1: 342-347. 1919.

1085. WALL, A. **The pronunciation of scientific terms in New Zealand, with special reference to the terms of botany.** Trans. and Proc. New Zealand Inst. 51: 409-414. 1919

BOTANICAL ABSTRACTS

A monthly serial furnishing abstracts and citations of publications in the international field of botany in its broadest sense.

UNDER THE DIRECTION OF

THE BOARD OF CONTROL OF BOTANICAL ABSTRACTS, INC.

BURTON E. LIVINGSTON, Editor-in-Chief
The Johns Hopkins University, Baltimore, Maryland

Vol. V SEPTEMBER, 1920 No. 2

ENTRIES 1086-2426

AGRONOMY

C. V. PIPER, *Editor*
MARY R. BURR, *Assistant Editor*

1086. ALWAY, F. J. **A phosphate-hungry peat soil.** Jour. Amer. Peat Soc. 13: 108-143. 1920.—Some Minnesota bogs are found to have a sufficient supply of lime and available nitrogen for the production of all crops suitable to the region. Phosphates, however, are very scant.—*G. B. Rigg.*

1087. ANONYMOUS. **Elephant-grass in elevated localities.** Agric. Gaz. New South Wales 31: 84. 1920.—Treats of *Pennisetum purpureum.*—*L. R. Waldron.*

1088. ANONYMOUS. **The department and elephant-grass.** Agric. Gaz. New South Wales 31: 143. 1920.—Treats of *Pennisetum purpureum.*—*L. R. Waldron.*

1089. ANONYMOUS. **Coffee in New South Wales.** Agric. Gaz. New South Wales 31: 133. 1920.—This crop (*Coffea* spp.) would be unsuited to New South Wales.—*L. R. Waldron.*

1090. ANONYMOUS. **Liming, cultivation and manurial experiments at Margam, Australia.** Australian Sugar Jour. 11: 679-681. 1920.

1091. ANONYMOUS. **Further reports on elephant grass.** Agric. Gaz. New South Wales 31: 244. 1920.

1092. ANONYMOUS. **Rice culture in New South Wales.** Agric. Gaz. New South Wales 31: 232. 1920.—Results so far not encouraging but further trials are advised.—*L. R. Waldron.*

1093. ANONYMOUS. **Weed seeds.** Sci. Amer. Monthly 1: 316. 1920.—Popular.—*Chas. H. Otis.*

1094. ANONYMOUS. **Paper from bagasse.** Sci. Amer. Monthly 1: 283. 1920. [Review of a paper in The Technical Engineering News. Feb., 1920.]—Describes the process for commercially making a special paper from bagasse, which is sugar cane from which the juice has been extracted.—*Chas. H. Otis.*

1095. ANONYMOUS. **Home-made syrup from sugar beets.** Sci. Amer. Monthly 1: 285-286. 1920.—This appears to be a brief of a paper by ORT and WITHROW in the Journal of Industrial and Engineering Chemistry. Feb., 1920.—*Chas. H. Otis.*

1096. Anonymous. **El zacaton como materia prima para papel. [Zacaton as a paper-making material.]** Revista Agric. [Mexico] 4: 107-111. *1 fig.* 1919.—A popular account based on: U. S. Dept. Agric. Bull. 309. 1919.—*John A. Stevenson.*

1097. Anonymous. **Origen, cultivo e industria del cacahuate. [Origin, cultivation and commercial aspects of the peanut.]** Jalisco Rural [Mexico] 2: 81-86. 1920.—Copied from El Boletin de la Camara Agric. de Leon [Mexico].—*John A. Stevenson.*

1098. Atkinson, Esmond. **Weeds and their identification.** New Zealand Jour. Agric. 19: 232-234. *1 fig.* 1919.—This is a continuation of a series of articles interrupted by the war in 1916. Plants known as "winter annuals" are under discussion. Spurrey (*Spergula arvensis*) is described in detail at various stages of growth. It is reported as a useful plant in some countries, but it can be considered only as a noxious weed in New Zealand. Its position as a weed, and possible control measures are discussed.—*N. J. Giddings.*

1099. Aumüller, F. **Nutation und Feinheitsgrad der Spelzen bei zweizeiliger Gerste. [Nutation and the degree of fineness of the glumes in two-rowed barley.]** Illustrierte Landw. Zeitg. 39: 430-431. *Fig. 332-333.* 1919.—The heads of varieties having fine glumes are shown by measurements to stand more nearly upright than those having coarser glumes. The former varieties are of higher quality but the latter are more productive.—*John W. Roberts.*

1100. Bancroft, Wilder T. [Rev. of: Peters, Charles A. **The preparation of substances important in agriculture.** *3rd. ed.* 19 x 14 *cm., vii + 81 p.* John Wiley and Sons, Inc.: New York, 1919. $0.80.] Jour. Phys. Chem. 23: 444. 1919.

1101. Banó, Jose de. **Dos cosechas de avena por una. [Two crops of oats for one.]** Rev. Agric. [Mexico] 4: 154-156. *2 fig.* 1919.—A ratoon crop secured under favorable weather conditions at small labor cost.—*John A. Stevenson.*

1102. Barber, C. A. **The growth of sugar cane.** Internat. Sugar Jour. 22: 198-203. 1920. —The fifth article of a series on the growth of sugar cane deals with the rate of maturing of the cane plant as a whole, the rate of early development, the average length and thickness of the mature joints, and the richness of the juice in branches of different ages. [See next following Entry, 1103.]—*E. Koch.*

1103. Barber, C. A. **The growth of sugar cane.** Internat. Sugar Jour. 22: 76-80. 1920. —The fourth article of a series on the growth of sugar cane deals with the formula for the branching of the cane plant. [See next preceding Entry, 1102.]—*E. Koch.*

1104. Barber, C. A. **Progress of the sugarcane industry in India during the years 1916 and 1917.** Agric. Res. Inst. Pusa Bull. 83. *46 p.* 1919.—The cane varieties in general use are poor, and the cultural practices and methods of handling the product primitive. The Department is endeavoring to introduce improvements along these lines, and the reports cover some of this work as carried out in the various provinces. Reports are given for Madras, Travancere, Mysore, Bombay, Central Provinces, Bengal, Bihar and Orissa, United Provinces, Punjab, North-west Frontier Province, Assam, and Burma.—*N. J. Giddings.*

1105. Besson, M. A., and Adrian Doane. **Darso.** Oklahoma Agric. Exp. Sta. Bull. 127. *20 p. Fig. 1-6.* 1919.—Darso is a new grain sorghum of unknown origin, possessing superior drouth resisting qualities. It is a dwarf variety of very uniform size, early maturing, leafy, red-seeded. The forage has a higher total sugar content than kafir or feterita. The feeding value of the seed is less than that of black-hulled white kafir. It is recommended as a grain sorghum in the drier regions of Oklahoma, Texas, and Kansas, but not in the more humid regions where other grain sorghums and corn make satisfactory yields.—*John A. Elliott.*

1106. Beverley, J. **Maize notes.** New Zealand Jour. Agric. 19: 242-243. 1919.

1107. Bolley, H. L. **Official field crop inspection.** Proc. Assoc. Official Seed Analysts 1919: 22-31. 1919.—Author believes "that the first step in cereal crop improvement rests in further extension of our state seed and weed laws and in the activity of the forces represented by them, to include proper control of seed crop production and of seed and grain distribution." Seed inspection laws alone have failed to insure seed and crop improvement since they inspect in the bin or bag after the goods has left the farm. Proposes for "every cereal producing state a law authorizing seed, field crop inspection, seed certification, seed standardization and seed sales lists" under the supervision of a competent officer, also providing for educational emphasis together with means for demonstrations and field work with seed plots.—*M. T. Munn.*

1108. Breakwell, E. **Popular description of grasses.** Agric. Gaz. New South Wales 31: 24-28. *Fig. 1-3.* 1920.—Habits of growth and seed production, palatability, behavior under irrigation and commercial possibilities are given for the genus Danthonia as found in New South Wales. *Danthonia longifolia, D. bipartita* and *D. pallida*, are figured. The Danthonias constitute 90 per cent of the grass herbage on the tablelands and slopes in New South Wales, and are common in western districts. Seed habits are fairly good. The Danthonias will be valuable in pastures in the future.—*L. R. Waldron.*

1109. Breakwell, E. **A remarkable fodder plant. Shearman's clover.** (*Trifolium fragiferum* var.) Agric. Gaz. New South Wales 31: 245-250. *4 fig.* 1920.—This report is given by the agrostologist. This clover was propagated vegetatively from an individual plant found growing alone several years previously. A taxonomic study indicates it to be unique, but closely allied to strawberry clover, *T. fragiferum.* The author suggests that it may have resulted from a cross between *T. fragiferum* and *T. repens* or *T. medium* or even between the two latter. Although under observation for over 20 years it has not been observed to produce viable seed. Compared with *T. fragiferum*, it is said to spread three times as quickly and to produce six times the amount of feed. Its palatability and nutritive quality are stated to be of the highest order. It thrives on marshy and slightly saline soils. It is not killed by frost. Chemical analyses are given.—*L. R. Waldron.*

1110. Breakwell, E. **Trials of Wimmers rye-grass. (Lolium subulatum.)** Agric. Gaz. New South Wales 31: 107-110. *2 fig.* 1920.—Conclusions as given are unfavorable to the grass both as to cultural results and palatability.—*L. R. Waldron.*

1111. Breakwell, E. **Bokhara Clover on the southern table-lands.** Agric. Gaz. New South Wales 31: 67. 1920.—Treats of *Melilotus alba.*—*L. R. Waldron.*

1112. Breasola, M. **La devitalizzazione dei semi di Cuscuta. [The killing of Cuscuta seeds.]** Staz. Sper. Agr. Ital. 52: 193-207. 1919.—This is a continuation of work which was reported upon in 1913. The purpose of the investigation was without screening to find a method of killing the seeds of Cuscuta in a lot of leguminous seeds. It was found that due to the different sizes of the seeds of *C. arvensis* and *C. Trifolii* screening would not separate the former from seeds of *Trifolium.* The method devised was that of heating the lot; incidentally it was found that the seeds of *Medicago sativa, Trifolium pratense, Trifolium repens* and *Lotus corniculatus* did not lose their vitality when exposed to the temperatures of experiment, i.e., 65°C. for one and two hours, 70°C. for one hour and 75°C. for one hour. In fact it was found that the number of seeds of these leguminosae germinating was in some cases greater after the treatment. The striking advantage was also found that the seeds of Cuscuta most easily screened out of seeds of the legume was the one that seemed to resist heat a little better (*C. Trifolii*) while the other (*C. arvensis*) was most easily killed. When tried in soil, the germinability of the two was found to decrease from 43.6 per cent to 11.8 per cent in *C. Trifolii* and from 55.6 per cent to 0.2 per cent for *C. arvensis* when heated for one hour at 75°C.—*A. Bonazzi.*

1113. Brown, Edgar. **Voluntary labeling by seedsmen.** Proc. Assoc. Official Seed Analysts 1919: 41–42. 1919.—Following a suggestion made by the Department of Agriculture, many large seed houses and firms pledged their support to the proposal that seedsmen label all farm seeds sold, giving on each lot of 10 pounds or more, purity, germination, and date when tested, and if imported, the country of origin. A series of purchases of seeds from seed dealers throughout the country showed that 78 per cent of the samples were not labeled, however, "a larger percentage of the seedsmen who specifically agreed to label their seeds were found to comply with the agreement than was the case with seedsmen who did not so express themselves."—*M. T. Munn.*

1114. Brown, W. H. **Philippine fiber plants.** Forestry Bur. Philippine Islands Bull. 19. *115 p., 28 pl.* 1919.—See Bot. Absts. 5, Entry 1304.

1115. Brunol, Gil Morice. **Algunos pastos naturales de Mexico.** [**Natural pastures in Mexico.**] Rev. Agric. [Mexico] 4: 58–62. *1 fig.* 1919.—Outlines the different types of pasture grasses in Mexico.—*John A. Stevens.*

1116. Burgess, J. L. **Relation of varying degrees of heat to the viability of seeds.** Proc. Assoc. Official Seed Analysts 1919: 48–51. 1919.—The author conducted experiments with corn, wheat, oats, rye, cowpeas, soy beans, and garden beans—seeds most liable to injury by insect pests, with a view of ascertaining the critical temperature above which the viability of each species is affected. The results of the experiments are given in tabular form.—*M. T. Munn.*

1117. Call, L. E. **Director's report.** Kansas Agric. Exp. Sta. 1917–18. *63 p.* 1918.—See Bot. Absts. 5, Entries 1466, 2024.

1118. Chambliss, Charles E. **Prairie rice culture in the United States.** U. S. Dept. Agric. Farmers Bull. 1092. *26 p., 13 fig.* 1920.

1119. Clayton, W. F. **The tea industry in South Africa. I.** South African Jour. Indust. 3: 112–120. *Pl. 1–2.* 1920.—Brief history of the tea industry in Natal, and of the cultural methods employed.—*E. M. Doidge.*

1120. Cockayne, L. **An economic investigation of the Montane tussock grassland of New Zealand.** New Zealand Jour. Agric. 19: 343–346. *2 fig.* 1919.—This is the fourth of a series of articles dealing with the Montane tussock grassland. The California thistle, *Cnicus arvensis*, is reported as becoming firmly established in some areas which were bare from overgrazing. It seems to be palatable to some animals, and may help to establish other useful plants, in which case it should not be considered a weed.—*N. J. Giddings.*

1121. Cowgill, H. B. **Cross pollination of sugar cane.** Jour. Dept. Agric. and Labor Porto Rico 3: 1–5. 1919.—See Bot. Absts. 5, Entry 1478.

1122. Crevost, C., and C. Lemarie. **Plantes et produits filamenteux et textiles de l'Indochine.** [**Fiber- and textile-producing plants of Indo-China.**] Bull. Econ. Indochine 22: 813–837. *Pl. 2.* 1919.—A continuation of the general paper on this subject, covering the families Asclepiadaceae, Ulmaceae, Urtricaceae, Scitamineae, Bromeliaceae, Amaryllidaceae, Liliaceae, and Pontederiaceae.—*E. D. Merrill.*

1123. Crocker, William. **Optimum temperatures for the after-ripening of seeds.** Proc. Assoc. Official Seed Analysts 1919: 46–48. 1919.—The author made a study of freshly harvested seeds of species of Crataegus, American linden, sugar maple, peach, and two species of Ambrosia. These seeds are typical of those having dormant embryos. The changes that go on and lead up to their normal germination are spoken of as after-ripening of the embryos. The embryos of these seeds must go through certain fundamental physiological changes before they sprout normally, since the embryos will not grow at all or only abnormally when they are naked and given all ordinary conditions favorable to germination. The optimum temperature for the process of after-ripening lies in the region of 4 to 5°C., and a constant temperature in these limits is very much more favorable than alternations between it and higher or lower temperatures. At freezing temperatures, after-ripening of these embryos progresses very slowly if at all, while temperature periods above 10°C. are especially detrimental to the process. The facts disclosed by the investigation raise the question whether nurserymen who layer their seeds to produce after-ripening would not do better to put the seeds in cold storage houses at optimum temperatures of 4 to 5°C., which would lead to a much more rapid and complete after-ripening than is attained in layering under fluctuating temperatures. It is the belief of the author that such methods should give returns in a greater percentage of seeds producing plants and in the general high vigor of the plants resulting from completed after-ripened embryos.—*M. T. Munn.*

1124. Cross, W. E. **The Kavangire cane.** Louisiana Planter and Sugar Manufacturer 63: 397–399. *1 fig.* 1919.—See Bot. Absts. 5, Entry 2113.

1125. Day, James W. **The relation of size, shape and number of replications of plats to probable error in field experimentation.** Jour. Amer. Soc. Agron. 12: 100–105. 1920.—Variation is reduced by increasing the size of the plat to one-twentieth of an acre or over. Most accurate results are obtained from plats that are long and narrow and extend in the direction of greatest variation of the soil. An increase in the number of replications of a plat of given size increases the accuracy of the results.—*F. M. Schertz.*

1126. Deem, J. W. **Pasture top-dressing test in Waipukuraw county.** New Zealand Jour. Agric. 19: 295–296. 1919.—Sheep were used in these experiments and the results for two seasons indicate that it is well worth while to top-dress.—*N. J. Giddings.*

1127. Descombes, Paul. **Le reboisement et le développement economique de la France.** [**Reforestation and the economic development of France.**] Mém. Soc. Sci. Phys. Nat. Bordeaux VII, 2: 103–217. *2 fig.* 1918.

1128. Descombes, Paul. **Installation d'expériences prolongées sur le ruissellement.** [**Protracted experiments upon stream-flow.**] Mém. Soc. Sci. Phys. Nat. Bordeaux VII, 2: 17–35. *2 fig.* 1918.

1129. Doblas, José Herrera. **El trigo tremesino.** [**Three-months wheat.**] Bol. Assoc. Agric. [España] 12: 47–52. 1919.—Discusses a variety of wheat known as "Tremesino" (three-months) secured by selection from the common fall type planted in Spain. Yields were much less than with the fall variety and it is not recommended for planting except where planting at the usual time has been impossible. The variety yielded in four experiments an average of 10.75 hectoliters per hectarea.—*John A. Stevenson.*

1130. Doblas, José Herrera. **Estudio sobre el cultivo de la almorta.** [**Studies in the cultivation of the grass pea (Lathyrus sativus).**] Bol. Assoc. Agric. [España] 11: 665–674. 1919.—Botanical classification, uses, varieties, cultivation and yields of *Lathyrus sativus* (grass pea).—*John A. Stevenson.*

1131. Duncan, J. **Noxious weeds.** New Zealand Jour. Agric. 19: 366–368. 1919.—It is urged that more attention be given to the destruction of noxious weeds. Weeds should be destroyed before seeding and the assistance of the public should be enlisted to destroy weeds as soon as they are observed. Methods of weed dissemination are discussed and means of prevention are indicated. It is suggested that in sowing to pasture the best of seed and plenty of it should be used in order to obtain a good close sod. This tends to choke out and prevent growth and spread of weeds. Farmers should not admit thrashing machines to their farms until the machines have been thoroughly cleaned.—*N. J. Giddings.*

1132. Duysen, F. **Ueber die Keimkraftdauer einiger landwirthschaftliche Wichtiger Samen.** [Concerning the vitality of certain agriculturally important seeds.] Illustrierte Landw. Zeitg. 39: 282–283. 1919.—As the result of germination experiments it was found that the seeds of wheat, rye, barley and oats possess greater vitality than is generally supposed. Seeds of wheat 8 years old were 80 per cent viable and those of 14 years old 10 per cent viable. Nearly 100 per cent of wheat seeds from 1 to 7 years old germinated. Similar results were obtained with seeds of rye, barley and oats.—*John W. Roberts.*

1133. Earle, F. S. **Varieties of sugar cane in Porto Rico.** Jour. Dept. Agric. and Labor, Porto Rico 3: 15–55. 1919.—One of the principal objects of this paper is to show that sugar cane varieties may be described, classified, keyed out and determined by ordinary methods of descriptive botany or taxonomy. Heretofore, remarkably few descriptions of the cane varieties have been published that would enable one to identify a variety. The cultural value and characteristics of the numerous varieties grown in Porto Rico are described in detail. A key for identification and a taxonomic description of a number of varieties is also contained in the article.—*Anthony Berg.*

1134. Evans, L. A. **Annual report of the acting-director of agriculture.** Tasmania Agric. and Stock Dept. Rept. 1918–19: 1–6. 1919.—Report giving statistics on production of principal crops. District reports are included.—*D. Reddick.*

1135. Fawcett, G. L. **The identity of canes grown in Argentina.** Internat. Sugar Jour. 22: 135–136. 1920.—The botanist of the Agricultural Experiment Station at Tucuman states that Java 36 is the true P. O. J. 36 as it is grown in Java today. The probable source of this incorrect designation is the description by Noel Deerr in his "Cane Sugar." Another inaccuracy is calling the variety J 228 (P. O. J. 228) by two names—its own and J 139, when in reality Java 228 is meant. Correspondence with the Java station and shipments of cane show that the Argentina canes of Javanese origin are identical with the varieties of corresponding names as grown in Java.—*E. Koch.*

1136. French, G. T. **Organization, development and activities of the Association of Official Seed Analysts of North America.** Proc. Assoc. Official Seed Analysts 1919: 15–20. 1919.

1137. Fruwirth, C. **Die Ansprüche der zur Körnergewinnung gebauten Lupinearten an Boden und Klima.** [The soil and climate requirements of lupine species grown for yield of seed.] Illustrierte Landw. Zeitg. 39: 199–200. 1919.—The soil and climate requirements of the following species are discussed: *Lupinus luteus*, *L. angustifolius*, *L. albus*, *L. cruikshanksii*, *L. mutabilis*, *L. hirsutus.*—*John W. Roberts.*

1138. Fruwirth, C. **Zur Frage des Verpflanzens der Luzerne.** [Concerning the question of transplanting alfalfa.] Illustrierte Landw. Zeitg. 39: 226. 1919.—Results obtained through three years of experimentation indicate that greater yields of forage and seed may be expected from a field in which the seed has been drilled in than from one in which a stand has been obtained by transplantation. The advantages and disadvantages of both methods are discussed.—*John W. Roberts.*

1139. Gajon, Carlos. **Cultivo del chicharo de vaca.** [Cultivation of the cowpea.] Rev. Agric. [Mexico] 5: 26–34. *5 fig.* 1919.—Explains the value of a green manure crop, the manner of fixation of nitrogen by legumes and outlines the culture of cowpeas, a green manure crop well adapted to Mexican conditions.—*John A. Stevenson.*

1140. Gammie, G. A. **Report of the imperial cotton specialist.** Sci. Rept. Agric. Res. Inst. Pusa 1918–19: 115–124. 1919.—The report summarizes the qualities of some of the various varieties of cotton grown in India, and outlines experiments either in progress or contemplated to improve the cotton yield.—*Winfield Dudgeon.*

1141. Gardner, H. A. **Research in the paint industry.** Sci. Amer. 122: 89. 1920.—Observations on the growing of soya beans and manufacturing of soya oil used in mixing paints.—*Chas. H. Otis.*

1142. Gillette, L. S., A. C. McCandlish, and H. H. Kildee. **Soiling crops for milk production.** Iowa Agric. Exp. Sta. Bull. 187: 33–59. 1919.—This bulletin treats of the utilisation of soiling crops for milk cows, discussing in this connection alfalfa, red clover, alsike, sweet clover, field peas, cowpeas, soy beans, maize, oats, rye, foxtail millet, sweet sorghum, Sudan grass, and the following mixtures: oats and peas, oats and vetch, barley and peas, rye and hairy vetch, cowpeas and corn, cowpeas and sorghum, clover and timothy. A résumé of work by other investigators is added.—*C. V. Piper.*

1143. Goss, W. L. **Greenhouse and germination-chamber tests of crimson clover seed compared.** Proc. Assoc. Official Seed Analysts 1919: 64. 1919.—The results of 164 comparative and simultaneous germination tests of crimson clover seed, made between folds of blotting paper and in the greenhouse in soil gave results as follows: "The average of these 164 samples in the germinator was 50 per cent. The average germination of these same samples tested in soil in the greenhouse was 42 per cent."—*M. T. Munn.*

1144. Griffiths, David. **Prickly pear as stock food.** U. S. Dept. Agric. Farmers' Bull. 1072. *24 p. 8 fig.* 1920.

1145. Guthrie, F. B., and G. W. Norris. **Note on the classification of wheat varieties.** Agric. Gaz. New South Wales 31: 243–244. 1920.—Classification based on milling values.—*L. R. Waldron.*

1146. Hadlington, James. **Poultry Notes. February.** Agric. Gaz. New South Wales 31: 137–141. 1920.—Notes on growing alfalfa, *Medicago sativa.*—*L. R. Waldron.*

1147. Hansen, W. **Degeneration und Saatgutwechsel. [Degeneration and seed variation.]** Illustrierte Landw. Zeitg. 39: 558–560. 1919.—The writer discusses the degeneration in the yield and quality of various field crops and strongly advises seed selection as a remedy therefor.—*John W. Roberts.*

1148. Harrington, Geo. T. **Comparative chemical analyses of Johnson grass seeds and Sudan grass seeds.** Proc. Assoc. Official Seed Analysts 1919: 58–64. 1919.—A brief account of the results of comparative microchemical and permeability studies, also, gross chemical analyses of the seeds of these two closely related grass plants are given. These studies were made to determine whether there are any differences in their chemical nature, which might serve as a basis for explaining their marked difference in dormancy, germinating and after-ripening.—*M. T. Munn.*

1149. Harrison, W. H. **Report of the Imperial Agricultural Chemist.** Sci. Rept. Agric. Res. Inst. Pusa 1918–19: 35–45. 1919.—See Bot. Absts. 5, Entry 2271.

1150. Haywood, A. H. **Elephant, Para, and Guinea grasses at Wollongbar.** Agric. Gaz. New South Wales 31: 6. 1920.—Growth results given for *Pennisetum purpureum*, *Panicum muticum* and *P. maximum*, respectively. Elephant grass gave largest bulk of feed, was drought resistant and stimulated milk yields. Para grass covered the ground forming succulent feed, which remained green throughout the winter.—*L. R. Waldron.*

1151. Heiduschka, A., and S. Felser. **Beitrag zur Kenntnis der Fettsäuren des Erdnussöles. [Fatty acids of peanut oil.]** Zeitschr. Untersuch. Nahrungs.- u. Genussmittel 38: 241–265. 1919.—The composition of the fatty acids of the peanut oil examined was: Arachidic 2.3 per cent, Lignoceric 1.9 per cent, Stearic 4.5 per cent, Palmitic 4.0 per cent, Oleic 79.9 per cent, Linoleic 7.4 per cent.—*H. G. Barbour.*

1152. Helweg, L. **Sale of Danish root seed with guarantee for genuineness.** Seed World 7[8]: 24–26. 1920.—This article deals with the Danish methods of growing seeds of carrots, mangels, rutabagas, and turnips and the guaranteeing of the genuineness of the varieties and strains, a method now adopted by nine of the important seed dealers.—*M. T. Munn.*

1153. Hilgendorf, F. W. **Methods of plant breeding.** New Zealand Jour. Agric. 19: 354–358. 1919.—The work of several investigators is briefly reviewed and the conclusion drawn that simple selection for the improvement of self fertilized plants, such as wheat, is not considered as very hopeful.—*N. J. Giddings.*

1154. Hillman, F. H. **Rhode Island bent seed and its substitutes in the trade.** Proc. Assoc. Official Seed Analysts 1919: 64–68. 1919.—In this paper the author reports recent investigations which show that there are certain seed characteristics peculiar to each of the species, by means of which the kinds of seed may be distinguished and to a certain extent their true proportions in a mixture determined. The source of the seed, shown or indicated by the kinds of weed seeds and extraneous crop seeds present, is also an aid in determining the kind of seed and liability of mixture due to condition of growth and trade practice. Attention is directed by the author to detailed and illustrative descriptions of the seeds of bent grasses found in Bulletin 692, Professional Series, U. S. Department of Agriculture.—*M. T. Munn.*

1155. Hite, Bertha C. **Forcing the germination of bluegrass.** Proc. Assoc. Official Seed Analysts 1919: 53–58. 1919.—Experiments designed to ascertain the effect of light, temperature, and nutrient solutions on the germination of Kentucky bluegrass and Canada bluegrass are discussed. The experiments lead to the conclusions that: A complete viability test of Kentucky blue grass can be obtained in the dark with an exact 20°–30°C. alternation. Under constant temperature conditions this grass gives a higher germination in the light.—An alternation of 20°–30°C. in a dark chamber does not give a complete viability test of Canada bluegrass.—Direct sunlight or diffuse light a few hours each day with approximately a 20°–30°C. alternation gives a complete viability test of both Canada blue grass and Kentucky bluegrass. —Nutrient solutions with 20°–30°C. alternation in the dark give a complete viability test of both Kentucky bluegrass and Canada bluegrass.—So far we have not been able to find an alternation of temperature alone that would give a complete viability test of all samples of Canada bluegrass.—*M. T. Munn.*

1156. Hodson, Edgar A. **Upland long staple cotton in Arkansas.** Arkansas Agric. Exp. Sta. Circ. 49: 1–4. 1920.—The conditions under which upland long staple cotton varieties may be expected to produce a profitable crop are given together with a map showing the regions suited to the culture of long staple, intermediate, and short staple cottons.—*John A. Elliott.*

1157. Hodson, Edgar A. **Cotton Club manual.** Arkansas Agric. Exp. Circ. 84: 1–26. *11 fig.* 1920.—A popular manual covering the history, physiology, histology, culture, and use of the cotton plant.—*John A. Elliott.*

1158. Hodson, Edgar A. **Lint frequency in cotton with a method for determination.** Arkansas Agric. Exp. Sta. Bull. 168: 1-12. 1920.—Lint frequency was determined for 100 seed samples from 10 plants each of 25 varieties of cotton under test. The length of lint was determined, also the percentage of lint by weight. The seed was delinted with sulphuric acid and the volume determined by displacement in alcohol. The weight of lint of a uniform length of 25 mm. was calculated to give an accurate comparison of weight of lint produced per square centimeter of seed surface. The lint index for a plant represents the average amount of lint produced on one seed. Six tables are given showing the lint index, lint percentage, lint length, and lint frequency of the varieties studied.—"High lint frequency is closely correlated with short lint, therefore, it is necessary in making selections for high lint frequency to consider length and per cent of lint."—*John A. Elliott.*

1159. Howard, A., and G. L. C. **Report of the Imperial Economic Botanist.** Sci. Rept. Agric. Res. Inst. Pusa 1918-19: 46-67. *Pl. 5-6.* 1919.—The report includes a summary of the progress of investigations during the year under report, a program for 1919-20, and a list of literature published. Improved wheats (*Triticum vulgare*) "Pusa 4" and "Pusa 12" have produced yields of 3350 pounds and 3000 pounds respectively per acre, under good cultivation, in contrast with the very low yields of ordinary Indian wheats under Indian methods of cultivation. These improved wheats are being sent to other countries for trial. Other work includes methods of culture and improvement of indigo (*Indigofera tinctoria*); sun-drying of vegetables; methods of packing fruit for shipment; pollination of Indian crop plants; and soil drainage. Poor drainage in the Gangetic Plains during the monsoon interferes with proper root development and promotes excessive denitrification. Actual crop production under improved methods of cultivation indicate that with small expenditure of organic fertiliser the fertility of alluvial soils may be maintained or improved.—*Winfield Dudgeon.*

1160. Howe, H. E. **The future of the cotton industry. What organized research promises to do for grower and manufacturer.** Sci. Amer. 122: 300. 1920.

1161. Hutchinson, C. M. **Report of the Imperial Agricultural Bacteriologist.** Sci. Rept. Agric. Res. Inst. Pusa 1918-19: 106-114. 1919.—See Bot. Absts. 5, Entry 2282.

1162. Hyde, W. C. **Orchard cover-crop experiments on the Mountere Hills.** New Zealand Jour. Agric. 19: 364-365. *1 fig.* 1919.—This is the final report of a 4-year series of experiments. Oats made a good growth and oats with partridge peas were particularly good. Blue lupine was the best of the legumes and it made much the strongest growth on limed area.—*N. J. Giddings.*

1163. Jones, Earl. **Northern grown seed wins in Massachusetts.** Potato Mag. 2^9: 24, 29. 1920.

1164. Jordan, W. H., and G. W. Churchill. **An experience in crop production.** New York Agric. Exp. Sta. [Geneva] Bull. 465. *20 p.* 1919.—An account of an experiment in which a 4-year rotation of crops (corn, oats, wheat, and hay) was carried through four rotations on plats fertilised in different ways—with farm manure, a complete chemical fertiliser, a partial chemical fertilizer, and no fertilizer. On some plats the hay crop was red clover; on others, timothy. The total amount of dry matter produced was somewhat greater on plats treated with farm manure than on plats receiving a complete chemical fertilizer; and about 56 per cent greater than on unfertilised plats. Especially noteworthy is the fact that crop production was maintained as efficiently on the timothy plats as on clover plats. The results of a series of soil analyses made in connection with the experiment show the unreliability of soil analysis as a means of measuring soil fertility.—*F. C. Stewart.*

1165. Jovino, S. **Osservazioni sull'aridocoltura italiana. [Observations upon dry farming in Italy.]** Stas. Sper. Agr. Ital. 52: 69-121, 125-192. 1919.—See Bot. Absts. 5, Entry 2328.

1166. Kellogg, James W. **Seed report, 1918.** Bull. Pennsylvania Dept. Agric. 2^2: 1-29. *5 pl.* 1919.—The bulletin includes a table giving standards of purity for various seeds; results of tests on special samples; average purity of official samples; results of inspection and analyses in tabular form; and illustrations of the noxious weed seeds found in farm seeds.—*C. R. Orton.*

1167. Kellogg, James W. **Seed report, 1920.** Bull. Pennsylvania Dept. Agric. 3^4: 1-28. 1920.—Standards of purity established by the Seed Law for 20 kinds of seeds are given; also the results of special samples tested for purity; the average purity of official samples and the results of inspection are discussed and the data arranged in tabular form.—*C. R. Orton.*

1168. Kerle, W. D., and R. N. Makin. **Farmers' experiment plots. Winter fodder trials, 1919.** Agric. Gaz. New South Wales 31: 77-83. 1920.—In the Upper North Coast district, trials of cereals and legumes with and without fertilizers were carried out by a number of farmers. Results showed the practice to be successful. In the South Coast district cereals were tried without manures, with success.—*L. R. Waldron.*

1169. Killer, J. **Über die Bewertung der Centaurea solstitialis als Charakterbegleitsame bei der Herkunftsbestimmung von Kleesaaten. [Concerning the value of Centaurea solstitialis as an indicator of the origin of clover seed.]** Jour. Landw. 67: 109–110. 1919.—*Centaurea solstitialis* has long been recognized as indicating a southern European origin of clover seed. As this plant in recent years has been growing in Alsace in increasing abundance its seed may also be found in clover seed from there.—*C. E. Leighty.*

1170. Koerner, W. F. **Auf welche Krankheitsformen ist beim "Durchsehen" und "Aushauen" der zur Saatgewinnung bestimmten Kartoffelfelder besonders zu achten. [What diseases are to be considered especially in going through and thinning out potato fields from which seed potatoes are to be selected.]** Illustrierte Landw. Zeitg. 39: 323–324. *Fig. 252–259.* 1919.

1171. Lansdell, K. A. **Some common adulterants found in agricultural seeds. I.** Jour. Dept. Agric. Union South Africa 1: 26–31. *Plates II–IV.* 1920.

1172. Lewis, A. C., and C. A. McLendon. **Cotton variety tests.** Georgia State Bd. Entomol. Circ. 29. *20 p.* 1920.—Outlines tests with twenty-eight varieties of cotton (Gossypium) for 1919 conducted in the following Georgia counties: Sumter, Stewart, Dooley, Burke, Wilks, Douglas and Habersham. In each test, from ten to twenty varieties were used. Summaries of the various tests and recommendations of the varieties for different sections and under different conditions are given. Lists are appended of coöperative cotton growers and of parties from whom cotton seed may be purchased.—*T. H. McHatton.*

1173. Macpherson, A. **Lucerne growing for seed.** New Zealand Jour. Agric. 19: 369–371. 1919.—This article discusses the preparation of the seed bed, general cultural methods, weather conditions, harvesting the seed crop, etc. Conclusions are drawn that good crops of lucerne seed may be produced on well drained soil of average fertility. Very rich land and soil supplied with an abundance of moisture produce herbage rather than seed. Thick stands of lucerne are not favorable for good seed production. During the period devoted to the seed crop, two crops of hay may be taken from thick stands, which will be found of more profit. Old stands that are thinning out will often produce good crops of seed. The best practice for seed production is to establish a special wide-spaced stand by sowing the seed in rows 28 inches or more apart and cultivating two or three times.—*N. J. Giddings.*

1174. Macpherson, A. **Lucerne-culture tests at Ashburton Experimental Farm.** New Zealand Jour. Agric. 19: 288–293. 1919.—Experiments were conducted to indicate the proper amount of seed; the best method of sowing, and the effects of lime and fertilizers. As a result of these tests it is recommended: Seed should be sown in drills from 14 to 21 inches apart, to admit of cultivation; that not less than 15 pounds of seed per acre should be used; and that lime should be used, but not fertilizers.—*N. J. Giddings.*

1175. Maiden, J. H. **Chats about the prickly pear. No. 1.** Agric. Gaz. New South Wales 31: 117–120. 1920.—A brief historical survey of *Opuntia* spp. as an Australian pest is presented.—*L. R. Waldron.*

1176. Maiden, J. H. **Chats about the prickly pear. No. 2.** Agric. Gaz. New South Wales 31: 195–199. 1920.—Remarks on possible minor uses of *Opuntia* spp.—*L. R. Waldron.*

1177. McDiarmid, R. W., and G. C. Sparks. **Farmers' experiment plots. Potato experiments, 1918–19.** Agric. Gaz. New South Wales 31: 37–42. 1920.—Yields are given for different varieties in the New England district and the southwestern slopes at different points, with different manures and for different cultural methods. Artificial manures proved to be valuable.—*L. R. Waldron.*

1178. McDiarmid, R. W. **Grain sorghums in northern districts.** Agric. Gaz. New South Wales 31: 17–18. 1920.—Satisfactory results were obtained at Pallamallawa and Tenterfield with 5 varieties of *Andropogon sorghum*, used both as green feed and for grain production. The maximum yield of grain was 28 bushels per acre from Kaoliang, which was also the earliest variety.—*L. R. Waldron.*

1179. McKay, J. W. **Assam Experiment Station.** Rept. Karimganj Agric. Exp. Sta. 1918–19: 1–16. 1919.—Annual report of Director of the Assam Experiment Station, recording progress in methods of cultivation and selection of promising varieties of commonly cultivated field crops.—*Winfield Dudgeon.*

1180. Menges, Franklin. **Report on soils and crops.** Bull. Pennsylvania Dept. Agric. 1[1]: 111–114. 1918.—Some brief considerations of the conditions favoring the conservation of food materials in the soil and what may be expected by a proper supplementation of them.—*C. R. Orton.*

1181. Miege, E. **Le desinfection du sol.** [The disinfection of the soil.] Prog. Agric. et Vitic. 74: 133–140. 1920.—See Bot. Absts. 5, Entry 2284.

1182. Miéville, R. **Note sur le théier sauvage du Phou-Sang Région du Tranninh (Haut-Laos).** [Note on the wild tea of Phou-Sang.] Bull. Agric. Inst. Sci. Saigon 2: 87–99. 1920.

1183. Mitscherlich, Eilh. Alfred. **Zum Gehalt der Haferpflanze an Phosphorsäure und seinen Beziehungen zu der durch eine Nährstoffzufuhr bedingten Ertragserhöhung.** [On the phosphoric acid content of the oat plant and its relation to the increased yield resulting from addition of nutrients.] Jour. Landw. 67: 171–176. *1 fig.* 1919.—The law which Pfeiffer and others believe they have established is not confirmed by these investigations.—*C. E. Leighty.*

1184. Münter, Dr. **Pflanzenanalyse und Düngerbedurfnis des Bodens.** [Plant analysis and fertilizer requirement of the soil.] Jour. Landw. 67: 229–266. 1919.—See Bot. Absts. 5, Entry 2275.

1185. Myers, C. H. **The use of a selection coefficient.** Jour. Amer. Soc. Agron. 12: 106–112. 1920.—See Bot. Absts. 5, Entry 1590.

1186. Nelson, Martin, and L. W. Osborn. **Report of oats experiments 1908–1919.** Arkansas Agric. Exp. Sta. Bull. 165. *32 p., 2 pl.* 1920.—Thirteen tables are given showing yields of 45 varieties of fall seeded and spring seeded oats under different dates of sowing and different rates of seeding. Tests were carried on in different sections of the state upon various types of soil. Recommendations are made of varieties adapted to different sections of the state and as to the cultural methods to be followed.—*John A. Elliott.*

1187. Nelson, Martin, and Edgar A. Hodson. **Varieties of cotton, 1919.** Arkansas Agric. Exp. Sta. Bull. 166. *8 p.* 1920.—Five tables are given showing the rank in seed cotton, lint production, seed production, and value of lint per acre of from 8 to 25 varieties, tested in various parts of the state, on different types of soil.—*John A. Elliott.*

1188. Olivares, Daniel. **Cultivo del lupulo.** [Cultivation of hops.] Revista Agric. [Mexico] 3: 374–378. *Ibid.* 4: 12–16, 62–64. *2 fig.* 1919.—An account of the importance and possibilities of hops as a crop in Mexico giving details, botanical description, varieties, cultivation, fertilizers, manner of harvesting and yields.—*John A. Stevenson.*

1189. Ortiz, Ruben. **Rotacion y alternacion de los cultivos.** [Rotation and alternation of crops.] Jalisco Rural [Mexico] 2: 61–64. 1920.—Popular résumé of reasons for crop rotations. A series of rotations suitable for Mexican conditions is given.—*John A. Stevenson.*

1190. Oswald, W. L. **Coöperation between the seed analysts and the seed trade.** Proc. Assoc. Official Seed Analysts 1919: 38–41. 1919.

1191. Pammel, L. H., and C. M. King. **An annual white sweet clover.** Proc. Iowa Acad. Sci. 25: 249-251. *Pl. 4-6.* 1920.—Origin and history of an annual strain of Melilotus alba found at Ames, Iowa.—*H. S. Conard.*

1192. Pammel, L. H., and C. M. King. **Test your clover and timothy seed.** Iowa Agric. Exp. Sta. Circ. 59. *8 p.* 1919.

1193. Pammel, L. H., and C. M. King. **Johnson grass as a weed in southwestern Iowa.** Iowa Agric. Exp. Sta. Circ. 55. *4 p., 3 fig.* 1919.—Johnson grass has become established in southern Iowa, and promises to become a menace to the farmers. A brief discussion is given, including a botanical description of the grass and seed, together with methods of extermination.—*Florence Willey.*

1194. Pavoni, P. A. **El cultivo de la higuerilla. [Cultivation of the castor bean.]** Jalisco Rural [Mexico] 2: 41-45. 1919.—A compiled account of the cultivation of the castor bean.—*John A. Stevenson.*

1195. Pieper, H. **Beschreibung einer Methode zur raschen Erkennung von Futterrübensamen im Zuckerrübensamen. [The description of a method for rapid differentiation between stock beet seed and sugar beet seed.]** Zeitschr. Vereins Deutsch. Zucker-Indust. 766: 409-418. 1919.

1196. Pitt, J. M. **Farmers' experiment plots. Winter green fodder exepriments, 1919.** Agric. Gaz. New South Wales 31: 7-12. *3 fig.* 1920.—Soiling crops are recommended for winter and spring in the Central Coast district, as dry weather invariably occurs. Cultural details and yield results are given for 10 localities (or less) for 8 varieties of wheat, 5 of oats and vetches and peas in combination with wheat or oats. The maximum yield of over 21 tons was secured from Thew wheat and peas.—*L. R. Waldron.*

1197. Pitt, J. M., and R. W. McDiarmid. **Farmers' experiment plots. Maize experiments, 1918-19.** Agric. Gaz. New South Wales 31: 99-105. 1920.—Different varieties, with and without phosphatic manures, were grown at various localities in the Central Coastal district. The use of manures generally showed profits. The Improved Yellow Dent gave a maximum yield of 125 bushels per acre. Light yields were secured in the Northern districts. —*L. R. Waldron.*

1198. Powers, W. L., and W. W. Johnston. **The improvement and irrigation requirement of wild meadow and tule land.** Oregon Agric. Exp. Sta. Bull. 167. *44 p., 25 fig.* 1920.—There are more than 515,000 acres of wild meadow and tule land in eastern Oregon, the former comprising more than one-third of the irrigated area of the state. The chief vegetation in the peat swamps consists of tules and flags, mingled with wire grass and sugar grass, while the chief meadow grasses are redtop, blue-joint, meadow grass and wild clover. In the Chewaucan Basin alsike clover and timothy have yielded 3½ tons an acre as compared to ¾ ton of native grass on adjoining land. Alfalfa in the Harney Basin has produced about 2 tons an acre, while native wild hay has averaged but ½ ton an acre. In the Fort Klamath region alsike clover and timothy have yielded more than double the amount of forage produced by native grasses. Results from 5 years experiments have shown that an average depth of 18 inches of water on the field could produce the maximum yield now obtained, while an average of 12 inches has given the largest yield per acre per inch of water used. The average cost for the production of wild hay has been nearly double that required for alsike clover and timothy. Marked increases in yield of alfalfa have been secured from an application of sulfur to swamp border soils.—*E. J. Kraus.*

1199. Ramsay, J. T. **Is change of seed necessary in the cultivation of potatoes?** Jour. Dept. Agric. Victoria 17: 651-657. 1919.—The selection of home grown seed potatoes has given as good results as imported seed potatoes.—*J. J. Skinner.*

1200. Ravaz, L. **Le nitrate d'ammoniaque. [Ammonium nitrate.]** Prog. Agric. et Vitic. 74: 33-34. *1 fig.* 1920.

1201. Rindl, M. **Vegetable fats and oils, II. Drying oils.** South African Jour. Indust. 3: 121-127. 1920.

1202. Robbins, W. W. **The organization of the Colorado seed laboratory.** Proc. Assoc. Official Seed Analysts 1919: 35-38. 1919.

1203. Robbins, W. W. **Research and seed testing.** Proc. Assoc. Official Seed Analysts 1919: 20-22. 1919.

1204. Robin, J. **Les differentes variétés de riz cultivées à la station de Cantho. [The different varieties of rice cultivated at the Cantho station.]** Bull. Agric. Inst. Sci. Saigon 2: 40-45. 1920.—Brief notes on the characters of 22 varieties of rice.—*E. D. Merrill.*

1205. Salmon, S. C. **Establishing Kanred wheat in Kansas.** Kansas Agric. Exp. Sta. Circ. 74. *16 p.* Aug., 1919.—Kanred wheat is a hard, red, winter wheat, resembling closely Turkey and Kharkof. It is resistant to winter killing, ripens early, yields more than any other commercial variety in Kansas and is very resistant to leaf rust and some forms of stem rust. It will probably be of commercial value in other states growing winter wheat.—*L. E Melchers.*

1206. Sanderson, T. **Value of Red Durum or D 5 wheat.** North Dakota Agric. Exp. Sta. Special Bull. 5: 507-517. 1920.—Deals with milling and baking values. There are presented coefficients of flour absorption, and also those for volume, color and texture of loaf. When these coefficients are applied to the data presented the D 5 wheat was found to be worth 23 cents per bushel less than No. 1 Amber Durum, and 38 cents less than No. 1 Hard Red Spring, for the years 1915-1919. The D 5 showed itself inferior in all loaf characters.—*L. R. Waldron.*

1207. Sayer, Wynne. **Report of the Imperial Agriculturist.** Sci. Rept. Agric. Res. Inst. Pusa 1918-19: 11-34. *4 pl.* 1919.—The report describes the results of experiments in crop rotation at the Agricultural Research Institute, Pusa, India, to determine the best methods of working the land of the Pusa farm, and field tests of new and improved varieties of commonly cultivated plants. A new variety of wheat (*Triticum vulgare*), "Hard Federation," stands up well in wind and rain, and yields up to 3300 pounds per acre.—*Winfield Dudgeon.*

1208. [Schule, N., and H. L. Maxwell.] **The oil in peanuts.** Sci. Amer. Monthly 1: 213. 1920. [Reprinted from Chemical News (London).]

1209. Scott, John M. **Bahia grass.** Jour. Amer. Soc. Agron. 12: 112-113. 1920.—A report of the promise of Bahia grass (*Paspalum notatum*), which has been introduced into the United States from South America and Mexico. Experiments in Florida have given very satisfactory results.—*F. M. Schertz.*

1210. Sparks, G. C. **Farmers' experiment plots. Potato experiments, 1918-1919.** Agric. Gaz. New South Wales 31: 251-254. 1920.—Different varieties were tried in several localities, with and without fertilisers. Fertilisers had a marked positive effect upon yield.—*L. R. Waldron.*

1211. Sparks, G. C., B. C. Meek, and R. W. McDiarmid. **Farmers' experiment plots. Wheat and oats experiments, 1919.** Agric. Gaz. New South Wales 31: 153-164. 1920.—Trials with wheat, also oats and barley, were carried out in three districts with a number of coöperators. The experiments dealt with the effect of fertilizing, early and late sowing, crop-harrowing, fallowing, rate of seeding and the effect of using graded and ungraded and acclimatized and unacclimatized seed. Yields and bushel weights of grain are given. Working the land after the rain gave growth and returns superior to that worked only prior to the rain and while the land was dry. The value of the properly compacted seed bed was demonstrated in the long and short fallowing plots and the May preparation with the spring-toothed cultivator only. The use of superphosphate with a quick maturing variety on the long and short fallowed land is unnecessary. Good yields on the long fallow plainly demonstrated the value of that system.—*L. R. Waldron.*

1212. STUCKEY, H. P. **Further studies in fertilizing and storing sweet potatoes.** Georgia Exp. Sta. Bull. 134: 77-87. 1920.—Bulletin 107 of the Georgia Experiment Station reports work on fertilizing sweet potatoes (*Ipomoea batatas*) which was begun in 1908, the first report being published in 1913. This Bulletin reports on the same work from 1914-1919 inclusive. The area utilized for the plats is Cecil clay loam, and the same kinds and amounts of fertilizer have been applied to the same plats from 1908 to 1919 inclusive. Plat No. 1, fertilized at the rate of 24 tons of stable manure per acre; plat No. 2, 2100 pounds 16-per-cent acid phosphate per acre; plat No. 3, 900 pounds sulphate of potash per acre; plat No. 4, 1500 pounds nitrate of soda per acre; plat No. 5, 1800 pounds of complete fertilizer. Results show that acid phosphate and sulphate of potash have increased the acidity of the soil. The complete fertilizer gave the largest total yield throughout the period of the test, stable manure coming second. Heavy nitrogenous fertilisation seemed to give potatoes a lighter color and somewhat poorer flavor. The variety of sweet potatoes used since 1913 has been Myers Early. The best quality potatoes were produced on the acid phosphate plat and the check. The potash seemed to have little influence in either color, flavor, or texture of the flesh. Potatoes from the experimental plats were tested in storage. Those from the check plat kept better through the winter than the others, but the data obtained were variable and a conclusion can hardly be drawn. In testing the influence of soil types on the keeping of sweet potatoes, potatoes grown on Cecil clay loam or red soil and on a gray phase of the Cecil clay loam were compared; it is concluded that under local conditions, potatoes grown on gray soil keep better than those grown on red soil. Potatoes from various plats were put in storage and loss of weight determined. The average loss of weight was 16.6 per cent. The loss of moisture from November 5th to March 1st was 3.73 per cent. The average total loss of weight was 16.6 per cent, and it is concluded that the percentage in loss of weight over the percentage of loss in moisture is doubtless due to the breaking down of carbohydrates and the giving off of carbon dioxide. In conclusion the author outlines a coöperative test on fertilizing sweet potatoes that is being carried on by several southern stations. It states results for one year. —*T. H. McHatton.*

1213. SYME, J. E. **Wheat plots at Narromine, 1919.** Agric. Gaz. New South Wales 31: 233-234. 1920.

1214. SYME, J. E. **Farmers' experiment plots. Wheat and oats experiments, 1919.** Agric. Gaz. New South Wales 31: 235-240. 1920.—Trials with wheat and oats were carried out with several coöperators with different varieties, under various cultural methods, with the use of manures, and with the use of home-grown and introduced seed. Yields of grain and wheat hay are given and rainfall data presented.—*L. R. Waldron.*

1215. TABOR, PAUL. **Permanent pastures for Georgia.** Georgia State Coll. Agric. Bull. 197. *36 p., 16 fig.* 1920.—Discusses the following pasture plats in Georgia: Japan clover (*Lespedeza stricta*), Bermuda grass (*Cynodon dactylon*), carpet grass (*Axonopus compressus*), Dallis grass (*Paspalum dilatatum*), white clover (*Trifolium repens*), Rhodes grass (*Chloris gayana*), Kudzu (*Pueraria thunbergiana*), bur clover (*Medicago arabica*), black medic (*M. lupulina*), red top or herds grass (*Agrostis alba*), orchard grass (*Dactylis glomerata*), tall oat (*Arrthenatherum elatius*), rescue grass (*Bromus unioloides*), arctic grass (*Bromus secalinus*), rye grass (*Lolium* sps.), Kentucky blue grass (*Poa pratensis*), The Paspalums (*Paspalum sps.*), giant carpet grass (*Axonopus furcatus*), broomsedge (*Andropogon sps.*), Indian oats (*Chrysopogon nutans*), wild rye (*Elymus sps.*), wire grass (*Aristida stricta*), lightwood-knot grass (*Sporobolus curtissii*), crab grass (*Syntherisma sps.*), crow foot (*Dactyloctenium aegyptium*), cane brake (*Arundinaria tecta, A. macrosperma*), maiden cane (*Panicum hemitomum*), smut grass (*Sporobolus berteroanus*), marsh bermuda (*Sporobolus virgatus*), Carolina clover (*Trifolium Carolinianum*), hop clover (*T. procumbeus*; *T. dubium*).—Directions for soil preparation and seeding are presented by the author and also mixtures of grass seeds suitable for various soils of the state.—*T. H. McHatton.*

1216. TAYLOR, H. W. **Tobacco culture, grading on the farm.** Rhodesia Agric. Jour. 17: 20–27. 1920.

1217. TRAN-VAN-HUU. **Note sur la variété de riz dite "Hueky."** [Variety of rice known as "Hueky."] Bull. Agric. Inst. Sci. Saigon 2: 75–78. 1920.

1218. TRAN-VAN-HUU. **Note sur la culture du riz flottant en Cochinchine.** [Cultivation of floating rice in Cochinchina.] Bull. Agric. Inst. Sci. Saigon 2: 46–52. 1920.—Notes on ten varieties and a description of the methods used in growing these forms of the rice plant which are peculiarly adapted to inundation.—*E. D. Merrill.*

1219. VAGELER, H. **Beziehung zwischen Parzellengrösse und Fehler der Einzelbeobachtung bei Feldversuchen.** [Relation between size of plot and error of the single observation in field experimentation.] Jour. Landw. 67: 97–108. *1 fig.* 1919.—Rye, oats, potatoes, and kohlrabi fields were each divided into 128 small rectangular plots, of which the yields were separately determined. The probable errors of the average yields of these plots considered singly and in different combinations were calculated. Different results were obtained according to the method and procedure followed, but when using the method considered least objectionable the probable error is not greatly reduced by enlarging the plots above about 50 square meters. —*C. E. Leighty.*

1220. VERNET, G., AND X. SALOMON. **Notes sur le Fourcroya gigantea Vent.** [Notes on Fourcroya gigantea Vent.] Bull. Agric. Inst. Sci. Saigon 2: 80–87. *Pl. 2.* 1920.

1221. WALDRON, L. R. **First generation crosses between two alfalfa species.** Jour. Amer. Soc. Agron. 12: 133–143. 1920.

1222. WALSTER, H. L. **Marquis versus durum wheats.** North Dakota Agric. Exp. Sta. Ext. Div. Circ. 34. *8 p.* 1920.—Summary of North Dakota yields.—*L. R. Waldron.*

1223. WEEKS, CHARLES R. **Growing alfalfa in western Kansas.** Kansas Agric. Exp. Sta. Circ. 73. *10 p.* July, 1919.—Information is given on soil requirements, seed bed preparation, date, rate and method of seeding, nurse crops, cultivation, time of cutting, seed crops, varieties and insects injurious to alfalfa in Kansas.—*L. E. Melchers.*

1224. WELTON, F. A. **Experiments with oats.** Monthly Bull. Ohio Agric. Exp. Sta. 5: 79–83. *7 tables.* 1920.—The article comprises tests of time, rate, manner, quality and varieties of seed.—*R. C. Thomas.*

1225. WENHOLZ, H. **Field peas as fodder. A substitute for wheat and oats.** Agric. Gaz. New South Wales 31: 167–170. 1920.

1226. WENHOLZ, H. **Soil improvement for maize. I. Manures and fertilizers.** Agric. Gaz. New South Wales 31: 29–35, 111–116, 117–183. 1920.

1227. WENHOLZ, H. **Fertilizers for green winter fodders.** Agric. Gaz. New South Wales 31: 241–242. 1920.

1228. WESTBROOK, E. C. **Tobacco culture. Bright leaf or flue-cured tobacco.** Georgia State Coll. Agric. Bull. 199. *36 p., 13 fig.* 1920.—Discusses a development in history of the bright tobacco (*Nicotiana tabacum*) industry in Georgia and considers advisability of increasing the crop. Discusses tobacco soils, crop rotation and general principles of tobacco culture, beginning with the preparation of the plant bed, and including transplanting, cultivating, insect enemies and diseases. Outlines directions for harvesting and curing, as well as for storage. Gives plans and suggestions for storage barns and curing sheds.—*T. H. McHatton.*

1229. Willey, Florence. **The vegetative organs of some perennial grasses.** Proc. Iowa Acad. Sci. 25: 341–367. *Fig. 121–144.* 1920.

1230. Williams, C. G. **Clipping tests of oats and wheat.** Monthly Bull. Ohio Agric. Exp. Sta. 5: 20–23. *4 tables.* 1920.

1231. Winters, S. R. **Paper from cottonseed waste.** Sci. Amer. 122: 299. *2 fig.* 1920.

1232. Wright, I. A. **The history of the cane sugar industry in the West Indies.** Louisiana Planter and Sugar Manufacturer 62: 414–415. *Ibid.* 63: 14–15, 108–109, 222–223, 237–239, 414–415. 1919.

1233. Young, J. P. **Report of Committee on the Cereal Crops.** Bull. Pennsylvania Dept. Agric. 1[1]: 11–13. 1918.—A report of the acreage, average yield per acre, estimated total production, average price per bushel, and estimated total value of the wheat, corn, rye, oats, buckwheat, potatoes, tobacco and hay crops in Pennsylvania for the year 1917. A comparative table with the yields per acre in 1916 is also given.—*C. R. Orton.*

BIBLIOGRAPHY, BIOGRAPHY AND HISTORY

Lincoln W. Riddle, *Editor*

1234. Anonymous. **Brief account of the life and works of Reginald Philip Gregory.** Jour. Botany 57: 47. 1919.

1235. Anonymous. **C. S. Harrison.** Florists' Exchange 47: 413. *1 fig.* 1919.

1236. Anonymous. **William J. Stewart.** Florists' Exchange 47: 413. *1 fig.* 1919.

1237. Anonymous. **Lewis S. Ware 1851–1918.** Internat. Sugar Jour. 21: 113. *1 pl.* 1919.—Lewis S. Ware, the distinguished sugar engineer, publisher, and author, of Philadelphia and Paris, made a special study of sugar beet industry and attempted unsuccessfully to establish it in the United States in 1873. In 1879 he established at Philadelphia a monthly publication, *The Sugar Beet*, which continued for 32 years. He also published pamphlets and books, his principal work being "Beet Sugar Manufacture and Refining," which is one of the standard works on this subject. Dr. Ware collected a sugar library of 12,000 volumes, which he has bequeathed to the Franklin Institute of Philadelphia.—*C. Rumbold.*

1238. Baker, C. F. **A contribution to Philippine and Malayan technical bibliography. Work fundamental to plant pathology and economic entomology.** Philippine Agric. 8: 32–37. 1919.—This bibliography gives mycological and entomological publications, each of which is based wholly or in part on the field results of the compiler, in the Philippines and Malaysia, during the period from 1913 to 1918, inclusive. The object of the index is to aid the investigator in obtaining the literature on these subjects, and to illustrate the great value of cooperation between scientists.—*S. F. Trelease.*

1239. Biggar, H. Howard. **The old and the new in corn culture.** U. S. Dept. Agric. Yearbook 1918: 123–137. *4 pl., 10 fig.* 1919.—See Bot. Absts. 4, Entry 28.

1240. Britten, James. **Bibliographical notes. LXXVI.—Henry W. Burgess's "Eidodendron."** Jour. Botany 57: 223–224. 1919.—A review of this work published in London in 1827 and bearing the full title "Eidodendron: Views of the general character and appearance of Trees, foreign and indigenous, connected with Picturesque Scenery." The work is of little or no botanical interest. Its only interest to the botanist is in connection with an essay headed "Botanical Diversions I" followed by a large title "Amoenitates Querneae." Here is included a comprehensive account of the oak in literature, history, poetry and commerce. The author of this essay was probably a more competent man than BURGESS. GILBERT BURNETT is often cited as the probable author. [See also next following Entry, 1241.]—*K. M. Wiegand.*

1241. BRITTEN, JAMES. **Bibliographical notes, LXXVII. John Ellis's directions for collectors.** Jour. Botany 57: 521. 1919.—This is an analysis of a damaged copy of this work published in 1771, which has lately been presented to the Department of Botany of the British Museum. It is entitled "Directions for bringing over Seeds and Plants from the East-Indies and other distant Countries in a State of Vegetation" and is anonymous. It proves to be a reissue of the first portion of the pamphlet published in 1770 by JOHN ELLIS, with some additional matter included. [See also next preceding Entry, 1240.]—*K. M. Wiegand.*

1242. COCKAYNE, L. **Presidential address.** New Zealand Jour. Sci. Technol. 2: 241-251. July, 1919.—Address delivered before the New Zealand Institute Science Congress, at Christchurch, 1919. Traces briefly the history of the New Zealand Institute, its activities, publications, equipment, influence, and aims. Urges the public support, financial and otherwise, of research in "pure" science, whether or not the given investigation has "an evident practical bearing." Notes the need of research in New Zealand in plant physiology and plant diseases.—*C. S. Gager.*

1243. FARR, BERTRAND H. **The peony and its people—from amateur to professional.** Flower Grower 6: 102. 1919.—References to the modern varieties of the peony and personal glimpses of those who produced them.—*W. N. Clute.*

1244. GAGNEPAIN, F. **Édouard Bureau. Sa vie et son oeuvre.** [Life and work of Édouard Bureau.] Rev. Gén. Bot. 31: 209-218. *Portrait.* 1919.—Edouard Bureau (1830-1918), entomologist, geologist and botanist, had a part in founding La Société Botanique de France. In 1874 A. DE JUSSIEU's chair of plant classification at the Paris Museum was reëstablished, and BUREAU was selected to occupy it. In this position he worked for more than 30 years in augmenting the great herbarium, developing the colonial floras, establishing a permanent exhibition of vegetable products, studying the palaeobotanical collections of BRONGNIART, and presenting courses in the Museum. A list of Bureau's 158 botanical contributions is appended.—*L. W. Sharp.*

1245. GUINET, A. **Auguste Schmidely. Sa biographie.** [The biography of August Schmidely.] Bull. Soc. Bot. Genève 10: 377-379. 1918.—SCHMIDELY is known for his study of the genera *Rosa* and *Rubus*. The results of his study from plants collected in the Swiss Alps are published mostly in the bulletin cited. He was born Jan. 26, 1838, and died Oct. 28, 1918.—*W. H. Emig.*

1246. HOLM, THEO. **The history of the popular name "Flower De Luce" or "Fleur De Lis" of the Iris.** Rhodora 21: 180-181. 1919.—A short discussion of the derivation of this name. It appears to have been first applied to the yellow iris growing on the shores of the river Lys in Flanders. The derivation dates back to the year 468 when the Franks left Flanders to invade and conquer Gaul, establishing the kingdom of France. In commemoration of their birthplace they selected this flower for their emblem. The name "Fleur de Lys" is therefore an abbreviation of "Fleur de la Lys."—*James P. Poole.*

1247. LEE, A. ATHERTON. **Plant pathology in Japan.** Phytopath. 9: 178-179. 1919.—The development of plant pathology in Japan commenced with Dr. Shirai's lectures at the Agricultural College, Tokyo, in 1886. Eighty pathologists now have a thriving society which publishes a journal with articles in English, German and Japanese. The latter are abstracted in English.—*R. E. Vaughan.*

1248. Meyer, Rud. **Heinrich Poselger.** Monatsschr. Kakteenkunde 29: 97-100. 1919.—There is given an account of the life of Poselger, his travels in Mexico in 1849-51, and his death in 1883.—*A. S. Hitchcock.*

1249. Nelson, J. C. **A little known botanist.** Amer. Botany 25: 129-133. 1919.—Juan Loureiro born in Lisbon, 1715. At the age of 20, visited Cochin China and later collected extensively there and in China proper, Cambodia, Bengal, and Malabar. He published Flora Cochinchinensis in 1790, and various shorter works in Portuguese.—*W. N. Clute.*

1250. Nicholson, Wm. Edw. **A reminiscence of the late Dr. Emil Levier.** Bryologist 21: 85-86. 1918.—The author gives an account of an evening spent with Dr. and Mme. Levier, and tells about the methods used by Dr. Levier in mounting specimens.—*Edward B. Chamberlain.*

1251. Peacock, Josiah C. **Franklin Muhlenberg Apple, Ph.G., Phar. D. Memoir.** Amer. Jour. Pharm. 91: 546-550. 1919.

1252. Petch, T. **Garcia da Orta's mongoose plants.** Ceylon Antiquary and Literary Register 4[2]: 143-149. 1919.—Discussion of the three plants of Ceylon, alleged to have been used as an antidote of snake poison, and described by the Portuguese physician Garcia da Orta, who lived at Goa from 1534 to about 1570. The first of these plants, which the ichneumon of fable seeks in order to protect itself against the bite of the cobra, is *Rauwolfia serpentina.* The second of Orta's species, the wood of which was formerly sent to Europe as *Lignum colubrinum*, was identified by Linné with *Strychnos nux-vomica.* In the author's opinion it is *S. trichocalyx.* The third species, hitherto unidentified, is determined as *Hemidesmus indicus* (Singhalese *iramusu*). None of these plants appears to be in use as a remedy for snake bite at the present day, nor are they enumerated in the recipes for snake-bite remedies, twenty in number, which Hoatson collected in Uva in 1822.—*B. Laufer.*

1253. Prain, (Sir) David. **"John" Roxburgh.** Jour. Botany 57: 28-34. 1919.—A discussion of the identity of "Roxburgh, junior," alluded to in Dr. William Roxburgh's *Flora Indica.*—*K. M. Wiegand.*

1254. Sewell, M. C. **Tillage: a review of the literature.** Jour. Amer. Soc. Agron. 2: 269-290. 1919.—See Bot. Absts. 3, Entry 1883.

1255. Stringer, H. B. **George Arnold.** Florists' Exchange 48: 521. *1 fig.* 1919.

1256. Vaupel, F. **Aus der alten Kakteenliteratur.** [On old cactus literature.] Monatsschr. Kakteenkunde 29: 25-31, 49-54, 61-66, 115-120. *5 fig.* 1919.—The author translates chapters from an old Spanish work published in 1547, Coronica de las Indias, by Gonçalez Hernandez de Oaiedo y Valdes. Chapter 23 describes the Pitahaya fruit; chapter 24 describes a columnar cactus called torches; chapter 25 concerns tunas and their fruits; chapter 1 of book 10 deals with tree cactuses.—*A. S. Hitchcock.*

1257. Whelpley, Henry M. **James Michenor Good.** Amer. Jour. Pharm. 91: 447-452. *Pl. 1.* 1919.—A review and appreciation of the life and work of the late James Michenor Good, one of the landmarks in American Pharmacy.—*Anton Hogstad, Jr.*

1258. Williams, Emile F. **George Golding Kennedy.** Rhodora 21: 25-35. *1 pl.* 1919.—Biographical sketch of the late George Golding Kennedy.—*James P. Poole.*

1259. Winslow, E. J. **Early days of the American Fern Society.** Amer. Fern. Jour 9: 33-38. 1919.

BOTANICAL EDUCATION

C. Stuart Gager, *Editor*
Alfred Gundersen, *Assistant Editor*

1260. Brown, Nelson Courtlandt. **The royal Italian forestry college.** Jour. Forestry 17: 807–812. 1919.—See Bot. Absts. 5, Entry 1303.

1261. Clute, Willard N. **Plant names and their meanings.—II. Ranunculaceae.** Amer. Bot. 26: 2–10. 1920.—The common names used for species of Ranunculaceae traced to their sources when possible.—*W. N. Clute.*

1262. Conard, H. S. **The general classification of higher plants.** Proc. Iowa Acad. Sci. 25: 237–240. 1920.

1263. Pammel, L. A. **State parks in Iowa.** Sci. Monthly 10: 516–521. 1920.—The plan proposes the preservation of some of the forests for the pleasure and education of all the people.—The parks are of different kinds. Lake parks which include enough of all lake shores to conserve animal and plant life; along streams where these have cut through ridges as the Devil's Backbone, and the forests associated with these; ledges on which most of the ferns of the state are found; mounds, palisades and similar areas suggest the plans.—It is far-sighted wisdom on the part of the state to establish these parks to preserve to future generations the natural history and geology and historic features of Iowa.—*L. Pace.*

1264. S., E. J. [Rev. of: Church, A. H. **Elementary notes on structural botany.** Oxford Botanical Memoirs No. 4. *27 p.* Oxford University Press, 1919.] Jour. Botany 58: 27. 1920.

CYTOLOGY

Gilbert M. Smith, *Editor*
George S. Bryan, *Assistant Editor*

1265. Balls, W. Lawrence. **The existence of daily growth-rings in the cell wall of cotton hairs.** Proc. Roy. Soc. London B 90: 542–555. *Pl. 14–16.* 1919.—Cellulose wall of Egyptian cotton swelled to five or ten times normal size by treatment with NaOH and CS_2 showed concentric layering. Correlated with Egyptian field crop conditions where growth is arrested each afternoon. Only one thin primary layer formed while cell is growing in length. When thickening sets in it proceeds to a maximum of 25 layers.—*Paul B. Sears.*

1266. Beer, Rudolph, and Agnes Arber. **On the occurrence of multinucleate cells in vegetative tissues.** Proc. Roy. Soc. London B 91: 1–17. *Pl. 1.* 1919.—Lists species in which multinucleate cells have been recorded in vegetative tissues, together with region of plant involved. List includes 177 species in 60 families of vascular plants. Theory of previous workers regarding amitotic origin of such multinucleate phases is questioned. No clear example of amitosis observed but numerous cases of mitosis normal up to cell plate stage observed. Instead of normal cell walls formation after mitosis Kinoplasm forms a hollow sphere around nucleus—"phragmosphere." This gradually enlarges until coextensive with cell cytoplasm. Suggested that numerous nuclei render available for use of cytoplasm valuable material (a) by increased nucleus surface (b) in certain cases by nuclear disintegration and resorption.—*Paul B. Sears.*

1267. Buscalioni, L. **Nuove osservazione sulle cellule artificiali.** [Further observations on artificial cells.] Malpighia 28: 403–434. *Pl. 11–12.* 1919.—This is a description and discussion of experiments with colloidal films. The plates are from photomicrographs of the results of experiments and show not only simulation of cell-walls, but also simulation of nuclei with chromatin-reticulum.—*L. W. Riddle.*

1268. Legrand, L. **Une conception biologique nouvelle de la cellule. [A new biological conception of the cell.]** Rev. Gén. Sci. Pures et Appliquées 30: 13. 1919.—Nothing essentially new, but a good review of the present situation.—*G. J. Peirce.*

1269. Mangenot, M. G. **Sur l'évolution du chondriome et des plastes chez les Fucacées. [The evolution of the chondriosome and of the plastids in the Fucaceae.]** Compt. Rend. Acad. Sci. Paris 170: 63-65. *1 fig.* 1920.—In the apical cells of *F. vesiculosus* and *F. platycarpus* mitochondria are to be found at some of the protoplasmic anastomoses in the cytoplasm, while at other anastomoses small phaeoplasts appear and elsewhere in these cells there are grains of fucosane. The adjacent peripheral cells also contain mitochondria, grains of fucosane and phaeoplasts, the last named being larger, having more pigment and reacting in a different fashion to the fixing solutions than those of the apical cell. Small phaeoplasts occur not only in the apical cells, but also in the cells of the central axis cut off from the apical cell on its proximal face and in the initial cells of adventitious shoots. The cells containing small phaeoplasts are considered to be embryonal in character.—*C. H. and W. K. Farr.*

FOREST BOTANY AND FORESTRY

Raphael Zon, *Editor*

J. V. Hofmann, *Assistant Editor*

1270. Adler, Friedrick v. d. **Aus dem Kubani Urwald. [The Kubani virgin forest.]** Oesterreich. Forst.- u. Jagdzeitg. 38: 23. 1920.—A short popular description of an 80 hectar area of virgin timberland in Bohemia. Trees 1 meter to 1.9 meters in diameter are found in contrast to the small sizes generally found in cut over forests in the same region.—*F. S. Baker.*

√ 1271. Aguilar, R. H., **The lumbang-oil industry in the Philippine Islands.** Philippine Jour. Sci. 14: 275-285. 1919.—Two kinds of lumbang nuts occur in the Philippines, lumbang bato (Aleurites moluccana) and lumbang banucalag (Aleurites trisperma), but when the word lumbang is employed it is taken to mean lumbang bato. The Bureau of Forestry is encouraging planting of the trees so that a sufficient supply of raw material may be assured. The nuts may be stored for a year or more without depreciable change. The oil is used in the calking of vessels, manufacture of soft soap, and in the manufacture of paints. The kernels may be separated from the shells and the oil expressed, or the whole nut ground up and the oil separated. The former is slower and more laborious but furnishes a larger percentage of oil and a cake of higher fertilizing value. The oil may be kept satisfactorily in copper containers.—*Albert R. Sweetser.*

1272. Ammon, W. **Ueber die Pflicht zum Unterholt subventionierter Aufforstungs und Verbauungs-Projekte. [The obligation to maintain subsidized forestation and construction projects.]** Schweiz. Zeitschr. Forstw. 71: 105-114. 1920.—One of the difficulties in maintaining a subsidized project is the change of ownership. When a change of title occurs the new owner accepts the subsidy as an obligation and fulfills it in so far as it is compulsory. Under the laws of Berne the acquisition of land carries with it the obligation to protect and continue any subsidized project although other cantons do not adequately provide for change of title. —A subsidy may consist of either a fixed sum or a per cent of the project undertaken. The State or Canton must have preference in the arrangement because in the event of non-fulfillment the project must be continued by the State or Canton.—Non-utilization of a tract for timber production or grazing constitutes a non-fulfillment of a subsidy agreement and leaves the present incumbent subject to a fine.—The regulations are still somewhat confused and it is recommended that the obligations of the State and land owner be more specifically defined and incorporated in the laws.—*J. V. Hofmann.*

1273. Anderson, J. **Ecuador contributes a wood that is lighter than cork.** Sci. Amer. 122: 281. *2 fig.* 1920.—Concerns *Ochroma lagopus*, balsa wood.—*Chas. H. Otis.*

1274. Anonymous. **Annual return of statistics relating to forest administration for the year 1917–18, British India.** *25 p., 1 diagram.* Simla, 1919.—The report contains summarised tabulated data on forest areas, improvement, protection, fires, grazing, planting, exports, expenditures, revenues, and other subjects for all the provinces. The present forest area under control of the Forest Department is 251,512 square miles or 23.3 per cent of the total area of all the provinces; 60,724 square miles, or 24 per cent of the forest area, are under approved working plans. 46.3 per cent of the entire forest area was under fire protection and 47,249 square miles, or 18.8 per cent, was entirely closed to grazing during the year. The financial statement shows a total revenue of 40,969,257 Rs, expenditure 21,157,063 Rs, leaving a surplus (cumulative) of 19,812,194 Rs. A final table gives the state of the finances by periods and years from 1869 to 1918, and the appended diagram shows graphically the relation of revenue, expenditure and surplus for the past ten years.—*E. R. Hodson.*

1275. Anonymous. **Automatic regulation of humidity in factories.** Sci. Amer. Monthly 1: 24–28. *6 fig.* 1920.—An article of interest to manufacturers of articles made from wood.—*Chas. H. Otis.*

1276. Anonymous. **Effect of decay on wood pulp.** Sci. Amer. Monthly 1: 247. 1920.

1277. Anonymous. **Fliegertätigkeit im Dienste des Forstschutzes.** [The use of air planes in forest protection.] Schweiz. Zeitschr. Forstw. 71: 82–83. *2 pl.* 1920.—Photographs taken from airplanes may be used for classification of areas in suitable regions for grazing, etc., also for topographic features and boundary locations of permanent forest areas. Photographs taken on a scale 1:25,000 bring out a great deal of detail. Often aerial patrol may bring out features that would be lost otherwise, such as snowslides and landslides in the initial stages. Taken in time, these may be prevented.—*J. V. Hofmann.*

1278. Anonymous. **Forests in Japan.** Amer. Forestry 26: 95. 1920.

1279. Anonymous. **Fra Dansk Skovforening. Handel og Priser i 1918–19.** [Business and prices, 1918–19.] Dansk Skovforenings Tidsskr. 4: 453–489. 1919.

1280. Anonymous. **Fuel value of wood.** Sci. Amer. Monthly 1: 425. 1920.

1281. Anonymous. **Holz als Ersatz der Kohle bei der Gaserzeugung.** [Wood as a substitute for coal in gas production.] Oesterreich. Forst.- u. Jagdzeitg. 38: 23. 1920.—Owing to the scarcity of coal in Zürich (Switzerland) wood was used in some of the retorts to eke out the coal supply. Mixtures of green cherry, oak, beech, alder, ash, willow, chestnut, hazel, birch were used. A yield of 27.5 per cent of gas was obtained of good quality running 29.2 per cent of hydrogen, 10.3 per cent methane and 2.9 per cent heavy hydrocarbons.—*F. S. Baker.*

1282. Anonymous. **Jaegersborg Dyrehave.** [The game reserve at Jaegersborg.] Dansk Skovforenings Tidsskr. 4: 4–8. 1919.

1283. Anonymous. **Kiln drying of green hardwoods.** Sci. Amer. Monthly 1: 247. 1920.

1284. Anonymous. **Lead pencils.** Sci. Amer. Monthly 1: 286. 1920.

1285. Anonymous. **Lumber salvage in France.** Sci. Amer. 122: 105. 1920.

1286. Anonymous. **Made of wood.** Sci. Amer. 122: 55. 1920. Some of the strange uses of wood and its by-products, as displayed in an exhibit prepared by the New York State College of Forestry.—*Chas. H. Otis.*

1287. Anonymous. **Paper famine if forests are wasted.** Amer. Forestry 26: 94–95. 1920.

1288. Anonymous. **Sodium fluoride as a wood preservative.** Sci. Amer. Monthly 1: 258. 1920.

1289. Anonymous. **The Southern Forest Conference.** Sci. Amer. Monthly 1: 286. 1920. —Notes on the meetings held in New Orleans, beginning Jan. 28, 1920.—*Chas. H. Otis.*

1290. Anonymous. **Die Sozialisierung des Forstwesens.** [The socialization of forestry.] Oesterreich. Forst.- u. Jagdzeitg. 37: 269-271. 1920.—During the war heavy cutting took place in Austrian forests and conditions are at present unsettled, the peasantry expecting a division and distribution of state forests and large estates. The future of sustained wood production and the very existence of many communities in the mountainous regions depends upon unification of management rather than further subdivision. The public value of the forests demands this. Formation of local voluntary associations of timber land owners, loggers, lumbermen and dealers is recommended, these associations to be united into a greater State association with large powers to govern forest management, lumber prices, export trade, and forest labor.—*F. S. Baker.*

1291. Anonymous. **Wohlfahrtseinrichtungen für Waldarbeiter.** [Housing conditions for forest laborers.] Schweiz. Zeitschr. Forstw. 71: 114-116. 1920.—Oberförster Schädelin advocated furnishing quarters in 1908 and Dr. Flury later pointed out that living conditions among the industries were better and more attractive than those of the forest laborers. This resulted in young men seeking other industries rather than the Forest Service.—The author describes the use of portable shelters built for 6 to 12 men that have proved successful in the Canton of Schaffhausen. The contentions in favor of a shelter equipped with a stove are that the men are more contented and willing to work in wet weather because they are able to dry their clothes when they return from work. Also the men do not use so much liquor in order to keep warm.—*J. V. Hofmann.*

1292. Ashe, W. W. **Notes on trees and shrubs in the vicinity of Washington.** Bull. Torrey Bot. Club. 46: 221-226. 1919.—See Bot. Absts. 3, Entry 2963.

1293. Baker, Hugh P., and Edward F. McCarthy. **Fundamental silvicultural measures necessary to insure forest lands remaining reasonably productive after logging.** Jour. Forestry 18: 13-22. 1920.—Silvicultural practice in the Adirondacks has not yet been fully settled and further work is needed in determining the limits of forest types, proper methods of slash disposal, and the requirements of the various species for establishment. A survey of forest lands and forests is needed.—*E. N. Munns.*

1294. Bang, J. P. F. **Lidt om Bjergfyrskovens Behandling.** [Notes on management of mountain fir.] Dansk Skovforenings Tidsskr. 4: 189-196. 1919.

1295. Bates, C. G. **A new evaporimeter for use in forest studies.** Monthly Weather Rev. 47: 283-294. *6 fig.* 1919.

1296. Bentley, J. B., Jr. **Municipal forestry in New York.** Amer. Forestry 26: 160-162. *4 fig.* 920. —Describes plantings made in Chenango County, N. Y.—*Chas. H. Otis.*

1297. Biilmann, H. H. **Nogle Tilvaekstoversigter fra Meilgaard Skovdistrikt.** [Some observations on growth in Meilgaard district.] Dansk Skovforenings Tidsskr. 5: 30-36. 1920.

1298. Blanford, H. R. **Financial possibilities of even-aged crops in Burma.** Indian Forester 46: 53-61. 1920.—Figures are presented which show possible returns from stands of teak and two other less important woods using 3 and 4.5 per cent as the interest rate. A rotation of around 75 years is forecasted.—*E. N. Munns.*

1299. Boas, J. E. V. **Det Nye Jagtlovsforslag og det Danske Skovbrug.** [The new game laws and Danish forestry.] Dansk Skovforenings Tidsskr. 5: 50-55. 1920.

1300. Bohn-Jespersen, J. F. W. **Sitkagranen i Klitten.** [Sitka spruce in Klitten.] Dansk Skovforenings Tidsskr. 4: 101-109. *Pl. 8.* 1919.

1301. Bowles, J. Hooper. **The California gray squirrel an enemy to the Douglas fir.** Amer. Forestry 26: 26. 1920.—A loss amounting to hundreds of thousands of dollars, caused by girdling of the trees by the squirrel.—*Chas. H. Otis.*

1302. Bridel, M. Marc. **Application de la méthode biochemique aux rameaux et aux écores de diverses espèces du genre Populus.** [Application of the biochemical method to the branches and barks of various species of the genus Populus.] Jour. Pharm. et Chim. 19: 429–434. Also *Ibid.* 20: 14–23. 1919.—See Bot. Absts. 3, Entry 2841.

1303. Brown, Nelson Courtlandt. **The royal Italian forestry college.** Jour. Forestry 17: 807–812. 1919.—A brief history of forest education in Italy is given with a description of the school at Vallombrosa. The school has a high scholastic requirement and courses and hours of work do not differ greatly from American practice.—*E. N. Munns.*

1304. Brown, W. H. **Philippine fiber plants.** Forestry Bur. Philippine Islands Bull. 19: 1–115. *28 pl.* 1919.—A general consideration of Philippine fiber producing plants with descriptions, occurrence, local names, methods of extracting fibers, and the uses to which the fibers are put. About 150 species are considered.—*E. D. Merrill.*

1305. Bruce, Donald. **Alinement charts in forest mensuration.** Jour. Forestry 17: 773–801. *15 fig.* 1919.—Alinement charts are adapted for formulae involving three variables. The development and principles underlying these devices with their application in problems of mensuration in determining the volume of trees is given in detail with illustrations as to their practical use. Advantages of much quicker computation and ease of construction are claimed over the use of slide rules and sets of curves employed in the past.—*E. N. Munns.*

1306. Butler, Ovid M. **Relation of research in forest products to forest administration.** Jour. Forestry 18: 275–283. 1920.—Silviculture cannot overlook the technical quality of the wood in its forest practice as the latter is influenced by silvicultural practices. Growth influences the technical properties of the wood greatly in seasoning, in strength and in use. Mechanical and physical qualities have already shown a close relation to rate and character of growth, and chemical uses may do likewise.—*E. N. Munns.*

1307. Cabrera, Teodoro. **La utilidad de los guayabos.** [Uses of the guava trees.] Revist. Agric. Com. y Trab. 2: 628. 1919.

1308. Carter, H. **Report on forest administration in Burma, for year ended June 30, 1918.** *114 p., 1 pl.* Rangoon, British India, 1919.—At the close of the year the aggregate area of the reserved forests was 29,116 square miles, about one-fifth of the total forest area of the province, and in addition there are large tracts proposed for reservation. The area under approved working plans is 10,832 square miles, or 37 per cent of the total reserved area. A system of cultivation called *taungya* (shifting cultivation, i.e., an area cleared and burned in hilly country for shifting cultivation) is practiced on areas aggregating 1,230 square miles of reserved forests by the wild hill tribes, comparatively low in the scale of civilization. When uncontrolled this system causes greater and more permanent damage than a fire. These wild tribes will not undertake permanent cultivation and are averse to settling in the plains. The problem is difficult but it is expected to regulate the *taungyas* by rotation in connection with the control of forest villages and also obviate local shortages of forest labor. By this plan the jungle tribes could be provided with all the virgin soil they require and the abandoned *taungyas* be stocked with a valuable forest crop. In a search for sites suitable for the extension of cinchona the following is reported of the damage by the *taungya* system: "Land with the necessary soil conditions has been very much to seek. Areas, some of which half a century or more ago would probably have afforded the requisite conditions, have been ruined by the practice of the jungle tribes of the pernicious system of shifting cultivation known in South India as *kumri*, in Burma as *taungya* and in Assam as *jhum*, by which enormous stretches of magnificent forest have been destroyed and the surface soil exhausted and more or less washed away by the unimpeded rush of rain water." And of an area west of the Upper Chindwin: "As regards cinchona prospects, the journey was disappointing. There was no need to go inland from the river for all along the outer ranges the ravages of shifting cultivation were only too evident. The evergreen forests are being rapidly destroyed." During the year 1,814 acres of *taungya* plantation were newly formed. Detailed tabulated data (72 pp.) is appended. In reviewing the year's work it is stated that the future before the Forest Department is one of the greatest activity; for not only has the better exploitation of the commercial forests to be undertaken, but the proper conservation of all that unclassed forest on which the agricultural demand is now concentrated can not be left in its present neglected condition. Such vast areas as the unclassed forests of Burma (74,707,834 acres) can not long be subjected to such profligate destruction as is now going on in many places for want of control and of staff to exercise it. *The conservation of these forests is not a matter of mere revenue,* but in the best interests of the whole population and most especially to the advantage of the agricultural classes.—*E. R. Hodson.*

1309. Cary, A. **Ticks and timber.** Amer. Forestry 26: 92–94. *5 fig.* 1920.—Concerns forest conditions in the Gulf states, U. S. A.—*Chas. H. Otis.*

1310. Chandler, B. A. **Financial loss to the community due to forest lands becoming wastes.** Jour. Forestry 18: 31–33. 1920.—Destructive lumbering is responsible not alone for the economic and financial loss due to the wasteful cutting and burning, but also for the degeneration of the people through loss of the vigorous stock, poor crops, whiskey and malnutrition. Such people need assistance from the outside and larger communities, as they are not self sustaining. In such regions, a peculiar type of degeneracy is developing.—*E. N. Munns.*

1311. Churchill, Howard L. **Approximate cost of private forestry measures in the Adirondacks.** Jour. Forestry 18: 26–30. 1920.—Cost of a forester and proper forest work in a lumber company was found to amount to an annual charge of 36 cents per thousand feet, while the charges due to conservative lumbering amount to 65 cents per thousand.—A comment by W. N. Sparhawk is to the effect that a number of items are not properly forestry but lumbering, thereby reducing the cost considerably.—*E. N. Munns.*

1312. Curtiss, C. F. **Forest parks and their relation to the rural community.** Rept. Iowa State Hortic. Soc. 53: 363–364. 1918.—See Bot. Absts. 3, Entry 3038.

1313. D'Aboville, P. **Detérmination du diamètre au milieu du tronc de l'arbre sur pied.** [**Determination of the middle diameter of a standing tree.**] Translated by S. T. Dana. Jour. Forestry 17: 802–806. *1 fig.* 1919.—By means of similar triangles based on known distances from the tree and the relation between the diameter of the tree at breast height and the intercepted diameter on a scale held at arms length, the diameter at half the height can be obtained. A formula is given for the practical application of this principle to field use.—*E. N. Munns.*

1314. Dalgas, J. M. **Døende Egeskov i Westfalen.** [**The dying oak forest: Westfalen.**] Dansk Skovforenings Tidsskr. 4: 64–72. 1919.

1315. Dalgas, J. M. **Gavntraeproduktionens Samfundsøkonomiske Betydning.** [**The economic importance of production of lumber.**] Dansk Skovforenings Tidsskr. 4: 446–453. 1919.

1316. Dalgas, J. M. **Nogle Oplysninger om Skove og Skovforhold i Nordslesvig.** [**Forest conditions in North Schleswig.**] Dansk Skovforenings Tidsskr. 4: 160–189. *1 fig.* 1919.

1317. Davis, R. N. **The winter aspect of trees.** Amer. Forestry 26: 87–91. *10 fig.* 1920.

1318. Dickie, F. **Discovery of sugar on Douglas fir.** Amer. Forestry 26: 84–86. *1 fig.* 1920.—The Indians of British Columbia knew of the existence of sugar on the Douglas fir long before the first white man came to North America. Only now the facts have been ascertained. Reporting upon the findings of Prof. Davidson and Mr. Teit, the writer states that "fir sugar" is occasionally formed during summer droughts or in dry-belt regions, sugar-bearing trees being most abundant between the 50th and 51st parallels and between 121°–122° longitude. The "manna" is a natural exudation from the tips of the needles, occurring as white masses ranging from ¼ inch to 2 inches in diameter on leaves and branches. A slight rain may quickly dissolve the sugar and it may be found recrystallized in patches at the base of the tree. At other times it remains in a semifluid condition. The sugar contains nearly 50 per cent of the rare trisaccharide, melezitose. Sugar-producing firs are chiefly those standing on gentle slopes facing east and north in comparatively open areas. In these situations, the leaves being exposed to the sun, an abundance of carbohydrates more than normal are formed during the day, which are not stored or carried to the growing tissues, as is the case with Douglas fir in heavily forested areas. The ground and atmosphere being dry, an increased root pressure and cessation of transpiration cause the leaves to become water-gorged. This water contains a sugar created by the reconversion of starch into sugar. By evaporation, the sugar is deposited on the leaf tips. By reason of the necessity for a succession of sunshiny days to produce the sugar, the Douglas fir does not yield a harvest that can annually be depended upon.—*Chas. H. Otis.*

1319. Dickie, F. **Sugar from the Douglas fir.** Sci. Amer. 122: 165, 174–175. *1 fig.* 1920. —The sugar-yielding firs are confined to the dry belt of British Columbia, and are chiefly found in the hottest parts of the interior of the province between parallels 50° and 51° and 121°–122° longitude. Trees standing on gentle slopes facing north and east and which are fairly wide spaced produce sugar in greatest abundance. The sugar occurs in white masses scattered over the foliage and branchlets, the accumulation of drops; drops of small size may appear upon the leaves at the tips and sometimes two or three tips will become imbedded in a very large drop. Analysis shows that the sugar yields about 50 per cent of the rare trisaccharide, known as melezitose. The Indians of the region have known of this occurrence of sugar on the Douglas fir for a long time and gathered it whenever available; but it is an uncertain crop, owing to reasons of climate.—*Chas. H. Otis.*

1320. Drolet, George. **Turpentine orcharding effect on longleaf timber.** Jour. Forestry 17: 832–834. 1919.—Turpentining with only slight damage to virgin longleaf timber has been successful in Alabama under a system where the crops are worked for only 2 years and then logged. Only healthy trees over 12 inches are tapped and not more than two cups are placed on a tree. Results of 4 years' work are given which show that there is a loss from turpentine operations which may be kept small, and that this loss increases with the length of the operation.—*E. N. Munns.*

1321. Dunbar, John. **Forty-two distinct forms of hickories.** [Rev. of: Sargent, C. S. **Notes on North American trees—II. Carya.** Bot. Gaz. 66: 229–258. 1918.] Amer. Nut. Jour. 10: 20–21. *1 fig.* 1919.

1322. Eldredge, I. F. **Management of hardwood forests in the southern Appalachians.** Jour. Forestry 18: 284–291. 1920.—An outline is given of a management plan for use in the hardwoods. The problem presented is one of area regulation with 6 age-classes to be considered in arriving at the volume of cut in any period in the working circle.—*E. N. Munns.*

1323. Eysselt, Joh. **"Weidwald."** **[Pasturewood.]** Oesterreich. Forst.- u. Jagdzeitg. 38: 1–2. 1920.—The present high value of grazing lands is leading to a demand for the extension of "pasture-woods" particularly in the alpine forests. This is considered contrary to public policy, however, as it would entail injury to exceedingly valuable protection forests, and lead to the extension of mountain torrents, avalanches and landslides, while experience as shown that the removal of the timber has also led to a deterioration of the pasturage as well. The segregation of all pasture-woods that have protection value is urged, to be managed on a strictly protective basis. Artificial extension should be practiced at least to the formation of clumps of trees, such as are naturally found in alpine meadow situations.—*F. S. Baker.*

1324. FABRICIUS, O. **Rødgran paa Fyn.** [Red spruce at Fyn.] Dansk. Skovforenings Tidsskr. 4: 317–372. 1919.

1325. FERNOW, B. E. [Rev. of: RECKNAGEL, A. B., AND JOHN BENTLY, JR. **Forest management.**] Jour. Forestry 17: 850–853. 1919.—See also Bot. Absts. 5, Entry 1373.

1326. FEUCHT, OTTO. **Zur Entstehung des Harfenwuchses der Nadelhölzer.** [On the formation of "harp-growth" on conifers.] Naturw. Zeitschr. Forst- u. Landw. 17: 137–139. *1 fig.* 1919.—S. KLEIN, and other authors, agree that the secondary stems, producing the so-called "harp" formation, are developed from the existing primary branches. The author, in the summer of 1917, discovered a white pine in the community of Würsbach (Wurtt, Black Forest), which exhibited a new sort of origin. On this tree, not a single branch has attempted to form a secondary stem, but some twenty young stems have arisen on the back of the tree below the upper third, evidently from dormant buds, either from the old whorls or between them.—*J. Roeser.*

1327. FLINT, HOWARD R. **A suggested departure in national forest stumpage appraisals.** Jour. Forestry 17: 823–831. 1919.—Present methods of stumpage appraisals on the national forests are deemed unsatisfactory and the proposal is made to change these by basing the price to be paid on the total receipts at stated intervals from lumber sales and costs of operation expressed in work hours of men, horses or machines.—*E. N. Munns.*

1328. GIRARD, JAMES W., AND U. S. SWARTZ. **A volume table for hewed railroad ties.** Jour. Forestry 17: 839–842. *1 fig.* 1919.—To overcome the recent change from two classes to five for railroad ties a volume table was prepared for Douglas fir and larch based on the diameter and number of ties per tree. The difference in form factors between the two species is not sufficient to affect the grades or number of ties.—*E. N. Munns.*

1329. GRAVES, H. S. **The extension of forestry practice.** Amer. Forestry 26: 50,51. 1920.

1330. GRAVES, HENRY S. **A policy of forestry for the nation.** Jour. Forestry 17: 901–910. 1919.—Present handling of forests in U. S. A. is not satisfactory and public interest requires public ownership of extensive areas and public participation in protection and management. A national policy demands action by the government, the states and by private owners of forest lands. National forest land should be increased, states should acquire and extend their holdings to assist in their economic and industrial life, and municipalities should have forest land to protect the water supply and to serve as a source of revenue.—On private lands, state and national aid should be given to prevent fires and legislation to this end should be undertaken by the states. Similar action by the states is necessary to require the forest owner to prevent lands becoming waste after lumbering and to assist the forest owner to secure the maximum production. In this, the states should be aided by the National government. Uniform taxation and a forest loan act are necessary, and a federal law is required to provide the government with authority to extend its influence and assistance to the states.—*E. N. Munns.*

1331. GREELEY, W. B. **The forest policy of France. The control of sand dunes and mountain torrents.** Amer. Forestry 26: 3–9. *7 fig.* 1920.—Material for this article has been taken largely from "*Cours de Droit Forestier,*" by CHARLES GUGOT, and from data prepared by G. GARBE, Engineer des Ponts et Chaussees. BREMONTIER is credited with having developed the methods which were successful in halting the destructive course of the Gascon dunes. These embraced the construction of a rampart along the coast, planting hardy herbs on the dunes within the rampart and planting seeds or seedlings of maritime pine. A national policy was adopted in 1810, and by 1864 the forestation of the 250,000 acres of dunes bordering the Landes was practically completed. Since that date the work has consisted largely in the care of the plantations established, the construction of new ramparts along the coast where dangerous dunes were forming, the extension of the successive zones of vegetation up to the limits of security thus established and the administration of the maritime pine forests which have been created. The successful reforestation of the dunes gave great impetus to the planting of maritime pine throughout the entire Landes. Today the Landes are a vast pinery, interspersed with little meadows and neat farms and traversed by a network of surfaced highways.—In the control of torrential erosion in the Alps and Pyrenees, France has been confronted with a far more difficult problem, which is, essentially, one in social economics. Following terrible floods in 1859, a reforestation law was passed in 1860, and by 1882 reforestation projects in the mountains had reached a total of some 350,000 acres. New laws passed at this time provided for more reduced areas for planting and other intensive methods, being limited to the immediate channels or slopes where erosion was taking place, and the establishment of large protection belts in the mountains, surrounding the limited water courses in which serious erosion was actually taking place. Further, the grazing of certain communal pasture lands was placed under public control. Human obstacles have prevented the perfect working of these measures. In controlling erosion, the line of attack is to reduce the trickling action of water on slopes, prevent the starting of gullies and hold loose soil or rock in place. This is accomplished by tree planting and by the employment of dams.—*Chas. H. Otis.*

1332. Greeley, W. B. **Private forestry in France.** Amer. Forestry 26: 139–143. *2 fig.* 1920.

1333. Greeley, W. B. **Self-government in forestry.** Jour. Forestry 18: 103–105. 1920.—Comment on national forest policy.—*E. N. Munns.*

1334. Griffin, Gertrude J. **Bordered pits in Douglas fir: a study of the position of the torus in mountain and lowland specimens in relation to creosote penetration.** Jour. Forestry 17: 813–822. *1 fig.* 1919.—Examination of the pits in Douglas fir showed a tendency in the torus of the mountain wood to aspirate (close) the pit while the opposite was true of the lowland woods, oven drying increasing the aspirated tori in both mountain and lowland varieties. In both sapwood and heartwood of the mountain variety, a large proportion of aspirated tori were found in air-dried wood, while only in the spring wood of the heartwood were the tori aspirated. Penetration of creosote was found to coincide directly with the number of aspirated tori. Subsequent treatments of air-dried material failed to open the tori when once aspirated, though soaking in alcohol before drying prevented their closing.—*E. N. Munns.*

1335. Gujer, A. **Zu unserer Titulaturfrage.** [The question of titles.] Schweiz. Zeitschr. Forstw. 71: 78–81. 1920.—The present titles are objectionable because they do not express the grade of the position and do not differentiate between the practical and technical positions.—It is proposed to replace "Förster" and "Oberförster" by "Förster" and "Förstmeister." "Förster" should apply to practical positions and "Förstmeister" to technical positions. The title could be used to cover all positions such as Kreis-, Bezirks-, Stadt-, Gemeinde- or Korporationsförstmeister. Such titles would eliminate the general usage of "Förster" for all employees in the profession of forestry.—*J. V. Hofmann.*

1336. Guthrie, John D. **Women as forest guards.** Jour. Forestry 18: 151–153. 1920.

1337. Hall, S. J. **Trees that are older than history.** Sci. Amer. 122: 303. *2 fig.* 1920.—Concerns the Sequoia.—*Chas. H. Otis.*

1338. Harvey, LeRoy H. **A coniferous sand dune in Cape Breton Island [Nova Scotia].** Bot. Gaz. 51: 417–426. *8 fig.* May, 1919.—See Bot. Absts. 4, Entry 288.

1339. HAUGH, L. A. **Klimaets Indflydelse Paa Udviklingen af Bøgens Sommerskud. [The influence of climate on the development of summer growth of beech.]** Dansk Skovforenings Tidsskr. 4: 13–28. *Fig. 4.* 1919.

1340. HAWES, A. F. **Raw material for the paper industry.** Amer. Forestry 26: 134–138. *5 fig.* 1920. The present paper shortage, U.S.A., is probably the result of the unusual amount of advertising carried by the newspapers, rather than of any scarcity of wood. The better grades of paper are still made from rags. While paper can be made from various plant fibers, straws and certain other materials, the collection of these materials in bulk is so costly that none of them can compete with wood. Spruce, hemlock and fir are the three main woods used in paper making. 95 per cent of the pulp and paper mills in the United States are located in the East, and the present supplies of these woods cannot be expected to last more than 25 years. Up to 1909 the country was self-supporting in respect to pulpwood, but since that date the consumption has exceeded the home product. Importations from Canada are constantly increasing. There are ample supplies of pulpwood for a great many years in Alaska and the Northwest. These may for several reasons become available.—*Chas. H. Otis.*

1341. HAWLEY, R. C. **Forestry in southern New England.** Amer. Forestry 26: 10–15. *7 fig.* 1920.—The territory embraced is roughly the states of Connecticut and Rhode Island. The region is primarily a manufacturing district. The forest area is now about 46 per cent of the total land surface. This forested area may be considered better suited for growing trees than for the production of agricultural crops. The forest is primarily hardwood in character. An upland hardwood type comprises over 80 per cent of the forest area, a swamp hardwood type less than 7 per cent, a pine (usually white) type about 2 per cent, an old field type (pine) 9 per cent and a hemlock type forms about 2 per cent of the area. As a whole the forests of southern New England are of second growth.—*Chas. H. Otis.*

√ 1342. HAY, R. DALRYMPLE. **Third annual report of the forestry commission, New South Wales, financial year ended June 30, 1919.** *38 p., 1 diagram, 8 pl.* Sydney, 1920.—The Forestry Act, passed by Parliament, November, 1916, created the Commission with powers to place the management of the forests on a business footing. Included in this plan is the systematic working of the forests with a view to regeneration and growth of future crops, and the disposal of timber and other forest produce to the best advantage. The Commission is exercising its powers with discretion and judgment in getting the new regime gradually under way, but is meeting with considerable opposition from the adherents of the old system of forest working, which was largely at the will of the operator. The forest area of New South Wales is estimated to be 11,000,000 acres, of which 5,043,800 acres have been proclaimed State forests and 566,730.5 acres are under working plans. It is stated that the available area of timber-bearing land of commercial value in the entire Commonwealth, previously estimated at 97,400,000 acres, can be reduced (on the basis of the past year's data) with certainty to about 24,500,000 acres. Of this area only about 18,000,000 acres had so far been protected from alienation in the interest of forestry. The estimated proportions in each State of the foregoing total (24,500,000 acres) are: New South Wales, 8,000,000 acres; Victoria, 5,500,000 acres; Queensland, 6,000,000 acres; Western Australia, 3,000,000 acres; Tasmania, 1,500,000 acres; and South Australia, 500,000 acres. At the instance of the Premier of New South Wales, the importance of ultimately appropriating a National forest area of about 30,000,000 acres for the whole Commonwealth, is being urged for the Commonwealth and the States' consideration. This area should comprise about 25,000,000 acres of indigenous forest country, and about 5,000,000 acres of coniferous plantation. During the year 98,372 acres of State forest area were released for settlement, 407¾ acres were planted to conifers, chiefly *Pinus insignis* and *P. pinaster*, and 23,707.5 acres were treated for natural regeneration and silvicultural improvement. A number of trees and fiber plants were tested for pulping material; the trees were mountain gum (*Eucalyptus goniocalyx*), coral tree (*Erythrina*), and mountain ash (*Eucalyptus sieberiana*). The algaroba bean (*Prosopis juliflora*) is being tested in a number of localities for fodder purposes. The outer sheathing of the gray ironbark (*E. paniculata*) has proven an excellent substitute for cork and cork waste, which is used largely in the manufacture of insulating material. Experiments undertaken to ascertain whether this sheathing could be removed without injury to the growing tree have resulted successfully. Mountain ash (*E. gigantea*) is being tested for veneer material. Many other investigations on a variety of subjects are also under way. Mistletoe is doing serious damage to the forests of the western districts. The following species are infested: *Acacia aneura, Eremophilla longifolia, E. creba, E. dealbata, E. rostrata,* and *C. luehmanni*. An area of 37,500 acres of Crown land in the vicinity of Buckenboura, on the South Coast was recently temporarily withdrawn from settlement for the growing of wattle trees for tanbark production. The principal species of wattle of tannic value (*Acacia decurrens*) is widely distributed on the area and appears well adapted to local climatic and soil conditions. It is expected therefore to set aside the better portions of the area as a National permanent reserve for the growth and preservation of wattle. Reference is made to an article by A. SHALLARD published in the October, 1918, issue of the *Australian Forestry Journal* which states that probably 20,000 people in Australia keep bees, and that the yield last season was between 5000 and 6000 tons of honey, the bulk of which came from the gum (eucalypt) trees, and among the principal varieties of honey value, the ironbarks, the stringybarks, the boxes, flooded gum, white mahogany, tallow wood, spotted gum, gray gum, and bloodwood, are given first place. In order to widen the use and productiveness of the state forests in this direction, the Commission has now made arrangements for the issuance of bee-farming permits, which convey to the holders certain privileges of occupation and use, and enable liberal areas of the state forests to be taken up as bee ranges.—*E. R. Hodson.*

1343. HELMS, JOHS. **Weymouthsfyrren paa Silkeborg Skovdistrikt.** [Pinus monticola at Silkeborg District.] Dansk Skovforenings Tidsskr. 4: 402–408. *Pl. 2.* 1919.

1344. HENKEL, J. S. **Afforestation in Zululand.** Rhodesia Agric. Jour. 17: 50–52. 1920.—Judging by the indigenous vegetation and the bad effects of strong winds, conditions at Empangeni appeared far from favorable for the growing of exotic timber trees. Quite a large number, however, have adapted themselves to the conditions, the outstanding successes being secured with eucalypts.—*E. M. Doidge.*

1345. HESSELMAN, HENRIK. **Iakttagelser över Skogsträdspollens Spridningsförmåga.** [Dissemination of pollen from forest trees.] Meddel. Statens Skogsförsöksanst. 16: 27–60. *3 fig.* 1919.—See Bot. Absts. 4, Entry 232.

1346. HODAL. **Fransk bergfuru (Pinus montana gallica).** [French mountain pine.] Tidsskr. Skogbruk 28: 1–12. *Pl. 2.* 1920.

1347. HOLE, R. S. **A new species of Ixora.** Indian Forester 45: 15–16. 1919.—See Bot. Absts. 3, Entry 2983.

1348. HOLTEN, JUST. **Gamle Ege i Christianssaedes Skove.** [Old oaks on Christian Manor.] Dansk Skovforenings Tidsskr. 4: 379–395. 1919.

1349. HOSMER, RALPH S. **One aspect of the national program of forestry: cost.** Jour. Forestry 18: 9–12. 1920.—The cost item has been left out of consideration in the discussion of a national forest policy. This is important because the antagonism of private owners is apt to result if the burden falls too heavily on them, and if the burden on the population is too heavy, there is apt to be trouble from the other side. In any case, the public pays the bills in the end.—*E. N. Munns.*

1350. HOSMER, R. S. [Rev. of: JUDD, C. S. **Report of the Division of Forestry, Territory of Hawaii, for biennial period ended Dec. 31, 1918.**] Jour. Forestry 17: 853–855. 1919.

1351. HUBAULT, E. **Efter krigen paa de britiske øer.** [The British Islands after the war.] [From Rev. Eaux et Forêts. Oct., 1919.] Tidsskr. Skogbruk 27: 276–291. 1919.

1352. Jessen, P. P. **En Ny Dansk Impraegneringsmetode.** [**A new Danish staining method called Teakin.**] Dansk Skovforenings Tidsskr. 4: 427-445. *Pl. 2.* 1919.—The process consists in pressing different kinds of liquids which contain coloring matter into the wood. These are either inorganic salts or aniline dyes. The color is taken up by the cells of the wood.—*J. A. Larsen.*

1353. Judd, C. S. **An historical mesquite tree.** Sci. Amer. 122: 165, 175. *1 fig.* 1920.—Descriptive of the algaroba (*Prosopis juliflora*), its occurrence in Hawaii, characteristics, uses and propagation.—*Chas. H. Otis.*

1354. Kellogg, R. S. **The news print paper situation.** Amer. Forestry 26: 147. 1920.

1355. King, H. E. **Tree planting in community, a suggested scheme.** South African Jour. Indust. 3: 161-163. 1920.

1356. Kinzel, Wilhelm. **Ueber eine neue Methode des Durchfrierens und die damit erzielten Erfolge bei zahlreichen bisher nicht oder kaum zur Keimung gebrachten Samen.** [**Concerning a new method of freezing and the results derived with numerous unfertile seed or seed with very low germinative power.**] Naturw. Zeitschr. Forst- u. Landw. 17: 139-142. 1919. —The author discusses the varying results obtained in the artificial treatment of seed either in light at 20° or in the dark under frost conditions. He cites a considerable number of examples. However, it is evident, that some species show little response to the methods hitherto employed. Treatment of seed by frost in conjunction with light has in the past been avoided, because where used, harmful results were obtained. This method, though, is very successful in many cases, and will yet become important in the case of many tree seeds. It cannot be used with seeds rich in chlorophyl, such as Acer and Fraxinus, or with frost sensitive seed, such as beech, hasel-nut, yew and others.—*J. Roeser.*

1357. Kirkland, Burt P. **Co-operation between national forests and adjacent private lands.** Jour. Forestry 18: 120-130. 1920.—To insure continuous forest production and the permanence of wood using industries, the owners of lands in units totaling more than 25,000 acres should consider the area as a whole. This would permit of better equipment and personnel, a permanent town-site and the development of practical forestry. Protection is to be paid for on an ownership basis, and the area to be restocked as cut by nature or planting. Careful cutting and trained supervision to follow the entire operation.—*E. N. Munns.*

1358. Kirkland, Burt P. **Economics of private forestry.** Jour. Forestry 18: 214-217. 1920.—The misconceptions of those who believe forestry uneconomic are due to misbeliefs in the rights of private property, interest returns and capitalisation and taxation.—*E. N. Munns.*

1359. Kitchin, P. C. **Preliminary report on chemical weed control in coniferous nurseries.** Jour. Forestry 18: 157-159. 1920.—Applications of copper sulphate, zinc chloride, and sulphuric acid to seed beds gave greatly reduced numbers of weeds, especially good were the results from the first two salts. Further work is in progress.—*E. N. Munns.*

1360. Knuchel, Hermann von. **Zur Praktikantenfrage.** [**The probation question.**] Schweiz. Zeitschr. Forstw. 71: 69-78. 1920.—A plea for better conditions for the probationer and more democratic relations between academic and applied forestry. The probationer should receive pay and should be allowed to serve under practical foresters on applied forest problems rather than the general system of working as a subordinate, without pay, under an instructor.—The state should encourage students to attend forest schools, but should not subsidise them. Enrollment at the forest schools should be limited to the number of men needed by the state. Foresters must receive better pay and be placed on social equality with other professions such as medicine, etc.—*J. V. Hofmann.*

1361. Koehler, Arthur. **Identification of mahogany.** [Review of several papers.] Jour. Forestry 18: 154-156. 1920.

1362. Kornerup A., and H. Mundt. **Aske-Gavnetra.** [Ash for lumber.] Dansk Skovforenings Tidsskr. 5: 1-29. *13 fig.* 1920.

1363. Kühl. **Traeets Kemiske Leknologi.** [The chemical composition of wood.] Dansk Skovforenings Tidsskr. 4: 28-64, 110-146. *45 fig.* 1919.

1364. Lee, Laurence. **Notes on the Parana pine of southern Brazil.** Jour. Forestry 18: 57-61. 1920.—The Parana pine has a stand of about 650 billion board-feet in Brasil. The wood is said to be superior to Swedish pine and even the southern longleaf pine of North America. There are no resin ducts and resin accumulates only at the base of knots. At the present time the lack of shipping facilities and the unfair taxes are keeping this timber from the market.—*E. N. Munns.*

1365. Leopold, Aldo. **Determining the kill factor for blacktail deer in the southwest.** Jour. Forestry 18: 131-134. 1920.—A method similar to that used in estimating cattle is proposed for obtaining data on the blacktail deer.—*E. N. Munns.*

1366. Maddox, R. S. **Reclamation work a vital forestry problem.** Amer. Forestry 26: 74-76. *5 fig.* 1920.—Relates particularly to conditions in Tennessee.—*Chas. H. Otis.*

1367. Maiden, J. H. **A critical revision of the genus Eucalyptus.** Vol. IV, Part 8. P. 201-237, *4 pl.* William Applegate Gullick: Sydney, 1919.—See Bot. Absts. 3, Entry 2995.

1368. Maxwell, Hu. **The uses of wood. Wood in agricultural implements.** Amer. Forestry 26: 148-155. *14 fig.* 1920.

1369. McLean, R. C. **Studies in the ecology of tropical-rain forest: with special reference to the forests of South Brazil. I. Humidity.** Jour. Ecology 7: 5-54. *1 pl., 21 fig.* 1919.

1370. Mell, C. D. **The mangroves of tropical America.** Sci. Amer. Supplem. 88: 388-389. *5 fig.* 1919.—The red mangrove (*Rhizophora mangle*) produces the bulk of the commercial bark used for tanning purposes. The bark is from three-fourths to one inch thick, of a dull reddish color, somewhat fibrous and covered with a grayish cork-like cuticle, and contains tannin superior to that of many other barks used for that purpose. The percentage of tannin is from 25 to 36. The gathering of the bark is a difficult task.—*Chas. H. Otis.*

1371. Metcalf, C. D. **Logging with belt tread tractors.** Sci. Amer. Monthly 1: 42-44. *5 fig.* 1920. [Reprinted from the *West Coast Lumberman.*]

1372. Minchin, A. F. **Annual rings in sal.** Indian Forester 46: 38-45. *2 fig.* 1920.—Annual rings in sal may be distinguished on a tangential cut when not possible on a radius. Fresh cut stumps only can be used and a clean smooth surface is essential. Stump counts and measurements of trees of known age show a very close relationship though based on a very small number of trees.—*E. N. Munns.*

1373. Moore, Barrington. [Rev. of: Recknagel, A. B., and J. Bentley, Jr. **Forest management.** *xiii + 269 p., 26 figs.* John Wiley & Sons: New York, 1919. Net $2.50.] Torreya 20: 34-35. 1920.—The book is written for owners of forest-lands who are not professional foresters. Four branches of forest management are treated: (1) mensuration; (2) regulation of cut; (3) finance; (4) administration. Both the forest-owner and professional forester will find the book valuable. [See also Bot. Absts. 5, Entry 1325.]—*J. C. Nelson.*

1374. Mulloy, G. A., and W. M. Robertson. **An analysis of logging costs in Ontario.** Jour. Forestry 17: 835-838. 1919.—Data on logging costs compiled from a large number of reports on operations in Ontario through several years is given for 11 divisions of cost covering 82 detailed items.—*E. N. Munns.*

1375. MUNNS, E. N. **Effect of fertilization on the seed of Jeffrey pine.** Plant World 22: 138–144. 1919.—Various crosses between thrifty, mistletoe-infested, insect-infested, and suppressed specimens of *Pinus jeffreyi* were made, with the result that thrifty trees produce larger and heavier seeds, with a higher germination percentage, higher rate of germination, higher real value per pound, and ability to produce stronger seedlings. Seeds borne on suppressed, malformed, and diseased trees are of inferior quality for planting. The author suggests forest management in which diseased and suppressed trees are removed, and only thrifty seed trees left for seed purposes. In collecting seed for forest tree nurseries, thrifty trees should be chosen as parents. [See also Bot. Absts. 5, Entry 1589.]—*Chas. A. Shull.*

1376. NELLEMANN, L. P. **Nogle Undersøgelser Over Arbejdstid og Arbejdsydelse. [Some investigations on working hours and working men's aid.]** Dansk Skovforenings Tidsskr. 4: 408–427. 1919.

1377. [NORDSTEDT, C. T. O.] [Swedish rev. of: OSTENFELD, C. H. **Bemerkninger om danske Traeer og Buskes Systematik og Udbredelse I. Vore Aelme-Arter. (Remarks on the systematics and distribution of Danish trees and shrubs. I. Our species of Elms.)** Dansk Skovforenings Tidsskr. 1918: 421–442. 1918.] Bot. Notiser 1919: 102. 1919.

1378. OPPERMANN, A. **Et Lovbuds Udviklingshistorie. [History of the development of a law.]** Dansk Skovforenings Tidsskr. 4: 146–160. 1919.

1379. OPPERMANN, A. **Vort Skovbrug Omkring Aar 1900. [Our forestry in 1900.]** Dansk Skovforenings Tidsskr. 4: 259–316. 1919.

1380. PAMMEL, L. H., AND C. M. KING. **The germination of some trees and shrubs and their juvenile forms.** Proc. Iowa Acad. Sci. 25: 292–340. *Fig. 45–120.* 1920.—One lot of seeds was placed in good greenhouse soil in the fall (1917) and stratified in a cold frame, from which they were removed to the greenhouse in March 1918. The second lot was planted in an open place covered with two inches of soil and leaves. Air temperature records were kept throughout the season; soil temperature records were kept in the fall until the ground was frozen, and again during the opening of the growing season of 1918. Tables of temperature and precipitation are given. Photographs or outline drawings of the leaves, and frequently outlines of trichomes, are given, with descriptive text, for the following species: *Juglans cinerea, J. nigra, Carya ovata, C. laciniosa, C. alba, C. glabra, C. cordiformis, Corylus americana, Ostrya virginiana, Betula lutea, B. alba papyrifera, Quercus coccinea, Q. ellipsoidalis, Q. falcata, Q. nigra, Q. imbricaria, Ulmus americana, U. fulva, U. pumila, Celtis occidentalis, Crataegus mollis, C. Crus-galli, Prunus padus, P. serotina, Gleditsia triacanthos, Gymnocladus dioica, Ptelea trifoliata, Acer saccharinum, A. saccharum, A. saccharum nigrum, A. negundo, Aesculus glabra arguta, Vitis vulpina, Tilia americana, Cornus alternifolia, Fraxinus pennsylvanica lanceolata, Catalpa speciosa.* A table gives number of seeds planted and total number germinated.—*H. S. Conard.*

1381. PAMMEL, L. H., AND C. M. KING. **A variation in the black walnut.** Proc. Iowa Acad. Sci. 25: 241–248. *Pl. 3, fig. 43–44.* 1920.

1382. PARNELL, RALPH. **Progress report on forest administration in the North-West Province for the year 1918–19.** *41 p., 1 map.* Peshawar, British India, 1919.—Incorporated with the annual report is a similar one covering the five-year period from 1914–15 to 1918–19. Since 1917 a beginning has been made in the departmental exploitation of timber. So far walnut, chil, and coniferous timber in one locality have been handled in this way. It is stated that the loss of revenue incurred by the government by leases for even relatively short periods in at all abnormal times, the difficulty of arranging for leases for long periods on a sliding scale of royalties on account of the vested interests involved and the friction inevitable in using the sliding scale, the importance of the Government's retaining its timber in its own hands for as long as possible in case of emergent needs and the public advantage obtained by the government's being in a position to use the profits of the timber trade for the benefit of the country as a whole instead of these profits going into the pockets of a few long-headed private firms, are believed to justify the abandonment of the system of sales of standing trees and the adoption of the system of departmental exploitation. During the year the department removed by this system 171,000 cubic feet of timber, or 14 per cent of the total timber outturn against 4 per cent the preceding year. Since the walnut supply is becoming exhausted and natural reproduction scarce, it is necessary to plant. A nursery has been established at Nagan and about ¼ acre sown with 21,000 walnuts. It appears the best method of restocking is to sow direct on the areas and fill in the gaps with trees raised in the nursery. Tests of bhan (*Rhus cotinus*) and garunda (*Carrissa spinarum*) leaves have shown a fairly satisfactory tannin content. However, the production from this source would only be sufficient to supplement the small local requirements of the province. Appended are numerous forms summarising detailed tabulated data and a map of the Hazara Division.—*E. R. Hodson.*

1383. Pabst, August. **Die Kienölgewinnung im Wald von Bialowies. [The production of pine-oils in the forest of Bialowies.]** Naturw. Zeitschr. Forst- u. Landw. 17: 105–137. *6 pl., 2 fig.* 1919.—The author briefly reviews the best known volatile oils obtained from conifers, under four headings: (1) those obtained from the bark and wood above ground, (2) through the distillation of needles and buds, etc., (3) through the distillation of cones and fruit, and (4) from the underground woody portion through extraction or dry distillation. The production of pine-oil, a variety of turpentine oil, is an important industry of that section of Europe lying between the Carpathians and the Baltic Sea, including the countries of Poland, Courland and Lithuania. The establishment founded by the writer in 1916 in the Forest of Nowi Most, after it was occupied by the Germans, is then described in considerable detail under the headings: (1) the raw material used in the process of distillation; (2) construction of the establishment including the retort, the heating chamber, the arrangement for carrying off the distillate, the cooling mechanism and the receiver of the pine-oil establishment; and the equipment of the tar and charcoal establishment; (3) the process of distillation; (4) the products resulting from the distillation, chiefly pine-oil, tar and charcoal; (5) cost accounting and profitableness; and (6) conditions necessary to establish the pine-oil industry in Germany. Numerous tables are included to illustrate topics (3), (4), and (5). The author believes that the industry can be successfully introduced, especially in North Germany, both on a small scale and on a large scale if a large supply of woody material can be obtained close at hand, and concludes, that since the Russian producer has made a success of it under very poor economic conditions, there is no ground for believing that success will not crown the efforts of the native contractor surrounded by an economic system organised and developed to the fullest extent. German forest culture is presented with a new prospect for increasing its forest revenue, and at the same time helping to break the economic bands now holding the country.—*J. Roeser.*

1384. Paschal, G. W. **A bigger tree.** Sci. Amer. 122: 61. 1920.—A letter concerning a poplar tree with a butt circumference of 39–40 feet.—*Chas. H. Otis.*

1385. Passler, Johannes. **Das Entrinden von Hölzern unabhängig von der Jahreszeit nach dem Gütschowschen Verfahren. [Bark-peeling independent of the season according to the Gütschow process.]** Schweiz. Zeitschr. Forstw. 71: 116–118. 1920.—It is well known that oaks and other trees do not peel easily except during the spring time when the sap is flowing freely, also the quantity or quality of tannin varies very little during the year. This makes it possible to peel only during a short season although it would be profitable to peel during the entire year. Methods of loosening the bark have been in use for a long time among which the Maitre method in use for the past fifty years is the most commonly used. By this method the wood is steamed at 100°C. before peeling.—A new method devised by Gütchow consists of steaming the wood for several hours at 30 to 40°C. This has the advantage of leaving the wood cooler and easier to handle. It may also be applied in the field by use of a wagon that Gütchow has constructed in which the steaming can be done and the bark dried.—His method applies to the pines also and is the most feasible for field conditions where the cutting is done during the winter season and the wood delivered to the industries later.—*J. V. Hofmann.*

1386. Perkins, G. W. **Forestry and recreation in the Palisades Interstate Park.** Amer. Forestry 26: 20–26. *8 fig.* 1920.

1387. Perrée, W. F. **Progress report of the Forest Research Institute for the year 1918–19.** *22 p.* Calcutta, British India. 1919.—The work of the Institute is organized in five branches: Silviculture, forest botany, forest economy, forest zoology, and forest chemistry. A silvicultural experiment in Thano forest indicates that two regeneration fellings are unnecessary where natural reproduction is already present in sufficient quantity. Sufficient overhead cover to protect from frost is also sufficient to suppress young Sal (*Shorea robusta*). Side protection is of greater value than overhead protection. In this forest the frost risk is slight and therefore it is believed that a clear felling in one operation followed by cleaning and cutting back will prove successful in regeneration. To test this point an experimental area of five acres has been marked for clear felling. Two other plots were laid out in this forest to determine the effect of severe thinning (1) at an early age, and (2) at maturity. The following is indicated in afforestation work at Zaberkhet Tappar: *Dalbergia sissoo* (less damaged by deer) and *Melia azedarach* are the most promising species; rooted cuttings of *Dalbergia sissoo*, *Bombax malabaricum*, *Eugenia jambolana*, and *Grewia vestita* have been successful, while *Terminalia tomentosa*, *Ougenia dalbergioides*, and *Mallotus philippinensis* have given fair results, and that Chir (*Pinus longifolia*) can be better raised from direct sowings than by transplanting. (July is best season for transplanting this species.) In the study of tan-yielding trees and shrubs *Anogeisus latifolia* is being tested to determine the best season for pollarding, *Cassia auriculata* for stimulation of germination and for methods of transplanting. *Phyllanthus emblica* was found frost hardy, and both direct sowings and transplants from nursery have proved successful; germination ranged from 70 to 90 per cent. *Elaeodendron glaucum*, also frost hardy, showed 70 per cent germination and both direct sowings and transplanting proved successful. In the branch of Forest Botany the problem of regenerating the Sal is believed solved by a series of recent investigations. The factors injurious to the establishment of the seedling, due to the interaction of a soil-covering of dead leaves, drought, and bad soil aeration, are eliminated more effectively by a complete removal of the overhead canopy than by either burning the soil covering, or by removal of undergrowth, with or without partial thinning of the overhead cover. Owing to the uncertainty of good seed years and for other reasons, the restocking of the area by artificial sowings is preferable to reliance on natural regeneration. It has further been proved that much better results are obtained from broadcast sowings in cleared patches and narrow strips with full overhead light than from sowings under the shade of a partial canopy. Therefore the system proposed for handling Sal is a combination of the group and strip methods, in which the size of the unit regeneration areas is determined by the average height of the forest at maturity, and their sequence and orientation by local requirements for shade. A number of woods have been investigated for industrial use. The branch of Forest Chemistry obtained from the leaves of *Cinnamomum glanduliferum* 0.20 per cent of camphor and 0.44 per cent of camphor oil. From the leaves of *Eucalyptus tereticornis* and *E. crebra* collected at Kaunli, Dehra Dun, were obtained oils which resembled those of similar species grown in Australia. The former contained a small percentage of eucalyptol but the oil from neither of these two species of eucalypts complies with the standard of the British Pharmacopoeia. *Artemisia maritima* was examined for santonin with negative results. The phenolic portion of the light Chir (*Pinus longifolia*) tar oil, a by-product in distilling this species for Stockholm tar, showed 8 per cent of guaiacol and 42 per cent of creosole. Kelp (*Saragosum species*) from the Bombay Coast contained 0.02 per cent of iodine and 1.14 per cent of potassium. The Institute library has increased its books and periodicals to 14,014. Appended is a list of the current year's publications and also a cumulative list from the beginning of the Institute. In general it is expected to develop the Research Institute, to serve not only the scientific and economic interests of the Forest Department, but also to function as the central bureau of information for the entire Indian scientific and commercial community.—*E. R. Hodson.*

1388. Pettis, C. R. **Legislative machinery for enforcement of private forestry measures.** Jour. Forestry 18: 6–8. 1920.—An attempt should be made to make lumbering operations and cut over lands more safe from fire. This may be done in New York by leaving strips and bands of uncut timber along roadways and creeks to create fire breaks, by the construction of fire lines, by burning the slash. Demonstration forests and foresters are needed to show what can be accomplished.—*E. N. Munns.*

1389. Pinchot, Gifford. **National or state control of forest devastation.** Jour. Forestry 18: 106–109. 1920.—State control does not offer the surest and strongest control of forest devastation; national control does and has proved its point in the past.—*E. N. Munns.*

1390. Pool, Raymond J. **The fuel situation in Nebraska and the need for greater wood production.** Publ. Nebraska Acad. Sci. 10: 17–28. 1920.—The author discusses the need of wood, the shortage of wood, and the value of woodlots in Nebraska. He urges thinning of groves and wind-breaks, and cutting off when the crop is mature.—*H. S. Conard.*

1391. Potts, H. W. **The honey locust tree.** Agric. Gaz. New South Wales 31: 85–90. *7 fig.* 1920. Gives chemical analysis of seeds.—*L. R. Waldron.*

1392. [Pratt, Geo. D.] **New York's forestry program.** Amer. Forestry 26: 51–52. 1920.

1393. Rafn, Johannes. **Skovfrøanalyser i Saesonen 1917–18. [Analysis of forest seed 1917–18.]** Dansk Skovforenings Tidsskr. 4: 8–12. 1919.

1394. Rafn, Johannes. **Skovfrøanalyser i Saesonen 1918–19, samt lidt om Egern. [Tests of forest seed, 1918–19, with notes on the oak.]** Dansk Skovforenings Tidsskr. 5: 55–64. 1920.

1395. Rao, B. Inamati Sham. **Brief note on the artificial raising of sandal in the Akola Division of the Berar Circle, Central Provinces.** Indian Forester 46: 1–10. *Pl. 1–2.* 1920.—Sandal seed was dibbled in the brush of Akola and in good years an excellent stand resulted. As the sandal coppices and spreads by root suckers, the future stands are well assured.—*E. N. Munns.*

1396. Recknagel, A. B. **Inspection, supervision and control of private forestry measures: methods and costs.** Jour. Forestry 18: 23–25. 1920.—There are nearly 300 timber land owners in New York with more than 500 acres in their holdings. To administer these properly would require technical supervision. Working plans for each tract should be prepared by a forester and filed with the Conservation Commission, failure to do so to be punished and violations of the plan carry fines. An office for handling these operations on 2,182,000 acres is needed with a mobile field force.—*E. N. Munns.*

1397. Record, S. J. **Possum wood.** Sci. Amer. 122: 569. 1920.—Descriptive of the tree and its wood, known by many common names, and botanically as *Hura crepitans.* This is one of the most recent introductions to the American timber market that seems certain to find a place.—*Chas. H. Otis.*

1398. [Ridsdale, P. S.] **A decade of progress in the Forest Service.** Amer. Forestry 26: 131–132. 1920.—An editorial, occasioned by the retirement of Henry S. Graves as head of the U. S. Forest Service, in which is reviewed the progress made during the ten years in which he has directed the forestry activities of the national government.—*Chas. H. Otis.*

1399. [Ridsdale, P. S.] **Increase in forest research necessary.** Amer. Forestry 26: 69–70. 1920.

1400. [Ridsdale, P. S.] **Light burning is a mistake.** Amer. Forestry 26: 68–69. 1920.—Light burning means nothing more nor less than the continuance of the frequent surface fire, which steadily and irresistibly destroys the western pine forests. At its best, the practice is simply a measure for the protection of old timber. An area cleaned by light burning has no advance young growth to replace the virgin timber after cutting. Light burning has no place in a system of forestry which seeks to perpetuate our western pine forests and make them continuously productive.—*Chas. H. Otis.*

1401. [Ridsdale, P. S.] **A national forest policy.** Amer. Forestry 26: 67–68. 1920.

1402. Skerrett, R. G. **Multiple production—a new slogan.** Sci. Amer. 122: 58–59, 72. *3 fig.* 1920.—Touches, among other things, on the waste of lumbering and some of the ways in which this waste may be lessened.—*Chas. H. Otis.*

1403. Skoien, Olaf. **Landsskogtakseringen.** [Taxation of the forests.] Tidsskr. Skogbruk 28: 12–15. *1 fig.* 1920.

1404. Smith, Annie Lorrain. **Hyphomycetes and the rotting of timber.** Trans. British Mycol. Soc. 6: 54–55. 1918.—See Bot. Absts. 3, Entry 2763.

1405. Smith, F. H. **Significant trends in lumber production in the United States.** Amer. Forestry 26: 143–147. *1 map, 2 tables.* 1920.

1406. Smith, F. H. **What our forests support.** Amer. Forestry 26: 16–17. 1920.—A consideration of the great value of forests and their economic importance to the wealth, independence and prosperity of U. S. A.—*Chas. H. Otis.*

1407. Sparhawk, William N., Donald Bruce, and Burt P. Kirkland. **Report of subcommittee on forest leasing, forest loans, and forest insurance.** Jour. Forestry 18: 260–274. 1920.—The details of a leasing plan are given whereby the government can lease forest land instead of buying it outright, the financial burden being distributed over a long period. Financial credit to forest users is at high interest rate because of the small units and a system of Federal Forest Loan Boards is described. To handle forest insurance properly an insurance organization is necessary and as a public necessity is at stake and a resource in danger, this work can best be accomplished by a national organization. To these ends, legislation by the states and by the government is essential.—*E. N. Munns.*

1408. Stevens, Carl M. **Rating scale for foresters.** Jour. Forestry 18: 143–150. 1920.

1409. Terry, E. I. **Further comment on a formula method of estimating timber.** Jour. Forestry 18: 160–161. 1920.

1410. Vestby, P. **Spredte træk fra en skogbefaring i Chili.** [Sketches from a trip to Chilean forests.] Tidsskr. Skogbruk 28: 17–27. *Pl. 2.* 1920.

1411. Vikhammer, P. **Om granen som fremtidig skogtre nordenfor polarcirklen.** [Norway spruce as a future tree north of the Polar Circle.] Tidsskr. Skogbruk 27: 253–276. *Fig. 4.* 1919.

1412. West, Erdman. **An undescribed timber decay of hemlock.** Mycologia 11: 262–266. 1919.

1413. Williams, I. C. **Report of forestry.** Bull. Pennsylvania Dept. Agric. 11: 119–122. 1918.—Remarks upon the loss of services of state foresters who entered war service and its effect upon forest protection. Brief statistics are given of plantings within the state forests and of the available seeds and seedlings for future planting. The number of forest fires recorded in 1917 was 2066 and the average area burned over 153.45 acres. The railroads within the state paid damages on 168 fires, the expense of extinguishing the same being $1674.80. Individuals made settlement for 81 fires, the expense of which amounted to $1016.73. During 1917 the state forests were increased by 5593 acres, bringing the total area to 1,017,773 acres.

At the present time there are 52 state forests. It is pointed out that the State Department of Forestry has to 1918 paid from its resources $148,052.33 to the State School Fund of Pennsylvania.—*C. R. Orton.*

1414. Wilson, Ellwood. **Use of seaplanes in forest mapping.** Jour. Forestry 18: 1–5. 1920.—Seaplanes in eastern Canada were found well adapted for forest use, the abundance of lakes and the absence of landing grounds making such a type of plane feasible. Hardwoods and softwoods can readily be distinguished and photographs with an aerial camera gave excellent results in mapping, 200 square miles a day being possible with a machine as against 50 square miles per month by a party of ten on foot.—*E. N. Munns.*

1415. Woodruff, George W. **Constitutionality of national laws to restrict forest devastation.** Jour. Forestry 18: 100–102. 1920.—The Supreme Court, U. S. A., has upheld previous legislation dealing with the control of forest lands because of the benefit to the public and liberty of posterity. The present scheme for control of devastation fits in with the past favorable decisions.—*E. N. Munns.*

1416. Woolsey, Theodore S., Jr. **Early Arizona problems.** Jour. Forestry 18: 135–142. 1920.

1417. Woolsey, T. S. **Natural regeneration of French forests.** Amer. Forestry 26: 77–81. *10 fig.* 1920.—In the Landes and the Gironde maritime pine matures in 70–80 years, at which time the trees are clear cut. The branches and unmerchantable tops are left on the ground; the sun opens the cones and the sand is quickly covered with a stand so dense as to require thinning. In the sapling stage the excess trees are tapped to death to produce resin and mine props and to favor the development of the crowns of the final stand. The sessile oak in the Adour, where there is an annual acorn crop, can be clear cut. Sessile and pedunculate oak stands (often mixed with beech in central France) must be regenerated by progressive cuttings. Oak matures in 180–240 years and the seedlings are intolerant, while the beech requires for a time a protective cover of older trees. Under these conditions there are 3 successive fellings; the seed felling aims at starting the seedlings, the development of the crowns of the seed trees and the partial removal of the merchantable crop; a secondary felling aims to gradually remove the seed trees and to gradually free the existing seedlings without causing too much damage; the final felling is made when the ground is seeded and the first seedlings have developed into saplings, and in this the seed trees that are left are removed at one stroke. In fir stands, where advance growth almost always exists, the seed felling is really a light secondary felling, designed to allow this advance growth to develop. Subsequent secondary fellings are also light; but the final felling should be complete. In the high mountains the treatment is different, since the objective is not solely the production of lumber, but the slopes must above all be protected to avoid damage by erosion. Group selection is the method practised. Soil preparation is often necessary, especially with spruce, since natural regeneration is hampered by (1) a dense vegetable cover which prevents the seed coming in contact with the mineral soil, (2) an excessive cover of undecomposed dead needles or (3) too compact surface of the soil.—*Chas. H. Otis.*

1418. Yates, Harry S. **The growth of Hevea brasiliensis in the Philippine Islands.** Philippine Jour. Sci. 14: 501–523. *1 fig.* 1919.—This paper has to do with the possibilities of cultivating Hevea in the Philippines on a commercial scale. The necessary conditions of climate, temperature, soil, and elevation are described. A comparison of these conditions with those of regions where *Hevea* is successfully cultivated indicates the suitability of the Islands for its cultivation, and the yield of rubber is satisfactory.—*Albert R. Sweetser.*

GENETICS

G. H. SHULL, *Editor*
J. P. KELLY, *Assistant Editor*

1419. A., D. **The doubling of the stock.** Gard. Chron. 66: 157. Sept. 20, 1919.—Author cites references contradicting Mr. Taylor, who states that Lothian growers succeed in obtaining double flowers from single-flowered plants without selection. It seems that seed selection must be made from plants showing tendency to doubling.—*A. C. Hildreth.*

1420. ABL [Zuchtinspektor, Halle, Sachsen]. **Unfruchtbare Zwillinge beim Rind. [Sterile twins in cattle.]** Deutsch. Landw. Tierzucht. 22: 34-35. 1918.—Author reviews briefly the theory of KELLER AND TANDLER in regard to the sterility and malformation of the free-martin heifer and describes two extreme examples.—*Sewall Wright.*

1421. ALLEN, EZRA. **Studies on cell division in the albino rat (Mus norvegicus, var. alb.). III. Spermatogenesis: the origin of the first spermatocytes and the organization of the chromosomes, including the accessory.** Jour. Morph. 31: 133-185. *58 fig.* June, 1918.—A technique which prevents clumping of the chromosomes is described. In the albino rat, the spermatogonial number of chromosomes is 37; the accessory divides in the second maturation division. Shapes of the chromosomes in spermatogonia are all curved rods; in first spermatocytes occur simple and compound rings, crosses, and one rod, the accessory; in the second spermatocytes, curved rods. The constitution of the first spermatocyte chromosomes is typically tetrad, with the four parts so organised that each may retain its individuality. The first spermatocyte chromosomes pass through clearly marked leptotene, pachytene, and diplotene stages without synizesis.—*Bertram G. Smith.*

1422. ALVERDES, F. [German rev. of: BOAS, J. **Zur Beurteilung der Polydaktylie des Pferdes. (Polydactyly in the horse.)** Zool. Jahrb. Anat. 4: 49-104. 1917.] Zeitschr. indukt. Abstamm. Vererb. 22: 287-288. May, 1920.

1423. ALVERDES, F. [German rev. of: LEBEDINSKY, N. G. **Darwins geschlechtliche Zuchtwahl und ihre arterhaltende Bedeutung. (Darwin's sexual selection and its significance for the maintenance of species.)** Habilitationsvortrag. *31 p.* 1918.] Zeitschr. indukt. Abstamm. Vererb. 22: 282-283. May, 1920.

1424. ALVERDES, F. [German rev. of: (1) NAEF, A. **Die individuelle Entwickelung organischer Formen als Urkunde ihrer Stammesgeschichte. (Kritische Betrachtungen über das sogenannte "biogenetische Grundgesetz.") (The individual development of organic forms as evidence of their evolutionary history.—Critical consideration of the so-called "biogenetic law.")** *77 p., 4 fig.* Jena, 1917. (2) *Idem.* **Idealistiche Morphologie und Phylogenetik. (Zur Methodik der systematischen Morphologie.) (Idealistic morphology and phylogeny.—On the method of systematic morphology.)** *77 p., 4 fig.* Jena, 1919.] Zeitschr. indukt. Abstamm. Vererb. 22: 279-282. May, 1920.

1425. ALVERDES, F. [German rev. of: PLATE, L. **Verbungsstudien an Mäusen. (Inheritance studies on mice.)** Arch. Entwicklungsmech. Organ 44: 291-336. *5 fig.* 1918. (See Bot. Absts. 3, Entry 658.)] Zeitschr. indukt. Abstamm. Vererb. 22: 284-285. May, 1920.

1426. ALVERDES, F. [German rev. of: (1) SCHAXEL, JULIUS. **Grundzüge der Theorienbildung in der Biologie. (Principles of theory formation in biology.)** G. Fischer: Jena, 1919. (2) SCHAXEL, JULIUS. **Über die Darstellung allgemeiner Biologie. (On the presentation of general biology.)** Abhandl. Theoret. Biol. 1919.] Zeitschr. indukt. Abstamm. Vererb. 22: 276-279. May, 1920.

1427. ANONYMOUS. **Report of the work of the plant breeding division for 1919.** Jour. Dept. Agric. Ireland 20: 102–107. 1920.—This report contains a brief summary of the work on wheat, barley, oats, flax and rye grass. It is stated that several new forms of spring wheat have been developed from a cross between Red Fife and April Red. It is planned to substitute one of these new forms for Red fife.—Hybrid barleys are compared with their parents and indicate slight increases in yield in some cases with deviations in others. Single plant selections were made in a crop sown with commercial Riga flax seed. The progeny of each of these selected plants was found to be remarkably uniform, not only in botanical characters but also in physiological characters such as resistance to frost, period of growth and vigor. The two progenies were found to be superior to the others and the propagation of them was continued. Twenty acres were sown from the two superior progenies and the plants showed great uniformity of growth.—In addition to these two selections, further selections were made from Riga flax and of these last selections two appear superior to the best two of the first selection. —Selections were also made of white-flowered and Kostroma flax. The results of these selections are not reported.—Single plant selections are being made in Perennial and Italian rye grass but no report of the success of this work is given.—*J. H. Kempton.*

1428. ANONYMOUS. **Daffodil breeding.** Florists' Exchange 49: 1082. May 8, 1920.—Notes on daffodil breeding in America and England. Finest English daffodils are raised by S. GOODELL of Seattle, Washington, from crossing English varieties. Some flowers measure 11 cm. and display exquisite coloring. Author describes choice collection of seedlings (red cups and red eyes) shown at Royal Horticultural Society's Daffodil show in London on April 13, raised by MRS. R. O. BACKHOUSE. Prices for best new seedlings range from $250 per bulb to $100 or less.—*Orland E. White.*

1429. ANONYMOUS. **A new dahlia of interest to plant breeders.** Jour. Heredity 11: 48. Jan., 1920.

1430. ANONYMOUS. **The heredity and environment of a great botanist.** Jour. Heredity 11: 6. Jan., 1920.

1431. ANONYMOUS. **University wants photographs of twin calves.** Jour. Heredity 11: 15. Jan., 1920.

1432. ANONYMOUS. **A genetic association in Italy.** Jour. Heredity 11: 45. Jan., 1920.

1433. ANONYMOUS. **New eugenics society in Hungary.** Jour. Heredity 11: 41. Jan., 1920.

1434. ANONYMOUS. **The birth rate in mixed marriages.** Jour. Heredity 11: 96. Feb., 1920.

1435. ANONYMOUS. **Eugenics in Germany.** Jour. Heredity 11: 110. Mar., 1920.

1436. ANONYMOUS. **Eugenics in Scandinavia.** Jour. Heredity. 11: 128. Mar., 1920.

1437. ANONYMOUS. **Eugenics and other sciences.** Jour. Heredity 11: 77–78. Feb., 1920.

1438. ANONYMOUS. **A common misconception concerning human heredity.** Jour. Heredity 10: 275. June, 1919.

1439. ANONYMOUS. **A factor influencing the sex-ratio.** Jour. Heredity 10: 256. June, 1919.

1440. ANONYMOUS. **Measuring intelligence.** Jour. Heredity 11: 86–87. *1 fig.* Feb., 1920.

1441. ANONYMOUS. **Deficiency in intellect found to be correlated with deficiency in the number of brain cells.** Jour. Heredity 10: 369. Nov., 1919.

1442. ANONYMOUS. **A supposed sheep-goat hybrid.** Jour. Heredity 10: 357-359. *2 fig.* Nov., 1919.

1443. ANONYMOUS. **Carriers of the germ plasm.** Jour. Heredity 10: 422. *Fig. 21.* Dec., 1919.

1444. ANONYMOUS. **To increase the birth rate.** Jour. Heredity 11: 64. Feb., 1920.

1445. ANONYMOUS. **An award of honor to Walter Van Fleet.** Jour. Heredity 11: 95-96. *1 fig.* Feb., 1920.

1446. ANONYMOUS. **The death of Richard Semon.** Jour. Heredity 11: 78-79. Feb., 1920.

1447. ANONYMOUS. **Systematic breeding.** Florists' Exchange 49: 986. April 24, 1920.—Popular discussion of breeding, with remarks on the importance of the F_1 generation in crossing work. Breeding problems of the carnation, rose, cyclamen and sweet pea are discussed.—*Orland E. White.*

1448. ANONYMOUS. **Historia de los metodos de seleccion. [History of the methods of selection.]** Jalisco Rural [Mexico] 2: 7-8. 1919.—Popular.

1449. ANTHONY, STEPHEN, AND HARRY V. HARLAN. **Germination of barley pollen.** Jour. Agric. Res. 18: 525-536. *2 pl., 2 fig.* Feb. 16, 1920.—Experiments with barley pollen were carried on: (1) with solutions, (2) with moist chambers, (3) fertilization in the field, (4) retention of viability in the laboratory, (a) when pollen is left in free air; (b) when pollen is kept over sulphuric acid; and (c) when pollen is kept in vacuo. No germinations were secured either with water or solutions of sugar, agar, or nutritive substances of various osmotic concentrations. Germination was finally obtained as follows: A slide containing pollen was placed inside a Van Tieghem cell; a piece of mesophyll from a leaf of garden pea was placed in the cell to supply water; the cell was covered with cover glass and placed outside on window ledge. Germination was thus obtained in five minutes. In field experiments receptivity of stigma and duration of viability of pollen were studied and results compared with those of laboratory experiments. Extreme delicacy of water adjustment is the most noticeable response of the pollen to treatment given in the experiments. Literature is reviewed. [See also Bot. Absts. 5, Entry 949.]—*W. E. Bryan.*

1450. BABCOCK, E. B. **Crepis—a promising genus for genetic investigations.** Amer. Nat. 54: 270-276. May-June, 1920.—It is desirable to find a genus with several crossable species, whose chromosome numbers are low and different; linkage groups corresponding to the chromosomes of each species should be understood. *Crepis* has 200 widely scattered and diversified species. Of these one is already known to have 3 chromosome pairs, 6 or 7 have 4, 4 have 5, one has 8, one has 9, and one has 20. Cytologically these are unusually favorable objects of study. *Crepis* is prolific, usually self-fertile, gives 2 or 3 generations a year, and probably its species are crossable. Disadvantage is smallness of flowers, making hybridization tedious though not impossible. Author has already commenced work on two species *virens* and *tectorum*, and urges other investigators to join in the attack, since an enormous mass of data will be necessary before the desired goal is reached.—*Merle C. Coulter.*

1451. BANCROFT, WILDER D. [Rev. of: JAEGER, F. M. **Lectures on the principles of symmetry.** *16 x 27 cm. xii + 333 p.* Elsevier Publ. Co.: Amsterdam, 1917.] Jour. Phys. Chem. 23: 516. 1919.—The book deals with the principles of symmetry in chemical substances, animals and plants. "While not easy reading, the book is an instructive one and contains a great deal that is of interest" to all morphologists, especially those in botany who are also interested in evolution.—*H. E. Pulling.*

1452. BANTA, ARTHUR M. **Sex and sex intergrades in Cladocera.** Proc. Nation. Acad. Sci. [U. S. A.] 4: 373-379. Dec., 1918.—Certain species of *Cladocera*, as *Daphnia pulex*, *Simo-*

cephalus serrulatus and three species of *Miona*, showed no intergradation of the secondary sex characters. In other species, however, as *Simocephalus vetulus*, sex-intergrades appeared very infrequently and in *Daphnia longispina* they were not very unusual. Frequently, in *Simocephalus vetulus*, there were many male intergrades produced with the female intergrades, but in *Daphnia longispina*, the intergrades were nearly all females. Sex intergrades appeared in certain cultures of *Simocephalus vetulus* in the 131st generation, in 1915, and have continued to appear throughout the 57 subsequent generations in the following three years. The females that showed only slightly developed intergrading sex characters reproduced with normal vigor but those with fully developed male characters were sterile.—*D. D. Whitney.*

1453. Barnils, Père. **Les éléments héréditaires dans le langage. [The hereditary elements in language.]** Compt. Rend. Soc. Biol. 82: 828–829. 1919.

1454. Bartlett, J. T. **A plant-breeder's opportunity.** Sci. Amer. 121: 372. 1919.—Desirable varieties of fresh vegetables and fruits are already available, but breeder now has notable opportunity in developing varieties adapted to such by-product industries as canning and evaporating. Special demands made, such as low water content, strawberries which husk easily, etc. Emphasises that canners and evaporators use first-quality produce, not produce unsuitable for shipment in fresh condition.—*Merle C. Coulter.*

1455. Bauin, P. **Sur la dimégalie des spermies dans certaines doubles spermatogénèse. Sa signification. [On dimegaly of sperms in certain cases of double spermatogenesis. Its significance.]** Compt. Rend. Soc. Biol. [Paris] 83: 432–434. Mar., 1920.

1456. Baumann, E. **Zur Frage der Individual- und der Immunitätszüchtung bei der Kartoffel. [On the question of individual selection in potatoes and the breeding for immunity.]** Fühlings landwirtsch. Zeitg. 67: 246–253. 1918.—Author points out the necessity of studying commercial potato varieties by means of clones. Data based on a number of individual selections vegetatively propagated from two varieties are presented. High yields are associated with an increase in number of tubers but a decrease in size. The percentage of starch in the tubers is lower in high yielders although the absolute amount of starch is greater.—Data on the influence of various leaf diseases in reducing yield is discussed. Author believes that the chief causes of "running out" in potatoes are leaf diseases.—*R. J. Garber.*

1457. Bishop, O. F., J. Grantham, and M. J. Knapp. **Probable error in field experiments with Hevea.** Agric. Bull. Federated Malay States 6: 596. 1918.

1458. Blaringhem, L. **Polymorphisme et fécondité du Lin d'Autriche. [Polymorphism and fecundity in Austrian flax.]** Compt. Rend. Soc. Biol. [Paris] 82: 756–758. 1919.

1459. Blaringhem, L. **Vigueur végétative compensatrice de la stérilité, chez les hybrides d'espèces de Digitales (D. purpurea et D. lutea). [Vegetative vigor compensating for the sterility in a species hybrid of Digitalis (D. purpurea and D. lutea).]** Compt. Rend. Acad. Sci. [Paris] 169: 481–483. 1919.—Reciprocal crosses of *Digitalis purpurea*, L., and *D. lutea*, L., give sterile progeny which surpass both parental species as follows:

	purpurea	*hybrid*	*lutea*
Height	50–150 cm.	150-185	40-80
Dry weight	150 g.	200–275	50
Duration of life	biennial	many years	triennial

First generation plants are very uniform. Reciprocal crosses do not differ in vegetative features but flowers differ in size, shape and color.—*D. F. Jones.*

1460. Bliss, A. J. **Hybridizing bearded Iris.** Gard. Chron. 67: 225. May 8, 1920.—Attempts to coördinate the results obtained by Bliss and by Sturtevant as to genetic composition of certain *plicatas*, basing an explanation on the results of Bateson and Punnett's experiment with Emily Henderson sweet pea. [See also Bot. Absts. 5, Entries 331, 1669.]—*J. Marion Shull.*

1461. Bonnevie, Kristine. **Polydaktyli i norske bygdeslegter. [Polydactyly in Norwegian peasantry.]** Norsk. Mag. f. Lägev. 6: 1–32. 1919.—In several families from different parts of Norway one and the same type of hereditary polydactyly occurs—a postaxial, asymmetrical polydactyly, mostly developed on the right side of the body. The extra finger (or toe) was always fixed at the base of the fifth finger, the metacarpalia showing no abnormalities. In all families the character in its occurrence follows the dominant type of inheritance, occurring in each of a series (2–5) of generations and in a relatively large number of individuals. The degree of development of the sixth finger (or toe) and its occurrence on one or both hands or feet, however, show considerable variation within each generation, from a well developed finger with three normal phalanges, down to a small soft knob at the side of the hand.—A genealogical investigation proved all the families in question to descend from one and the same parish of Norway and also to have at least one ancestor in common.—*Kristine Bonnevie.*

1462. Bonnevie, Kristine. **Om tvillingsfödslers arvelighet. Undersökelse over en norsk bygdeslegt. [On the inheritance of twin births. Investigations on Norwegian peasantry.]** Norsk. Mag. f. Lägev. 8: 1–22. 1919.—Hereditary disposition of twin births is stated within certain branches of a large country family (counting about 5000 individuals), the multiple births making in these branches no less than 7.7 per cent of all births, while the percentage of twin births within the whole country makes only 1.3–1.4 per cent. Through the "difference method" of Weinberg (subtraction of all twin "pairs" from the number of one-sexed twins) it is proved that about 80 per cent of all multiple births investigated should be considered as two-egged twin births, while probably only 20 per cent of multiple births have been from one egg. Younger mothers (below 30 years old) seem to give rise to one-egg and two-egg twin births in about equal number, while the number of one-egg twin births rapidly decreases among older mothers. The inheritance of two-egg twin births which must depend upon some hereditary character of the ovary is investigated through a genealogical study of the ancestry of twin mothers. Among 88 twin mothers 73 are shown to belong to twin-producing branches of the families investigated, while the ascendence of 15 twin-producing mothers is unknown. 67 twinning mothers whose ascendence is known through several generations on one (30 cases) or on both sides (37 cases) are without exception shown to descend from twin-producing families through both parents, or through the one of them whose ascendence is known. The type of inheritance seems, therefore, to be that of a recessive character demanding for its manifestation that the twinning mother should receive her disposition in a double dose, through both her parents. The investigations are being continued on other families and all results should as yet be considered as preliminary.—*Kristine Bonnevie.*

1463. Boulenger, G. A. **Un cas intéressant de dimorphisme sexuel chez un serpent africain (Bothrolycus ater Günther). [An interesting case of sexual dimorphism in an African snake.]** Compt. Rend. Acad. Sci. Paris 168: 666–669. 1919.—Sexes are distinguished by number of rows of scales, 19 in female, 17 in male. Variations in other species mentioned in literature are not related or are only indefinitely related to sex.—*A. Franklin Shull.*

1464. Burch, D. S. **Heredity and economical production of food.** Jour. Heredity 11: 7–11. *2 fig.* Jan., 1920.

1465. Burt, B. C., and N. Haider. **Cawnpore-American cotton: An account of experiments in its improvement by pure line selection and of field trials. 1913–1917.** Agric. Res. Inst. Pusa Bull. 88. *32 p., 10 pl., 1 fig.* 1919.—Describes effort to isolate pure lines adapted to Indian conditions from a badly mixed stock of an American upland variety.—*T. H. Kearney.*

1466. Call, L. E. **Director's report.** Kansas Agric. Exp. Sta. 1917–18. *63 p.* 1918.—Author states breeding parthenogenetic *Appotettix* indicates certain characters may be affected by temperature and moisture. Of several thousand parthenogenetic offspring, all were females except four. Parthenogenesis occurs among homozygotes and heterozygotes. "Crossing over" and "linkage" also occur.—Corn leaf aphis: *Aphis maidis*, reared at temperature of 84° to 90°F. produced no winged forms; reared at 72°F. one winged form appeared among many hundred wingless ones; reared at temperature of 60° to 70°F. large numbers of winged forms appeared. "In entire 55 generations no males appeared."—Cereal crops: Author states Kanred winter wheat is markedly resistant to cold and certain strains of stem rust. Kansas Nos. 2414 and 2415 exhibit similar resistance.—Hessian fly seldom lays eggs on "oats, barley, einkorn, spring emmer, and durum wheat, and less abundantly on soft than on hard winter wheats." Very few "flax seeds" were developed on wheat varieties, Illini Chief, Dawson Golden Chaff, Beechwood Hybrid, and Currell Selection, although eggs were laid on them "in abundance."—Swine: Following tendencies have been noted: (1) Wide Berkshire forehead is dominant over medium forehead of Duroc Jersey and narrow forehead of Tamworth and wild hog, (2) Berkshire dish of face is recessive to straight face of Tamworth and wild hog, (3) Berkshire short face is completely recessive to Tamworth long face, (4) Erect ear of Berkshire is dominant over drooping ear of Duroc Jersey.—Apparently there are distinct hereditary differences between Berkshire and Duroc Jersey with respect to size, rate of growth and early maturity."—*Fred Griffee.*

1467. CARD, W. H. **Originating and standardizing a new variety of Cornish.** Reliable Poultry Jour. 26: 647, 672, 725, 748, 749, 817, 857, 858, 927, 975, 976. *8 fig.* 1919.—An account of the origin of the White Laced Cornish fowl, by its originator, a practical breeder.—*H. D. Goodale.*

1468. CARLE, E. **Sélection pédigrée appliquée à la variété de riz "Nàng Mèo."** **[Pedigreed selection applied to the variety of rice known as "Nang Mèo."]** Bull. Agric. Inst. Sci. Saigon. 2: 73–78. 1920.

1469 COHEN-STUART, C. P. **A basis for tea selection.** Bull. Jard. Bot. Buitenzorg. III, 1: 193–320. 1919.—A comprehensive study of the origin, distribution and cultivation of tea. The systematic treatment of the genus *Camellia* is thoroughly discussed and a synoptic key is given for the determination of the various species. There is appended also a list of the specimens contained in the herbaria of Kew, Buitenzorg, Singapore and Berlin. This article comprises the first of three sections of a paper on selection of tea.—*J. H. Kempton.*

1470. COLE, LEON J., AND HEMAN L. IBSEN. **Inheritance of congenital palsy in guinea-pigs.** Amer. Nat. 54: 130–151. Mar.–Apr., 1920.—A definite neurosis (congenital palsy), characterized by clonic spasms, particularly of the legs, appeared in stock of normal guinea-pigs. All affected animals die at or before two weeks after birth. Defect is due to Mendelian recessive. DR × DR gave 183 normal, 63 palsied. Tested normals from this mating gave 7 DD and 15 DR. Variations of symptoms are noted and discussed. Defect is due to a factor mutation, cause unknown. Comparison is made with certain hereditary motor disturbances in pigeons, mice, rats, rabbits, goats, sheep, man and progeny of alcoholized guinea-pigs, none of which cases are considered identical with congenital palsy observed by the writers.—*C. C. Little.*

1471. COLE, LEON J. **An early family history of color blindness.** Jour. Heredity 10: 372–374. *1 fig.* Nov., 1919.

1472. COLLINS, G. N., AND J. H. KEMPTON. **Heritable characters of maize. I. Lineate leaves. Description and classification of lineate plants—value of maize as material for investigation, and economic importance of discovering latent variations.** Jour. Heredity 11: 3–6. Jan., 1920.

1473. COOK, O. F., AND ROBERT CARTER COOK. **Biology and government. Further discussion of Alleyne Ireland's article on democracy and the accepted facts of heredity.** Jour. Heredity 10: 250–253. June, 1919.

1474. COOK, O. F. **A disorder of cotton plants in China: Clubleaf or cyrtosis.** Jour. Heredity 11: 99–110. *9 fig.* Mar., 1920.

1475. COOLEY, CHARLES H. **A discussion of Popenoe and Johnson's "Applied eugenics" and the question of heredity vs. environment.** Jour. Heredity 11: 80–81. Feb., 1920.

1476. CORRENS, C. **Fortsetzung der Versuche zur experimentellen Verschiebung des Geschlechtsverhältnisses. [Continuation of experiments on artificial shifting of sex relations.]** Sitzungsber. Preuss. Akad. Wiss. Berlin 1918: 1175–1180. *3 fig.* 1918.

1477. COULTER, MERLE C. **Inheritance of aleurone color in maize.** Bot. Gaz. 69: 407–425. May, 1920.—An attempt was made to test the certainty with which predicted aleurone ratios would be fulfilled in complicated crosses. Crosses were made involving the *Rr Cc* and *Pp* factors in such a way as to require eight different ratios. The general conclusion is reached that the expectation in these cases is reasonably fulfilled. Seeds of different shades of color were separated and planted to determine whether it was possible to recognize genotypes by the intensity of the color. The author concludes that with experience genotypes may be separated by this method, particularly among red seeds. The inheritance of faintly colored or parti-colored seeds was studied. It is assumed that such seeds lack the aleurone factor *C* but have some partial substitute which is very erratic in its effect on the expression of color. An unusual case is reported where a plant known to have the factorial composition *Pp rr Cc* gave, when selfed, an ear with a perfect ratio of 9 colored to 7 white seeds. It is believed in this case that some unusual condition is present which produces purple aleurone when combined with the factors *PC* but colorless aleurone in combination with *C* only. Practically all the grains on this ear had irregularly split pericarps and when planted germinated slowly or not at all with a subsequent slow and stunted growth, suggesting that the aleurone ratio may be due to pathological causes. Crosses in which EMERSON'S *R*-tester was used as the male parent and *C*-tester as the female parent (*PPRRcc* × *PPrrCC*) were found to have only self purple seeds but when the parentage was reversed (*PPrrCC* × *PPRRcc*) all the seeds were mottled. This confirms the results of EMERSON from whom the material was received. In various crosses of EMERSON'S *C* and *R* testers with material obtained from East, the author concludes that these investigators have given similar symbols to the same set of factors. A study of mottling led to the conclusion that it can appear only when the *R* aleurone factor enters the seed from the male parent and then only when some other condition is present. This other condition was found in EMERSON'S *C*-tester. A very small percentage of mottled seeds is obtained where no mottling is to be expected, in some crosses involving *R*-tester. Such mottled seeds are believed to differ genetically from the mottling in the crosses involving *C*-tester.—It was found that there were no differences in the inheritance of aleurone color between inflorescences on the main stalk and suckers, but there was evidence, not given, that differences might be expected in the inheritance of plant colors, particularly chlorophyll, between the main culm and lateral branches.—A further test of the variability in inheritance which may occur between different parts of the same plant was obtained by self-pollinating both ears of two-eared plants. In most cases the two ears were reasonably alike but in some instances significant differences were found. The agreement between the two ears of the same plant is especially poor where faint aleurone color is involved.—The chance distribution of the different-colored seeds on the ear was tested and found to hold for starchy-sweet and colored-colorless but on ears where less than 10 per cent of the grains were particolored the majority of spotted grains were found in groups of 4 or 5, indicating the influence of local conditions. With respect to this phenomenon the author believes that local conditions on the ear do not determine but merely limit the appearance of particolored aleurone.—*J. H. Kempton.*

1478. COWGILL, H. B. **Cross-pollination of sugar cane.** Jour. Dept. Agric. Porto Rico 3: 1–5. Jan., 1919.—Method used at Insular Experiment Station of Porto Rico is satisfactory and many seedlings are produced. Bags are made of cheese cloth 48 inches long and 18 wide, held extended by heavy wire rings sewed into them. Rings placed one at top and other 16 inches from bottom so that a skirt of 16 inches is left to be drawn in and tied about stems of panicles. Bags are supported over panicles by means of bamboo poles set in ground with cross-bar at top. Poles are set to windward side of stools just before panicles "shoot;"

when panicles shoot, the bag is immediately suspended over each panicle and tied around its stem so that it is protected from undesirable pollen before any florets open. Cane blossom is hermaphrodite but some varieties are almost completely self-sterile, making it possible to cross-pollinate with another variety with assurance that nearly all offspring will be hybrids of the two chosen varieties. Pollinating is done by placing panicles of desired variety in bag, in such position that pollen will be shed or carried by wind or insects to florets of other variety as they open. One or two panicles are used at a time, allowed to remain in bag two or three days, being renewed as often as necessary. It is found advantageous to cut stems 4 to 6 feet long and put cut end in joint of bamboo filled with water, thus keeping fresh 2 or 3 days. —Results: 1915–1916. Ten crosses attempted, eight produced seedlings, majority of which showed characteristics of both parents. About 1500 seedlings produced, one panicle yielding over 1000.—1916–1917. Thirty crosses made comprising nine different combinations, of which nineteen were successful. From one combination 1309 seedlings were obtained and in all 2589 were produced.—1917–1918. Thirty crosses were attempted, comprising nine combinations. Fifteen were successful and 1794 seedlings were produced, 157 from one combination, 735 from another.—Effect of crossing: In 1915–1916 and 1916–1917 pollinator was dark-colored cane while seed-parent was medium light, and dark color of pollen parent was seen in many of offspring.—At least two of old standard varieties are nearly pollen-sterile here (Crystalline and Rayada).—*E. E. Barker*.

1479. Cunningham, J. T. **Results of a Mendelian experiment on fowls, including the production of a pile breed.** Proc. Zool. Soc. London 1919: 173–202. *1 pl.* Sept., 1919.—A male black-red *Gallus bankiva* was crossed to a silky hen. Data on inheritance of plumage, skin pigmentation, comb, booting and crest are given. The production of a pile race from the cross, which bred true, is described. "The simplest explanation" of its origin "is that segregation is not complete or perfect . . . ." Attempts to increase amount of pigmentation in the piles by repeated back-mating to normals did not result in any consistent increase. —*H. D. Goodale*.

1480. Danforth, C. H. **Resemblance and difference in twins.** Jour. Heredity 10: 399–409. *Frontispiece, fig. 1–14, 20, 22–30.* Dec., 1919.

1481. Daniel, L., and H. Teulié. **Extension des limites de culture de la vigne au moyen de certains hybrids. [Extension of the limits of culture of the grape by means of certain hybrids.]** Compt. Rend. Acad. Sci. Paris 166: 297–299. 1918.

1482. Davenport, C. B. **A strain producing multiple births.** Jour. Heredity 10: 382–384. Nov., 1919.

1483. Delage, Y., and M. Goldsmith. **Le Mendelisme et le mécanisme cytologique de l'hérédité. [Mendelism and the cytological mechanism of heredity.]** Rev. Sci. Paris 57: 97–109, 130–135. 1919.—Part I is a brief summary of Mendelism, "Neo-Mendelism" and the chromosome theory of heredity, including the factorial hypothesis, the phenomena of linkage, crossing over and non-disjunction and the chromosomal mechanism of sex determination. Mendelism is compared with Weismannism. Credit Naudin with many discoveries attributed to Mendel. Mention influence of environment and cytoplasmic inheritance. Part II is a critique of Mendelism (or Neo-Mendelism). Acknowledge great advances and brilliant achievements in this field but think Mendelians are blinded to the uncertainties, defects, lacunae and improbabilities of the theory and the fragility of the objective bases upon which it rests. Illustrate (1) by questioning continuity of chromosomes because these are not visible in resting stage, (2) by questioning linear arrangement of genes because chemical differentiation of chromatin within individual chromosomes has not been demonstrated, (3) by contending that a force which will bring homologous chromosomes into such intimate and accurate alignment as necessitated by crossover hypothesis will not permit them to lie X-wise and give crossovers, and (4) by maintaining that Mendelian conception gives no explanation of successive appearance of characters in ontogeny or, (5) of the origin of new characters during evolution. Predict downfall of Mendelism from weight of accessory hypotheses needed to explain special cases.—*C. W. Metz*.

1484. Demoll, R. **Zur Frage nach der Vererbung vom Soma erworbener Eigenschaften. [On the question of the inheritance of acquired characters.]** Arch. Entwicklungsmech. Organ. 46: 4–11. *3 fig.* 1920.

1485. Detjen, L. R. **A mutating blackberry—dewberry hybrid.** Jour. Heredity 11: 92–94. *4 fig.* Feb., 1920.

1486. Detlefsen, J. A., and W. W. Yapp. **The inheritance of congenital cataract in cattle.** Amer. Nat. 54: 277–280. May–June, 1920.—On mating the F_1 son of Holstein-Friesian bull 62924 to the F_1 daughters of this bull 8 F_2 offspring (2♀ and 6♂) with well-defined congenital cataracts of the stellate type to 55 F_2 normal offspring were produced. Ninety-three normal F_1 offspring of 62924 were produced. Pedigree studies of bull 62924 reveal no ancestors which had cataracts. Assuming the bull 62924 heterozygous the F_2 expectation is 55.125 normal + 7.875 cataractous. 62924 mated to his own daughters produced 7 offspring, 3 (1♂ + 2♀) of which were cataractous. It is concluded that congenital cataract in cattle is a simple recessive Mendelian character.—*John W. Gowen.*

1487. De Vries, Hugo. **Oenothera Lamarckiana erythrina, eine neue Halbmutante. [Oenothera Lamarckiana erythrina, a new half-mutant.]** Zeitschr. indukt. Abstamm. Vererb. 21: 91–118. 1919.

1488. Doncaster, L. **The tortoiseshell tomcat. A suggestion.** Jour. Genetics 9: 335–338. Mar., 1920.—Author criticizes Little's hypothesis of mosaic character of tortoiseshell tomcat and on basis of work of Chapin, Lillie, and Magnusson on free-martin and of Cutler and Doncaster on histology of testis of sterile tortoiseshell tomcat, suggests that latter be a masculised female.—*P. W. Whiting.*

1489. Doncaster, L., and H. G. Cannon. **On the spermatogenesis of the louse (Pediculus corporis and P. capitis), with some observations on the maturation of the egg.** Quart. Jour. Microsc. Sci. 64: 303–328. *1 pl., 1 fig.* Mar., 1920.—*P. corporis* has 12 chromosomes in somatic cells of both sexes. In the testis certain large cells, supposed to be follicular, also have 12. Other cells of testis, believed to be spermatogonia, have 6, apparently double, chromosomes. Spermatocytes, also with 6 chromosomes, pass through growth period followed by a very asymmetrical division, giving one large cell which develops into a spermatid and one small "polar cell" which degenerates. A conspicuous mitochondrial body remains in the large cell. No second spermatocyte division occurs. Centrosomes of spermatids are double and there are two axial filaments. No oögonial or oöcyte divisions were found. Author did not observe unisexual broods or sex-ratio disturbances described by Hindle. Spermatogenesis of *P. capitis* apparently agrees with that of *P. corporis.*—*C. W. Metz.*

1490. Duerden, J. E. **Methods of degeneration in the ostrich.** Jour. Genetics 9: 131–193. *Pl. 5–6, 8 fig.* Jan., 1920.—Author describes type of degenerative changes observed in coverts, wing quills, down feathering, wing digits and toes, and regards these as suggestive of the manner in which degeneration proceeds, and as favorable data for throwing light on the nature of variation and method of evolution generally.—In his discussion of relation of the degenerative changes to adaptation, author concludes that, compared with other factors, such losses have little or no bearing upon the welfare of the ostrich; and hence, that natural selection has been inoperative in directing their course. "Natural selection may wipe out the race, but cannot guide its evolution."—Referring to ontogenetic and phylogenetic degeneration, author believes process of degeneration is in no way affected during the life of the individual, but only with the formation of the zygote; in plumes, scales and claws of embryos and chicks the degenerative changes are found expressed just as in the adult. "Degeneration may be defined as the somatic expression of a phylogenetic degradation and loss of genetic factors."—As to cause of degeneration, author acknowledges our ignorance on this point but believes they are certainly intrinsic as opposed to environmental. "The influence is so slowly acting . . . as to call for an aloofness, an independence, of external vicissitudes. Only something in the organism itself, and beyond all varying somatic responses, could meet demands so continuous and so consistent." According to the author the agency at work possesses a strong determinate influence; and the evidence is of such a nature as to remind one of Nägeli's conception of a mystical, internal, vitalistic force. In the ostrich, it is suggested that the changes may be interpreted in terms of "a germinal senescence, perhaps expressing itself in factorial fractionation and loss." The author believes that the ostrich race may present us with an example of "mass mutation."—In conclusion, author discusses the possibility of factorial changes, but this point, with reference to the bearing of the ostrich data, is left inconclusive. —*P. B. Hadley.*

1491. Elderton, Ethel M. [Rev. of: Whipple, George Chandler. **Vital statistics: An introduction to the science of demography.** *12 x 18 cm., v + 517 p., 63 fig.* John Wiley & Sons, Inc.: New York, 1919.] Science Progress 14: 696-697. April, 1920.—See Bot. Absts. 3, Entry 2212.

1492. Ellinger, Tage. [German rev. of: Punnett, R. C., and the late Major P. G. Bailey. **Genetic studies in poultry. I. Inheritance of leg feathering.** Jour. Genetics 7: 203-213. May, 1918. (See Bot. Absts. 1, Entry 492.)] Zeitschr. indukt. Abstamm. Vererb. 22: 288. May, 1920.

1493. Ellinger, Tage. [German rev. of: Rasmuson, Hans. **Über eine Petunia-Kreuzung. (On a petunia cross.)** Bot. Notiser 1918: 287-294. 1918. (See Bot. Absts. 3, Entry 2181.)] Zeitschr. indukt. Abstamm. Vererb. 22: 289. May, 1920.

1494. Ellinger, Tage. [German rev. of: Rasmuson, Hans. **Zur Genetik der Blütenfarben von Tropaeolum majus. (On the genetics of the flower colors of Tropaeolum majus.)** Bot. Notiser 1918: 253-259. Nov., 1918. (See Bot. Absts. 3, Entry 2180.)] Zeitschr. indukt. Abstamm. Vererb. 22: 288-289. May, 1920.

1495. Ellinger, Tage. [German rev. of: Raunkiaer, C. **Om Løvspringstiden hos Afkommet af Bøge med forskellig Løvspringstid. (On leaftime in the descendants of beeches with different leaf times.)** Bot. Tidsskr. 36: 197-203. 1918. (See Bot. Absts. 2, Entry 42.)] Zeitschr. indukt. Abstamm. Vererb. 22: 289. May, 1920.

1496. Emerson, R. A. **Heritable characters of maize. II. Pistillate flowered maize plants.** Jour. Heredity 11: 65-76. *8 fig.* Feb., 1920.

1497. Emoto, Y. **Über die relative Wirksamkeit von Kreuz- und Selbstbefruchtung bei einigen Pflanzen. [On the relative effectiveness of cross- and self-fertilization in several plants.]** Jour. Coll. Sci. Imp. Univ. Tokyo 43: 1-31. *2 pl., 6 fig.* Mar. 15, 1920.

1498. Erdmann, Rhoda. **Endomixis and size variations in pure bred lines of Paramaecium aurelia.** Arch. Entwicklungsmech. Organ. 46: 85-148. *12 fig.* 1920.

1499. Erikson, J. **Platanthera bifolia × montana i Blekinge. [Platanthera bifolia × montana in Blekinge.]** Bot. Notiser 1918: 59-62. 1918.

1500. Euler, K. **Ein bemerkenswerter Fall von Knollen-Farbabänderung der Kartoffel. [A remarkable case of change of color in potato tubers.]** Deutsch. Landwirtsch. Presse 1919: 161-162. 1919.

1501. Fairchild, David. **Twins.** Jour. Heredity 10: 387-396. *Frontispiece, fig. 1-14, 20, 22-30.* Dec., 1919.

1502. Fleischmann, R. **Die Auslese bei der Maiszüchtung. [Selection in maize breeding.]** Zeitschr. Pflanzenzücht. 6: 69-96. 1918.—Selection has been practiced since 1909 on the yellow horse-tooth variety of maize. The characters used were yield of grain, length and number of rows on the ear, per cent of grain to cob, weight of 100 seeds, and time of maturity. —It was found that in selecting for yield of grain the best results were obtained when the progeny row was taken as the unit of selection rather than the individual plant, although positive results were obtained in either case.—Selection for number of rows was ineffective since the progenies regressed to a fourteen-rowed type regardless of whether the selection was made for a greater or less number of rows.—The per cent of grain to cob was found to be readily changed by selection but it was found also that the size of the cob was directly associated with the yield of grain. Care, therefore, must be exercised in selecting for an increased ratio of grain to cob, not to reduce the absolute size of the cob.—The author questions the value of many-eared strains and restricted selection to single-eared plants.—*J. H. Kempton.*

1503. FLORIN, RUDOLF. **Zur Kenntnis der Fertilität und partiellen Sterilität des Pollens bei Apfel- und Birnensorten. [On the fertility and partial sterility of the pollen of different varieties of apple and pear.]** Acta Horti Bergiani 7: 1–39. 1920.—If there is self-sterility or insufficient power of germination of the pollen of a variety of fruit trees it is not advisable to grow the variety in question alone in great closed groups, but other sorts should be grown among them which produce plenty of pollen with great efficiency. Author has examined the power of germination of the pollen (in solutions of sugar of variable concentration) of 102 apple and 14 pear varieties, which are cultivated in Sweden. He gives a tabulated summary of 405 experiments, wherein he states date, time of examination, temperature, per cent of germination and maximum and minimum length of the measured pollen tubes.—Of the apples 24 sorts showed 0–30 per cent of germination; 13 showed 31–70 per cent; and 65 showed 71–100 per cent. The last group is of course the most preferable for use as pollenizers. A list of literature is given containing 27 citations.—*K. V. Ossian Dahlgren.*

1504. FOOT, KATHARINE. **Determination of the sex of the offspring from a single pair of Pediculus vestimenti.** Biol. Bull. 37: 385–387. Dec., 1919.—A pair of fleas produced 143 fertilized eggs. Of these 125 hatched and the sex was determined for 115 of the young or 92 per cent of the total. There were 62 males and 53 females. The earlier-produced eggs yielded a higher percentage of females than males. Later the proportion of the sexes became equal and then, as the last eggs were produced, the earlier sex ratio was reversed—more eggs developing into males than females.—*D. D. Whitney.*

1405. FRASER, ALLAN CAMERON. **The inheritance of the weak awn in certain Avena crosses and its relation to other characters of the oat grain.** Cornell Univ. Agric. Exp. Sta. Mem. 23: 635–676. June, 1919.—A study is made of the inheritance of the weak awn in *Avena* crosses. Burt oats were used as parent for the weak awn and Sixty Day for awnless. The reciprocal crosses indicated an approach to dominance of awnlessness. In F_2 generations, two distinct classes of the weak awn and awnless appeared with a variation between the two types of about all the possible differences between the parent sorts. These intermediate forms could not be separated into classes on a multiple factor basis. If all these intermediate forms were thrown into one class, there would be a close approximation to the 1:2:1 ratio. The fully awned type is evidently pure recessive. Data in F_2 or F_3 generations did not include the entire plant, the center spikelet only being used. This method was based upon results of Love and McRostie on the tendencies of the plant to agree in its characteristics with the terminal spikelet. The data seemed to show that both parents contain a factor for awning, but that the Sixty Day parent possesses an inhibitor linked with yellow color. The inhibitor seems to be affected in its power of inhibition by environmental factors. The partly awned plants in F_3 generations are shown to be heterozygous in successive progeny types. Spikelets with two awns on a kernel are found only on completely awned spikelets. Increase in soil moisture and nitrogen seems to decrease number of awns.—The appearance of strong and intermediate awns in F_2 and F_3 progenies is considered to be a reversion. There is strong linkage shown between medium long basal hairs and the awned condition. Short basal hairs or no hairs are dominant over long basal hairs.—With respect to color, the F_1 plants are intermediate. On account of the difficulty of determining color under weather conditions, the F_2 is not considered well classified. The Burt oat possesses a red factor and a yellow factor, which are quite distinct from the Sixty Day factor. The Sixty Day yellow factor inhibits awning. The Burt yellow carries no such inhibitor. The F_3 generation bears out most of the conclusions reached in F_2. The appearance of brown berries is attributed to mutation or reversion.—*Alvin Kezer.*

1506. FRATEUR, J. L. **La robe sauvage du lapin.** [The wild coat of the rabbit.] Réunion Soc. Belge Biol. 1919: 941–943. 1919.

1507. FRETS, G. P. **De polymerietheorie getoetst aan de erfelijkheid van den hoofdvorm.** [Theory of polymery tested in the inheritance of head-form.] Genetica 2: 115–136. Mar., 1920.

1508. FRUWIRTH, C. **Neunzehn Jahre Geschichte einer reinen Linie der Futtererbse.** [Nineteen-year history of a pure line of field peas.] Fühlings landw. Zeitg. 69: 1–28. 1920.—Study of variations in a pure line, in sense of Johannsen, of field peas breeding absolutely true for three years to pink flowers and yellowish-green seed-coats. In succeeding years, "spontaneous variations" occurred from time to time such as plants with red-purple flowers and maple seed-coats, purple specked and purple-striped seed-coats, albino foliage, variegated yellow and green or more rarely green and white foliage, and plants that either died prematurely or set no pods or set pods, but matured no seeds. Detailed data given including tables, of selection and crossing experiments with some of the variants of this pure line. Only negative results obtained with selection lines. Variants may be regarded as phases of eversporting races, the variations arising either in vegetative cells or in sexual cells. In latter case parents of variants are hybrids, giving segregation ratios of a Mendelian type although these may be irregular. Some spontaneous variations such as red-purple flowers and maple seed-coats are dominants, while others such as albinism and other foliage-chlorophyll defects are recessive. Albino foliage variations appear first in a ratio of 3 green: 1 white, but the variation must have arisen in the sex cells two generations back, but since green foliage is dominant, did not appear except as members of an F_2 generation. Albinism and other chlorophyll defects appeared only in F_2 and later generations of cross of the "pure line" with a white-flowered green-foliage variety. Literature of chlorophyll defects is reviewed. "Disassociation" and "association" concept of TSCHERMAK is discussed; also "pluripotency" concept of HAECKER. Variations occurring in sex cells uniting with the unvarying sex cells appear as hybrids. Variations taking place in vegetative cells later give rise to sex-cells which unite and produce pure races of hereditary variations at once. Eversporting proclivity may express itself rarely in some races and as regards some characters.—*Orland E. White.*

1509. GAINES, E. F. **The inheritance of resistance to bunt or stinking smut of wheat.** Jour. Amer. Soc. Agron. 12: 124–132. 1920.—Bunt resistance to wheat is not a simple Mendelian unit character, but resistance, if Mendelian, is composed of multiple factors, for a continuous series ranging from complete immunity to complete susceptibility has been obtained. Different wheat varieties possess different kinds of resistance. Linkage between resistance and morphological characteristics is not sufficient to prevent the selection of a resistant strain of any morphological type desired.—*F. M. Schertz.*

1510. GALLOWAY, BEVERLY T. **Some promising new pear stocks.** Jour. Heredity 11: 25–32. *8 fig.* Jan., 1920.

1511. GAUGER, MARTIN. **Die Mendelschen Zahlenreihen bei Monohybriden im Lichte der Dispersionstheorie.** [The Mendelian ratios in monohybrids in the light of the dispersion theory.] Zeitschr. indukt. Abstamm. Vererb. 22: 145–198. Mar., 1920.

1512. GOLDSCHMIDT, RICHARD. **Intersexualität und Geschlechtsbestimmung.** [Intersexuality and sex determination.] Biol. Zentralbl. 39: 498–512. Nov., 1919.

1513. GOWEN, J. W. **Appliances and methods for pedigree poultry breeding at the Maine Station.** Maine Agric. Exp. Sta. Bull. 280: 65–88. *13 fig.* 1919.—This is a revision of an earlier bulletin on the same subject.—*H. D. Goodale.*

1514. Grantham, J., and M. D. Knapp. **Field experiments with Hevea.** Agric. Bull. Federated Malay States 6: 596–597. 1918.

1515. Grantham, J., and M. D. Knapp. **Field experiments with Hevea.** Arch. Rubbercultuur 2: 614–630. 1918.

1516. Green, Heber. **The application of statistical methods to the selection of wheat for prolificacy. Agricultural research in Australia.** Advisory Council Sci. and Ind. Commonwealth of Australia Bull. 7: 49–56. 1918.—Author discusses application of familiar biometric methods and points out their limitations in wheat breeding. Experiments have been conducted for seven generations in selecting the heavy-, medium-, and light-yielding plants of wheat. Progress in both directions resulted, though apparently much more rapid, in the direction of high yield.—In an attempt to develop a wheat suitable for semi-arid climates an unusually severe season destroyed all but three plants in a plot. One of these three was a giant, the progeny of which has given rise to a valuable strain.—*J. H. Kempton.*

1517. Haecker, V. **Eine medizinische Formulierung der entwicklungsgeschichtlichen Vererbungsregel. [A medical formulation of the developmental law of heredity.]** Deutsch. Med. Wochenschr. 44: 124–126. 1919.—The author's "developmental law of heredity" [See Bot. Absts. 4, Entry 588] is briefly explained and illustrated. In general the clearness with which a trait segregates in heredity is a function of the autonomy of that trait in development. Hereditary defects occurring in organs with a high degree of developmental autonomy tend to follow simple Mendelian rules in heredity while those dependent for their manifestation on disharmonies in several organs or systems (e.g., diabetes) do not do so. Cases in which the same organ shows different defects in various members of the same family are interpreted as indicating an early autonomy of the organ in question with a more or less generalized weakness of that organ in the particular family concerned.—*C. H. Danforth.*

1518. Harlow, H. V., and H. K. Hayes. **Breeding small grains in Minnesota. II. Investigations in barley breeding.** Minnesota Agric. Exp. Sta. Bull. 182: 45–56. *4 fig.* Mar., 1919.—Two lines of investigation (pure-line and hybridisation) are discussed as methods of barley improvement. From selections of domestic and foreign sorts it was found that almost as wide variations in yield were found within a variety as in different varieties. By means of several crosses between Lion, a smooth-awned black barley, and Manchuria, a smooth-awned barley of high yielding ability has been produced. Other promising crosses have also been obtained. Sixty-eight selections, crosses and new introductions are compared on the basis of the yearly production. A method for discarding in elimination tests based on the probable error is presented.—*W. E. Bryan.*

1519. Harper, R. A. **Inheritance of sugar and starch characters in corn.** Bull. Torrey Bot. Club 47: 137–186. *3 pl.* April, 1920.—Work of Correns and of East and others on the inheritance of sugar and starch characters in corn endosperm (*Zea*) is reviewed to show that intermediate sweet-starchy types result from crossing these two forms. Original experiments with crosses of different sweet and starchy endosperm varieties carried to the fourth filial generation are described and illustrated. Dominance of starchiness is shown in first cross but in segregating generations intermediate kernels ranging from practically pure sweet to pure starchy in appearance were obtained in varying proportions and degree along with other cases in which more definite segregation occurred. The different grades of kernels are classified and tabulated. Marked tendency shown for intermediate types to breed true but with more of an inclination to revert to sweet type than to starchy type. Practically pure starchy ears, in appearance, were obtained from a cross of two sweet varieties. Continuity of variation in both sexually and asexually reproduced types is taken as an indication of mutual modification of germplasm where contrasting characters are brought together. The main features of chromosome individuality and of reduction phenomena are considered as established but the physiological nature of the chromatin is thought to permit mixing of hereditary materials resulting in intergradations between parental forms.—*D. E. Jones.*

1520. Hendrickson, A. H. **Plum pollination.** California Agric. Exp. Sta. Bull. 310. *28 p., 5 fig.* July, 1919.—Experiments show 13 varieties self-sterile, 3 self-fertile and 1 doubtful. Early-blooming Japanese varieties produce little pollen and are not efficient pollenizers. Late-blooming varieties produce abundant pollen. Except for the self-fertile French and sugar prunes interplanting of varieties is recommended to increase yields. No evidence of intersterility among plum or prune varieties was found. Experiments show that bees are efficient agents of cross-pollination. Set of fruit is also influenced by climatic factors.—*J. L. Collins.*

1521. Herre, Albert C. **Hints for lichen studies.** Bryologist 23: 26–27. 1920.—See Bot. Absts. 5, Entry 1949.

1522. Hertwig, P. [German rev. of: Boveri, Theodor. **Zwei Fehlerquellen bei Merogonieversuchen und die Entwicklungsfähigkeit merogonischer und partiellmerogonischer Seeigelbastarde. (Two sources of error in investigations of merogony and the ability of merogonic and partially merogonic sea-urchin hybrids to develop.)** Arch. Entwicklungsmech. Organ. 44: 417–471. *3 pl.* 1918.] Zeitschr. indukt. Abstamm. Vererb. 22: 216–218. Mar., 1920.—See also Bot. Absts. 3, Entry 600.

1523. Hertwig, P. [German rev. of: Hertwig, Günther. **Kreuzungsversuche an Amphibien. (Hybridization studies on amphibians.)** Arch. Mikrosk. Anat. 91: 203–271. *2 fig.* Aug. 20, 1918. See Bot. Absts. 3, Entry 1005. Zeitschr. indukt. Abstamm. Vererb. 22: 219–221. Mar., 1920.

1524. Hilgendorf, F. W. **Methods of plant breeding.** New Zealand Jour. Agric. 19: 354–358. 1919.—Popular. [See Bot. Absts. 5, Entry 1153.]

1525. Holländer, Eugen. **Familiäre Fingermissbildung (Brachydaktylie und Hyperphalangie). [Familial abnormalities of the fingers (brachydactyly and hyperphalangy).]** Berlin Klin. Wochenschr. 55: 472–474. 1918.—A man and his son, and probably also his sister, are characterized by a shortening of the fingers accompanied by an extra bony element in the basal phalanx of digits two and three. Evidence is brought forth to show that the extra element is an ununited epiphysis, the inhibition of normal union being in these cases apparently an hereditary trait.—*C. H. Danforth.*

1526. Holmberg, O. R. **Carex dioica×paniculata, en för Skandinavien ny hybrid. [Carex dioica×paniculata, a hybrid new for Scandinavia.]** Bot. Notiser 1918: 249–252. *3 fig.* 1918.

1527. Honing, J. A. **Selectie-proeven med Deli-tabak. II. [Selection experiments with Deli-tobacco. II.]** Meded. Deli-Proefstation, Medan, Sumatra, 2: 84. *1 pl.* 1918.—Gives results of selection experiments at Deli Proefstation for 1917. The tobacco was harvested separately, tied in bundles with specially colored twine, fermented in bulk with the other tobacco, and finally separated for testing. In general the results of 1917 were inferior to those of 1916 due to less favorable weather. Both large- and small-scale trials were made. In the small-scale trials there were 467 lots, most of these containing 800–1200 plants. These repres ented 150 seed-numbers belonging to 81 lines. Of the large-scale trials, with from 90,000 to 560,000 plants per lot, there were 34. These trials were distributed over 17 estates and were supervised by 5 assistants. Figures for production, percentages of various qualities, estates' grading and manufacturers' grading, leaf measurements, numbers of leaves per plant, burning tests, etc., are given for most of these lines. The writer does not agree with Koch (Koch, L. Algem. 1528 Landbouwincekblad voor Med. India, Dec. 7, 1917) that mixed seed is to be preferred to that from pure lines, so far as tobacco culture is concerned. [See also next following Entry, 1528.]—*Carl D. La Rue.*

1528. Honing, J. A. **Selection experiments with Deli tobacco. III.** Meded. Deli-Proefstat. Medan 2: 25. 1919.—See also next preceding Entry, 1527.

1529. Hottes, Alfred C. **Our American originators.** Florists' Exchange 48: 933. *3 fig.* Dec. 27, 1919.—The work of the A. W. Livingston Seed Co., of Columbus, Ohio, is discussed somewhat flatteringly and information is given as to the source or point of origin of nineteen commercial varieties of potatoes.—*H. F. Roberts.*

1530. Houwink, R. Hzn. **Erfelijkheid. Populaire beschouwingen omtrent het tegenwoordige standpunt der erfelijkheid, verzameld uit theorie en practijk. [Heredity. Popular presentation of the present status of heredity compiled from theory and practice.]** Assen. Stoomdrukkerij Floralia 1919: 1-62. *5 pl.* 1919.

1531. Howe, Lucien. **The relation of hereditary eye defects to genetics and eugenics.** Jour. Heredity 10: 379-382. Nov., 1919.

1532. Hume, A. N. **Corn families of South Dakota.** South Dakota Agric. Exp. Sta. Bull. 186: 114-134. Aug., 1919.—A plan of corn breeding is described in which a 96-ear-row breeding plot is employed. The plot is divided into four independent quarters of twenty-four rows each and alternate rows are detasseled in order to insure against the most extreme forms of inbreeding. Thus far the system follows that devised by the Illinois Agricultural Experiment Station. An important modification, however, lies in the fact that instead of planting the tasseled or "sire" rows from different individual ears, all of the twelve "sire" rows of each quarter are planted from kernels of a single ear. This not only permits a more intense selection for high yield but also makes possible the establishment of a definite ear pedigree along both lines of parentage. Data are given to show the tendency of yielding capacity of seed ears to follow lines of ancestry.—*L. H. Smith.*

1533. Hume, A. N. **Yields from two systems of corn breeding.** South Dakota Agric. Exp. Sta. Bull. 184: 70-86. Jan., 1919.—Two systems of corn breeding are compared, both of which are based upon the ear-row plan of continuous selection. The essential difference between the two systems is that in the one, alternate rows of the breeding plot are detasseled and seed is taken only from detasseled plants thereby insuring a certain degree of crossing while in the other system this precaution is omitted. The results based upon several seasons' data indicate no significant difference in effectiveness in increasing yield. The working details of a plan of corn improvement intended to meet the demand for simplicity and practicability are appended.—*L. H. Smith.*

1534. Ikeno, S. **Études d'hérédité sur la réversion d'une race de Plantago major. [Hereditary studies on reversion in a race of Plantago major.]** Rev. Gén. Bot. 32: 49-56. 1920.

1535. Ireland, Alleyne. **Democracy and heredity—A reply.** Jour. Heredity 10: 360-367. Nov., 1919.

1536. Janssens, F. A. **À propos de la chiasmatype et de la théorie de Morgan. [Concerning the chiasmatype and Morgan's theory.]** Réunion Soc. Belge Biol. 1919: 917-920. 1919.

1537. Janssens, F. A. **Une formule simple exprimant de qui se passe en réalité lors de la "chiasmatypie" dans les deux cinèses de maturation. [A simple formula expressing what really takes place in chiasmatypy in the two maturation divisions.]** Réunion Soc. Belge Biol. 1919: 930-934. 1919.

1538. Johannsen, W. **Weismanns Keimplasma-Lehre. [Weismann's germplasm theory.]** Die Naturwiss. 6: 121-126. 1918.

1539. Johannsen, W. **Om Weismann's Kimplasma-Laere. [Weismann's germplasm theory.]** Vidensk. Meddelelser fra Dansk Naturhist. Foren i Kjøbenhavn. 69: 153-164. 1918.

1540. Johnson, Charles W. **Variation of the palm weevil.** Jour. Heredity 11: 84. Feb., 1920.

1541. Johnson, James. **An improved strain of Wisconsin tobacco. Connecticut Havana No. 38.** Jour. Heredity 10: 281-288. *Fig. 8-10.* June, 1919.

1542. Jones, D. F., and W. O. Filley. **Teas' hybrid catalpa. An illustration of the greater vigor of hybrids; increased growth and hardiness as a result of crossing; illustrating definite principles of heredity.** Jour. Heredity 11: 16-24. *6 fig.* Jan., 1920.

1543. Jones, D. F. **Selection in self-fertilized lines as the basis for corn improvement.** Jour. Amer. Soc. Agron. 12: 77-100. 1920.—Selection in self fertilized lines makes possible a reliable estimation of hereditary values of both sexes and is suggested for corn improvement. —*F. M. Schertz.*

1544. Kappert, H. **Über das Vorkommen volkommener Dominanz bei einem quantitativen Merkmal. [The occurrence of complete dominance in a quantitative character.]** Zeitschr. indukt. Abstamm. Vererb. 22: 199-209. *1 fig.* Mar., 1920.

1545. Kempton, J. H. **Heritable characters of maize. III. Brachytic culms.** Jour. Heredity 11: 111-115. *4 fig.* Mar., 1920.

1546. Klatt, B. **Experimentelle Untersuchungen über die Beeinflussbarkeit der Erbanlagen durch den Körper. [Experimental investigations on the modifiability of the hereditary factors through the soma.]** Sitzungsber. Ges. Naturf. Freunde. 1919: 39-45. 1919.—Writer experimented with three races of gypsy moth (*Lymantria dispar*). The caterpillars of one of these had an unusually broad yellow stripe along the back, dominant on the whole over the narrow yellow stripe of the normal race. The third race had a black longitudinal stripe, dominant over yellow and normal and clearly differing by a unit factor. He extirpated the ovaries of individuals dominant in one or both factors (yellow or black) and transplanted in their place ovaries from recessive individuals. These females were mated with recessive males. The caterpillars appeared to be pure recessives, showing no trace of the dominant characters of the foster mothers. [See also Bot. Absts. 5, Entry 1579.]—*Sewall Wright.*

1547. Klatt, Berthold. [German rev. of: Dürken, Bernhard. **Einführung in die Experimentalzoologie. (Introduction to experimental zoology.)** *16 x 23 cm., x + 446 p., 224 fig.* Julius Springer: Berlin, 1919.] Zeitschr. indukt. Abstamm. Vererb. 22: 275-276. May, 1920.

1548. Klatt, B. [German rev. of: (1) Palmgren, Rolf. **Till Kännedomen om Abnormiteters Nedärfning hos en del Husdjur. (Inheritance of abnormalities in certain domestic animals.)** Acta Soc. pro fauna et flora fennica 44: 1-22. 1918. (2) Palmgren, Rolf. **Tvenne bastarder mellan getbock och fartacka, födda i Högholmes zoologiska trädgard. (Two hybrids between sheep and goats produced in Högholm zoological gardens.)** Med. pro fauna et flora fennica 44: 124-125. 1918.] Zeitschr. indukt. Abstamm. Vererb. 22: 283-284. May, 1920.

1549. Klatt, B. [German rev. of: Pézard, M. A. **Transformation expérimentale des caractères sexuels secondaires chez les Gallinacés. (Experimental transformation of secondary sexual characters in Gallinaceae.)** Compt. Rend. Acad. Sci. Paris 160: 260-263. 1915.] Zeitschr. indukt. Abstamm. Vererb. 22: 284. May, 1920.

1550. Koch, L. **Verdere Onderzoekingen betreffende de praktijkwaarde van de lijnenselectiemethode, mede in verband met het gemengd planten van varieteiten. [Further observations on the practical value of the line-selection method and a comparison of it with the mixed planting of varieties.]** Teysmannia 29: 389-423. 1918.—Author has made comparative tests of planting in (a) pure lines, (b) mixed populations and (c) populations made up of definite mixtures of pure lines of the following crops: rice, katjang tanah, kedelee, corn, potatoes, and cassave, and finds that in rice and katjang, line selection gives no satisfactory results. Varieties of rice when in mixed plantings influence each other greatly. The results of such influences depend upon the kind and proportion of the varieties in the mixture. It is possible to get mixtures that produce a higher average yield than any of the varieties of which the mixture is composed. [See also next following Entry 1551.]—*W. H. Eyster.*

1551. KOCH, L. Onderzoekingen betreffende de praktijkwaarde van de lijnenselectie-methode voor verschillende éénjarige landbouwgewassen. [Researches concerning the practical value of the line selection method for various annual tropical crops.] Teysmannia 29: 1–36, 96–127, 156–191, 389–423. 1918.—The line-breeding method was first practised in 1907 by VAN DER STOK, then assistant at the botanical section of the Experiment Station for Rice and other Annual Crops at Buitenzorg, Java. A great deal of line breeding had been performed before 1915, the selected crops being specially rice, ground-nuts and soy beans. During the years when most breeding took place (1911–1915) some peculiarities were noticed, which gave birth to the idea that line breeding was by no means a method for securing high-producing rice strains, etc. In the trials (almost all of them with 8 or more control plots) it was observed that the population (mixture of all strains, high- and low-producing) gave in most cases an unexpectedly high yield, higher than most selected pure strains. Breeding did meet with success where immunity for certain diseases or qualitative peculiarities were aimed at. As most breeding was for increasing the yield, a series of trials was undertaken to determine whether line breeding should be continued or not, and to investigate the reason why there was so little success.—In the years 1914–1916 selection took place for 6 rice varieties. In only 2 of 16 trials did the selected rice strains give a fairly good yield in comparison to the unselected mixture. As a rule, a strain that gave one year the highest yield, failed to do so in the next. More than once such a strain yielded much less than some others had that been much inferior the previous year.—As the climate at Buitenzorg is somewhat peculiar, and results might perhaps be influenced by the great rainfall or the moist atmosphere, trials were made at the same time at the experimental farms at Ngandjoek and at Sidoardjo, these places being situated respectively in the central and the eastern part of Java. Out of six trials at Ngandjoek, the pure strains and the unselected mixture were alike; at Sidoardjo, in 2 out of 3 cases, the strains failed to give a higher yield than the population.—The supposition arose that the high yield of the population might be caused by the fact that the mixture is, generally speaking, more suited for uneven circumstances than is a pure variety.—In order to investigate this matter author began, in 1915, a series of trials wherein mixed-up pure strains were compared with the same races unmixed. The same was done by mixing up pure varieties. In most trials the varieties or strains were compared in this way: (1) variety A, 100 per cent; (2) variety B, 100 per cent; (3) A, 75 per cent + B 25 per cent; (4) A 50 per cent + B 50 per cent; (5) A 25 per cent + B 75 per cent.—Not only the yielding but also the stooling power was examined. When the paddy was ripe the ears were cut by hand and afterwards all the product in the trials where pure varieties had been mixed up was separated by hand so that one could know exactly which part of the yield had been provided by variety A, and what part by B. All heads were counted, so that the average weight was determined. The result of 4 trials with 8 controls showed that the pure strains and varieties did, on the whole, not so well as the mixtures. The stooling power shown by weekly counts, was in most cases higher than the pure strains; in one of the four cases, however, all the counts were remarkably lower with the mixtures than with the pure strains. Of two varieties, the highest producer (singly planted) did not always give the greatest proportion of the product of the mixture. In most cases the heads of the varieties that suppressed the other one became heavier and the heads of the suppressed one became lighter.—Trials of the same order were made with maize, soy beans and peanuts. With maize, yellow Menado corn and Saipan corn, singly planted, were compared with mixtures of these varieties. The mixtures yielded as much as 12 per cent more than the highest-producing variety separately planted. With soy beans the same was to be observed: 70 per cent of black mixed with 30 per cent of white soy beans yielded 12 per cent more than black alone, and 28 per cent more than white alone. With peanuts, 9 out of 10 mixtures gave a higher yield than might have been expected from the yield for the pure strains.—In the year 1916–1917, out of 4 trials comparing pure strains with mixtures of the same strains, no conclusions could be reached as to which should be preferred, strains or mixtures.—Out of 5 other such trials made at Sidoardjo, only in one case did the strains yield more than the mixtures.—The same was done for peanuts, the strains producing a little more than the mixtures.—The conclusion could be reached that: (1) Mixed planting of rice or peanuts does not necessarily raise the production. (2) Line selection with paddy gives wholly unsatisfactory results.—In 15 other trials, made in 1916–1917, where mixed-up pure varieties of paddy had been compared (8 controls) with the same varieties unmixed, the following conclusions were reached: (1) The yield of a mixture of pure varieties is, on the whole, higher than the calculated yield based on the production of the varieties planted singly. (2) The stooling power in a mixture is generally higher than the calculated.—(3) The percentage of stalks bearing heads is somewhat less in mixtures than in pure varieties. (4) The mean head-weight of different varieties in a mixture exhibits greater variation, and may differ greatly from the weight of the same variety not mixed. (5) In a mixture one variety may suppress another. (6) The suppressing variety is not necessarily the highest yielding when planted singly. (7) The suppressing variety is generally the race that stools most, when other characters are the same. (8) As a rule, the mean weight of the head increases with the suppressing variety and decreases with the suppressed one. (9) Perhaps it may be possible to find empirically mixtures that are well suited to certain circumstances.—Mixing trials have also been made with sweet potatoes (14 trials) and cassava varieties (1 trial). With sweet potatoes no conclusions could be made as to the yielding power; with cassava the mixture proved to be better than the best pure race. [See also next preceding Entry, 1550.]—*L. Koch.*

1552. KOHLBRUGGE, J. H. F. **De erfelijkheid van verkregen eigenshappen. [Inheritance of acquired characters.]** Genetica 1: 347–386. 1919.

1553. KRAFKA, JOSEPH, JR. **The effect of temperature upon facet number in the bar-eyed mutant of Drosophila.** Part I. Jour. Gen. Physiol. 2: 409–432. *10 fig.* Mar. 20, 1920. Part II. *Ibid.*, 433–444. *4 fig.* May 20, 1920. Part III. *Ibid.*, 445–464. May 20, 1920.—Breeding experiments with the bar-eyed mutant of *Drosophila melanogaster* at constant temperatures between 15°–31°C. have shown that the mean facet number varies inversely with the temperature at which the larvae develop, though no such variation occurs in the normal wild stock. The temperature coefficient for the variation in facet number of bar eye is of the same order as that for chemical reactions, and the variation may be plotted as an exponential curve. The greatest percentages of increase per degree centigrade come at the upper and lower temperatures. The temperature curve for rate of development of the immature stages of the fly corresponds with the facet curve from 15°–27°C., but drops above that point. The rate of development may be interpreted as the resultant of a number of different processes having different temperature coefficients. Temperature is effective in determining facet number during a relatively short period in larval development only, i.e., at a stage when about 36 per cent of immature development is completed. This period is about 18 hours long, and the temperature either before or after that time has no effect on facet number. The time at which this period is reached is dependent on the rate of development, but the facet number is not influenced by the length of the immature stage. The correlation between the two curves is therefore only apparent. It is suggested that the decrease in facet number in the bar-eyed flies may be accounted for by the presence of an inhibitor in the mutant stock, the temperature coefficient of which differs from that of the normal facet-producing reaction.—It is shown also that the coefficient of variability of the facet number in bar-eyed flies increases with temperature, while the standard deviation apparently decreases. The effect of temperature on facet number in bar-eyed stock is not inherited.—*H. H. Plough.*

1554. KUIPER, K. **Onderzoekingen over kleur en teekening bij runderen. Naar experimenten van R. Houwink Hzn. [Studies on color and color pattern in cattle. Based on experiments of R. Houwink Hzn.]** Genetica 2: 137–161. *5 pl.* Mar., 1920.

1555. KÜSTER, E. **Über mosaikpanaschierung und vergleichbare Erscheinungen. [Mosaic variegation and comparable phenomena.]** Ber. Deutsch. Bot. Ges. 36: 54–61. 1918.

1556. Küster, E. **Über sektoriale Panaschierung und andere Formen der sektorialen Differenzierung.** [On sectorial variegation and other forms of sectorial differentiation.] Monatshefte f. d. natw. Unterr. 12: 84-87. 1919.

1557. Lebedinsky, N. G. **Darwins geschlechtliche Zuchtwahl und ihre arterhaltende Bedeutung.** [Darwin's sexual selection and its significance for the maintenance of species.] Habilitationsvortrag. *31 p.* 1918.—See Bot. Absts. 5, Entry 1423.

1558. Lehmann, Ernst. **Zur Terminologie und Begriffsbildung in der Vererbungslehre.** [Terminology and formation of genetical concepts.] Zeitschr. indukt. Abstamm. Vererb. **22:** 236-260. May, 1920.

1559. Lehmann, E. [German rev. of: (1) Sperlich, Adolf. **Die Fähigkeit der Linienerhaltung (phyletische Potenz), ein auf die Nachkommenschaft von Saisonpflanzen mit festen Rhythmus ungleichmässig übergehender Faktor. (Capacity to maintain lines (phyletic potency) a factor distributed irregularly to the offspring of plants with fixed seasonal rhythm.)** Sitzungsber. Akad. Wiss. Wien 128: 379. 1919. (2) Sperlich, Adolf. **Über den Einfluss des Quellungszeitpunktes von Treibmitteln und des Lichtes auf die Samenkeimung von Alectorolophus hirsutus All. Charakterisierung der Samenruhe. (On the influence of the time of application of forcing-agents and of light on the germination of seeds of Alectorolophus hirsutus. Characterization of seed rest.)** Sitzungsber. Akad. Wiss. Wien 128: 477. 1919.] Zeitschr. indukt. Abstamm. Vererb. 22: 299-301. May, 1920.

1560. Leighty, Clyde E. **Natural wheat-rye hybrids of 1918.** Jour. Heredity 11: 129-136. *4 fig.* Mar., 1920.

1561. Levine, C. O. **The water buffalo—A tropical source of butter fat.** Jour. Heredity 11: 51-64. *9 fig.* Feb., 1920.

1562. Levine, C. O. **Swine, sheep, and goats in the orient.** Jour. Heredity 11: 117-124. *6 fig.* Mar., 1920.

1563. Lewis, A. C. **Annual report of the State Entomologist for 1918.** Georgia State Bd. Ent. Bull. 55: 1-31. *Fig. 2.* 1919.—The cotton breeding work is along three main lines; to improve the wilt resistant varieties which have already been developed, breeding for earliness in Sea Island cotton, and to improve the varieties of cotton which are especially adapted to central and north Georgia. Breeding for wilt resistance is being done with three varieties, Lewis 63, Council Toole and DeSoto, all of which now give satisfactory results under wilt conditions. Efforts are being made to stabilise the length of lint in the hybrid Dix-Afifi, a long staple upland wilt-resistant variety. Selections are being made to improve ten varieties of cotton adapted to north and central Georgia. A strain of Sea Island cotton known as No. 33 has been developed which is much earlier than the ordinary varieties. This strain is also very prolific and produces a small stalk.—*D. C. Warren.*

1564. Lienhart. **De la possibilité pour les eleveurs d'obtenir a volonté des males ou des femelles dans les races gallines.** [On the possibility for the raiser to obtain males or females at will in the races of poultry.] Compt. Rend. Acad. Sci. Paris 169: 102-104. 1919.

1565. Lindhard, E., and Karsten Iversen. **Vererbung von roten und gelben Farbenmerkmalen bei Beta-Rüben.** [Inheritance of red and yellow color characters in beets.] Zeitschr. Pflanzenzücht. 7: 1-18. June, 1919.—Crosses were made between red, yellow and white types of beets (*Beta*) and carried through the F_4 generation in some cases. A provisional factorial hypothesis is presented in which *R G* denotes red; *r G*, yellow; and *R g* and *r g* white. This presupposes a 9:3:4 ratio when a plant *RrGg* is self-pollinated. A large F_2 generation approximates such a ratio rather poorly and the author suggests a linkage between *R* and *G* with a gametic ratio of 1.8:1 which fits the F_2 results closely. This linkage relation, however, does not apparently hold in the only two back-crosses listed, although the total number of individuals is slightly less than 400. The author then suggests the presence of a lethal factor (*T*) but does not develop this idea.—*E. W. Lindstrom.*

1566. LIPSCHÜTZ, A. **Bemerkung zur Arbeit von Knud Sand über experimentellen Hermaphroditismus.** [**Comments on the work of Knud Sand on experimental hermaphroditism.**] Pflüger's Arch. 176: 112. 1919.

1567. LITTLE, C. C. **A note on the origin of piebald spotting in dogs.** Jour. Heredity 11: 12–15. *1 fig.* Jan., 1920.

1568. LITTLE, C. C. **Is there linkage between the genes for yellow and for black in mice.** Amer. Nat. 54: 267–270. May–June, 1920.—Discussion of recent paper of DUNN's referring to a deficiency of black young in a family of yellow mice. Because of small number of offspring involved, it is pointed out that the deviation from normal expectation may be entirely a matter of chance. DUNN states that yellow and black may possibly be linked. Author calls attention to the fact that yellow and agouti are allelomorphic and that agouti has been shown not to be linked to black. Author gives alternative explanation for observed facts, viz., assumption is made that a lethal factor is linked to black in the family above noted, and that this lethal is effective in a heterozygous condition in non-yellow mice but not in yellow mice. —*H. L. Ibsen.*

1569. LITTLE, C. C. **The heredity of susceptibility to a transplantable sarcoma (J. W. B.) of the Japanese waltzing mouse.** Science 51: 467–468. May 7, 1920.—In a cross between a Japanese waltzing mouse one hundred per cent susceptible to a transplantable sarcoma (J. W. B.) and the common non-waltzing mouse not susceptible to the sarcoma, the F_1 generation hybrids were all susceptible to the sarcoma, but the F_2 hybrids gave a total of twenty-three susceptible to sixty-six non-susceptible animals thus supporting the expectations on the three-, four-, five-, and seven-factor hypotheses.—To determine more closely the number of factors involved F_1 hybrid mice,—themselves susceptible,—were crossed back with the non-susceptible parent race. The numbers obtained were twenty-one susceptible to 208 non-susceptible which indicates that from three to five factors—probably four— are involved in determining susceptibility to the mouse sarcoma (J. W. B.).—Simultaneous presence of these factors is considered necessary for susceptibility. None of these factors is carried in the sex (X) chromosome since all the "X" chromosomes in the resulting animals, of the back-cross, if the original mating is a non-susceptible female with a susceptible male, will be derived from the common non-susceptible mice.—*Mary B. Stark.*

1570. LO PRIORE, G. **Sulla ereditarietà della fasciazione nelle spighe del mais.** [**On the inheritance of a fasciation in the maize ear.**] Staz. Sper. Agr. Ital. 51: 415–430. 1918.—Four fasciated ears of maize were found in 1902. A progeny of these, grown from open-pollinated seed, produced fasciated ears on one-third of the plants. The second year 40 per cent of the plants bore fasciated ears, while in the third year the progeny of a better-fasciated ear produced such ears on 60 per cent of the plants. The plants with fasciated ears showed no other abnormalities and yielded exceptionally well. The author concludes that a fasciated race of maize can be developed by selection although the abnormal form is transmitted to only a part of the offspring and according to laws of heredity not yet formulated.—The relation of traumatic and chemical treatment to the development of fasciations and other abnormalities as well as the relation of fasciation to the origin of the normal maize ear is discussed.—*J. H. Kempton.*

1571. LOSCH, HERMANN. **Ascidienbildung an Staubfäden vergrünter Blüten von Tropaeolum majus.** [**Ascidia formation on stamens of virescent flowers of Tropaeolum majus.**] Ber. Deutsch. Bot. Ges. 37: 369–372. Dec., 1919.—Describes on virescent stamens of *Tropaeolum majus* ascidia in various stages of development. Inner side of ascidium is foliar under side.—*James P. Kelly.*

1572. Lotsy, J. P. **Heribert Nilsson's onderzoekingen over soortsvorming bij Salix met opmerkingen mijnerzijds omtrent de daarin en in publicaties van anderen uitgeoefende kritiek aan mijn soorts-definitie. [Heribert-Nilsson's investigation on species formation in Salix with remarks of my own on the author's critique, and that of others on my taxonomic definitions.]** Genetica 2: 162–188. Mar., 1920.

1573. Lotsy, J. P. **Cucurbita-Strijdpagen. De soort-quaestie.—Het gedrag na kruising.—Parthenogenese? II. Eigen onderzoekingen. [Cucurbita problems. The species question. The result of crossing. Parthenogenesis? II. Investigations by the author.]** Genetica 2: 1–21. *1 3-colored plate, 9 fig.* Jan., 1920.

1574. Löhning. **Die erbliche Geschlechtsverknüpfung. [Hereditary sex linkage.]** Deutsch. Landw. Tierzucht. 22: 77–78. 1918.

1575. Malinowski, Edmund. **Die Sterilität der Bastarde im Lichte des Mendelismus. [The sterility of hybrids in the light of Mendelism.]** Zeitschr. indukt. Abstamm. Vererb. 22: 225–235. May, 1920.

1576. Mass, J. G. J. A. **Field experiments with Hevea.** Agric. Bull. Federated Malay States 6: 561–613. 596–597. 1918.

1577. Masui, Kiyoshi. **The spermatogenesis of domestic mammals. I. The spermatogenesis of the horse (Equus caballus).** Jour. Coll. Agric. Imperial Univ. Tokyo 3: 357–376. *2 pl., 2 fig.* 1919.

1578. Masui, Kiyoshi. **The spermatogenesis of domestic mammals. II. The spermatogenesis of cattle (Bos taurus).** Jour. Coll. Agric. Imperial Univ. Tokyo 3: 377–403. *3 pl., 1 fig.* 1919.

1579. Matthaei, R. [German rev. of: Klatt, B. **Experimentelle Untersuchungen über die Beeinflussbarkeit der Erbanlagen durch den Körper. (Experimental investigations on the modifiability of the hereditary factors through the soma.)** Sitzungsber. Ges. Naturf. Freunde 1919: 39–45. 1919. See Bot. Absts. 5, Entry 1546.] Zeitschr. Allg. Physiol. 18: 46–47. 1920.

1580. McAlpine, D. **Immunity and inheritance in plants.** Advisory Council Sci. Indust. Australia Bull. 7: 76–86. 1918.—A general discussion of the inheritance of disease resistance in wheat. Author recommends crossing susceptible varieties with resistant ones as means of developing rust immunity.—*J. H. Kempton.*

1581. Mendel, Kurt. **Familiäre peripherische Radialislähmung. [Familial peripheral paralysis of the radial nerve.]** Neurol. Centralbl. 39: 58–59. 1920.—It is recognised that heredity often plays a rôle in cases of facial paralysis, but it has not been determined whether the manifestation in these cases is due to the indirect effect of some hereditary anatomical anomaly, such for example as an unusually acute bend in the facial canal, or to a heightened susceptibility inherent in the nerve itself. The author now reports a family in which the father and two sons suffered from paralysis of the hand following very trivial injuries to the radial nerve at the elbow or near the shoulder. From these cases the author is led to believe that in hereditary paralysis involving the radial, probably the facial, and possibly other peripheral nerves, the underlying factor is to be sought not in any gross anatomical variation of the related parts, but in an hereditary condition of increased vulnerability of the particular nerve involved in the paralysis.—*C. H. Danforth.*

1582. Metz, Chas. W. **Correspondence between chromosome number and linkage groups in Drosophila virilis.** Science 51: 417–418. April 23, 1920.—Whereas in *Drosophila melanogaster* there are three large pairs and one very small pair of chromosomes, and three large groups and one very small group of linked genes, there are in *D. virilis* five large pairs, and one very small pair of chromosomes, and five known groups of linked genes. Author points out that only twenty-seven mutant characters, of which fourteen are sex-linked, have thus far been investigated in this species, and that the failure to detect the sixth (and presumably small) group, is not surprising in view of the small number of characters investigated. He promises full data on this case in a future publication.—*John S. Dexter.*

1583. Mitscherlich, Eilh. Alfred. **Über künstliche Wunderährenbildung.** [The artificial production of abnormal heads of cereals.] Zeitschr. Pflanzenzücht. 7: 101–109. *8 fig.* Dec., 1919.

1584. Mohr, Otto L., and Chr. Wriedt. **A new type of hereditary brachyphalangy in man.** Carnegie Inst. Washington Publ. No. 295. *64 p., 7 pl., 4 fig.* 1919.—A careful study based on personal examinations, authentic records, photographs and X-ray plates has been made of the hands of nearly 100 members of a Norwegian family in which an unusually clear-cut type of brachyphalangy occurs in at least six generations. The trait behaves as a simple dominant and is not sex-linked. In heterozygous individuals the manifestation is confined exclusively, so far as can be determined, to the middle phalanx of the index finger (and the comparable phalanx of the corresponding toe). The affected phalanx may be shortened to a moderate degree or reduced almost to the point of elimination in which case it is sometimes subluxated toward the ulnar side causing the terminal phalanx to bend radial-ward giving a "crooked" finger which is not (in this family) genetically different from a "short" finger. Of especial interest is the fact that the manifestations of the trait do not fluctuate around a single mode but arrange themselves in two distinct groups without any overlapping. The authors, therefore, postulate a second, modifying, gene which intensifies the effect of the main gene. This modifier is one of presumably many such genes which may be widely distributed in the human germ plasm without often having an opportunity to manifest themselves. Certain individuals who have married into the family have been heterozygous for the modifier, others have lacked it altogether. Failure to recognise the existence of this gene might easily have led to erroneous conclusion as to "dilution" of the main gene. In reality no dilution has taken place in the course of six generations. Of possibly great importance is the result of the marriage of two affected individuals. A single marriage of this sort yielded three children, one of whom lacked all fingers and toes and died at the age of a year. The authors are inclined to regard this case as the one instance of an individual homozygous for brachyphalangy and to look upon the gene as one which, when heterozygous, produces relatively inconsequential effects, but which when homozygous produces very serious, perhaps lethal, results.—*C. H. Danforth.*

1585. Mohr, Otto L. **Mikroskopische Untersuchungen zu Experimenten über den Einfluss der Radiumstrahlen und der Kältewirkung auf die Chromatinreifung und das Heterochromosom bei Decticus verruccivorus (♂). [Microscopic studies in experiments on the influence of radium rays and effect of cold on the maturation and the heterochromosome of Decticus verruccivorus (♂).]** Arch. mikrosk. Anat. 92: 300–368. *6 pl.* 1919.

1586. Morgan, T. H. **Variations in the secondary sexual characters of the fiddler crab.** Amer. Nat. 54: 220–246. *6 fig.* May–June, 1920.—Two variations are described that are shown not to be due to regeneration. Whether due to genetic change, to infection, or to some embryonic "slip" could not be determined. Literature relating to sex-intergrades in crustacea is reviewed.—*T. H. Morgan.*

1587. Mottet, S. **Digitalis hybride de Lutz. [The Lutz Digitalis hybrid.]** Rev. Hortic. 91: 396–397. 1919.—See Bot. Absts. 5, Entry 1827.

1588. Mumford, H. W. **Famous Angus cows of Scotland.** Breeder's Gaz. 76: 462–463. 1919.—Author discusses briefly the records of the foundation cows of certain famous families of the Aberdeen Angus breed.—*Sewall Wright.*

1589. Munns, E. N. **Effect of fertilization on the seed of Jeffrey Pine.** Plant World 22: 138–144. 1919.—Author reports on results of various cross- and self- pollinations among 8 trees of *Pinus Jeffreyi*, three of which were thrifty, two mistletoe-infected, one insect infected, and two "suppressed trees." On basis of observations author recommends that seed should be collected from localities with strong winds at time of flowering so that cross-pollinated seeds may be secured; and that for heavy seeds and consequent stronger seedlings collections should be from thrifty parents; poor trees gave largest number of seeds to pound but produced smallest nursery trees; and that in timber-sale practice only thrifty trees should be left. [See also Bot. Absts. 5, Entry 1375.]—*James P. Kelly.*

1590. Myers, C. H. **The use of a selection coefficient.** Jour. Amer. Soc. Agron. 12: 106-112. 1920.—$\frac{\text{Number of ripe ears}}{\text{Total number of ears}}$ = percentage of maturity. The average yield per stalk of maize was determined in pounds. It was desirable to combine the yield and the maturity into a single expression which would serve as a basis for selection. The average yield per stalk times the percentage of maturity gives the "selection coefficient."—*F. M. Schertz.*

1591. Nachtsheim, Hans. **Crossing-over-Theorie oder Reduplikationshypothese? [The crossover theory or the reduplication hypothesis?]** Zeitschr. indukt. Abstamm. Vererb. 22: 127–141. *4 fig.* Jan., 1920.

1592. Nachtsheim, Hans. **Zytologische und experimentelle Untersuchungen über die Geschlechtsbestimmung bei Dinophilus apatris Korsch. [Cytological and experimental studies on the sex determination of Dinophilus apatris Korsch.]** Arch. Mikrosk. Anat. 93: 17–140. *4 pl., 5 fig.* Nov., 1919.

1593. Naef, A. **Idealistische Morphologie und Phylogenetik. (Zur Methodik der systematischen Morphologie.) [Idealistic morphology and phylogeny. (On the method of systematic morphology.)]** *77 p., 4 fig.* Jena, 1919.

1594. Noack, Konrad. [German rev. of: Stomps, Theo. J. **Gigas-mutation mit und ohne Verdoppelung der Chromosomenzahl. [Gigas-mutation with and without doubling of the chromosome number.** Zeitschr. indukt. Abstamm. Vererb. 21: 65–90. *3 pl., 4 fig.* July, 1919. (See Bot. Absts. 4, Entry 778.)] Zeitschr. Bot. 12: 36–39. 1920.

1595. Noyes, Hilda H. **The development of useful citizenship.** Jour. Heredity 11: 88–91. Feb., 1920.

1596. Nuttall, G. H. F. **The biology of Pediculus humanus.** Parasitology 2: 201–220. *1 pl., 1 fig.* 1919.—Lice reared on white backgrounds developed very little pigment and appeared whitish or translucent but those reared on black backgrounds became very darkly pigmented thus showing that pigmentation is not inherited. In some lots taken from their host as high as 9 per cent of the adult individuals were hermaphrodites.—*D. D. Whitney.*

1597. O., A. **Zonal Pelargoniums.** Gard. Chron. 66: 157. Sept. 20, 1919.—Maxime Kavolsky, a comparatively new variety is briefly described.—*A. C. Hildreth.*

1598. Palmgren. Rolf. **Till Kännedomen om Abnormiteters Nedärfning hos en del Husdjur. [Inheritance of abnormalities in certain domestic animals.]** Acta Soc. pro fauna et flora fennica 44: 1-22. 1918.—See Bot. Absts. 5, Entry 1548.

1599. Palmgren, Rolf. **Tvenne bastarder mellan getbock och fartacka, födda i Högholmes zoologiska trädgard. [Two hybrids between sheep and goats produced in Högholm zoological gardens.]** Med. pro fauna et flora fennica 44: 124–125. 1918.—See Bot. Absts. 5, Entry 1548.

1600. Pammel, L. H., and C. M. King. **An annual white sweet clover.** Proc. Iowa Acad. Sci. 25: 249-251. *Pl. 4-6.* 1920.—See Bot. Absts. 5, Entry 1191.

1601. PAMMEL, L. H., AND C. M. KING. **A variation in the black walnut.** Proc. Iowa Acad. Sci. 25: 241-248. *Pl. 3, fig. 43-44.* 1920.

1602. PATTERSON, J. T. **Polyembryony and sex.** Jour. Heredity 10: 344-352. *2 fig.* Nov., 1919.

1603. PELLEW, CAROLINE. **The genetics of Campanula carpatica.** Gard. Chron. 66: 238. *3 fig.* Nov. 8, 1919.—A brief consideration of investigations of the author more completely discussed in "Types of segregation," Jour. Genetics 6: 1917.—In *Campanula carpatica* hermaphrodites occur with male and female organs fully developed. In other plants the anthers fail to develop beyond a rudimentary stage while in others, still, development of the anthers is partial. Self-sterility is general in this species. In certain strains of *C. carpatica pelviformis* crosses between hermaphrodites or between females and hermaphrodites invariably gave mixed families consisting both of females and hermaphrodites, often with a preponderance of females. The hermaphrodites appear to produce more gametes carrying the female character than gametes carrying the hermaphrodite character. There is no consistent difference in this respect between the ovules and pollen of a single plant. In other strains the pollen and ovules differ. Two hermaphrodites were found, pollen of which, when used on females, gave rise exclusively to females, whereas ovules of the same plant fertilized by other hermaphrodites gave rise exclusively to hermaphrodites. A flower-color factor pair in this species also follows this unusual type of segregation by which the ovules and pollen are differentiated. Normal segregation of the color factor occurs on the female side resulting in equal numbers of ovules bearing blue or white allelomorphs. Ninety-seven per cent of the pollen grains, however, carry the white allelomorph and three per cent only the blue allelomorph.—Power of transmitting this unusual mode of segregation from parent to offspring is apparently limited to the ovules, for no plant similar to *C. carpatica pelviformis* has been derived from its male side. This type of segregation may be compared with the double-throwing variety of stock.—*C. B. Hutchison.*

1604. PÉZARD, A. **Castration alimentaire chez les coqs soumis au régime carné exclusif. [Alimentary castration in a cock subjected to an exclusive meat diet.]** Compt. Rend. Acad. Sci. Paris 169: 1177-1179. 1919.

1605. PITT, FRANCES. **Notes on the inheritance of color and markings in pedigree Hereford cattle.** Jour. Genetics 9: 281-302. *4 pl.* Feb., 1920.—Notes and photographs on which this paper is based come chiefly from the breeding of pure bred Herefords owned by W. J. PITT.—Excessive white on the sides of the belly and down the spine behaved to well marked animals as a recessive factor. The ratios were: heterozygote to heterozygote, 25 well marked: 10 excessive white; heterozygote to pure dominant well marked, 52 well marked; heterozygote to recessive excessive white, 7 well marked to 9 excessive white.—Dark neck or extension of the pigment area to include the neck, the crest, and to encroach on the white area on the tail behaved nearly as a dominant to the desired white markings. In the presence of the factor for excessive white the "dark-necked" factor apparently may be inhibited in its action.—A ring of red around the eyes is dependent on a single dominant factor. The ratios for heterozygote × heterozygote were 42 red-eyed to 12 white-eyed. The mating of the heterozygote × the recessive white-eyed gave 12 heterozygote to 15 complete recessive. It appears that the factor for red pigment around the eyes is independent of the other factors.—Pigment on the nose behaves as a dominant to clean nose, pigmented × non-pigmented giving 4 pigmented in F_1; pigmented heterozgyous × non-pigmented, giving 3 pigmented to 3 not pigmented in the F_1.—Two pigment factors control coat color. Pale brown coat is dominant over the deep rich purple or claret coat. The factors for coat color apparently behave independently of the rest save with the possible exception of the pigmented nose.—The observation is made that the "claret"-coated animals may not feed as rapidly as the pale brown.—The history of the breed is cited to show that the factors discussed were present in early times.—*John W. Gowen.*

1606. Pleijel, C. **Valeriana excelsa Poir × officinalis L. nova hybrida. [Valeriana excelsa Poir × officinalis L. a new hybrid.]** Bot. Notiser 1918: 295–296. 1918.

1607. Popenoe, Paul. **Inbreeding and outbreeding.** [Rev. of: East, E. M., and D. F. Jones. **Inbreeding and outbreeding.** *14 x 21 cm., 285 p., 46 fig.* J. B. Lippincott: Philadelphia, 1919. (See Bot. Absts. 4, Entry 571; 5, Entries 437, 1695.)] Jour. Heredity 11: 125–128. Mar., 1920.

1608. Popenoe, Paul. **World-power and evolution.** Jour. Heredity 11: 137–144. Mar., 1920.

1609. P[openoe], P. **Lock's last work.** [Rev. of: Lock, R. H. **Recent progress in the study of variation, heredity, and evolution.** *4th ed., 336 p.* E. P. Dutton & Co.: New York, 1916.] Jour. Heredity 11: 110. Mar., 1920.

1610. P[openoe], Paul. **Morgan on heredity.** [Rev. of: Morgan, Thomas Hunt. **The physical basis of heredity.** *14 x 21 cm., 300 p., 117 fig.* J. B. Lippincott Co.: Philadelphia, 1919.] Jour. Heredity 11: 144. Mar., 1920.

1611. P[openoe], P. [Rev. of: Punnett, Reginald Crundall. **Mendelism.** *5th ed., 13 x 19 cm., 219 p., 7 pl., 52 fig.* Macmillan & Co.: London, 1919.] Jour. Heredity 11: 115. Mar., 1920.

1612. Pridham, J. T. **Oat and barley breeding, agricultural research in Australia.** Advisory Council Sci. and Ind. Commonwealth of Australia, Bull. 7: 22–38. 1918.—Cross was made between the Algerian variety of oats and Carter's Royal Cluster. The F_2 consisted of 1,092 plants. There was great diversity among the young plants, some having coarse, broad leaves, while others had leaves almost like rye-grass in their fineness. There was also great diversity in character of stooling, foliage color, and habit of growth (erect or prostrate). On approaching maturity some plants showed pink or reddish color at base of stalk, a characteristic of the Algerian parent. 32.48 per cent of the plants exhibited the reddish straw, a percentage considered by the author to conform with a Mendelian ratio. The grain was of varying shades of brown, except in a few plants which produced yellow seeds, but no plants were found with white seeds like those of the male parent.—Four crosses were made between varieties of the Algerian type and those of the tree "class" and one cross was made between Algerian and a "side" oat. The F_1 plants were intermediate in character and of pronounced vigor. In subsequent generations from oat crossbreds of the "tree" or branching type, no individuals of the "side" type were found.—Attempts were made to cross *Avena fatua* with the Algerian variety and also with Chinese skinless, but without success.—A cross was effected between a "false wild oat" resembling *A. fatua* and white Bonanza. The progeny had slender straw, pale foliage and the open thin head with drooping branches of the wild oat. The line was not pursued further as no individuals of promise were found.—The most successful cross from the standpoint of the production of new varieties is white Ligowo × Algerian. From this cross sprang "Guyra," "Lachlan," and other strains of merit which have not yet been named. It is stated that the most productive varieties are those with stout awns and the value of skinless varieties is deprecated.—Seeds of various oat varieties and crossbreds were taken from Cowra and planted at Longerenong College, Victoria. In selections of these grown again at Longerenong striking variations were found in the Algerian oats. Among these were several plants with very coarse awns, very tall straw, white, large grain, and a limited number of stalks. These plants ripened unusually early. The possibility of the seed having been mixed was considered but no plants resembling these were found in other plots. This variation with a few individual exceptions bred true in succeeding years and was named "Sunrise" on account of its earliness.—A remarkable plant was found in Chinese skinless oats at Cowra in 1913. In addition to being much earlier than the other plants the early stools bore heads on which the upper flowers were like the skinless oat (three to five flowers to a spikelet) while the lower flowers resembled Algerian (two flowers to a spikelet with stiff glumes). The early stools had a darker foliage than the late ones, which latter bore flowers typical of the skinless oat. The straw, when mature, was reddish like that of Algerian. Some of the seeds were naked and some were black or dark brown hulled. Progenies of individual plants have been grown for several seasons and continue to be quite variable, some having wholly naked, some half and half and some yielding only hulled seed. The hulled seed germinated best and also yielded best. Crosses between this oat and Dun and Ruakura have given no promising material.—An oat resembling *A. fatua* was found in a progeny of the natural crossbred of the Sunrise variation. In this progeny most of the plants resembled Sunrise and seed from such plants bred true, but the wild oat type split up remarkably. The plants varied in seed color, degree of awn, stoutness of straw and hairiness of grain, some being thickly felted while others were smooth. Since none of these types were of economic importance they were not persevered with.—Author states that well-marked variations which bred true have been found in the Kelsall's, Black Bell, Ruakura, and Winter Turf varieties. The characteristics of several varieties are given and the technique of oat crossing described.—Under the heading of Barley Breeding the author records having found a few plants of wild barley *Hordeum spontaneum* in a sample of wild wheat *Triticum dicoccum dicoccoides*. The wild barley was crossed with the Standwell and Kinver varieties. The F_1 crossbreds were more vigorous than the cultivated parents. They were uniformly of the Chevalier type and scattered easily. Plants in which the grain adhered more or less firmly to the rachis and resembled malting barley were selected. In the F_4 these selections compared favorably in productiveness with Kinver, Standwell and a two-row selection from Chilian. The straw is stronger, the awns stouter, the grain larger and the plants more drought-resistant than the ordinary malting barleys.—Author's assistant crossed a two-row naked-awned barley with ordinary skinless, also Kinver malting barley with the two-row and naked type. Among other variations the latter cross gave rise to a six-rowed bearded type.—Author states that a MR. PEACOCK of the Bathhurst Experiment Farm found a natural crossbred in the Standwell barley which gave rise to a two-rowed awnless, six-rowed awnless and a six-rowed awned, all of which bred true.—*J. H. Kempton.*

1613. PRZIBRAM, HANS. **Ursachen tierischer Farbkleidung.** [Causes of animal coat colors.] Arch. Entwicklungsmech. Organ. 45: 199–259. 1919.

1614. PUNNETT, R. C. **The genetics of the Dutch rabbit.—A criticism.** Jour. Genetics 9: 303–317. *1 pl., 2 fig.* Mar., 1920.—Author recognizes three true-breeding grades with reduced pigmentation and frequent heterochromia iridis, viz., White Dutch, Spotted Dutch, and Typical Dutch. Self-color is *PPTTSS* and White Dutch is *ppttss*. *S* raises White Dutch to Spotted Dutch and if *T* is also added pigmentation is increased to Typical Dutch. *P* produces darker types and eliminates heterochromia. The various combinations of these factors are fitted to Castle's data and this multiple factor theory is considered to agree better than Castle's hypothesis of multiple allelomorphism of the four types, Self-color, Dark Dutch, "Tan" Dutch, and White Dutch, and to make unnecessary his conception of "mutual modification."—*P. W. Whiting.*

1615. PYE, H. **Wheat breeding in its incidence to production. Agricultural research in Australia.** Advis. Council. Sci. and Ind. Commonwealth of Australia Bull. 7: 10–22. 1918. —General discussion of the application of breeding to improvement of wheat. Author records having noticed in the past few years many more natural crosses in wheat than heretofore. This increase in crossing is attributed to lack of pollen, some varieties having been lost through a failure to fertilize the ovules. An emasculated bearded wheat left to wind or insect pollination produced nine seeds, eight of which germinated, six producing bald ears and two bearded. Author lists four features which influence prolificacy and thirteen qualities which are associated with prolificacy in its relation to inherency and economical harvesting.—*J. H. Kempton.*

1616. R. [German rev. of: TROW, A. H. **On "albinism" in Senecio vulgaris L.** Jour. Genetics 6: 65–74. 1916. (See Bot. Absts. 1, Entry 947.)] Zeitschr. Pflanzenzücht. 7: 141. Dec., 1919.

1617. REIGHARD, JACOB. **The breeding behavior of the suckers and minnows. I. The suckers.** Biol. Bull. 38: 1–32. Jan., 1920.—The white sucker (*Catostomus commersionii*), the red-horse (*Moxostoma aureolum*), and the hogsucker (*Catostomus nigricans*) breed in the swift water of small streams on gravel bottom. In all, the breeding males bear pearl organs, and in the hogsucker the female also bears them. In spawning, those surfaces of the male that are rendered rough by the pearl organs are brought into contact with the female, and aid the fish in maintaining their relative positions. In the white sucker and the red-horse, two males pair with the female at one time, one on either side of her. In the hogsucker, six or eight males may pair with the female at one time. In each species, the female repeats the spawning act in many places and with different groups of males. The male does not enter into combat with other males, but coöperates with them. The relation of the sexes is thus promiscuity, not polyandry or polygamy; this promiscuity is not found in fishes in which combat takes place between the males.—*Bertram G. Smith.*

1618. RENNER, O. **Zur Biologie und Morphologie der männlichen Haplonten einiger Önotheren. [Biology and morphology of the male haplonts of some Oenotheras.]** Zeitschr. Bot. 11: 305–380. *39 fig.* 1919.

1619. RENNER, O. **Bemerkungen zu der Abhandlung von Hugo de Vries: Kreuzungen von Oenothera Lamarckiana mut. velutina. [Comments on the paper by Hugo de Vries: Crosses of Oenothera Lamarckiana mut. velutina.]** Ber. Deutsch. Bot. Ges. 36: 446–456. 1918.

1620. RICHARDSON, A. E. V. **Production of cereals for arid districts. Agricultural research in Australia.** Advisory Council Sci. and Ind. Commonwealth of Australia Bull. 7: 57–77. 1918.—Following a general discussion of location of arid regions, progress of cultural methods, differences between species and their ability to withstand drought, and relation of the migration ratio (i.e., ratio of grain to stalk) to drought-resistance, author describes the Hays centgener-plot system of wheat breeding. Cross-breeding as a method of producing new types is considered with brief summary of Mendel's principles. In this connection a list of dominant and recessive characters in wheat and barley is given.—Attempt was made to determine whether high and low yielding power are Mendelian characters. A high-yielding variety of wheat such as Federation or Yandilla King was crossed with one of low yield such as Huguenot. In the F_2 the plants were grown in centgener plots and each plant harvested separately. While the parental varieties give normal frequency curves the F_2 appears to show segregation into two distinct groups, one consisting of high yielding plants (several of which outyield the best parent) and one of low yielding plants. Progenies of both groups were grown and the results indicate that the observed differences were inherited.—By propagating the extreme plants found in F_2 of a cross between a two-rowed bearded and a six-rowed skinless variety of barley a new race has been obtained which is six-rowed and bearded and exceeds the parents in migration-ratio as well as in yield.—*J. H. Kempton.*

1621. ROBERTS, HERBERT F. **The founders of the art of breeding.** Jour. Heredity 10: 99–106. *4 fig.* Mar., 1919. *Ibid.* 10: 147–152. *1 fig.* Apr., 1919. *Ibid.* 10: 229–239. *1 fig.* May, 1919. *Ibid.* 10: 257–270. June, 1919.—See also Bot. Absts. 5, Entry 90.

1622. ROLFE, R. A. **The pre-Mendelian age.** Gard. Chron. 66: 288. Dec. 6, 1919.—Author takes somewhat positive attitude regarding MENDEL and the supposed sanctification of his results, basing his objections upon the fact that Goss, SETON, KNIGHT and GÄRTNER all experimented with peas, obtaining concurrent results as to the uniformity in the F_1, and diversity in the F_2 generations, the overlooking of which data by MENDEL and his commentators, seems to the author curious, and a manifest fault subject to criticism. Author thinks that MENDEL has blinded all investigators to the merits of those who preceded him.—*H. F. Roberts.*

1623. ROMELL, LARS-GUNNAR. **Något om artbildningsproblem. [On problems of the origin of species.]** Skogsvårdsföreningens Tidskr. 18: 92–100. 1920.—After brief description of different theories concerning origin of species author discusses rather particularly the treatise of van der Wolk, "Onderzoekingen ober blijvende modificaties en hun betrekking tot mutaties" [Researches on persistent modifications and their relation to mutations]. Cultura 1919. *K. V. Ossian Dahlgren.*

1624. S., W. [Rev. of: Rignano, Eugenio. **Upon the inheritance of acquired characters: A hypothesis of heredity, development, and assimilation.** *413 p.* Open Court Publishing Co.: Chicago, 1911.] Science Progress 14: 514–515. Jan., 1920.

1625. Salisbury, E. J. **Variation in Anemone apennina, L., and Clematis vitalba, L., with special reference to trimery and abortion.** Ann. Botany 34: 107–116. *9 fig.* Jan., 1920. —Author presents further data on his views relative to essential trimery of Ranunculaceae. In *A. apennina* perianth segments ranged from 9 to 21 with 34 per cent of the flowers departing from trimerous condition in perianth. Distribution of variations tends to be symmetrical about mode in contrast to condition in *A. nemorosa* where skewness was associated with lower modal value. Stamen number in *A. apennina* ranged from 48 to 111 (multiples of 3). Curve was multimodal with succession of trimerous modes, greatest frequency being at 72, 81 and 87; in 55.3 per cent of flowers staminal number was multiple of three; departure from modes explainable on basis of fission or fusion. In 57.3 per cent of flowers carpel number was multiple of 3, largest modes being at 60, 63, 51, and 57 with limits of range 27 and 87. One instance of a carpel with two stigmas suggested fission as cause of departure from trimery.—In *Clematis vitalba* the gynaecium of 1202 specimens furnished again a many-peaked curve with modes at multiples of three. There was a tendency for number of abortive carpels to increase as total carpel number increased. Abortion seemed to depend on conditions of nutrition and development and not on idiosyncrasies of pollinating agent.—*James P. Kelly.*

1626. Schaffner, John H. **A remarkable bud sport of Pandanus.** Jour. Heredity 10: 376–378. *1 fig.* Nov., 1919.

1627. Schaffner, J. H. **The expression of sexual dimorphism in heterosporous sporophytes.** Ohio Jour. Sci. 18: 101–125. *25 fig.* 1918 —"The sexual condition is simply a state of the living substance which may continue for a greater or less length of time before a neutral state or the opposite sex condition is set up." Author maintains inadequacy of sex-chromosome mechanism for most plants, even suggesting that Allen's work on *Sphaerocarpus* is not conclusive. Body of paper involves examples of various stages in development of dioecious condition. No original monosporangiate flowers exist; few seeming examples show direct relationship to groups with opposite structures present. Usually dioecious condition comes directly from bisporangiate; sometimes monoecism is intermediate. Carpellate flowers more likely to retain vestiges of stamen structures, than are staminate to retain carpel parts. *Zizania aquatica* has staminate spikelets awnless, carpellate long-awned, bisporangiate short-awned; latency of awn factor caused by presence of male condition. *Cannabis sativa* normally an extreme example of dioecism, but plants grown under unusual conditions may show reversal of certain parts to opposite sex. Discusses genera (*Acer, Rumex, Fraxinus*) and larger groups which themselves show many gradations in the development of dioecism. Suggests inadequacy of sex-chromosome idea even in animal kingdom, though in some cases "hereditary factors may arise in a special chromosome which may assist in retaining and intensifying a male or female state already established." Sex-linked transmission can be readily explained without sex chromosomes. With assumption of sex chromosomes greater part of sexual phenomena becomes unexplainable and contradictory. Adds list of 41 plant species which are promising for investigation, describing general sexual condition of each.—*Merle C. Coulter.*

1628. Schaxel, Julius. **Über die Darstellung allgemeiner Biologie.** [On the presentation of general biology.] Abhandl. Theoret. Biol. *62 p.* 1919.—See Bot. Absts. 5, Entry 1426.

1629. Schaxel, J. **Grundzüge der Theorienbildung in der Biologie.** [Principles of theory formation in biology.] *221 p.* G. Fischer: Jena, 1919.—See Bot. Absts. 5, Entry 1426.

1630. Schellenberg, G. **Über die Verteilung der Geschlechtsorgane bei den Bryophyten.** **[On the distribution of sex organs in the bryophytes.]** Beih. z. Bot. Zentralbl. 37: 1–39. 1919.—See Bot. Absts. 5, Entry 1639.

1631. Schermers, D. **Erfelijkheid en rasverbetering. [Heredity and race-improvement.]** Schild en Pijl 10: 1–26. 1919.

1632. Schiemann, E. **Zur Frage der Brüchigkeit der Gerste—eine Berichtigung. [To the question of brittleness in barley—a correction.]** Zeitschr. indukt. Abstamm. Vererb. **21:** 53. May, 1919.

1633. Schiemann, E. [German rev. of: Baerthlein, K. **Über bakterielle Variabilität, insbesondere sogenannte Bakterien-mutationen. (On bacterial variation, especially the so-called Bacteria mutations.)** Centralbl. Bakt. 81: 369–475. 1918.] Zeitschr. indukt. Abstamm. Vererb. 22: 303–304. May, 1920.

1634. Schiemann, E. [German rev. of: Bateson, W., and Ida Sutton. **Double flowers and sex linkage in Begonia.** Jour. Genetics 8: 199–207. *Pl. 8.* June, 1919. (See Bot. Absts. 3, Entry 2081.)] Zeitschr. indukt. Abstamm. Vererb. 22: 296–297. May, 1920.

1635. Schiemann, E. [German rev. of: Collins, E. J. **Sex segregation in the Bryophyta.** Jour. Genetics 8: 139–146. *Pl. 6, 5 fig.* June, 1919. (See Bot. Absts. 3, Entry 2103.)] Zeitschr. indukt. Abstamm. Vererb. 22: 296. May, 1920.

1636. Schiemann, E. [German rev. of: Correns, C. **Fortsetzung der Versuche zur experimentellen Verschiebung des Geschlechtsverhältnisses.** Sitzungsber. Akad. Wiss. **1918:** 1175–1180. 1918.] Zeitschr. indukt. Abstamm. Vererb. 22: 293. May, 1920.

1637. Schiemann, E. [German rev. of: Kajanus, Birger. **Kreuzungsstudien an Winterweizen. (Studies on crossing winter wheat.)** Bot. Notiser 1918: 235–244. 1918. (See Bot. Absts. 4, Entry 622.)] Zeitschr. indukt. Abstamm. Vererb. 22: 292. May, 1920.

1638. Schiemann, E. [German rev. of: (1) Nilsson-Ehle, H. **Untersuchungen über Speltoidmutationen beim Weizen. (Experiments on speltoid mutations in wheat.)** Bot. Notiser 1917: 305–329. *1 fig.* 1917. (2) Kalt, B., and A. Schulz. **Über Rückschlagsindividuen mit Spelzweizeneigenschaften bei Nacktweizen der Emmerreihe des Weizens. (On atavists with spelt characters in naked wheat of the Emmer series.)** Ber. Deutsch. Bot. Ges. 36: 669–671. 1918. (See Bot. Absts. 4, Entry 624.)] Zeitschr. indukt. Abstamm. Vererb. 22: 291–292. May, 1920.

1639. Schiemann, E. [German rev. of: Schellenberg, G. **Über die Verteilung der Geschlechtsorgane bei den Bryophyten. (On the distribution of sex organs in the bryophytes.)** Beih. Bot. Zentralbl. 37: 1–39. 1919.] Zeitschr. indukt. Abstamm. Vererb. 22: 298. May, 1920.

1640. Schiemann, E. [German rev. of: Thellung, A. **Neure Wege und Ziele der botanischen Systematik erläutert am Beispiele unserer Getreidearten. [New methods and purposes of botanical taxonomy illustrated by examples of our cereal species.]** Naturw. Wochenschrift 17: 449–458, 465–474. *3 fig.* 1918.] Zeitschr. indukt. Abstamm. Vererb. 22: 293–295. May, 1920.

1641. Schmidt, J. **Experimentelle Konstanz og Arvelighedsundersogelser med Lebistes reticulatus (Peters) Regan. [Experimental studies on constancy and heredity in Lebistes reticulatus.]** Meddel. Carlsberg Lab. 14: 8. 1919.

1642. Schultz, W. **Gleichlauf von Verpflanzung und Kreuzung bei Froschlarven. [Parallelism between transplantation and crossing in frog larvae.]** Arch. Entwicklungsmech. Organ. 43: 361-380. *1 pl.* 1918.

1643. Seiler, J. [German rev. of: Goldschmidt, Richard. **Crossing over ohne Chiasmatypie? (Crossing over without chiasmatype?)** Genetics 2: 82–95. 1917.] Zeitschr. indukt. Abstamm. Vererb. 22: 215–216. Mar., 1920.

1644. Semon, Richard. **Über das Schlagwort "Lamarckismus." [On the catch-word "Lamarckism."]** Zeitschr. indukt. Abstamm. Vererb. 22: 51–52. Dec., 1919.

1645. Shamel, A. D. **Origin of a new and improved French prune variety.** Jour. Heredity 10: 339–343. *Frontispiece, 3 fig.* Nov., 1919.

1646. Shamel, A. D. **A bud variation of the Le Grande Manitou dahlia.** Jour. Heredity 10: 367–368. *1 fig.* Nov., 1919.

1647. Sheppard, Hubert. **Hermaphroditism in man.** Anat. Rec. 18: 259–260. April 20, 1920.—Author's abstract of paper read before American Association of Anatomists April 1–3, 1920:—In 1911 Gudernatsch asserted that "hermaphroditism in the sense that separate testicles and ovaries are found has not been demonstrated in man, nor even in other mammals beyond a doubt." In so far as we are able to determine, this assertion has not been questioned. We thought it worth while, in the light of this and other investigations, to report a study of the anatomical structures of an extreme case of hermaphroditism which came to the dissecting room.—The testicles in this individual were located in the scrotum and the ovaries in the pelvic cavity. The tissue from both organs proved to be normal in structure under a close microscopic examination. The broad ligament was thicker and wider than is usually found in a female subject, due to the fact that the uterus was a little lower in the pelvis than normal. The uterus measured about 5 cm. in length, 4 cm. in width and 2 cm. in thickness. A muscular wall, as well as a lumen which opened downward into the vagina, could be easily seen by both microscopic and macroscopic examinations. The oviduct took a normal course to the lateral angle of the uterus. A microscopic examination of the tube showed a lumen with walls containing the usual tunics. The cervix of the uterus passed into the inferior portion of the prostate about one-half inch below the urethra. The position of the organs might be described as follows: The bladder was superior and anterior to the uterus, with the prostate almost below the bladder, and a little anterior to the inferior portion of the uterus. Both are connected to the prostate, the urethra entering the prostatic substance near its superior anterior surface, the cervix of the uterus occupying the lower two-thirds. The cervix of the uterus held almost the exact position of the utriculus prostaticus of the male.—Externally the genitalia featured decidedly as a male. However, upon a closer examination of the region, and palpation of the organs, certain irregularities could be observed. The penis was small with a urethral orifice three-fourths as large as the organ itself. The opening gradually increased in size until it terminated at the cervix of the uterus. This portion of the urethra was in all respects a vagina attached to the inferior surface of the penis. Both the lumen of the uterus and the urethra opened directly into the vaginal opening.—It has been found in all true cases of hermaphroditism that there is always a sharp distinction between the male and female genital tissue and never an indefined mixing of the two elements (true ovitestis). In this unusual case we found the same phenomenon with a wider separation of the two kinds of tissue, the testes and ovaries in the exact position of a normal individual.—*Hubert Sheppard.*

1648. Shull, George H. **A third duplication of genetic factors in shepherd's-purse.** Science 51: 596. June 11, 1920.—Author's abstract of paper read before American Philosophical Society, April 23, 1920.—In the third generation of a cross between a wild biotype of the common shepherd's-purse (*Bursa bursa-pastoris*) from Wales and Heeger's shepherd's-purse (*B. Heegeri*) there appeared a small number of plants of unique type, having a more coriaceous texture than in the plants of either of the two original strains involved in the cross. This new type has been designated *coriacea*. It differs from the common form, not only in texture, but the lobing of the leaf is reduced and simplified and the angles of the lobes are almost spinescent. The proportion of *coriacea* to the typical sibs in this F_3 family was 12:187 or almost exactly a 1:15 ratio. This suggested at once the presence of two independently inherited factors for the normal texture, the *coriacea* type being produced only when these two factors *K* and *L* were absent. Subsequent breeding has shown that *coriacea* breeds true when selfed, and has also confirmed the interpretation of this as a third case of duplication of factors in this species. The two characters previously shown to be thus constituted are the triangular form of capsule, and the division of the leaf to the midrib which brings to light the characteristic lobing found in the form designated *rhomboidea*. The duplication of the capsule determiners is practically universal while that of the leaf-lobe factor is less frequently found. Studies on the *coriacea* character are still too limited in extent to justify a statement as to the prevalence of duplication of the factor for the usual texture of the leaves.—*George H. Shull.*

1649. SIEMENS, H. W. **Rashygienens biologiska grundvalar. [Biological foundations of race hygiene.]** *98 p.* Gleerup: Lund, 1918.

1650. SIEMENS. [German rev. of: HAECKER, V. **Die Erblichkeit im Mannesstamm und der vaterrechtliche Familienbegriff. (Inheritance in man and the male-line concept of the family.)** *32 p.* Gustav Fischer: Jena, 1917.] Zeitschr. indukt. Abstamm. Vererb. 22: 213. Mar., 1920.

1651. SIRKS, M. J. **Verwantschap als biologisch vraagstuk. [Relationship as a biological problem.]** Genetica 2: 27–60. Jan., 1920.

1652. SIRKS, M. J. **De analyse van een spontane boonenhybride. [Analysis of a spontaneous bean hybrid.]** Genetica 2: 97–114. Mar., 1920.

1653. SIRKS, M. J. **Uit het Instituut voor Veredeling van Landbouwgewassen. Vergelijking van gerst en tarwerassen, van het Instituut afkomstig met andere voortreffelijke rassen van deze gewassen 1915–1917. [From the Institute for the Improvement of Agricultural Plants. Comparison of barley and wheat varieties originating from the Institute with other superior races of these plants 1915–1917.]** Med. Landb.-Hoogeschool Wageningen 14: 1–34, 210–232. 1918.

1654. SIRKS, M. J., AND J. BIJHOUWER. **Onderzoekingen over de eenheid der linneaansche soort Chrysanthemum leucanthemum L. [Investigation of the homogeneity of the Linnean species Chrysanthemum leucanthemum L.]** Genetica 1: 401–442. Sept., 1919.

1655. SOLER, RAFAEL ANGEL. **Cultivo del tomate. [Tomato culture.]** Revist. Agric. Com. y Trab. 2: 479–483. *8 fig.* 1919.

1656. SPERLICH, ADOLF. **Die Fähigkeit der Linienerhaltung (phyletische Potenz), ein auf die Nachkommenschaft von Saisonpflanzen mit festem Rhythmus ungleichmässig übergehender Faktor. [Capacity to maintain lines (phyletic potency), a factor distributed irregularly to the offspring of plants with fixed seasonal rhythm.]** Sitzungsber. Akad. Wiss. Wien 128: 379. 1919. —See Bot. Absts. 5, Entry 1559.

1657. SPERLICH, ADOLF. **Über den Einfluss des Quellungszeitpunktes, von Treibmitteln und des Lichtes auf die Samenkeimung von Alectorolophus hirsutus All.: Charakterisierung der Samenruhe. [On the influence of the time of application of forcing-agents and of light on the germination of seeds of Alectorolophus hirsutus. Characterization of seed rest.]** Sitzungsber. Akad. Wiss. Wien 128: 477. 1919.—See Bot. Absts. 5, Entry 1559.

1658. SPRAGG, FRANK A. **The spread of Rosen rye.** Jour. Heredity 11: 42–44. *1 fig.* Jan., 1920.

1659. STEIN, E. [German rev. of: KLEBAHN, H. **Impfversuche mit Pfropfbastarden. (Infection experiments with graft hybrids.)** Flora 11–12: 418–430. 1918.] Zeitschr. indukt. Abstamm. Vererb. 22: 304. May, 1920.—See also Bot. Absts. 3, Entry 2124.

1660. Stein, E. [German rev. of: van Herwerden, M. A. **De invloed van radiumstralen op de ontwikkeling der eieren van Daphnia pulex. (Effects of the rays of radium on the oogenesis of Daphnia pulex.)** Genetics 1: 305–320. July, 1919. (See Bot. Absts. 3, Entry 1044.)] Zeitschr. indukt. Abstamm. Vererb. 22: 286–287. May, 1920.

1661. Steinach, E. **Histologische Beschaffenheit der Keimdrüse bei homosexuellen Männern. [Histological condition of the gonads in homosexual men.]** Arch. Entwicklungsmech. Organ. 46: 29–37. *Pl. 3–5.* 1920.—Interstitial cells characteristic of the ovary were found in the testes of several homosexual men, associated with degeneration of male interstitial cells, and of the seminal tubules. As reported elsewhere (Steinach und Lichtenstern, Münch. med. Wochensch. Nr. 6, 1918), these testes were removed and cryptorchid testes with normal puberty gland implanted, restoring normal sexual instincts to the homosexuals.—*H. D. Goodale.*

1662. Steinach, E. **Künstliche und natürliche Zwitterdrüsen und ihre analogen Wirkungen. [Artificial and natural hermaphroditic glands and their analogous functioning.]** Arch. Entwicklungsmech. Organ. 46: 12–28. 1920.—A discussion of castration, feminisation, masculinisation, and hermaphroditisation, some of it based on work previously unpublished, with particular reference to the similarity between homosexuals and certain artificial hermaphrodites. Two instances of homosexual goats are described.—*H. D. Goodale.*

1663. Stockard, Charles R., and G. N. Papanicolaou. **Variations of structural expression in the inheritance of polydactyly.** Anat. Rec. 18: 262–263. April 20, 1920.—Author's abstract of paper read before American Association of Anatomists, April 1–3, 1920.—The inheritance of polydactyly in a strain of guinea-pigs has been studied for the past several years. This character when it appears in the race is inherited as a Mendelian dominant.—The expression of the character in a series of individuals presents a most striking condition. The extra toe on the hind foot may be a perfectly developed functional toe in one animal, while in others the toe presents varying degrees of imperfect development and structure until in some it is represented by only a minute toe-nail attached to the foot by a thread-like filament. This poorly formed toe is frequently broken off or lost shortly after birth, and would often escape notice if not carefully looked for. Other animals inherit the extra toe, but fail to develop it sufficiently to show any evidence of its existence at birth. The fact that these have the character for extra toes is demonstrated by their offspring which may exhibit the toe as frequently as do offspring from parents with well-expressed polydactyly.—These normal variations in the expression of this dominant character renders it a most uncertain quantity for judging the influences of experimental treatments on its inheritable behavior in different groups of animals.—*Charles R. Stockard and G. N. Papanicolaou.*

1664. Stomps, Theo. J. **Über zwei Typen von Weissrandbunt bei Oenothera biennis L. [On two types of white margins in Oenothera biennis L.]** Zeitschr. indukt. Abstamm. Vererb. 22: 261–274. May, 1920.

1665. Stout, A. B. **The aims and methods of plant breeding.** Jour. New York Bot. Gard. 21: 1–16. Jan., 1920.—Author notes rise of subject during past three decades which has culminated in the present day development of genetics. Broadly considered, plant breeding, through selection of seed parents, is older than written history, but modern methods of plant breeding are based on a knowledge of sexuality in plants. Notes work of early investigators in study of sexuality, hybridization and selection including the early pedigree methods of Le Couteur and Shirreff, also the early work on sugar beets.—Importance of hybridization as a means of inducing variation is noted. Also development and importance of chromosome theory of inheritance and Mendel's Law. The rise of the mutation theory, linkage, multiple and modifying factors is also noted. Author notes "the germ plasm is the seat in which most of the hereditary changes occur." Cites the case of the 6,500 varieties of *Dahlia* in cultivation in the United States, all of which have descended from a single American species during the past 130 years. Similarly with *Phlox*, 200 varieties of which have descended from a single wild Texan species which was introduced into cultivation in 1835.—Importance of bud sports is noted in the case of the Sword fern and variegated *Coleus* and also with the citrus fruits.—Discusses modern method of plant breeding with respect to pedigree culture, and summarizes: "In practical application, the methods of plant breeding are (1) to maintain in a highly productive condition races whose qualities make their cultivation desirable, (2) to recognise and preserve new characteristics which may lead to further improvement, (3) to combine qualities of different strains into one strain through crossing, and (4) to induce hereditary variation through hybridization. Plant breeding aims to regulate, to control, to direct, and to utilize the processes of heredity and variation."—*C. E. Myers.*

1666. STREETER, GEO. L. **Formation of single-ovum twins.** Bull. Johns Hopkins Hospital 30: 235–238. *4 fig.* 1919.—The mature ovum here referred to is the one previously described by the same author (Carnegie Inst. Washington Publ., 272.) The ovum, which is about 17 days old, contains two embryos. One of these is considerably more advanced than the other. The primary embryo is in the primitive-groove stage, and has an embryonic plate 0.92 mm. long by 0.78 mm. wide. Two small vesicles slightly separated from each other are found in the loose mesenchyme in the posterior region of the body-stalk. These two vesicles represent the Amniotic vesicle and the yolk-vesicle, respectively, of the smaller twin. This small embryo is undoubtedly abnormal. By comparing this ovum with the Miller specimen and the Bryce-Teacher specimen, the author is able to indicate how in all probability monozygotic, or identical twins are formed. The ovum is one of unusual interest, in that it shows the youngest stage of twinning so far recorded for the human species.—*J. T. Patterson.*

1667. STRONG, LEONELL C. **Roughoid, a mutant located to the left of sepia in the third chromosome of Drosophila melanogaster.** Biol. Bull. 38: 33–37. Jan., 1920.—New mutant, characterised by roughened eyes, found to lie to the left of sepia, which had been furthest to the left of known third-chromosome loci. Roughoid sepia crossover value of 24.9 was obtained. —*A. H. Sturtevant.*

1668. STUCKEY, H. P. **Work with Vitis rotundifolia, a species of Muscadine grapes.** Georgia Agric. Exp. Sta. Bull. 133: 60–74. *4 pl. (colored), 8 fig.* Dec., 1919.—Work with *rotundifolia* was started at the Experiment Station in 1909. A history of workers with this species of grape is given. The work at the Station bears out the fact that *Vitis rotundifolia* is self-sterile, though the fruiting vines produce pollen. This pollen is infertile due to a degeneration of the generative nuclei. Work with more than two thousand seedlings which have been brought into bearing shows that approximately one-half are males and one-half are females. The male vines are more vigorous in growth and a larger percentage of these produce flower before the female vines. In more than one thousand bearing vines, it was found that the color of the tendrils and new growth correspond to the color of the fruit; vines having red or reddish green tendrils bear black or reddish black grapes, while those with green tendrils, internodes and new growth, produce light or amber-colored fruit, as the Scuppernong. Male vines fall into these two groups just as the females except they bear no fruit. Black is dominant over white and latter color is pure recessive. A formula showing crosses between plants heterozygous for black, red and white is given. Thomas × dark male produced only plants with dark fruit, but seedlings from Scuppernong × dark male produced plants of different colors of fruits. Certain male vines were prepotent for quality. In Flowers × light male no. 1, out of 41 seedlings, only one produced fruit inferior in flavor to the Flowers; the others were superior. From nine vines of Flowers × Black No. 1, only one produced fruit equal in flavor to the Flowers. New varieties described are Hunt, Irene, November, Qualitas, Spalding and Stuckey. It is further stated that pruned vines growing by the trellis system, when in good bearing should produce 50 to 60 pounds or about a bushel of fruit per year. Test of various strains of Scuppernongs, which is the most common variety of *Vitis rotundifolia*, demonstrated that nurserymen have made some effort to eliminate poor-bearing types and those untrue to name. Last page of the bulletin details methods of planting and pruning, and uses of the fruit.—*T. H. McHatton.*

1669. Sturtevant, Robert S. **Hybridizing bearded irises.** Gard. Chron. 67: 184. April 10, 1920.—Refers to a number of crosses of horticultural varieties and presents evidence to show that *plicata* characters are not due to a recessive factor as proposed by Bliss [see Bot. Absts. 5, Entry 1460]; also indicates that in the *Iris*, venation acts as a simple Mendelian dominant but that in regard to color and its disposition in other ways a more complex explanation is needed.—*J. Marion Shull.*

1670. Sumner, Francis B. **Geographic variation and Mendelian inheritance.** Jour. Exp. Zool. 30: 369–402. *7 fig.* April 5, 1920.—Paper is continuation of earlier biometric and genetic work on geographic races of deer-mouse (*Peromyscus maniculatus*) found in California. Characters chosen for study were length of tail, foot, ear, pelvis, femur and skull, width of dorsal tail stripe, color of pelage, pigmentation of foot and number of caudal vertebrae. Fewer grades for any one character have been found than number of localities from which material was collected. Members of same subspecies collected from different localities often differ widely. To a certain extent and for certain characters gradations considered follow geographic and climatic sequence. Degrees of difference in characters are, however, not proportional to geographic intervals between races and there are other incongruities which greatly complicate the situation. Characters which vary together, when geographic sequence is considered, may or may not vary together within any single local collection and *vice versa*. It seems that special factors, operating locally, must be responsible for modification of parts which do not ordinarily vary together. Animals from coastal stations, which probably present graded series in respect to both temperature and humidity, show similar gradation in respect to mean width of tail stripe and mean length of tail, foot and ear. Suggestion is made that environment in course of time has modified characters of animals dwelling at various points. Variations within each race are partly hereditary and partly 'somatic' in origin. Differences between local races do not act, in crossing, as simple Mendelian factors although theory of multiple factors would undoubtedly be invoked by many geneticists. Author prefers theory of contamination of genes. Deviations from type of various characters in F_1 and F_2 generations have been compared. Conclusion is made that variation is slightly greater in F_2. Incidentally differences have been observed between sexes, viz., smaller feet and larger pelvis in females. These are attributed to presence of at least two hormones varying independently.—*P. W. Whiting.*

1671. Sutton, Arthur W. **Brassica crosses.** Gard. Chron. 67: 20. Jan. 10, 1920.—Issue is taken with the statements of a writer in a recent issue of the journal in regard to crosses between cauliflower and kohl-rabi. Cauliflower crosses readily with any other type of *Brassica* but the resulting forms are worthless.—*C. B. Hutchison.*

1672. Tammes, T. **De leer der erffactoren en hare toepassing op den mensch. Rede, uitgesproken bij het aanvaarden van het ambt van buitengewoon hoogleeraar aan de Rijks-Univ. te Groningen. [The theory of hereditary factors and its applicability to man. Address, delivered on assumption of the office of Professor Extraordinarius in the State University at Groningen.]** *24 p.* Wolters: Groningen., 1919.

1673. Th., G. **Systematic breeding.** Florists' Exchange 49: 882. April 10, 1920.—Popular account of the value of systematic breeding based on a knowledge of the laws of heredity, especially as applied to carnation breeding. Lack of such knowledge may result in an occasional striking prize in commercial breeding, but no continuous series of successes. Describes some work of carnation breeders.—Dorner & Sons, Ward, and Fisher. Systematic breeding in the hands of these men brought the five-inch carnation and many other improvements. Dorner & Sons' promising new carnation productions are tested out by Samuel Goddard, Framingham, Mass. Carnation breeding is toward better keeping qualities, better form and color, larger number of blooms per plant without decrease in flower size and production of a good yellow type.—*Orland E. White.*

1674. Th., G. **Systematic breeding organization proposed.** Florists' Exchange 49: 1089. May 8, 1920.—Discussion of the advantages of systematic breeding and of the formation of a society of growers interested in applying theoretical knowledge to their own problems. Records of practical breeders' work should be kept by a central body. Work is often repeated through lack of an organisation through which knowledge can be distributed. Many a valuable discovery has probably been made by individual workers and then lost to the world because the records have not been passed on. Author says "Darwin's theories today are repudiated to a large extent, while Mendel's law is recognised."—*Orland E. White.*

1675. Thomson, J. Arthur. [French rev. of: Macleod, J. **The quantitative method in biology.** *15 x 23 cm., v+228 p., 27 fig.* Longmans, Green & Co.: New York, 1919. (See also Bot. Absts. 4, Entry 758.)] Scientia 27: 244–246. 1920.

1676. Tischler, G. [German rev. of: (1) Renner, O. **Ueber Sichtbarwerden der Mendelschen Spaltung im Pollen von Oenotherabastarden. (On the visibility of Mendelian segregation in hybrids of Oenothera.)** Ber. Deutsch. Bot. Ges. 37: 129–135. 1919. (2) Idem. **Zur Biologie und Morphologie der männlichen Haplonten einiger Önotheren. (Biology and morphology of the male haplonts of some Oenotheras.)** Zeitschr. Bot. 11: 305–380. *39 fig.* 1919.] Zeitschr. indukt. Abstamm. Vererb. 22: 221–223. Mar., 1920.

1677. Tornau, Dr. **Einige Mitteilungen über variabilitätsverhältnisse in einem konstanten Weizenstamm. [Some communications concerning variability relations in a constant wheat strain.]** Jour. Landw. 67: 111–149. 1919.—A biometrical study of variability and correlation in a pure line of wheat, the constants for different years being compared.—*C. E. Leighty.*

1678. Van Fleet, W. **Rose-breeding notes for 1918.** Amer. Rose Ann. 1919: 29–35. 1919. —Description of results from crossing numerous species and types of roses. Considerable improvement is seen in newer hybrids of *Rosa rugosa, R. Hugonis, R. Soulieanea, R. Moyesii.* Color range in *R. rugosa* hybrids covers single and double, constant-blooming forms in clear whites to glowing crimsons. No pure yellows. Creams, common and bright yellows may be expected in time. Main ideal for *R. rugosa* hybrids is high class blooms of Hybrid Perpetual and Hybrid Tea types combined with vigorous, hardy, disease-resistant plants. Premier English rose of 1918 is Mermaid, said to be result of *R. bracteata* crossed with a tea-scented variety. Efforts are being made to secure hybrids of *R. bracteata* able to endure climate of northern plains region, hybrids of *R. bracteata* × *R. carolina* giving promising results, and enduring zero weather. They have beautifully-formed pink buds. No success has been attained in attempts to cross Harison's Yellow for over 20 years. More success with seedlings of this variety, especially one similar to one of reputed parents of Harison's Yellow. Out of many thousand seeds of Harison's Yellow sown, only three grew so far. There is possibility of fragrance of the sweetbrier being intensified through breeding work.—*Orland E. White.*

1679. van Wisselingh, C. **Über Variabilität und Erblichkeit. [Concerning variability and heredity.]** Zeitschr. indukt. Abstamm. Vererb. 22: 65–126. *10 fig.* Jan., 1920.—Emphasizes importance of a study of the lower and simpler plants in the attempt to get at fundamentals of heredity and variation. Many illustrations are cited from author's and Gerassimoff's extensive studies of *Spirogyra.* Variations in the form and size of the cell, thickness and markings of cell walls, number and form of chromatophores, presence or absence of pyrenoids and method of starch-formation, number of nuclei, abnormalities in nuclear and cell-division, number of chromosomes, and nature and development of nucleoli are described in detail, and the causes giving rise to them are discussed. Variations in cell length, rate of starch formation and cell division may be induced through alterations in amount of light, but are not heritable. Thickness and markings on cell walls are heritable even in cells without nuclei. Chromatophores without pyrenoids form starch in a diffuse manner and are passed on through innumerable cell generations regardless of environmental conditions. Binucleate cells may be produced through anaesthesia, low temperatures, or centrifuging. This condition is perpetuated by cell division and so is heritable. Instead of two nuclei there may be one giant nucleus. In either case the cells assume a much larger diameter which is inherited both through cell division and conjugation. Author concludes that the nucleus is not the sole bearer of hereditary factors but that on the contrary heritable variations may arise in and be transmitted by the chromatophores and the cytoplasm.—*Leonas L. Burlingame.*

1680. Vestergaard, H. A. B. **Observations on inheritance in lupines, wheat, and barley.** Tidsskr. Planteavl. 26: 491–510. *7 fig.* 1919.

1681. Vigiani, D. **Sulla selezione del frumento "Gentil Rosso." [Upon the selection of the wheat "Gentil Rosso."]** Staz. Sper. Agr. Italiane 52: 5–13. 1919.

1682. Vogt, A. **Der Altersstar, seine Heredität und seine Stellung nach exogener Krankheit und Senium. [Senile cataract, its heredity and its place in exogenous disease and senile degeneration.]** Zeitschr. Augenheilkunde 40: 123. 1918.

1683. von Graevenitz. [German rev. of: Crane, M. B. **Heredity of types of inflorescence and fruits in tomato.** Jour. Genetics 5: 1–10. 1915.] Zeitschr. indukt. Abstamm. Vererb. 22: 223–224. Mar., 1920.

1684. von Ubisch, G. **II. Beitrag zu einer Faktorenanalyse von Gerste. [Contribution to a factorial analysis of barley.]** Zeitschr. indukt. Abstamm. Vererb. 20: 65–117. *7 fig., 11 diagrams.* Jan., 1919.

1685. von Wettstein, Fritz. **Vererbungserscheinungen und Systematik bei Haplonten und Diplohaplonten im Pflanzenreich. [Genetical phenomena and taxonomy in haplonts and diplonts in the vegetable kingdom.]** Zeitschr. indukt. Abstamm. Vererb. 21: 233–246. Nov., 1919.

1686. W., B. C. A. [Rev. of: Punnett, R. C. **Mendelism.** *5th ed.*, Macmillan & Co.: London, 1919.] Jour. Botany 57: 357–358. 1919.

1687. W., F. A. **The meaning of continuous variation in color.** Jour. Heredity 11: 84–86. *1 fig.* Feb., 1920.

1688. Waldron, L. R., and J. A. Clark. **Kota, a rust resisting variety of common spring wheat.** Jour. Amer. Soc. Agron. 11: 187–195. *1 pl.* 1919.—A variety of bearded hard red spring wheat, named Kota, has been found to possess resistance to the form or forms of stem rust of wheat present at Fargo, North Dakota, Brookings, South Dakota, and St. Paul, Minnesota, in 1918. Preliminary tests with Kota show it to have yielding ability. In baking tests it ranked high when compared with other bread wheats.—*H. K. Hayes.*

1689. Waldron, L. R. **First generation crosses between two alfalfa species.** Jour. Amer. Soc. Agron. 12: 133–143. 1920.—A report on the weight of plants of the first generation hybrids, secured by crossing *Medicago sativa* (common alfalfa) with *Medicago falcata* (yellow-flowered). The hybrids showed 47.5 per cent more weight than the parents. No significant differences were observed in the heights of the hybrid and the non-hybrid plants. Increased weight was then probably due to an increased number of stems per plant. Plants of *M. falcata* showed less winter-killing than the other groups.—*F. M. Schertz.*

1690. Wangerin, W. **Der Generationswechsel im Tier- und Pflanzenreich. [The alternation of generations in the animal and plant kingdoms.]** Schrift. Naturf. Ges. Danzig 15: 1–13. 1918.

1691. Warren, Don C. **Spotting inheritance in Drosophila busckii Coq.** Genetics 5: 60–110. *1 pl., 4 fig.* Jan., 1920.—Variation was noted among males of *D. busckii* in number of spots on tergum of fifth abdominal segment. Selection isolated two types, the two-spot and the six-spot, although the germinal behavior of the three separate strains was distinct. Crosses indicate that (1) the same high factor has been isolated in all three strains; (2) the high or low is neither completely dominant to the other; (3) the female may transmit the factor for specific configuration although she is always of the six-spot type.—Stock 501 gave a mutant with an exceptionally large outer spot. Tests with two-spot line indicate that the factor for the middle spots may be sex-linked in this particular strain.—Temperature has a differential effect on spotting. Low temperatures (11–15°C.) emphasize outer spots and reduce the middle ones, even in the two-spot selected lines.—Six females appeared simultaneously in one stock, lacking the middle spots. When mated to brothers, these gave rise to a variable abnormal strain. Selection purified the stock. Crosses show that male can transmit the factor, although not showing the character himself.—To conclude, inheritance of spotting in *D. busckii* is complicated. The same spot in the female and in different strains of males are due to different factors. Environment, particularly temperature, has a differential effect on the development of the various spots, and is important in the interpretation of selection.—*Joseph Krafka, Jr.*

1692. Weatherwax, Paul. **The origin of the intolerance of inbreeding in maize.** Amer. Nat. 54: 184–187. Mar.–Apr., 1920.—In regard to androgyny and to protogyny of individual inflorescences maize presents no fundamental difference from other American representatives of Maydeae. This fact together with reduction in number of inflorescences due to the mode of long continued cultivation and hence widespread cross-pollination make it unnecessary to assume the introduction of intolerance of self-pollination from another group.—*D. F. Jones.*

1693. Weimer, J. L. **Variations in Pleurage curvicolla (Wint.) Kuntze.** Amer. Jour. Bot. 6: 406–409. 1919.—Data on the extent of variation in certain characters due to differences in substratum upon which a pure strain of *Pleurage curvicolla* was grown, indicates unreliability of taxonomic criteria for species formation in fungi. Spore size was found to be relatively constant but size of perithecia showed greater variation and secondary spore appendages, a recognized character for this species, were not seen. Observations of author and others indicate that this species may have 128, 256, or 512 spores in ascus as a result of 7, 8 or 9 mitoses. [See also Bot. Absts. 5, Entry 694.]—*T. H. Goodspeed.*

1694. Wenholz, H. **Maize breeding. Agricultural research in Australia.** Advisory Council Sci. and Ind. Commonwealth of Australia Bull. 7: 39–48. 1918.—Author believes that improvement of maize can be accomplished largely by selection within a variety and therefore the experiment farms of New South Wales have been restricted to the use of one variety which previous experience has shown to be the best for the district.—Study of ear characters had led to the conclusion that some of them are associated with yield. These characters are length and shape of ear, weight and percentage of shelled grain, space between the rows, filling and character of the butts and tips, depth of grain and size of core. The ideal ear with many of these desirable characters highly developed has not been found by experiment to be positively correlated with yielding capacity under all conditions.—Data are being collected to discover what visible characters in the ear are associated with yielding capacity. Thus far it is found that although depth of grain is correlated with yield in a late-maturing variety on the coast, this correlation does not exist with the early variations of the tablelands. In regions of good rainfall, moderate-sized core is correlated with yielding capacity while in regions of scanty rainfall smallness of core is a character somewhat related to drought resistance but not to very high yields.—Another measurable ear character found to be related to yield is the weight. Author states that uniformity in the appearance, size, shape of ear, and character of the indentation of the grain gives a greater uniformity in the maturing of the crop and in consequence a greater uniformity in flowering which latter has been found to be directly associated with a smaller percentage of barren stalks.—Ear-to-row breeding is highly recommended and in ear-to-row tests author notes having made some very careful observations which have thrown considerable light on maize breeding and selection. It has been found, for instance, that some rows from individual ears contain a high percentage of barren stalks while other rows have practically none. It has also been found that many of the highest-yielding rows in the tests have been most uniform in the type of ears produced. Author considers that this observation supports the practice of breeding for uniformity in ear type.—In breeding for early maturity author recommends selecting early-maturing plants in the field instead of the longer process of elimination of the late-maturing types in the variety by gradual acclimatization.—In breeding for drought-resistance the greatest difficulty to be overcome is the "blasting" effect of hot, dry winds on pollen viability, although in some districts this is obviated by planting at the proper time. It is stated that while breeding may produce a drought-resistant pollen it must be borne in mind that the limitation of moisture in the soil is also a contributing factor in low yields.—*J. H. Kempton.*

1695. White, Orland E. [Rev. of: East, Edward M., and Donald F. Jones. **Inbreeding and outbreeding.** *14 x 21 cm., 285 p., 46 fig.* J. B. Lippincott: Philadelphia, 1919.] Torreya 20: 32–34. Mar.–Apr., 1920.—See also Bot. Absts. 4, Entry 571; 5, Entries 437, 1607.

1696. Wiggans, C. C. **Some factors favoring or opposing fruitfulness in apples.** Missouri Agric. Exp. Sta. Res. Bull. 32: 1–60. *6 fig.* 1918.—Studied individual fruit spurs of six commercial varieties of apples. Three varieties were known as annual bearers and these developed fairly high percentage of blossoms each year while three were classed as alternate bearers. Two of the annual bearers were able to develop blossoms in successive seasons on the same spur in much greater proportion than other varieties observed. Bearing spurs ranged from 2 to 8 years in age, 3 to 6 or 7 years being most effective fruiting age.—Found slightly higher concentration of sap (freezing point method) in bearing than in non-bearing spurs and noted marked decrease in sap concentration in late June or early July. Sugar and starch were shown by chemical methods to be present in slightly greater quantities in bearing than in non-bearing spurs. Determined effect of girdling, fertilizers, cultural treatments, and etherization on concentration of cell sap.—*H. K. Hayes.*

1697. Wilder, Harris Hawthorne. **Physical correspondence in two sets of duplicate twins.** Jour. Heredity 10: 410–420. *Fig. 15–19.* Dec., 1919.

1698. Winters, A. Y. **Eugenics, the war instinct and democracy.** Jour. Heredity 10: 254–256. June, 1919.

1699. Woods, Frederick Adams. **Twins prove the importance of chromosomes.** Jour. Heredity 10: 423–425. Dec., 1919.

1700. Woods, Frederick Adams. **A definition of heredity—"Nature vs. nurture" not a good expression.** Jour. Heredity 10: 426–427. Dec., 1919.

1701. Wriedt, Chr. **The brindle colour in cattle in relation to red.** Jour. Genetics 9: 83. Dec., 1919.—Author concludes from records on Telemark breed in Norway that brindle is dominant to red (and not a heterozygote between red and black as J. Carlson had concluded), on the basis of the following: Brindle × brindle or brindle × red gives both brindle and red, but red × red gives only red. Black is said to be very rare in this breed, the characteristic colors being brindle and red.—*J. A. Detlefsen.*

1702. Zeleny, Charles. **A change in the bar gene of Drosophila melanogaster involving further decrease in facet number and increase in dominance.** Jour. Exp. Zool. 30: 293–324. *9 fig.* April 5, 1920.—Author, who has for some time been studying the effect of selection upon the physical appearance and hereditary determiners ("bar gene") of the barred eye of *Drosophila melanogaster*, reports several mutants that have arisen in respect to this character. Bar gene, which is sex-linked, is concerned with the production of an eye with a greatly reduced number of facets (an average of about seventy-five, instead of the usual eight hundred of normal "full-eye" flies). The F_1 generation of bar by full-eye is nearly intermediate between the parents. To avoid the effects of varying temperature, the flies of these data were reared at uniform temperature. Though considerable variability occurs in facet number, one male appeared, having only nineteen facets, a number markedly lower than the lowest otherwise known for bar eye. This fly produced a race with average of twenty-two or twenty-three facets. The gene concerned is named ultra-bar, and shows a marked dominance over both bar and full-eye, so that the F_1 generation has eyes almost as small as those of ultra-bar. Crossing-over tests seem to show that ultra-bar is an allelomorph of bar. Author calls attention to this evidence of mutation in a gene during selection, but thinks the direction of mutation probably not significant inasmuch as mutations toward full-eye have also occurred.—*John S. Dexter.*

HORTICULTURE

C. H. Gourley, *Editor*

FRUITS AND GENERAL HORTICULTURE

1703. Allen, W. J. **Orchard notes. February.** Agric. Gaz. New South Wales 31: 142–143. 1920.

1704. Allen, W. J. **Apricot growing in New South Wales.** Agric. Gaz. New South Wales 31: 201–207. *1 fig.* 1920.

1705. Allen, W. J. **Peach growing in New South Wales.** Agric. Gaz. New South Wales 31: 127–133. *2 fig.* 1920.

1706. Allen, W. J., and W. C. G. Brereton. **Orchard notes. January.** Agric. Gaz. New South Wales 31: 65–67. 1920.

1707. Allen, W. J., and W. le Gay Brereton. **Orchard notes.** Agric. Gaz. New South Wales 31: 294–295. 1920.

1708. Allen, W. J., and S. A. Hogg. **Cherry growing in New South Wales.** Agric. Gaz. New South Wales 31: 277–279. 1920.

1709. Allen, W. J., and S. A. Hogg. **Orchard notes. March.** Agric. Gaz. New South Wales 31: 221–222. 1920.

1710. André, G. **Sur l'inversion du sucre de canne pendant la conservation des oranges. [The inversion of sucrose in oranges during storage.]** Compt. Rend. Acad. Sci. Paris 170: 126–128. 1920.—See Bot. Absts. 5, Entry 2193.

1711. Anonymous. **The cocoanut raft.** Sci. Amer. 122: 339. *1 fig.* 1920.

1712. Anonymous. **Lime sulphur spray following Bordeaux.** New Zealand Jour. Agric. 19: 371–374. 1919.—See Bot. Absts. 5, Entry 2001.

1713. Anonymous. **The most valuable crop.** Sci. Amer. Monthly 1: 316. 1920.—A note concerning the value of the cocoanut palm.—*Chas. H. Otis.*

1714. Anonymous. **Liming fruit trees.** Jour. Dept. Agric. Victoria 17: 699. 1919.—The following formula is given for washing tree trunks: 10 pounds of fresh quicklime in 50 gallons of water, enough water being added at first to cover the lime, add 8 pounds of flowers of sulphur, allow to boil for 20 minutes, and add the remaining quantity of water.—*J. J. Skinner.*

1715. Anonymous. **Conference on fruit growing.** Jour. Roy. Hortic. Soc. 45: 60–80. 1919.—This is a report of a discussion of the distribution, varieties, disease control, and grading of deciduous fruits.—*J. K. Shaw.*

1716. Anonymous. **Revival of indigo.** Sci. Amer. Supplem. 88: 271, 279. 1919. [Abstract.]

1717. Anonymous. **Spraying programs for the orchard and fruit garden.** Monthly Bull. Ohio Agric. Exp. Sta. 5: 67–78. 1920.

1718. Baker, C. F. **Coöperative seed exchange.** Philippine Agric. 8: 19–21. 1919.—This paper gives a list of tropical plants, seeds of which are desired by the College of Agriculture (Philippine Islands) in exchange for seeds of the College stock.—*S. F. Trelease.*

1719. Ballou, F. H., and I. P. Lewis. **Horticultural notes from the county experiment farms of Ohio.** Monthly Bull. Ohio Agric. Exp. Sta. 5: 52–57. *3 pl.* 1920.—Plans for pruning, fertilizing, landscaping and management are given.—*R. C. Thomas.*

1720. Ballou, F. H., and I. P. Lewis. **Culture and feeding of the apple orchard.** Monthly Bull. Ohio Agric. Exp. Sta. 5: 43–48. *2 pl.* 1920.—The article includes a comparison of the value of fertilizers used respectively with the grass mulch and tillage systems of culture. —*R. C. Thomas.*

1721. Ballou, F. H., and I. P. Lewis. **Pruning tests in young apple orchards.** Monthly Bull. Ohio Agric. Exp. Sta. 5: 85–90. *5 pl.* 1920.—This is a report of tests made in orchards of County Experiment Farms in Ohio. Seven methods are discussed briefly, viz., (1) Light dormant pruning. (2) Heavy dormant pruning. (3) Light summer pruning. (4) Heavy summer pruning. (5) Light dormant pruning with summer clipping of new shoots. (6) Heavy dormant pruning with summer clipping of new shoots, and (7) No pruning.—*R. C. Thomas.*

1722. Balme, Juan. **El olivo y su porvenir en Mexico.** [The olive and its future in Mexico.] Rev. Agric. [Mexico] 3: 379–383. *2 fig.* 1919.—History of olive culture in California and other parts of the new world, and the possibilities of growing the tree in Mexico.—*John A. Stevenson.*

1723. Beckwith, Charles C. **The effect of certain nitrogenous and phosphatic fertilizers on the yield of cranberries.** Soil Sci. 8: 483–490. 1919.—As a result of one year's studies on the effect of fertilizers on the yield of cranberries, the optimum amount of a mixed fertilizer consisting of sodium nitrate, 75 pounds; dried blood, 75 pounds; rock phosphate, 300 pounds; potassium sulfate, 50 pounds, was found to be 800 pounds. A mixture of mineral and organic nitrogen did not prove superior to sodium nitrate alone. Calcium cyanamid and barium phosphate proved unsatisfactory sources of nitrogen and phosphorus respectively.—*W. J. Robbins.*

1724. Bernard, Charles. **La culture du thé aux Indes néerlandaises.** [Tea-culture in the Dutch East Indies.] Rev. Gén. Sci. Pures et Appliquées 30: 17–18. 1919.—This paper, by the Director of the Tea-Experiment Station in Buitenzorg, Java, covering the industry indicated by the title, is of such conciseness as not to lend itself to further condensation into an abstract.—*G. J. Peirce.*

1725. Blair, W. S. **Orchard cultivation.** Fruit Growers' Assoc. Nova Scotia Ann. Rept. 55: 18–27. 1919.—Early plowed land contained 5.6 per cent more moisture in August than land plowed two weeks later. In another experiment sod land contained 5.9 per cent moisture in August while land cultivated six times and seeded to a cover crop on July 20 contained 14.1 per cent. Of the cover crops used crimson clover depleted the soil moisture least and millet most.—*Paul A. Murphy.*

1726. Boulger, G. S. [Rev. of: Bedford, Duke of, and Spencer Pickering. **Science and fruit growing: Being an account of the results obtained at the Woburn Experimental Fruit Farm since its foundation in 1894.** *xxii+351 p.* Macmillan & Co.: New York, 1919.] Jour. Botany 58: 28–29. 1920.

1727. Boyer, G. **Études sur la biologie et la culture des champignons superieurs.** [Biology and culture of mushrooms.] Mém. Soc. Sci. Phys. Nat. Bordeaux VII, 2: 233–344. *4 pl., 20 fig.* 1918.—See Bot. Absts. 5, Entry 1931.

1728. Cabrera, Teodoro. **La utilidad de los guayabos.** [Uses of the guava-trees.] [Revist. Agric. Com. y Trab. 2: 628. 1919.

1729. Call, L. E. **Director's report.** Kansas Agric. Exp. Sta. Ann. Rept. 1917-18. *63 p.* 1918.—See Bot. Absts. 5, Entry 1466, 2024.

1730. Calvino, Mario. **Reseño general sobre la arboricultura frutal de Mexico.** [Fruit trees of Mexico.] Rev. Agric. [Mexico] 5: 34-42. *6 fig.* 1919.—Lists the fruits of Mexico both for the tropical and the temperate belts, giving uses and possibilities of development of each. Fruits belonging to the following genera are discussed: *Crataegus, Carasus, Persea, Juglans, Casimiroa, Diospyros, Lucuma, Citrus, Musa, Theobroma, Annona, Spondias, Carica, Achras, Psidium, Chrysophyllum, Mangifera, Cocos, Cudonia, Phoenix, Vitis*, and *Olea*.—*John A. Stevenson.*

1731. Condit, I. J. **Caprifigs and caprification.** Univ. California Agric. Exp. Sta. Bull. 319: 341-375. 1920.—Figs which drop may be of the Smyrna class, the fruits of which require caprification in order to set and remain on the tree; they may be of the caprifig class, the fruits of which drop unless inhabited by the fig insect; or they may be common figs which drop because of unsuitable climatic conditions. Varieties of caprifigs which consistently bear quantities of polleniferous figs year after year, should be discarded, as they are of no value in caprification. A list of commercially grown varieties is given.—*A. R. C. Haas.*

1732. Ducomet, M. V. **Par quel moyen peut-on assurer a l'obtenir la propriété des variétés nouvelles de plants cultivées.** [How can the ownership of new varieties of cultivated plants be assured to the owner.] Jour. Soc. Nation. Hortic. France 20: 120-121, 139-144, 173-177. June, July and August, 1919.—The writer calls attention to the fact that the originator of a new and worthy plant is not protected in his rights in the same way that an inventor or writer is. He thinks that a man who has spent years in developing a worthy plant should be protected by law so that no one else would be allowed to propagate and disseminate it without paying a royalty to the originator. The writer recommends for France:—(1) That an association of French plant breeders be formed.—(2) That one or more government establishments, open to the public, be instituted for the acceptance and preservation of new varieties.—(3) That committees of acceptance and control be appointed.—(4) That every request for entry be accompanied by a detailed description of the new variety; a supply of seeds, bulbs, roots, buds or grafts; as exact an account as possible of the parentage of the new form; and a promise to send periodically fresh supplies of seeds, roots, etc., and to permit visits to the plantations in the event of controversy.—(5) That the request for registration of the new variety be publicly announced.—(6) That in the case of annuals a provisional certificate be given after one year and a final certificate after not less than two years and that certificate in the case of perennials be granted in as short a time as the nature of each permits.—(7) That the certificate guarantee only the authenticity of the plants not their productivity or any other quality.—(8) That the certificate be revoked if the variety prove unstable or is shown not to be a novelty.—(9) That during the period of certification no sale of the variety be allowed without the authorisation of the originator.—No recommendation is made as to the length of the period of protection for the originator.—*H. C. Thompson.*

1733. Ellenwood, C. W. **Bearing habits of the Delicious apple.** Monthly Bull. Ohio Agric. Exp. Sta. 5: 27-28. *2 tables.* 1920.

1734. Enfer, V. **L'ensachage des fruits.** [Bagging fruits.] Rev. Hortic. 91: 294-295. June, 1919.—The enclosing of fruits in sacks has long been practiced as a protection against various insects and hail, and because fruits thus protected are improved in texture and size. Sacks of a size appropriate to the fruit to be enclosed are chosen, the deformed and excess fruits removed, and those remaining enclosed when they are the size of a small nut, or at least by June 15 before the egg-laying period of the codling moth. Small holes are cut near the bottoms of the sacks in order that air may be admitted and excess moisture drained out. The fruit may remain covered until harvested, but the more highly colored varieties should be gradually uncovered by cutting out parts of the sack about September 10. It may be removed entirely several days later, after the skin has hardened somewhat. Bits of paper should be left attached to the peduncles of the fruits, in order to prevent attacks by birds.—*E. J. Kraus.*

1735. Enfer, V. **Sélection des jeunes fruits. [The selection of young fruits.]** Rev. Hortic. 91: 333-334. August, 1919.—In spite of the fact that many fruit buds are removed by pruning or are destroyed by cold or unfavorable weather, still, more generally remain than can be matured into good fruits. It is advisable, therefore, to remove all deformed and small fruits as early in the season and as rapidly as possible. When the fruit spurs are close together the fruits from half of them should be removed entirely in order that there may be a crop the following year. Later, selection is to be made of those which are to be sacked. The number of fruits to be preserved on each tree will vary with the vigor of the tree and the final volume of the fruit when mature. If a variety is the more valuable because of its extraordinary size, very few fruits should be allowed to remain even on vigorous trees.—*E. J. Kraus.*

1736. Fenzi, E. O. **Le culture ortive in Tripolitania. [Vegetable culture in Tripolitania.]** Bull. R. Soc. Toscana Orticult. 44: 105-109. 1919.—A discussion of the crops cultivated in this Italian colony.—*W. H. Chandler.*

1737. Ginarte, Benjamin Muñoz. **Algo más sobre el cultivo de la piña. [More about pineapple culture.]** Revist. Agric. Com. y Trab. 2: 592-593. *Fig. 1-2.* 1919.—The opinion of Rossi that the pineapple is a native of Brazil is recorded. The qualities of the fruits of different varieties of pineapple and closely related plants are described. A classification by Rossi is given.—*F. M. Blodgett.*

1738. Gladwin, F. E. **A test of methods of pruning the Concord grape in the Chautauqua grape belt.** New York Agric. Exp. Sta. [Geneva] Bull. 464: 189-213. *10 pl.* 1919.—Experiments covering a period of eight years were conducted at Fredonia, N. Y. Seven methods of training were tested and early winter pruning compared with late winter pruning. So far as yield is concerned, the single-stem Kniffin, the Munson, and the Chautauqua methods of training proved about equal; while fruit from the high-renewal and two-stem Kniffin methods was smaller in quantity and poorer in quality. Considering all of the advantages and disadvantages of the several methods, the single-stem Kniffin outranked all other methods of training. On the whole, late winter pruning made a slightly better showing than early winter pruning; but the difference in yield, wood growth, and maturity of fruit was too slight to warrant the definite conclusion that either method of pruning is to be preferred to the other.—*F. C. Stewart.*

1739. Green, W. J. **Smudging to prevent frost.** Monthly Bull. Ohio Agric. Exp. Sta. 5: 63. 1920.

1740. Gruebber, Charles. **Annual report of the senior fruit inspector.** Tasmania Agric. and Stock Dept. Rept. 1918-19: 10-11. 1919.—Administrative report on enforcement of various regulations at the port of entry and departure. The "apples and pears standaridization act" was not complied with satisfactorily. Many growers preferred to ship ungraded stock and some such shipments sold as well as stock marked "Fancy."—Shipments from Hobart for the year were over one million cases of fresh fruit.—*D. Reddick.*

1741. Hatton, Ronald G. **Paradise apple stocks; their fruit and blossom described.** Jour. Roy. Hortic. Soc. 44: 89-94. *Fig. 26-33.* 1919.—The author lists nine types of dwarf apple stocks grown at the Wye College Fruit Experiment Station, England. These have been compared with a series of "free" or standard stocks and there appears to be no strict dividing line between the two series. Eight of the dwarf types have fruited and tabular description of the flowers and fruit are given.—*J. K. Shaw.*

√ 1742. HAYWOOD, A. H. **The rice bean (Phaseolus calcaratus) or so-called Jerusalem pea (P. trinervis).** Agric. Gaz. New South Wales 31: 289–290. *1 fig.* 1920.—Notes are given on the plant as a cover crop for bananas. Its use is recommended.—*L. R. Waldron.*

1743. HODGSON, R. W. **Pruning the navel oranges.** California Citrograph 5: 138, 160. 1920.

1744. HONNET, G. **Les hybrides en 1919.** [The hybrids in 1919.] Rev. Vitic. 52: 53–59. 1920.—The oldest hybrids most resistant to drought are: Oberlins, Gaillard Number 2, Couderc 202 × 75, 146 × 51, Seibel 1000, 2859, Bertille-Serve 450; those less resistant are: S. 2003, 2006, G. 194, 157, S. 2734, 880, C. 272 × 60. Two black grapes, B-S, 413 and C. 106 × 46, have grown and produced well. The new black direct producers are: Baco Number 1, B-S 872, 893, 1129, Malegue 829 × 6, M. 2049 × 3, S. 4121, 4643, 4636 and 5212. Among the white varieties are: C. 162 × 5, S. 2638, 4681, 4955, 4986, 5279, M. 1647 × 8, 1157 × 1, Baco 22A, B. 43 × 23. A certain number of these new varieties appear to be very promising. They are more resistant to fungous diseases than *Vinifera* varieties.—*L. Bonnet.*

1745. HOUSER, J. S. **Recent tests of materials to control San Jose scale.** Monthly Bull. Ohio Agric. Exp. Sta. 5: 49–51. 1920.

1746. HOWARD, A., AND G. L. C. **Report of the Imperial Economic Botanists.** Sci. Rept. Agric. Res. Inst. Pusa 1918–19: 46–67. *Pl. 5 and 6.* 1919.—See Bot. Absts. 5, Entry 1159.

1747. HYDE, W. C. **Orchard cover-crop experiments on the Mountere Hills.** New Zealand Jour. Agric. 19: 364–365. *Fig. 1.* 1919.—See Bot. Absts. 5, Entry 1262.

√ 1748. JONES, J. **Plant importations.** Report on the Agricultural Department, Dominica, 1918–19: 2–3. [Imp. Dept. Agric. Barbados, 1919.]—Notes are given on the following plants: Mexican apple (*Casimiroa edulis*), Rambutam (*Nephelium lappaceum*), *Poutaria suavis*, Jaboticaba (*Myrciaria cauliflora*), Brazil nut (*Bertholetia excelsa*), Sapucaia nut (*Lecythis Zabucajo*), Suwarri nut (*Caryocar nuciform*) and Chicle gum tree. Other plants under trial are Sarawak bean (*Dolichos Hosei*), from St. Lucia, and *Cytisus Palmensis*, *C. stenopetalus* and *C. pallidus*, plants used in the Canary Islands for forage purposes. Mention is also made of *Momordica cochin-chinensis*, the seeds of which contain an oil of remarkable drying properties.—*J. S. Dash.*

1749. JONES, J. **Plot experiments with orchard cultivation.** Report on the Agricultural Department, Dominica, 1918–19: 18–23. [Imperial Department of Agriculture, Barbados. 1919.]—The author treats in a full and interesting manner the difficulties encountered while carrying on manurial and other experiments with such permanent crops as cacao and limes. Many useful suggestions are given.—*J. S. Dash.*

1750. KIRBY, R. S., AND J. S. MARTIN. **A study of the formation and development of the flower beds of Jonathan and Grimes Golden in relation to different types (clover sod, blue grass sod, cover crop, and clean tillage) of soil management.** Proc. Iowa Acad. Sci. 25: 265–290. *Pl. 7.* 1920.—Experiments made at Council Bluffs, Iowa, indicate that flower buds of apple form earlier and in greater numbers where soil moisture is less, and that nitrogen added by clover sod induces earlier formation of flower buds. The flowers are differentiated during a period of about four weeks on each tree, at some time between July 1 and September 15, according to variety and location.—*H. S. Conard.*

1751. KELLEY, W. P., AND E. E. THOMAS. **The effects of alkali on citrus trees.** California Agric. Exp. Sta. Bull. 318: 305–337. 1920.—The bulletin aims to enable citrus growers to recognize the effects of alkali, to appreciate the seriousness of alkali in citrus culture, to apprehend the relationships between irrigation and the accumulation of alkali, and to see that the application of certain fertilizers, especially nitrate of soda, may bear an important relation to the accumulation of alkali. The discussion is confined mainly to the effects of excessive salt concentration. Alkali content of the soil may ultimately reach a harmful concentration where irrigation water is applied that contains only a relatively low concentration of alkali salts. The rate of salt accumulation varies in different soils, depending on (1) the composition of the water, (2) the amounts applied, and (3) the freedom with which it penetrates into the subsoil. There exists a close relationship between the composition of irrigation water and the accumulation of alkali salts, and the condition of the citrus trees.—*A. R. C. Haas.*

1752. LAFFER, H. E. **The pruning of the vine.** Agric. Gaz. New South Wales 31: 47-55, 121-126. *Fig. 5-13.* 1920. [Continued from: *Ibid.* 30: 808. 1919.]

1753. LARUE, P. **Taille du Pineau à Chablis. [Pruning the Chablis Pineau grape.]** Rev. Vitic. 52: 7-11. *2 fig.* 1920.

1754. LEWIS, C. I., A. E. MURNEEK, AND C. C. CATE. **Pear harvesting and storage investigations in Rogue River Valley.** (Second report.) Orgeon Agric. Exp. Sta. Bull. 162: 1-39. *Fig. 1-12.* 1919. Fruits of Bartlett pears increase gradually in size, but at an accelerated rate in volume, throughout the growing season, apparently independently of climatic or cultural conditions. A distinct correlation appears to exist between the degree of maturity of Bartlett pears and the resistance offered by the cortical and epidermal tissues to pressure as measured by the amount of pressure required to force into them a $\frac{7}{16}$ inch steel ball up to one half its diameter. There is no direct relationship between such resistance to pressure and the diameter of the fruit. Storage investigations showed that, in the case of Bartlett pears, the size of the fruit was not a factor in time of ripening or decay. Fruits picked during the middle or latter part of the season of development kept longer than those picked early, and were superior in quality, and those picked exceptionally late were superior both in keeping and eating qualities. No difference in rate of maturity in storage was noted when a change in temperature of 10° to 15°F. was registered, provided the same approximate percentage of humidity was maintained. In the case of Bosc pears it was determined that both relatively high temperature with low humidity and low temperature with high humidity were harmful to proper ripening, that fruit picked very early in the season must be allowed to ripen partially before being placed at low temperatures, and that at least two weeks should elapse before putting the fruit into cold storage, though this time may be decreased under conditions of higher humidity.—*E. J. Kraus.*

1755. LODIAN, L. **Strange things to eat.** Sci. Amer. 122: 302, 312, 314. *9 fig.* 1920.—A popular enumeration of seeds, bulbs and flowers used by cosmopolitan New York City for food, which are out of the ordinary for that region.—*Chas. H. Otis.*

1756. MACOUN, W. T. **The commercial varieties of apples of Canada and the United States.** Fruit Growers' Assoc. Nova Scotia Ann. Rept. 55: 119-137. 1919.

1757. MANARESI, A. **Sulla biologia florale del pesco. 2a nota. [On the floral biology of the peach. 2nd note.]** Staz. Sperim. Agrarie Italiane 52: 42-67. 1919.—A study of the structure of the flower, its various parts and functions in a large number of varieties. Statistical study of the size of the various types of buds in different varieties, of the shape of the flower as connected with the character and adherence of the stone; the classification of the varieties into two groups characterized by a campanulaceous perianth in one case and a rosaceous perianth in the other case. A study of the flowering period and its daily periodicity; the action of meteorological conditions upon the functions and longevity of the various floral parts; form and dimension of pollen, and its relation to varietal classification. Study of the germination of the pollen of seventy varieties, of the size of the pollen tube, its morphological characteristics and speed of germination when tested in solutions of the following sugars: lactose, saccharose, maltose, glucose, laevulose, and galactose in solutions ranging in concentrations from 5 to 30 per cent. Distinct differences were obtained with the different sugars, saccharose being the most generally useful in concentrations ranging from 10 to 20 per cent; maltose in a concentration of 10 to 15 per cent may give results that approach and sometimes surpass those obtained with saccharose; lactose and glucose gave relatively good results only in concentrations varying from 5 to 15 per cent while galactose gave passable results at this concentration the optimum being between 5 and 10 per cent. Laevulose gave very poor results. Distinct differences were to be observed in the pollen tubes germinated in the different sugars, and in the different concentrations. Accidental differences were observed in some varieties in the position of the style with respect to the position of the anthers, and differences in the number of styles and ovules in the pistils. Anthesis was found to take place exclusively in day time, and mostly in the forenoon, the petals first expanding being the ones first touched by the sun. Cleistogamy was often observed in good seasons, and dehiscence took place mostly in the early forenoon under the direct guidance of the sun. Anthesis appears to follow a centrifugal path along the branch. The influence of the position and altitude of the tree and of grafting upon the time of flowering are also touched upon. A bibliography is appended.—*A. Bonazzi.*

1758. MARSHALL, ROY E. **Pruning fruit trees.** Virginia Polytech. Inst. Ext. Bull. 38. *37 p., 29 fig.* 1919.—A popular discussion of the training and pruning of apple, peach, pear, cherry, and plum trees with special emphasis on those phases of the subject of most practical importance in eastern United States.—*F. D. Fromme.*

1759. MARTIN, J. N., AND L. E. YOCUM. **A study of the pollen and pistils of apples in relation to the germination of the pollen.** Proc. Iowa Acad. Sci. 25: 391–410. *Fig. 163–166.* 1920. —The pollen of the five varieties of apples studied contains proteins or amino-acids, some pectin, and occasionally small amounts of sugars at the time of pollination. Pollen grains germinate in sugar solutions from pure water to 70 per cent, but most successfully at 2½ per cent. A temperature of 22°–25°C. was best. The stigma is papillate; pollen germinates when caught between the papillae. The styles contain much cane sugar at some distance below the stigma. No secretion was found on the stigma at the time favorable for pollination.—*H. S. Conard.*

1760. MATSUSHIMA, T. **Untersuchungen über die Wasseraufnahme bei abgeschnittenen Zweigen.** [Investigation of the water-absorption of amputated branches.] Jour. Coll. Sci. Imp. Univ. Tokyo 43[2]: 1–27. 1919.—After an abstract of the literature of the subject, Matsushima considers the relations of water-absorption to the Japanese art of arranging bouquets, and reports the methods and results of his experiments. In these he used water, both tap and distilled, and dilute acid and alkaline solutions, branches or sprays cut in the air and others under water, and still others the cut ends of which were deliberately burned. His results, as summarized at the end of the paper, are that in twigs cut off in air the decrease in water-absorption is insignificant if the twigs have abundant wood, but that in plants with much slime, milk or gum it is considerable: that burning the cut ends is especially favorable in the slime, milk and gum carrying plants as thereby the stopping of the water-carrying vessels is prevented: and that acids, especially the organic acids, increase the water-absorption while alkalies decrease it in ordinary plants, whereas in milk, slime and gum containing twigs the reverse is true.—*G. J. Peirce.*

1761. MUNN, M. T. **Spraying lawns with iron sulfate to eradicate dandelions.** New York Agric. Exp. Sta. [Geneva] Bull. 466: 21–59. *Pl. 1–6.* 1919.—Experiments made at Geneva, New York, demonstrate that dandelions (*Taraxacum officinale* and *T. erythrospermum*) may be eradicated from lawns at small expense and without material injury to the grass by spraying four or five times during the season with a solution prepared by dissolving 1.5 to 2 pounds of iron sulfate in one gallon of water. Spraying should be supplemented by the use of fertilizers and the application of grass seed in the spring and fall of each year. With proper management a lawn may be kept practically free from dandelions by spraying every third year. The cutting-out method of fighting dandelions is laborious and ineffective unless the greater part of the root is removed. A study of seed production in *Taraxacum officinale* shows it to be parthenogenetic.—*F. C. Stewart.*

1762. Pellett, Frank C. **American honey plants, together with those which are of special value to the beekeeper as sources of pollen.** *15 x 23 cm. 297 p., 155 illust.* American Bee Journal: Hamilton, Illinois, 1920.—The book is primarily an annotated and illustrated list of a very large number of plants that are of more or less interest to beekeepers. Each plant is listed in alphabetical order by some "common" name, but the Latin name is added in each case and a cross-referenced index makes it possible to find a given entry by either name. Names of states in the United States and of provinces in Canada are inserted in the alphabetical arrangement, each with a brief consideration of the honey-plants of that area. Many other names besides those of plants and regions find place in the list; under P is found a discussion of about seven pages on "Physiology of nectar secretion." The annotations are mainly with reference to the production of honey or pollen, but many facts of plant distribution are stated. Numerous quotations, with their sources are given. The geographical region considered is practically the United States, Alaska and Canada, rather than the whole of even North America. The illustrations are mostly reproductions of photographs. The book contains much to interest gardeners and plant-lovers generally, as well as beekeepers.—*B. E. Livingston.*

1763. Pickford, Verne C. **Control of air conditions in lemon storage rooms.** California Citrograph 5: 139, 164. 1920.

1764. Quisumbing y Arguelles, E. **Studies of Philippine bananas.** Philippine Agric. Rev. 12[3]: 1–73. *30 pl.* 1919.

1765. Ravaz, L. **Obtention des greffes-soudés. [Method of handling callused grape grafts.]** Prog. Agric. et Vitic. 74: 173–182. 1920.

1766. Ravaz, L. **Plantation des bouturos soins speciano. [Planting of cuttings—special precautions.]** Prog. Agric. et Vitic. 74: 21–32. 1920.—In planting grape cuttings "in situ" it is recommended that only the cuttings from the base of canes be planted; to mound them to avoid injury by frost, or to stratify them in a reverse position and plant after roots have appeared. The vineyard should be carefully cultivated.—*L. Bonnet.*

1767. Rawes, A. N., and F. J. Chittenden. **Effect of grass on apple trees.** Jour. Roy. Hortic. Soc. 45: 116–119. 1919.—Twenty-five trees of five varieties on crab and Paradise stocks were grown at Wisley, England under continuous cultivation, under grass with a space around the tree 2 to 3 feet in diameter kept bare and stirred at intervals, and under a grass mulch. Growth and yield were largest where the trees were cultivated.—*J. K. Shaw.*

1768. Rives, Louis. **Affinité des hybrides pour les porte-greffes. [Affinity of hybrids for resistant stocks.]** Prog. Agric. et Vitic. 74: 13–14. 1920.—The direct producers are generally very exigent in water and give good results on vigorous resistant stocks. The Vinifera Americans, 1202, the Aramon × Rupestris, the 93-5, show a sufficient affinity for them. The Rupestris St. George gives varying results. The author concludes that affinities must be studied experimentally in various soils and localities.—*L. Bonnet.*

1769. Rivière, C. **Le Jardin d'Essai d'Alger. [The Experimental Garden at Algiers.]** Rev. Hortic. 91: 340–342. Sept., 1919.

1770. Rivière, G. **De la progression de la maturation dans les poires a couteau. [Progression of ripening in table pears.]** Jour. Soc. Nation. Hortic. France 20: 306–307. Dec., 1919. —The author states that ripening of pears begins at the stem end and proceeds toward the calyx end. Analyses were made to determine the sugar content of different parts of the pear as it begins to ripen and it was found that the stem end section contained a larger percentage of sugar than the middle section and the latter more than the calyx end section. From this the author concludes that ripening progresses from the stem end toward the calyx end.—A table showing the percentage of sugar in three cross sections of three varieties of pears is given.—*H. C. Thompson.*

1771. Sanders, J. G., and L. H. Wible. **List of owners of commercial orchards and licensed nurserymen in Pennsylvania, including list of registered dealers in nursery stock.** Bull. Pennsylvania Dept. Agric. 1[10]: 1-56. 1918.

1772. Shamel, A. D. **Control of humidity conditions in lemon storage rooms.** California Citrograph 5: 137, 170. *3 fig.* 1920.

1773. Shamel, A. D. **Results of individual tree performance record studies with pruned and unpruned Marsh grapefruit trees.** California Citrograph 5: 248, 268. *4 fig.* 1920.—Experiments showed that heavy pruning of middle-aged vigorous grapefruit trees reduced the crop with no compensating benefits. The author recommends the training of young trees to proper form; the renewal by heavy pruning of old worn out trees; and the pruning out of dead brush and conflicting branches at any time. Aside from this he would not prune Marsh grapefruit trees.—*J. E. Coit.*

1774. Stuckey, H. P. **Work with Vitis rotundifolia, a species of muscadine grapes.** Georgia Agric. Exp. Sta. Bull. 133: 60-74. *4 pl. (colored), 8 fig.* 1920.—See Bot. Absts. 5, Entry 1668.

1775. Thayer, Paul. **Selecting nursery stock.** Monthly Bull. Ohio Agric. Exp. Sta. 5: 58-62. *4 pl.* 1920.—Particular attention is called to the supreme merit of standard varieties and the importance of selecting clean, disease-free stock.—*R. C. Thomas.*

1776. Thayer, Paul. **The Bartlett plum.** Monthly Bull. Ohio Agric. Exp. Sta. 5: 26. *1 pl.* 1920.—This variety combines beauty and utility and is recommended for semi-formal plantings.—*R. C. Thomas.*

1777. Thomas, P. H. **Annual report of the Assistant Fruit and Forestry Expert.** Tasmania Agric. and Stock Dept. Rept. 1918-19: 19-20. 1919.—Brief notes on the following: a fruit tree disease, caused by *Armillaria mellea*, can be prevented in early stages by cutting away all diseased tissue and painting over the wound with Bordeaux paste; a treatment of apples with nitrate-caustic soda spray did not have appreciable effect on fruit production or growth. Experiments are in progress with stocks for fruit trees, cold storage of fruit, etc.—*D. Reddick.*

1778. Tribolet, I. **Olives. I.** South African Jour. Indust. 2: 1160-1167. 1919.

1779. Tribolet, I. **Olives. II.** South African Jour. Indust. 3: 42-49. 1920.

1780. Truelle, A. **La vente des pommes de calville blanc, de Méran (Tyrol). [Marketing white Calville apples from Meran (Tyrol).]** Rev. Hortic. 91: 378-380. Nov., 1919.—Special methods of grading and packing are followed in marketing the fruit in order to utilise it to the best advantage and to derive the largest profits from its sale. The fruits are first graded into three classes. Class A contains perfect fruits only; Class B those which have slight defects; and Class C those which have more pronounced imperfections. These groups are again divided according to the form, beauty, and particularly the weight of the fruits. After grading the finest fruits are wrapped first in a white and then in a colored paper, and packed into boxes whose gross weight does not exceed 5 kilos. Usually such boxes contain from 12 to 20 fruits, though there may be as few as 9 or as many as 32. The less choice fruits are packed into cases which contain from 400 to 500 fruits arranged in 6 layers, with a gross weight of 55 to 60 kilos, and into half-cases which contain from 180 to 200 fruits arranged in 5 layers, with a gross weight of 25 to 30 kilos. There are also special cases for special purposes. The price received for the fruit varies with the quality and quantity available. Over a ten-year period, for the best fruits this has ranged from 10, to more than 50 cents a pound; very large individual specimens selling readily for 80 cents each. The inferior grades sell for less.—*E. J. Kraus.*

1781. Turney, A. G. **Report of the Horticulturist.** Province of New Brunswick Rept. on Agric. 1918: 86-109. 1919.

1782. Tyson, Chester J. **Report of the Pomologist.** Bull. Pennslyvania Dept. Agric. 1[1]: 27-29. 1918.—Includes general remarks upon exporting, market conditions, prices, packing and marking bill, new plantings, and the future of apple-growing industry in Pennsylvania.—*C. R. Orton.*

1783. Verdié, H. **Enquête sur les producteurs directs dans le Gers.** [Inquiry on direct producers in the Gers.] Rev. Vitic. 52: 78-19. 1920.—The direct producers giving the best results in that region are: Seibel 128, 138, 880, 1000, 1020, 2859, Condere 235-120, 272-60, 71-20, Mategue 829-6, 1100-2, 1157-1, Gaillard-Girerd 157, 194, Castel 1028, Bertille-Seyve 450, Baco 22A. These hybrids vary in regard to their adaptation to soils and their resistance to fungous diseases.—*L. O. Bonnet.*

1784. Ward, J. M. **Annual report of the fruit and forestry expert.** Tasmania Agric. and Stock Dept. Rept. 1918-19: 16-19. 1919.—Administrative report of work among fruit growers in connection with enforcement of grading and packing laws and the like.—Apple trees are found to do very much better when worked on seedling stock than when on root-graft stock. Jonathan, Fameuse, Dunn's, King David and Alfriston seem to do well on root-graft stock.—*D. Reddick.*

1785. Webber, Herbert John. **Selection of stocks in citrus propagation.** California Agric. Exp. Sta. Bull. 317: 269-301. 1920.—The influence of the character of the stock on the variation in yield of trees, forms the basis of the bulletin, although other important factors are considered. Differences in size of nursery trees of transplanting age is largely due to the fact that the seedling root-stocks on which the trees are budded are of different types, possessing inherent characters that react differently on the growing buds. Recommends that all small seedlings in the seed bed and nursery, regardless of what has caused their dwarfness, be discarded.—*A. R. C. Haas.*

1786. Weidner, A. I. **Report of committee on fruit and fruit culture.** Bull. Pennsylvania Dept. Agric. 1[1]: 22-24. 1918.—Some general remarks upon the condition of orchard fruits in Pennsylvania during the year 1917. Includes brief notes on San Jose scale, dusting, cultivation and fertilization, pruning, borers and spraying.—*C. R. Orton.*

1787. Whitehouse, W. E. **Cold storage for Iowa apples.** (Third progress report.) Iowa Agric. Exp. Sta. Bull. 192. 181-216. *14 fig.* 1919.—Methods of control of disease of cold stored apples are set forth. A study was also made of the factors relating to the control of common diseases of the apple in cold storage, such as temperature, humidity, maturity of fruit, size of apples, wrapping paper used in packing and rate of cooling.—*Florence Willey.*

1788. Woglum, R. S. **Is it safe to fumigate while trees are in bloom?** California Citrograph 5: 190. *Fig. 1.* 1920.—Experiments show that citrus blossoms and blossom buds are more resistant to hydrocyanic acid gas than tender leaves and shoots. It is suggested that this may be due to higher concentration of reducing sugars in the former. Ordinary fumigation if carefully done will not injure the bloom.—*J. E. Coit.*

1789. Young, Floyd D. **Rate of increase in temperature with altitude during frosty nights in orange groves in southern California.** California Citrograph 5: 136, 160. *3 fig.* 1920.

FLORICULTURE AND ORNAMENTAL HORTICULTURE

1790. Anonymous. **Spring-flowering stocks tried at Wisley, 1916-17.** Jour. Roy. Hortic. Soc. 44: 117-122. 1919.—This is a report of trial of 180 varieties of spring flowering stocks at Wisley, England. A classification with brief descriptive notes is given.—*J. K. Shaw.*

1791. Anonymous. **Oriental poppies at Wisley, 1917.** Jour. Roy. Hortic. Soc. 45: 120-125. 1919.—Report is made of the trials of 80 stocks of oriental poppies at Wisley, England, together with a classification and brief description and the awards of the Floral Committee.—*J. K. Shaw.*

1792. Anonymous. [Rev. of: Farrer, Reginald. **The English rock-garden.** *2 vol., 4to, lxiv, 504+viii, 524, 102 pl.* T. C. & E. C. Jack: London and Edinburgh.] Jour. Botany 57: 354-357. 1919.—The plants concerned are treated at considerable length in part from a botanist's standpoint, and the nomenclature is discussed.—*K. M. Wiegand.*

1793. B, D. **Fougères utiles.** [Useful ferns.] Rev. Hortic. 91: 330-331. Aug., 1919.—In addition to the specific mention of certain ferns enumerated by Prince Bonaparte (in: Notes Pteridologiques, Fasc. VII, p. 19, October, 1918), several others of medicinal or decorative value are listed.—*E. J. Kraus.*

1794. Bellair, G. **Comment economiser. Note sur le Verbena venosa.** [Economic comment. Note on Verbena venosa.] Rev. Hortic. 91: 387-388. *Fig. 119.* December, 1919.—Although this plant was introduced from La Plata in 1830, it is still grown but little as an ornamental because of the uncertainty connected with the germination of the seeds. In order to obviate this difficulty various methods of seed treatment were tested. Good results were secured from the following methods: (1) Immersion in water at 100°C. for 60 seconds; (2) immersion in water at 55°C. for 5 minutes; (3) immersion in a 2/1000 solution of nitrate of soda for 48 hours; (4) immersion in a 2/1000 solution of sulfate of ammonia for 48 hours; (5) stratification for 5 months. Poor results followed the following methods: (1) immersion in water at 70°C. for 4 minutes; (2) immersion in a 5/1000 solution of sodium nitrate for 48 hours; (3) immersion in a 5/1000 dilution of wood ashes, (4) immersion for 48 hours in pure water, though this latter result was fairly good. With the exception of the hot water treatments, the seeds were bathed in the solution indicated, rather than immersed in them. The stratified seeds, sown in March, germinated rapidly and completely. Seedlings may be carried over winter or grown in the spring and planted out of doors from the middle to the end of May.—*E. J. Kraus.*

1795. Berthault, P. **La réorganisation du Jardin d'Essais du Hamma.** [The reorganization of the Experimental Garden at Hamma.] Rev. Hortic. 91: 292-294. *Fig. 92-93.* June, 1919.—The experimental garden established in 1832, was later (1867) taken over by the Algerian Company with the three-fold view of making it a public park, a garden for scientific investigations, and a center for the introduction of exotics as well as for the propagation and distribution of indigenous plants. In 1913 the government again took possession of it and has declared its intention of maintaining it for these same purposes. Much has already been done in the way of removing the Garden from the state of neglect into which it had fallen, and it is confidently expected that it will soon assume high rank as a laboratory for various phytological investigations.—*E. J. Kraus.*

1796. Blot, F. **Corbeilles de chrysanthèmes.** [Beds of chrysanthemums.] Rev. Hortic. 91: 355-356. *Fig. 109.* Oct., 1919.—The chrysanthemum is preëminent among autumn flowers. Many types, forms and colors are available. Cutting prepared during the early part of March or the early part of April, or even the end of April for the varieties to be used as borders, should be cut back several times during the summer in order to secure stocky, bushy plants. Some of the single flowered varieties are especially desirable as budding plants. A classified list of more than forty varieties is given.—*E. J. Kraus.*

1797. Bois, D. **La rose "Los Angeles."** [The rose, "Los Angeles."] Rev. Hortic. 91: 296. *1 pl. (colored).* June, 1919.—This rose, exhibited by Howard and Smith of Los Angeles, California, at Bagatelle, where it received a gold medal, is the result of a cross between Lyon Rose (Pernetiana) and Madame Segond-Weber (hybrid tea). It much resembles the former in bud, flower, and color, and is said to be more hardy, more floriferous and less subject to fungous diseases.—*E. J. Kraus.*

1798. Bontrager, W. E. **What shade and ornamental trees shall we plant?** Monthly Bull. Ohio Agric. Exp. Sta. 5: 35-41. *5 pl.* 1920.—A discussion is given of the relative merits of types most suitable for lawn and shade planting, including those which will survive unfav-

orable conditions occasioned by gas, smoke and restricted areas, also with reference to distinctly ornamenental species. Directions for the care and planting of young trees are outlined.—*R. C. Thomas.*

1799. Brodie, Ian. Seedling daffodils selected to grow on at Brodie Castle. Jour. Roy. Hortic. Soc. 45: 113–155. 1919.

1800. Brooks, A. J. Work in the gardens and observations on plants. Report on the Agricultural Department, St. Lucia, 1917–18: 1–5. [Imp. Dept. Agric., Barbados.] 1918.—Contains a list of economic and ornamental plants introduced. Notes are also given on several plants which are under trial. These include: hybrid hibisci, dracaenas, ixoras, bougainvilleas, *Swietenia mahogani*, *S. macrophylla*, *Carum copticum* and *Hyoscyamus muticus*.—*J. S. Dash.*

1801. Burnham, Stewart H. Commercial fern gathering. Amer. Fern Jour. 9: 88–93. 1919.—The author gives accounts of the commercial gathering of ferns, especially the shield fern, in Vermont. It seems that the ferns are bearing up under the strain of annual pickings, but it is hoped that some one with the opportunity will make accurate observations of the real effect of commercial picking.—*F. C. Anderson.*

1802. Clute, Willard N. An unknown honeysuckle. Amer. Bot. 26: 17. *Fig. 1.* 1920.—The plant sent out by the Foreign Seed and Plant Introduction Division of the U. S. Bureau of Plant Industry as No. 39697 from Nanking, China, collected by Joseph Baillie has flowered at Joliet, but the Division was unable to supply the name. [This plant has since been named *Lonicera Maackii* var. *podocarpa* by Dr. C. S. Sargent.]—*W. N. Clute.*

1803. Clute, Willard N. [Editorial.] Amer. Bot. 26: 34. 1920.—Attention called to variations of commercial importance or of unusual interest in the writer's grounds. Red-leaved peaches, red forms of *Rubdeckia hirta*, a *Podophyllum* with multiple fruits, single-leaved locust, and various giant races mentioned.—*W. N. Clute.*

1804. Constantin, L. L'Epiphora de Pobeguin de Finet (Epiphora Pobeguini). Rev. Hortic. 91: 398–399. *1 pl. (colored).* Dec., 1919.—This species is an introduction from the Nenkan plateau, French Guinea. The plants are small, entirely epiphytic, and should be kept at a temperature of 18° to 22°C. throughout the year. In its native habitat it blooms in February or March, but certain plants which were brought into the greenhouses showed a progressive modification of the time of flowering as follows: June 14, 1910; September 31, 1912, and October 15, 1913.—*E. J. Kraus.*

1805. Crawford, Mrs. Wm. My experience with the peony. Flower Grower 7: 24–25. 1920.—Observations on the cultivation and propagation of the peony.—*W. N. Clute.*

1806. Cummings, Alex., Jr. Hardy roses for the garden. Gard. Chron. Amer. 24: 135. 1920.—Methods of cultivating and pruning described. [See also next following Entry, 1807.]—*W. N. Clute.*

1807. Cummings, Alex., Jr. Hardy roses for the garden. Gard. Chron. Amer. 24: 94–96. 1920.—Garden roses considered as tea roses, hybrid teas, dwarf polyantha or baby ramblers, and pernetiana. A list of 14 new or comparatively new roses is given. [See also next preceding Entry, 1806.]—*W. N. Clute.*

1808. Cushman, L. B. Aegopodia podagraria variegata. Amer. Bot. 26: 13–14. 1920.—This plant identified as a familiar form with variegated leaves in old gardens.—*W. N. Clute.*

1809. Daveau, J. Ficus Saussureana et F. eriobotryoides Kunth et Bouché. Rev. Hortic. 91: 389. Dec., 1919.—In 1840 A. P. de Candolle described *Ficus Saussureana*, basing his description upon a specimen then known as a *Galactodendron*, growing in the greenhouses of

Th. de Saussure, at Geneva. Later, in 1846, Kunth and Bouché described *Ficus eriobotryoides*. The descriptions of these two species are almost identical except that in the former species the petiole is said to be hirsute, in the latter, glabrous. In the Botanic Garden at Montpellier is a tree, also listed under the name of *Galactodendron*, but in reality it is a *Ficus* and corresponds to the descriptions of both of the foregoing species, since the petioles are at first hirsute, but gradually become more nearly glabrous, and finally completely so the second year. It is practically certain that the two species are synonymous and therefore the name assigned by de Candolle should be retained. The tree is a beautiful one and should find a place among collections of exotics.—*E. J. Kraus.*

1810. Denis, F. **Quelques iris nouveaux.** [Some new irises.] Rev. Hortic. 91: 362. Oct., 1919.—A number of new varieties have been obtained during the past several years by hybridizing various species or by crossing forms which in themselves are hybrids. The variety John Wister is a valuable hybrid between *I. aurea* and *I. fulvala.* The latter, itself a hybrid between *I. fulva* and *I. hexagona Lamancei*, is intermediate in flower color and is self fertile. Hybrids somewhat lacking in vigor have been obtained between *I. Ciengialti* and *I. tectorum*, and between *I. Edina* and *I. tectorum.* The floral characters of the seedlings are intermediate for the most part. It is possible, also, to secure plants by hybridizing *I. tectorum* and *I. pallida dalmatica*, but no seeds are secured when the former is crossed with any variety from the groups *germanica neglecta, amoena.* The name *Iris filifolia* is applied to two distinct forms. The one commonly listed in floral catalogues is really an early flowering vigorous form of *Xiphium.* The other is the true *I. filifolia Boissier*, and it is found but rarely in various collections. The two species *I. Xiphium* and *I. filifolia* have been successfully hybridized—the resultant seedlings are intermediate in the color of the flower and the length of the tube.—*E. J. Kraus.*

1811. Hirscht, Karl. **Epiphytische Kakteen im Zimmergarten** [Epiphytic cacti in window gardens.] Monatsschr. Kakteenkunde 29: 74–80. 1919.—A popular account is given of species suitable for window gardens and hints as to successful culture.—*A. S. Hitchcock.*

1812. Jackson, T. P. **Plant importations.** Report on the Agricultural Department, Antigua, 1917–18, 4–5. [Imp. Dept. Agric., Barbados, 1919.]—Interesting notes are given on trials with certain new plants at the Botanic Station, Antigua, notably the "Guada" bean (gourd), *Trichosanthes anguina*, useful as a vegetable, and several grasses.—*J. S. Dash.*

1813. Jacob, J. **Freesias and Lachenalias.** Jour. Roy. Hortic. Soc. 45: 29–38. 1919.—These two plants were introduced into England from South Africa more than a hundred years ago and have recently been the object of renewed interest. Discussions of cultural methods are given and a list of varieties to which awards have been given by the Royal Horticultural Society.—*J. K. Shaw.*

1814. Jarmillo, P. J., and F. J. Chittenden. **On double stocks.** Jour. Roy. Hortic. Soc. 44: 74–82. *Fig. 22, 23.* 1919.—Selecting the most vigorous seedlings gave a higher percentage of double stocks than were secured from selection of medium and weak seedlings. Such selection appears to have practical value in securing a high percentage of double flowering plants.—*J. K. Shaw.*

1815. Jahandiez, E. **Mesembryanthemum a formes étranges.** [Mesembryanthemums of unusual form.] Rev. Hortic. 91: 372–374. *Fig. 112–113.* Nov., 1919.—Many species of this genus are especially unsuitable for growing in the open air in the more southern regions, where they are able to accommodate themselves to conditions of dryness, poor soil and salt air. It is possible to make excellent borders by using several species of varying height and flower color which ranges from violet to red, orange-red, and yellow. *M. acinaciforme* L., having broad violet flowers and *M. edule* L. which has large white or yellow flowers have become naturalized in southern France. *M. Bolusii* Hook. fil. from South Africa is one of the more striking species because of the close resemblance of its leaves to pebbles. Two related species *M. simuland* Marloth and *M. testiculatum* Jacq. which has white, glaucous leaves, are equally remarkable. *M. pseudotruncatellum* Berger, has its leaves reduced to flattened balls, while those of *M. concinnum* N. E. Brown, from Damaraland, are covered with small, white tubercles. The leaves of *M. tigrinum* are marked with white and are bordered with long hairs, whereas those of *M. felinum* Haw are denticulate. *M. digitatum* Ait. resembles a very large finger, and *M. Barklyi* N. E. Brown is eaten by animals because of the large leaves which are filled with a watery sap. There are three native European species, *M. angulatum* Thunb., *M. cordifolium* L. and *M. crystallinum*, the leaves from all of which may be used in the same way as is spinach.—*E. J. Kraus.*

1816. LANTES, ADELAIDE. **El alamo. [The pipal tree.]** Revist. Agric. Com. y Trab. 2: 612-613. *3 fig.* 1919.—It is pointed out that the pipal tree (*Ficus religiosa*) is undesirable for common planting in parks and along roads. Its roots injure cement work, the leaves fall continuously, the fruits fall in quantities, and the trees are favorite retreats of birds. Other trees are mentioned which are preferred.—*F. M. Blodgett.*

1817. MANRIN, G. **Support rotatif pour plantes d'appartement. [A rotary support for house plants.]** Rev. Hortic. 91: 331. *Fig. 102-103.* August, 1919.—A brief description and working drawings are given.—*E. J. Kraus.*

1818. MARIE-VICTORIN, FR. DES E. C. **Le "Micrampelis lobata." "Une Plante lance-torpilles." [Micrampelis lobata (Michx.) Greene.]** Naturaliste Canadien 46: 172-174. Feb., 1920.—A graphic popular sketch of an interesting cucurbitaceous plant used for veranda decoration, found growing native in fertile soil along water courses in southern Canada.—*A. H. MacKay.*

1819. MEYER, RUD. **Kulturregeln aus alter Zeit. [Culture rules of ancient times.]** Monatsschr. Kakteenkunde 29: 37-41. 1919.—In this chapter, which is a continuation from the volume for 1917, page 120, are discussed the choice, packing, and shipping of cactus specimens.—*A. S. Hitchcock.*

1820. MILLARD, ALBERT. **Natural effects in landscape work.** Gard. Chron. Amer. 24: 103. 1920.—Numerous plants named for use in the natural style of planting.—*W. N. Clute.*

1821. MOREL, F. **Le clematis montana et ses dérivés. [Clematis montana and its derivatives.]** Rev. Hortic. 91: 358-360. *Fig. 110.* 1919.—The hybrid offspring of *C. montana grandiflora* and *C. montana rubens* were intermediate in color of flower, and generally more vigorous than the red form. When the former species was crossed with *C. repens*, individuals were secured which both preceded and followed the parent varieties in period of flowering, and possessed flowers which were larger than those of *repens* and of greater consistency than those of *grandiflora*. By careful selection, it was possible to interhybridize some of the latest appearing flowers of *C. repens* with some of the earliest flowers produced during the second period of bloom of *C. montana rubens*. From these crosses plants of unusual vigor and substance, bearing flowers of large size, good form and of various shades of rose or with red pencilings, were secured. A succession of blossoms may be secured by growing the following varieties: April—*C. montana rubens*; May—*C. montana grandiflora*, then *C. repente-montana rubens* with variously colored flowers, and finally *C. repente-montana grandiflora* with white flowers; June—*C. repens*; July and August—*C. montana rubens* and *C. repente-montana rubens* commence at this time a second period of flowering which may be prolonged into September and October. It may be possible to select a free-flowering, everblooming race from among the individuals disposed to flower more than a single time during the year.—*E. J. Kraus.*

1822. MOTTET, S. **Les leucanthèmes. [The leucanthemums.]** Rev. Hortic. 91: 312-313. *1 pl.* July, 1919.—It seems most probable that *L. lacustre Brot.* and *L. maximum D. C.* have contributed principally in the development of the large-flowered marguerites or Shasta daisies, though it is probable that several other species have been concerned also. Although these large flowered forms were first introduced into Europe from America in 1902 or 1903, little is known definitely concerning their origin. The "Shasta Daisy" of Luther Burbank is thought to have been derived by a vigorous selection from the seedlings of *Chrysanthemum leucanthemum* crossed with an American species; this progeny in turn having been crossed with *C. nipponicum*, a Japanese species. Whatever may have been the origin of the various large flowered forms, it is certain that great variation now eixsts, and they are among the most generally useful decorative plants.—*E. J. Kraus.*

1823. Mottet, S. **Paederia tomentosa.** Rev. Hortic. 91: 298–300. *Fig. 95.* June, 1919.—This species was first introduced into Europe from China in 1806, and again in 1907, through seeds collected by E. H. Wilson for the Arnold Arboretum. It is recommended as a suitable covering for walls and trellises. A description and synonymy are given.—*E. K. Kraus.*

1824. Mottet, S. **Un rhododendron à fleurs jaunes. (R. campylocarpum.)** [A yellow flowered rhododendron.] Rev. Hortic. 91: 328–329. *1 pl.* August, 1919.—This species was collected by Hooker in Himalaya and introduced into England in 1856. Though it has been overlooked for a long time, there is little doubt that it is really a desirable, hardy form with persistent foliage and clear yellow flowers. It should serve, also, as valuable material for crossing with other forms. A detailed description of the species is given. Another yellow flowered species *Rhododendron lutescens* Franch, is mentioned as having been recently introduced from China by Wilson.—*E. J. Kraus.*

1825. Mottet, S. **Nouveaux oeillets remontants grandiflores.** [New large flowers, ever-blooming carnations.] Rev. Hortic. 91: 360–361. *1 pl. (colored).* Oct., 1919.—Attention is directed to seven varieties of carnations which represent the progress made in the last several years in breeding for flowers of large size and special colors. Although perpetual blooming carnations have been known since about 1845, it was not until near the end of the last century that the large flowered forms appeared, several varieties having been exhibited in 1900. New varieties have been introduced with considerable rapidity since that time. Most of these varieties may be placed in one of five or six type classes, each of which possesses distinctive characters of stem, foliage and flower. Intercrossing between the classes has been frequent, however, so that as a result several of the various types may be represented among any particular lot of seedlings.—*E. J. Kraus.*

1826. Mottet, S. **Un nouveau chamaecyparis (Ch. formosensis).** [A new chamaecyparis.] Rev. Hortic. 91: 342–344. *Fig. 105.* Sept., 1919.—The two Japanese species, *Ch. obtusa* Sieb. and Zucc. and *Ch. pisifera* Sieb. and Zucc., together with *Ch. sphaeroidea* Spach have produced many varieties commonly known under the name *Retinospora.* Two other forms are known from North America, namely *Ch. nutkaensis* Spach. and *Ch. Lawsonia*, Parlat. Each of these has given rise to several varieties. To this list of species should be added *Ch.* formosensis Matsum. which, on the island of Formosa, is said to attain a great size, one specimen having measured 22 meters in circumference at the base. The species was described by Matsumura in the Botanical Magazine for 1901. Seeds were introduced into England in 1911. It is highly recommended as a decorative tree, since the branches are as light and graceful as certain ferns, and they assume an attractive, bronze tint at the beginning of winter. Young trees are not entirely hardy in the vicinity of Verrières, though this defect may be overcome when the trees have grown older. The species may be propagated by grafting or from seeds.—*E. J. Kraus.*

1827. Mottet, S. **Digitale hybride de Lutz.** [The Lutz digitalis hybrid.] Rev. Hortic. 91: 396–397. Dec., 1919.—From seeds of an apparently spontaneous hybrid between *Digitalis purpurea* and *D. lutea*, the following types of plants were obtained: (1) Flowers clear chamois, spotted, foliage very downy. (2) Flowers purple, stems brown, and foliage smooth. (3) Flowers yellow-white, spotted.—Seeds were secured from plants of the first two types. From the first, five plants were obtained, three of which produced purple flowers and two chamois flowers. From the second, 37 plants were obtained, but only five of them were sufficiently sturdy to bloom; all bore chamois, spotted flowers. One of the plants of the latter type was then chosen for seed production, but was not isolated, though the plants which produced purple flowers were destroyed. From this plant 300 individuals were secured. Of these, 13 produced purple flowers, the remainder yellow flowers. A few of the plants were weak. The variety probably will prove to be of value as an ornamental. Another hybrid between *Digitalis purpurea* and *D. ambigua* is more or less sterile and can not be propagated with sufficient ease to make it of horticultural importance.—*E. J. Kraus.*

1828. PEREZ, G. V. **Vitalité des racines de Bougainvillea.** **[Vitality of the roots of Bougainvilleas.]** Rev. Hortic. 91: 380. Nov., 1919.—Cuttings of this plant, put out in 1916, although they have not produced roots, are still alive and have not decayed. Small pieces of roots which were split lengthwise are also well preserved. Ordinary cuttings of conifers are preserved an equally long time in the open air, those of *Juniperus Cedrus* may not start roots for more than a year after they are planted out.—*E. J. Kraus.*

1829. PINELLE, A. **Robinia Kelseyi Hort.** Rev. Hortic. 91: 339. *Fig. 104.* Sept., 1919. —It is still uncertain whether this form is a true species of a hybrid between *R. hispida* and *R. pseudoacacia.* It is a shrub or small tree and bears a superficial resemblance to both forms. The flowers are pink and appear earlier in the season than those of either of the species mentioned. It is said to have arisen spontaneously in the nursery of a MR. KELSEY, of Boston, from seeds secured in the southern Alleghany Mountains. It is readily propagated by grafting on *R. pseudoacacia*, but it is unknown whether it will reproduce true to type from seed.—*E. J. Kraus.*

1830. POLE-EVANS, I. B. **Our aloes. Their history, distribution and cultivation.** Jour. Bot. Soc. South Africa 5: 11–16. *Pl. 2–3.* 1919.—Aloe rockeries and gardens are becoming fashionable in South Africa as they did in Holland and Britain at the beginning and in the middle of the eighteenth century. There are many aloes of reputed South African origin which have been under cultivation in Holland and England for at least one or two centuries, but which today are unknown in South Africa. The first to be cultivated in European gardens was *A. succotrina* Lam.—*E. P. Phillips.*

1831. QUEHL, L. **Auswahl der Arten zu einer Kleinen Kakteensammlung. [Choice of species for a small cactus collection.]** Monatsschr. Kakteenkunde 29: 54–55. 1919.

1832. RAGIONIERI, ATTILIO. **Un bel problema per i biologi: Sulla comparsa dell'odore nel fiore delle "Rosseline di Firenze" (Ranunculus asiaticus var.). [A good problem for biologists: on the appearance of odor in the flowers of the Florentine "rosseline" (Ranunculus asiaticus).]** Bull. R. Soc. Toscana Orticult. 44: 87–94. 1919.—He reports an experience with *Ranunculus asiaticus*, that had a marked rose odor not characteristic of the variety. Seedlings resulting from selfing the flowers of this plant showed this odor to a reduced extent. The strain had been grown on the same land since 1844 producing both vegetatively and as seedlings. He thinks that there was no chance for the odor to have been introduced by crossing with another variety, and that it is the reappearance of an ancestral character.—*W. H. Chandler.*

1833. RICCOBONO, VINCENZO. **La prima fioritura in Europa del Pilocereus Dautwitzii Fr. A. Haage. [The first flowering in Europe of Pilocereus dautwitzii Fr. A. Haage.]** Bull. R. Soc. Toscana Orticultura 44: 94–96. 1919.—Description of *Pilocereus dautwitzii*, introduced into Italy from northern Peru. Observations on its behavior.—*W. H. Chandler.*

1834. RINGELMANN, M. **Murs garnis de Lierre. [Ivy-covered walls.]** Rev. Hortic. 91: 363. *Fig. 111.* Oct., 1919.—It is believed by many that climbing plants, especially English ivy, are destructive to the walls upon which they grow. As a matter of fact, if young plants of English ivy are originally planted about 1½ or 2 feet from the base of the wall, when they have grown and covered it the overlapping leaves will tend to shed water and also aid in keeping out the cold. The clinging rootlets, stem and branches of this vine aid in holding together the pieces of which the wall is constructed, rather than forcing them apart. Many other vines, however, which lose their leaves in winter, actually do tend to hold moisture against the wall that supports them.—*E. J. Kraus.*

1835. Sheward, T. **The dracenas.** Gard. Chron. Amer. 23: 61. *1 fig.* 1920.

1836. Smith, Arthur. **Twelve most desirable shrubs for gardens.** Gard. Chron. Amer. 24: 141. 1920.

1837. Smith, Arthur. **A lesson on seed sowing and germination.** Gard. Chron. Amer. 24: 108–110. 1920.

1838. Sturtevant, Robert Swan. **The garden plus irises.** Gard. Chron. Amer. 24: 97–98. *Fig. 2.* 1920.—Mention of various named varieties for garden planting.—*W. N. Clute.*

1839. Van den Heede, A. **Une superbe plante annuelle.** [A superb annual plant.] Rev. Hortic. 91: 393. Dec., 1919.—*Salpiglossis sinuata* Ruiz and Pavon, also known as *S. straminea* Hooker, *S. atropurpurae* Graham, *S. picta* Sweet, *S. Barclayana* Sweet, *S. hybrida* Hort. and *S. variabilis* Hort., is a native of Chili and was introduced into Europe about 1830. Several other Chilian species, *S. fulva*, *S. integrifolis*, *S. intermedia*, and *S. linearis* were also introduced at about the same period, but these, together with *S. sinuata coccinea* and *S. straminea picta* have disappeared from cultivation, so that at the present time *S. sinuata* and its dwarf variety alone persist. The plants are readily grown out of doors and the flowers possess a wide range of harmonious colors.—*E. J. Kraus.*

1840. Von Oven, F. W. **Perpetuating our native flora.** Amer. Bot. 26: 24–27. 1910.—The great individual differences that exist in the botanical species are pointed out and the proposal made that the best of these should be selected and propagated. The writer is a nurseryman and will undertake to grow variations that may be called to his attention.—*W. N. Clute.*

1841. Vorwerk, W. **Beitrag zur Kultur der Asclepiadaceae-Gattungen Trichocaulon und Hoodia.** [Contribution to the culture of the asclepiad genera Trichocaulon and Hoodia.] Monatsschr. Kakteenkunde 29: 41. 1919.—This includes remarks upon the cultivation of *T. keetmanshopense* and *H. Currori.*—*A. S. Hitchcock.*

1842. Weingart, W. **Aussaat von Cereus formosus S.-D.** [Seed of Cereus formosus.] Monatsschr. Kakteenkunde 29: 105. 1919.—Seed of *C. formosus* obtained by Haage and Schmidt from Los Angeles gave four forms: *C. formosus monstruosus*, *C. variabilis* Pf. (*C. Pitahaya* DC.), *C. formosus*, *C. obtusus.*—*A. S. Hitchcock.*

1843. Whitten, James. **The public parks of Glasgow.** Jour. Roy. Hortic. Soc. 45: 39–55. 1919.

1844. Williams, W. L. **The beet sugar industry.** Jour. Dept. Agric. Victoria 17: 722–730. 1919. *Ibid.* 17: 15–24, 65–74. 1920.—Sugar beet growing in Victoria is discussed.—*J. J. Skinner.*

VEGETABLE CULTURE

1845. Anonymous. **Runner beans at Wisley, 1918.** Jour. Roy. Hortic. Soc. 44: 95–100. 1919.—Report is made on sixty varieties of *Phaseolus multiflorus*, giving recommendations of the judging committee and a classification and description of the varieties.—*J. K. Shaw.*

1846. Anonymous. **Climbing French beans, 1918.** Jour. Roy. Hortic. Soc. 44: 101–110. 1919.—A report on seventy-nine climbing varieties of *Phaseolus vulgaris* with recommendations of the Vegetable Committee concerning their value. A classification with description of varieties is given.—*J. K. Shaw.*

1847. ANONYMOUS. **Vegetable marrows at Wisley, 1918.** Jour. Roy. Hortic. Soc. 44: 114–116. 1919.—Tests of fifty-seven stocks of vegetable marrows, at Wisley, England, are reported, with the awards of the judges and brief discriptions of the different varieties.—*J. K. Shaw.*

1848. ANONYMOUS. **Leeks tried at Wisley, 1917–18.** Jour. Roy. Hortic. Soc. 44: 111–113. 1919.—Brief description of 31 varieties of leeks are given with brief notes on cultural method and the awards of the judging committee.—*J. K. Shaw.*

1849. ANONYMOUS. **Brussels sprouts at Wisley, 1918.** Jour. Roy. Hortic. Soc. 45: 125–127. 1919.—Brief descriptions of 64 stocks of Brussels sprouts and the awards of the Fruit and Vegetable Committee are given.—*J. K. Shaw.*

1850. ANONYMOUS. **Carrots at Wisley, 1918.** Jour. Roy. Hortic. Soc. 45: 128–130. 1919. —Report is made of the trial of 61 stocks of carrots together with a classification, brief description and the awards of the Vegetable Committee.—*J. K. Shaw.*

1851. BLIN, H. **L'exploitation rationelle des cressonnières. [The rational utilization of cress-beds.]** Rev. Hortic. 91: 313–316. *Fig.* 99. July, 1919.—The growing of cress is a profitable industry in the vicinity of large cities. The number of beds which may be formed is directly dependent upon the flow of water available; 70 to 75 litres a minute will supply 240 square meters as a maximum. Each bed should not exceed 80 meters in length and should be so arranged that there is a slow but continuous flow of water through it, the amount of such flow being regulated by an adjustable dam. New plantings are established either by sowing the seeds or transplanting cuttings, which may be put out at any season, though if this is done in August or September a good stand for the more valuable winter harvest will be secured. Successive plantings will furnish a supply throughout the year. Decomposed stable manure is an excellent fertilizer. It should be carefully applied when new beds are established and further application should be made after each cutting. In winter it is advisable completely to submerge the plants to protect them from cold. Such inundation or spraying will aid in the controlling of insect pests. It is possible to harvest a crop from the beds within 3 months following the sowing of the seed, or within one month after transplanting the cuttings. During the rapid growing season the beds may be cut over every 15 to 20 days, and during the winter every six or seven weeks. The shoots should be from 15 to 20 cm. in length before being cut, and care should be used to avoid disturbing the roots. Though the beds would last for many seasons, better results are secured by renewing them each year. The shoots, after being cut, are tied into bunches weighing at least 275 grams each, and these are then packed into oval baskets holding from 15 to 20 dozen bunches. In order to prevent yellowing a space is left in the center of the basket. From an area of 100 square meters about 300 dozen bunches may be harvested, which would yield a gross return of 200 to 280 francs.—*E. J. Kraus.*

1852. FISHLOCK, W. C. **Sweet potatoes.** Report on the Agricultural Department, British Virgin Islands, 1918–19: 3–5. [Imp. Dept. Agric., Barbados. 1919.]—Results of experiments with 31 varieties are recorded, with descriptions of each variety. Bourbon heads the list over a period of 8 years, with a yield of 7600 pounds per acre.—*J. S. Dash.*

1853. LEVY, E. BRUCE. **Swede variety types and their perpetuations by pure seed.** New Zealand Jour. Agric. 19: 284–287. 1919.—A rough classification of Swede types (of turnips) has been drawn up. Three varieties, as listed by seedsmen, were tested and great variation was found. It is urged that more effort be exercised to select and breed pure strains.—*N. J. Giddings.*

1854. LIVVENTAAL, A. **The crop factory.** Sci. Amer. 122: 563, 582. *1 fig.* 1920.—An attempt to solve the problem—can gardening be made a standardized industry, independent of the elements? By the novel equipment pictured, heat, moisture, light and other conditions are made constant and labor is reduced to a minimum.—*Chas. H. Otis.*

1855. Meunissier, A. **De quelques idées sur la selection des légumes.** [Some ideas on the selection of vegetables.] Rev. Hortic. 91: 300–303. June, 1919.—This is a discussion of the ideas of variation in general with specific emphasis on the necessity for recognizing pure lines, as defined by Johannsen, as the real basis for selection in crop improvement.—*E. J. Kraus.*

1856. Rogers, Stanley S. **Methods for marketing vegetables in California.** California Agric. Exp. Sta. Circ. 217: 1–19. 1920.—A survey of the probable causes for success or failure in the production and marketing of vegetables in California.—*A. R. C. Haas.*

1857. Stokes, Fred. **The food value of vegetables.** Jour. Roy. Hortic. Soc. 44: 21–30. 1919.—The author has devised a formula for calculating the "economic value" of a crop. This formula applies, however, only when the produce is not sold for profit.—

$$\frac{\text{Caloric value} \times \text{yield in pounds per rod}}{\text{Cost of crop in shillings} \times \text{Number of weeks the ground is occupied}} = \text{Economic value}$$

According to the formula the economic value of potatoes is 69.5, carrots 31.6, kidney beans (dry) 28, peas (shelled) 18, parsnips 15.5, onions 4.3, and cabbage 3.—The various vegetables may not only be valuable because of the amount of proteid, carbohydrate, fat and salts they contain, but also because they yield bulk and furnish the indispensable vitamines. Especially valuable are the green vegetables like spinach, cabbage, celery, etc., which give the body the necessary salts and vitamines and also add the necessary bulk to the diet. The bulbs, especially the onion and leek "are remarkable for their beneficial action upon inflamed mucus membrane and for their germicidal powers." The onion is valuable for its salts and essential oil and no doubt contains "a potent vitamine as well." Roots are of value chiefly because of their salts and carbohydrates and the legumes because of their richness in protein and carbohydrates.—*H. A. Jones.*

1858. Sutton, Arthur W. **How amateurs may secure three successive crops of vegetables in twelve months without the aid of glass houses or of heat.** Jour. Roy. Hortic. Soc. 44: 13–20. 1919.

1859. Woolsey, C. **Sweet potato culture in Arkansas.** Arkansas Agric. Ext. Circ. 90. *20 p., 10 fig.* 1920.—A popular discussion on bedding, cultivating, digging, grading, storing and marketing the sweet potato. Directions are given for seed selection and the common varieties are briefly described.—*John A. Elliott.*

1860. Woolsey, C. **The home vegetable garden in Arkansas.** Arkansas Agric. Ext. Circ. 89. *32 p., 9 fig.* 1920.—A popular presentation of gardening methods suited to the conditions of the state, giving dates of planting, culture and rotation of garden crops.—*John A. Elliott.*

1861. Zimmerley, H. H. **Greenhouse tomato growing in Virginia.** Virginia Truck Experiment Station Bull. 26. *23 p., 2 fig.* 1919.—Methods of growing tomatoes in the greenhouses under Virginia conditions are given. The seed for the winter crop is sown in August and the plants shifted to the beds in September. The seed for the spring crop is sown in November and the plants shifted in December. Discussions of varieties, soil treatment and the control of diseases are given.—*T. C. Johnson.*

HORTICULTURE PRODUCTS

1862. Aguila, Isidoro. **Notas sobre la elaboracion de aceite de oliva.** [Notes on the preparation of olive oil.] La Informacion Agric. [Madrid] 9: 318–322. 1919.—Lists defects occurring in olive oil and gives the causes and approved manner of avoiding them. Proper methods of obtaining high grade oils are discussed.—*John A. Stevenson.*

1863. Anonymous. **A new vegetable ivory.** Sci. Amer. Monthly 1: 346. 1920.—Descriptive of a substance produced from the kernel of an edible fruit growing upon the palm, *Borassus ethiapicum.*—*Chas. H. Otis.*

1864. Bancroft, Wilder D. [Rev. of: Peters, Charles A. **The preparation of substances important in agriculture.** 3rd ed. *19 x 14 cm.* *vii + 81 p.* John Wiley and Sons, Inc.: New York, 1919. $.80.] Jour. Phys. Chem. 23: 444. 1919.—See Bot. Absts. 5, Entry 1100.

1865. Bredemann, G., and Chr. Schätzlein. **Über Herstellung und Zusammensetzung kleinasiatischer Traubensaftkonserven.** [Preparation and composition of grape-juice preserves from Asia Minor. Zeitschr. Untersuch. Nahrungs- u. Genussmittel 38: 16-24. 1919.

1866. Carles, P. **La prune d'ente et les pruneaux d'Agen: Explication scientifique de leur préparation et des moyens de les conserver temporairement pour l'Europe et de façon indéfinite pour l'exportation mondiale.** [A scientific account of methods used in preparing "prunes of Agen" for foreign and domestic consumption.] Mém. Soc. Sci. Phys. Nat. Bordeaux VII, 2: 219-232. 1918.—The preparation of the fruit consists of two phases, (1) a chemical phase during which it is subjected to temperatures of from 40°-50°C. to facilitate the action of a soluble ferment (oxydase) and (2) a physical phase during which the temperatures are increased to 75°-80°C. to produce desiccation. The author discusses various methods of packing and sterilisation.—*I. W. Bailey.*

1867. Crevost, C., and C. Lemarie. **Plantes et produits filamenteux et textiles de l'Indochine.** [Fiber- and textile-producing plants of Indo-China.] Bull. Econ. Indochine 22: 813-837. *Pl. 2.* 1919.—See Bot. Absts. 5, Entry 1122.

1868. Davis, R. A. **Fruit and fruit products in South Africa. III. The canning, drying and preserving business.** South African Jour. Indust. 2: 1138-1148. 1919.

1869. Fernandez, O., F. Bustamenta. **Estudio analitico de los aceites de oliva aspanoles.** [Analytical study of the Spanish olive oils.] Rev. R. Acad. Cienc. Exactas, Fisecas y Nat. [Madrid] 17: 281-286. 1919.

1870. Hartmann, Wilhelm. **Über Gärversuche mit Zuckerrüben.** [Fermentation experiments with sugar beets.] Zeitschr. Untersuch. Nahrungs- u. Genussmittel 28: 287-290. 1919.

1871. Laborde, J. **Recherches sur le vieillissement du vin.** [Aging of wine.] Mém. Soc. Sci. Phys. Nat. Bordeaux VII, 2: 37-75. *Tables 1-15.* 1918.

1872. Mach, F., and M. Fischeer. **Die Zusammensetzung der Moste des Jahres 1918 in Baden.** [Musts of 1918 in Baden.] Zeitschr. Untersuch. Nahrungs- u. Genussmittel 38: 93 99. 1919.

1873. Maxwell, Harold L., and Nicholas Knight. **The oil in cherry pits.** Proc. Iowa Acad. Sci. 25: 451-455. 1920.—Oil was extracted from seeds of "the common cherry *Prunus erratus*" [doubtless *P. cerasus*]. It was found to be essentially the same as almond oil, having a saponification equivalent of 276.8.—*H. S. Conard.*

1874. Roettgen, Theodore. **Zur Bestimmung der Milchsäure im Weine.** [Determination of lactic acid in wines.] Zeitschr. Untersuch. Nahrungs- u. Genussmittel 38: 99-100. 1919.

1875. Rothéa, and De Bon, F. **Essay industriel de fabrication d'huile d'amandes d'abricots. Resultats analytiques des matières premières et des products obtenus.** [An industrial experiment in regard to the manufacture of oil from apricot seeds. Analytical results of the original material and of the products obtained.] Bull. Sci. Pharm. 26: 505-514. *1 fig.* 1919. —As the title indicates, a description of apricot kernels, of the process of obtaining the oil by expression together with the chemical and physical constants of the oil are given.—*H. Engelhardt.*

1876. STERN, J. **Moste des Jahres 1918 aus den Weinbeugebeiten der Nahe, des Glans, des Rheintales unterhalb des Rheingaues, des Rheingaues, des Rheins, Mains und der Lahn.]** [Musts of 1918 of the Rhine Valley, etc.] Zeitschr. Untersuch. Nahrungs- u. Genussmittel 38: 91-93. 1919.

1877. TEVIS, MAY. **Cutting the cocoanut cake.** Sci. Amer. Monthly 1: 404-407. *4 fig. and frontispiece.* 1920.—Concerns the cocoanut tree, *Cocos nucifera*, its growth, products and their preparation.—*Chas. H. Otis.*

MORPHOLOGY, ANATOMY AND HISTOLOGY OF VASCULAR PLANTS

E. W. SINNOTT, *Editor*

1878. BANCROFT, WILDER D. [Rev. of: JAEGER, F. M. **Lectures on the principles of symmetry.** *27x16 cm., xii+333 p.* Elzevir Publishing Co.: Amsterdam, 1917.] Jour. Phys. Chem. 23: 516. 1919.—See Bot. Absts. 5, Entry 1451.

1879. BETTS, M. WINIFRED. **Notes on the autoecology of certain plants of the Peridotit Belt, Nelson: Part I. Structure of some plants (No. 2).** Trans. and Proc. New Zealand Inst 51: 136-156. *27 fig.* 1919.

1880. BREWSTER, A. A. **Aerating roots or pneumatophores of mangroves (Avicennia).** Australian Nat. 4: 136. 1920.—These plants have an aerating system strongly suggesting that of the cypress of the southern United States.—*T. G. Frye.*

1881. BREWSTER, A. A. **Germination of choko seed.** Australian Nat. 4: 121. 1920.

1882. BREWSTER, A. A. **Leaf of the grasstree (Xanthorrhoea).** Australian Nat. 4: 135. 1920.—Paper deals with the leaf structure of this xerophyte. The most striking features are the abundance of sclerenchyma, and the occurrence of numerous crystals in the cells of the epidermis.—*T. C. Frye.*

1883. BUCHHOLZ, JOHN T. **Embryo development and polyembryony in relation to the phylogeny of conifers.** Amer. Jour. Bot. 7: 125-145. *89 fig.* 1920.—The author has here summarized all published work on the proembryo and early embryo of conifers, in an endeavor to throw light on the phylogeny of this group by a comparative study of their embryogeny and in particular of the manner in which polyembryony occurs within them. Cleavage polyembryony—the separation of the zygote into a number of smaller units which compete with each other—is distinguished from simple polyembryony, which results from the fertilization of several eggs. The phylogenetic values of these two types of polyembryony and of various other embryological characters are discussed, and the affinities suggested by embryogeny among the 22 genera of conifers studied are represented by a diagram. The occurrence of cleavage polyembryony, together with the presence of an apical cell, of rosette embryos and rosette cells, and the direct organization of embryo initials from the free nuclei of the proembryo are regarded as primitive features. On the other hand, a return to simple polyembryony, the presence of a proembryo that fills the entire egg with cells, an archegonial complex and an embryo cap, together with the organization of embryo initials after walls form in the embryo, are regarded as specialized features characteristic of more recent types.—*E. W. Sinnott.*

1884. BUSCALIONI, L., AND G. MUSCATELLO. **Studio anatomo-biologico sul Gen. Saurauia Willd.** [Anatomical-biological studies on the genus Saurauia.] Malpighia 28: 331-370. *Pl. 5-10.* 1918.—This is the concluding part of a detailed anatomical study, the publication of which was begun in earlier numbers of the journal cited.—*L. W. Riddle.*

1885. Chamberlain, Charles J. **The living cycads and the phylogeny of seed plants.** Amer. Jour. Bot. 7: 146–153. *Pl. 6.* 1920.—The position of the living cycads in the evolution of the seed plants is considered. A general resemblance is noted between the living cycads and the Bennettitales and Cycadofilicales. The last named group is undoubtedly the most primitive. The living cycads are so different from the Bennettitales that there is little likelihood that the former have been derived from the latter. The origin of the living cycads is quite unknown.—Living cycads are also evidently not ancestral to any of the other great groups of seed plants, since they differ so radically from Cordaitales, Ginkgoales, Coniferales, Gnetales and Angiosperms. They are evidently a terminal group on the road to extinction. The author brings forward evidence that it is the Coniferales and the Gnetales, rather than the cycad-like plants, to which we should look for ancestors of the Angiosperms.—*E. W. Sinnott.*

1886. Collins, Marjorie I. **On the leaf-anatomy of Scaevola crassifolia, with special reference to the epidermal secretion.** Proc. Linnean Soc. New South Wales 43: 247–259. *Pl. 27–28, 6 fig.* 1918.—This plant, one of the Goodeniaceae, a xerophyte, with special adaption for sand dune existence (where it will survive burial by elongation and the production of adventitious roots) was found to be characterized by the development of peltate glandular hairs which secrete yellow resin in great quantity. This activity was at a maximum in buds and young leaves and decreased in older leaves, where the resin dried, producing a lacquered appearance on the leaf surface. Mature leaves appeared succulent, the glands shrunken, but active in the region of the leaf base; the resin serving there to protect axillary buds. Other xerophilous adaptations noted were the secondary increase in the size of epidermal cells, massive development of palisade tissue and production of special water storage cells.—*Eloise Gerry.*

1887. Frucht, Otto. **Zur Entstehung des Harfenwuchses der Nadelholzer.** [On the formation of "harp-growth" in conifers.] Naturw. Zeitschr. Forst. u. Landw. 17: 137–139. 1919. —See Bot. Absts. 5, Entry 1326.

1888. Fletcher, J. J., and C. T. Musson. **On certain shoot-bearing tumors of Eucalypts and Angophoras, and their modifying influence on the growth habit of the plants.** Proc. Linnean Soc. New South Wales 43: 191–233. *Pl. 4–26.* 1918.—The nodules and tumors produced in the axils of the cotyledons and early leaves of Eucalypts and Angophoras are illustrated and discussed with reference to their occurrence, external characteristics and development. These growths are also noted in ten species of other genera. The fully developed tumors, though subject to much variation, are said to be generally characterized by the following stages: (1) Axillary shoot-bearing stem nodules; (2) Composite shoot-bearing, stem-encircling tumors; (3) Composite, stem-encircling, shoot-bearing, root-incorporating (but not root-emitting) tumors. Seedlings of the non-Mallee or tree forms of Eucalypts, where tumors usually persist for a limited period only, and do not seriously interfere with growth were especially studied. Six species, apparently exempt from tumors, were found. The Mallee or shrubby forms of Eucalypts (where the tumors incorporate the water-storing roots, persist throughout the life of the plant and appear to cause stunting) and the Angophoras, were also examined. The tumors are considered attributable to parasitic soil organisms, which produce proliferation of the cambium, and not to insects. Related work in Australia and the United States is discussed.—*Eloise Gerry.*

1889. Fyson, P. F. **Note on the oecology of Spinifex squarrosus L.** Jour. Indian Bot. 1: 19–24. *3 fig.* 1919.—This plant and other strand-formation species are not halophytes, but rather xerophytic psammophytes; they depend for their water supply on rain water and dew retained by the sand. Further, the air blown over these plants from the sea is always damp. —*A. J. Eames.*

1890. Griffin, Gertrude J. **Bordered pits in Douglas fir: a study of the position of the torus in mountain and lowland specimens in relation to creosote penetration.** Jour. Forestry 17: 813–822. *1 fig.* 1919.—See Bot. Absts. 5, Entry 1334.

1891. Hamilton, A. A. **Root fasciation in cycads.** Australian Nat. 4: 134. 1920.—All cycadean genera produce root nodules primarily caused by infection with *Bacillus radicicola*.—*T. C. Frye.*

1892. Holloway, J. E. **Studies in the New Zealand species of the genus Lycopodium: Part III. The plasticity of the species.** Trans. and Proc. New Zealand Inst. 51: 161–261. *Pl. 9–14, 16 fig.* 1919.—Eleven species of *Lycopodium* occur in New Zealand. A comparative study of these, character by character, shows that there is a great range of variability in the plants, but at the same time a distinct interdependence of characters. The author concludes with a discussion of the relationships and phylogeny of the species of *Lycopodium* in the light of his observations.—*L. W. Riddle.*

1893. Jivanna Rao, P. S. **The formation of leaf-bladders in Eichornia speciosa Kunth (water hyacinth).** Jour. Indian Bot. 1: 219–225. *5 fig.* 1920.—Bladder formation near the base of the petiole is the result of high water content in the plant. All gradations are found from well developed bladders on plants growing in an abundant supply of fresh water, to bladderless leaves on plants growing in pools that are drying up or in mud. An account of the structure of the bladder is given.—*Winfield Dudgeon.*

1894. Kashyap, S. R. **Abnormal number of needles in the spurs of Pinus longifolia.** Jour. Indian Bot. 1: 115–119. 1919.—The number of leaves on spur shoots of mature trees is quite constantly 3, but an examination of 100 4-year-old nursery seedlings revealed 57 bearing spurs with from 2 to 5 leaves. The number of leaves was 4 in 83.8 per cent of the abnormal shoots, from which the author concludes that "a 3-leaved spur has been derived from a spur with more leaves, and that pines with a small number of needles in their spurs are more specialised than species with a larger number of needles."—*Winfield Dudgeon.*

1895. Kenoyer, L. A. **Dimorphic carpellate flower of Acalypha indica L.** Jour. Indian Bot. 1: 3–7. *21 fig.* 1919.—The carpellate flowers on the lower branches of the inflorescence are trilocular; those at the tips of the staminate cymes are unilocular. In the latter there are no traces of other carpels.—*A. J. Eames.*

1896. Kirby, R. S., and J. S. Martin. **A study of the formation and development of the flower buds of Jonathan and Grimes Golden in relation to different types (clover sod, blue grass sod, cover crop, and clean tillage) of soil management.** Proc. Iowa Acad. Sci. 25: 265–290. *Pl. 7.* 1920.—See Bot. Absts. 5, Entry 1750.

1897. Manaresi, A. **Sulla biologia fiorale del pesco. 2 nota.** [On the floral biology of the peach. 2nd note.] Staz. Sperim. Agrarie Italiane 52: 42–67. 1919.—See Bot. Absts. 5, Entry 1757.

1898. Mascre, M. **Sur le rôle de l'assise nourriciere du pollen.** [The rôle of the tapetum.] Compt. Rend. Acad. Sci. Paris 168: 1120–1122. *4 fig.* 1919.—An account of the changes taking place in the cytoplasm of the tapetal cells during maturation and spore formation of *Datura arborea L.* At tetrad formation the cytoplasm contains numerous mitochondrial threads and granules, together with tannin corpuscules. The cells are usually multinucleate. In older stages the nuclei disappear, after fusing in pairs; the mitochondria also disappear. As the cytoplasm becomes vacuolate numerous deutoplasmic vesicles appear, as well as some starch.—*F. B. Wann.*

1899. Metcalf, Woodbridge. **A precocious youngster.** Amer. Forestry 26: 15. *1 fig.* 1920.—A demonstration of the fact that coniferous cones are simply modified branches, the leaves of which are changed in shape to form the cone scales.—*Chas. H. Otis.*

1900. Miller, Robert B. **The wood of Machaerium Whitfordii.** Bull. Torrey Bot. Club 47: 73–79. *8 fig.* 1920.—See Bot. Absts. 5, Entry 218.

1901. PAMMEL, L. H., AND C. M. KING. **The germination of some trees and shrubs and their juvenile forms.** Proc. Iowa Acad. Sci. 25: 292–340. *Fig. 45–120.* 1920.—See Bot. Absts. 5, Entry 1380.

1902. POLE-EVANS, I. B., AND K. LANSDELL. **The weeds of South Africa. Notes on the Canada thistle (Cnicus arvensis).** Jour. Dept. Agric. Union South Africa 1: 73–75. *1 fig.* 1920.

1903. RONCAGLIOLO, M. **Descrizione anatomica e comparata degli organi epigei di cinque specie di mimosa. [Comparative anatomy of the aerial organs of five species of Mimosa.]** Malpighia 28: 435–457. 1919.

1904. SABNIS, T. A. **The physiological anatomy of the plants of the Indian desert.** Jour. Indian Bot. 1: 33–43. *16 fig.* 1919.—The author has studied the structure of the leaf and stem of 165 species, 125 genera, and 50 orders of xerophytic plants of the Indian desert. This is the introductory section of his paper and contains chiefly a discussion of the physical aspects of the desert, including tables of meteorological data. The anatomy of a few forms in the Menispermaceae and Capparidaceae is described and illustrated. Herbarium specimens were used, and were sectioned unembedded. [See also Bot. Absts. 6, Entry 771.]—*A. J. Eames.*

1905. SCHAFFNER, JOHN H. **The dioecious nature of buffalo-grass.** Bull. Torrey Bot. Club. 47: 119–124. 1920.—The buffalo-grass, *Bulbilis dactyloides* (Nutt.) Raf., has been variously considered, and even in our present manuals inconsistent statements are made as to its dioecism. Field observations in Kansas and experimental results indicate that the dioecious condition is the normal one, it being the only one found in the course of this investigation.—*P. A. Munz.*

1906. SHIRLEY, JOHN, AND C. A. LAMBERT. **The stems of climbing plants.** Proc. Linnean Soc. New South Wales 43: 600–609. *Pl. 60–66.* 1918.—The results of the examination of 53 climbing plant stems are given. A grouping of the structures according to natural orders was found impossible, for similar characteristics were common to plants of many different families, especially among dicotyledons. Therefore, classes were created and are discussed in some detail, illustrated, and type species indicated. Under Subclass I: Dicotyledones, are seven classes; (1) *Normales*, single cambium, wood and bast of each bundle lying along the same radius; (2) *Chiastoxylon*, single cambium, in young stems four rays of alternate wood and bast; (3) *Astroxylon*, single cambium, bundles separated by stellate arrangement of pluriseriate rays; (4) *Endophloia*, second bast occurring at inner margin of wood ring (bicollateral); (5) *Exocycla*, besides normal cambium, new cambium;—zones appear successively centrifugally; (6) *Phloiocycla*, new bast zones are produced in centripetal order; (7) *Polycycla*, oldest bundles in pith, then a normal zone of wood and bast, or alternating rings may be formed. Under Subclass II: Monocotyledones, are two classes; (1) *Vulgares*, usual rind and scattered closed bundles; (2) *Abnormales*, differing from subclass (1) in one or other of the above characters. The authors conclude that these abnormal stem structures in climbers assist the free flow of elaborated sap in the bast.—*Eloise Gerry.*

1907. SHREVE, FORREST. **Proliferation in cacti.** [Rev. of: JOHNSON, DUNCAN S. **The fruit of Opuntia fulgida; a study of perennation and proliferation in the fruits of certain Cactaceae.** Carnegie Inst. Wash. Publ. 269. *Pl. 12.* 1918.]—Plant World 2: 182–183. 1919.

1908. STEIL, W. N. **The distribution of the archegonia and the antheridia on the prothallia of some homosporous leptosporangiate ferns.** Trans. Amer. Microsc. Soc. 38: 271–273. *2 fig.* 1919.—In ordinary *Polypodiaceae*, the archegonia are formed on the so-called cushion directly back of the apical notch, and the antheridia on the posterior portion of the prothallium; but in some species the antheridia are produced on the lobes and margins. Under favorable conditions of nutrition male prothallia became monoecious. In Osmundaceae the archegonia are produced on the sides of the midrib from the notch to the posterior end where the antheridia are borne. A peculiar arrangement of the sex organs was found on the prothallia of *Pteris ensiformis* Burm. var. *Victoria*. On the prominent and highly developed cushion the archegonia occupy only the highest portions while the antheridia are found on the lower parts from the notch to the posterior end. In some cultures a large number of prothallia produced antheridia only, on both surfaces, especially when the prothallia were equally illuminated on both surfaces. In other cultures when the dishes were about half filled with sphagnum and nutrient solution, several species were grown which produced both archegonia and antheridia on both surfaces. It was observed that prothallia may be grown in weak light indefinitely, but under such conditions antheridia only are produced. When the light is sufficiently strong, archegonia will form with the continued growth of the prothallium, provided fertilisation is prevented.—*S. H. Essary.*

1909. VIELHAUER, [—] **Vierblätteriger Klee.** [Four-leaved clover.] Illustrierte Landw. Zeitg. 39: 373–374. 1919.—The formation of four or more leaflets is encouraged by conditions favoring luxuriant growth. It is to be regarded as a condition of robustness or hypertrophy, or as a certain form of fasciation; and it diminishes the fruitfulness of the plant. Whether the property of forming four leaflets is hereditary or not is not known.—*John W. Roberts.*

1910. VÖCHTING, HERMANN. **Untersuchungen zur experimentellen Anatomie und Pathologie des Pflanzenkörpers. II. Die Polarität der Gewächse.** [Experimental anatomy and pathology of the plant body. II. Polarity.] *vi*+333 *p., 12 pl., 113 fig.* Tübingen, 1918.—Review by O. VON K[IRCHNER] in: Zeitschr. Pflanzenkr. 29: 242–249. 1919 (1920).

1911. VON K[IRCHNER], O.. [Rev. of: VÖCHTING, HERMANN. **Untersuchungen zur experimentellen Anatomie und Pathologie des Pflanzenkörpers. II. Die Polarität der Gewächse.** (Experimental anatomy and pathology of the plant body. II. Polarity.) *vi*+333 *p., 12 pl., 113 fig.* Tübingen, 1918.] Zeitschr. Pflanzenkr. 29: 242–249. 1919 (1920).—See also next preceding Entry, 1910.

1912. WEATHERWAX, PAUL. **The ancestry of maize—a reply to criticism.** Bull. Torrey Bot. Club. 46: 275–278. 1919.—H. J. KEMPTON's criticism of author's paper of September, 1918, on the evolution of maize make necessary a brief presentation of the present status of the question. Errors were made in the paper in question in confusing "bracts" with "prophylla" and in substituting "one-rowed" for "single-rowed"; these are to be corrected. The theories of the origin of maize by hybridization and of the ear by fasciation are discussed, and the importance of the use of comparative morphology in explaining the origin of *Zea*, *Euchlaena* and *Tripsacum* from common ancestry is re-emphasized.—*P. A. Munz.*

1913. WIELAND, G. R. **Distribution and relationships of the cycadeoids.** Amer. Jour. Bot. 7: 154–171. *Pl. 7, 3 fig.* 1920.—See Bot. Absts. 5, Enry 1999.

1914. WILLEY, FLORENCE. **The vegetative organs of some perennial grasses.** Proc. Iowa Acad. Sci. 25: 341–367. *Fig. 121–144.* 1920.

MORPHOLOGY AND TAXONOMY OF BRYOPHYTES

ALEXANDER W. EVANS, *Editor*

1915. ALLEN, C. E. **Sex inheritance in Sphaerocarpos.** Proc. Amer. Philos. Soc. 58: 289–316. *27 fig.* 1919.—See Bot. Absts. 4, Entry 486.

1916. ANDREWS, A. LE ROY. **Dicranoweisia crispula in the White Mountains.** Rhodora 21: 207–208. 1919.—See Bot. Absts. 4, Entry 313.

1917. ANDREWS, A. LEROY. **Hymenostomum in North America. I. Delimitation of the genus.** Bryologist 23: 28–31. 1920.—The author maintains that the mosses usually classified under *Astomum*, *Hymenostomum*, and *Weisia* are so closely related that generic separation is unwarranted; that the revival of *Kleioweisia* is wholly needless; and that *Tetrapterum* should not be included in *Astomum*. The understanding of the genus has been further obscured by the inclusion of many unrelated tropical and south-temperate forms, as well as by careless identifications. The genus, as here delimited, corresponds with Lindberg's *Mollia*, subgenus *Hymenostomum*; it may be naturally divided into the three subgenera *Astomum*, *Euhymenostomum* and *Weisia*.—*E. B. Chamberlain.*

1918. ARMITAGE, ELEANORA. **On the habitats and frequencies of some Madeira bryophytes.** Jour. Ecol. 6: 220–225. 1918.—See Bot. Absts. 4, Entry 274.

1919. DOUIN, CH. **Le capitule du Marchantia polymorpha expliqu paré Leitgeb et ses disciples. [The receptacle of Marchantia polymorpha explained by Leitgeb and his disciples.]** Rev. Gén. Bot. 32: 57–71. 1920.—A criticism and refutation of the interpretation of Leitgeb who held that growing points in the angles between the original 8 fused thalli (rays) grew into additional archegonium-bearing thalli, which folded underneath and fused with the lower surface of the receptacle.—*L. W. Sharp.*

1920. EVANS, ALEXANDER W. **The North American species of Asterella.** Contrib. U. S. Nation. Herb. 20: 247–312. 1920.—In this revision of the North American species of the liverwort genus *Asterella* Beauv. (including the Mexican and West Indian representatives) 15 species are recognized and very fully described, and the following new species and names occur: *Asterella saccata* (Wahl.) Evans, *A. venosa* (Lehm. & Lind.) Evans, *A. rugosa*, *A. reticulata*, and *A. versicolor*. Five species described by Stephani from Mexico are referred to a list of doubtful species. The systematic treatment is preceded by a discussion of the nomenclature of the genus, which is generally known in Europe under the name *Fimbriaria*, and by notes on its morphological characters.—*S. F. Blake.*

1921. HOLZINGER, JOHN M. **Dr. Correns's investigations and sterile mosses.** Bryologist 23: 27–28. 1920.—Few bryologists, when determining sterile material, seem to use the keys given in the chapter upon Systematic Determinations in Correns's "Vermehrung der Laubmoose durch Brutorgane und Stecklinge." Two examples of the usefulness of these keys are given.—*E. B. Chamberlain.*

1922. INGHAM, W. **Mosses and hepatics of the magnesium limestone of West Yorkshire (continued).** Rev. Bryologique 41: 77–82. 1914. [Issued in 1919.]—See Bot. Absts. 4, Entry 340.

1923. SCHACKE, MARTHA A. **A chromosome difference between the sexes of Sphaerocarpos texanus.** Science 49: 218–219. 1919.—See Bot. Absts. 3, Entry 1034.

1924. WATSON, W. **The bryophytes and lichens of fresh water.** Jour. Ecol. 7: 71–83. 1919.—See Bot. Absts. 4, Entry 310.

MORPHOLOGY AND TAXONOMY OF FUNGI, LICHENS, BACTERIA AND MYXOMYCETES

H. M. FITZPATRICK, *Editor*

1925. ANONYMOUS. **Index to American mycological literature.** Mycologia 12: 112–114. 1920.

1926. BAL, S. N., AND H. P. CHAUDHURY. **Commentationes Mycologicae. 7. A short study of Plicaria repanda (Wahl.) Rehm on Borassus flabellifer Linn.** Jour. Dept. Sci. Calcutta Univ. 2: 35–36. *1 pl.* 1920.—The authors record the occurrence of the fungus at Calcutta, and give a short description.—*Winfield Dudgeon.*

1927. Bal, S. N. **Commentationes Mycologicae. 5. Vermicularia Jatropha Speg., on Jatropha integerrima.** Jour. Dept. Sci. Calcutta Univ. 2: 31–32. *1 pl.* 1920.—This is a record of the occurrence of the fungus at Calcutta. A short description is given.—*Winfield Dudgeon.*

1928. Beardslee, H. C. **A new species of Amanita.** Jour. Elisha Mitchell Sci. Soc. 34: 198–199. *Pl. 30–31.* 1919.—*Amanita mutabilis* is described, growing on white sand along the coast (Davis Island, North Carolina). In a note by W. C. Coker the same species is also reported in similar soil from Charleston, South Carolina.—*W. C. Coker.*

1929. Börgesen, F., and Raunkiaer, C. **Mosses and lichens collected in the former Danish West Indies.** Dansk Bot. Ark. 2[9]: *18 pl.* 1918.—See Bot. Absts. 6, Entry 150

1930. Bose, S. R. **Descriptions of fungi in Bengal. (Agaricaceae and Polyporaceae.)** Proc. Indian Assoc. Cultivation Sci. 4: 109–114. *Pl. 1–11.* 1918.—The following species, collected near Calcutta, Hooghly, and neighboring places, are described, and with the exception of the first are figured: *Schizophyllum commune, Lentinus praerigidus, L. caespitosus, L. irregularis, Lepiota ermineus, Collybia mimicus, C. ambustus, Daedalea quercina, Favolus scaber, Polystictus sanguinus,* and *Hexagonia sub-tenuis.* The author states that he expects to publish similar descriptions of the Polyporaceae in Bengal at frequent intervals, and will cover the group in two or three years.—*H. M. Fitzpatrick.*

1931. Boyer, M. G. **Études sur la biologie et la culture des champignons supérieurs. [Biology and culture of mushrooms.]** Mém. Soc. Sci. Phys. Nat. Bordeaux VII, 2: 233–344. *4 pl., 20 fig.* 1918.—The work is divided into two parts: 1. Experiments on the germination of spores and culture of mycelia of edible Basidio- and Ascomycetes. 2. Special researches on *Morchella esculenta* and *Psalliota campestris.*—The author attempted to obtain the germination of many kinds of spores but had only a few positive results. He was thus unsuccessful with *Boletus, Russula* and *Amanita.* Contrary to the findings of Matruchot, de Lesparre, and others, the author has never observed the germination of *Tuber* spores. He attempted without success also the germination of spores which had gone through the digestive tract of animals. In contact with oak leaves or rootlets, spores remain equally inert. The author studied in particular a group of fungi neither saprophytic nor apparently parasitic, found in the vicinity of trees. He believes them to be always symbiotic with trees through mycorrhiza. This fact has been satisfactorily proved for several Agaricineae and for *Tuber.* The direct connection between fungus and mycorrhiza is difficult to establish in the species that do not form rhizoids. Symbiotic forms are apparently capable of adopting parasitic habits and vice versa. The author found *Hypholoma fasciculare* and *Trametes pini* growing on earth in contact with their host through mycorrhiza only, and a normally mycorrhizal form (*Boletus*) growing parasitically on tree trunks.—Aseptic Mycelia: Constantin and Matruchot saved the industry of mushroom culture in France, attacked by *Mycogona perniciosa,* when they introduced in the market aseptic mycelia, raised from spores. The author does not obtain satisfactory results with this method. He recommends another which he believes to be new. It consists simply in growing mycelia not from the spores but from fragments of pseudotissue taken from the pileus or stipe. Most of these cuttings grow vigorously. Those of *Boletus* are of weak growth, and those of *Morchella, Amanita,* and *Tuber,* do not grow at all. This fact the author considers as further proof of the semi-parasitic nature of these latter fungi. The saprophytic mycelia of *Morchella* can easily be obtained from the spores, but it remains permanently sterile. The author believes that in order to produce carpophores *Morchella* must become parasitic or symbiotic. All attempts to bring about this condition have, however, failed. The mycelium remains sterile in field, garden, or orchard. When inoculated on live tubers or rootlets of Jerusalem artichoke, it does not penetrate the living tissue. His special studies on *Psalliota campestris* seemed to prove that cultural characters are preserved by the mycelia arising from cuttings.—*Mathilde Bensaude.*

1932. Bronfenbrenner, J., and M. J. Schlesinger. **Carbohydrate fermentation by bacteria as influenced by the composition of the medium.** [Abstract.] Absts. Bact. 3: 8. 1919.

1933. Chaudhury, H. P. **Commentationes Mycologicae. 6. Phyllosticta glycosmidis Sydow and Butler, on Glycosmis pentaphylla Corr.** Jour. Dept. Sci. Calcutta Univ. 2: 33–34. *1 pl.* 1920.—This is a record of the occurrence of the fungus at Calcutta. A short description is given.—*Winfield Dudgeon.*

1934. Clark, Paul F. **Morphological changes during the growth of bacteria.** [Abstract.] Absts. Bact. 3: 2. 1919.—"In some instances as early as two hours after transplanting, and in practically all cases by the fourth or sixth hour of growth, the majority of the organisms in any given smear were approximately twice as large as the organism we have considered the average, namely, the organisms from a twenty-four-hour culture." Members of the diphtheria group are a marked exception. In cultures 4 to 6 hours old the individuals are smaller, less variable and stain more readily than those from cultures 24 hours old. [From author's abst. of paper read at scientific session, Soc. Amer. Bact.]—*D. Reddick.*

1935. Coker, W. C. **The Hydnums of North Carolina.** Jour. Elisha Mitchell Sci. Soc. 34: 163–197. *Pl. 1–29.* 1919.—Twenty-eight species of the larger Hydnums, including *Hydnum, Manina, Steccherinum, Hydnellum, Phellodon* and *Hydnochaete* are reported. Of these, *Hydnellum carolinianum* Coker and *Phellodon Cokeri* Banker are reported as new. Resupinate species are not treated. Of the plates two are in color, three are high power drawings of the spores, the remainder are photographs.—*W. C. Coker.*

1936. Coker, W. C. **Craterellus, Cantharellus, and related genera in North Carolina with a key to the genera of gill fungi.** Jour. Elisha Mitchell Sci. Soc. 35: 24–48. *Pl. 1–17 (in color).* 1919.—Twenty-six species are reported, belonging to the following genera: *Eomycenella, Trogia, Nyctalis, Craterellus, Cantharellus* and *Plicaturella.* Plate 17 gives the spore characters.—*W. C. Coker.*

1937. Darnell-Smith, G. P. **The occurrence of an inverted hymenium in Agaricus campestris.** Proc. Linnean Soc. New South Wales 43: 883–887. *Pl. 91–93.* 1918.—The article records teratological observations on *Agaricus campestris* derived from a particular sample of spawn imported from France. The under surface was normal but the upper surface was broken by one or more black protuberances composed of irregular, sinuous, labyrinthiform lamellae having the appearance of small inverted caps without stipe. These were quite separate from the normal hymenium. Spores were borne upon enlarged cells provided with 1–4 sterigmata. These abnormal mushrooms are considered as a partial reversion to an ancestral, cylindric, dome-shaped form having semi-alveolar or labyrinthiform gill formation over the exposed upper surface. In the evolution of the normal cap the hymenium is considered as having been relegated to the lower surface, the gills having developed from the original pore or alveolar structure.—*C. J. Humphrey.*

1938. De Mello, Froilano. **Contribution to the study of the Indian Aspergilli.** Jour. Indian Bot. 1: 158–161. 1920.—The author describes *Aspergillus (Sterigmatocystis) polychromus* as a new species, and records its behavior on a number of different culture media.—*Winfield Dudgeon.*

1939. Donk, P. J. **Some organisms causing spoilage in canned foods, with special reference to flat sours.** [Abstract.] Absts. Bact. 3: 4. 1919.—See Bot. Absts. 5, Entry 2164.

1940. Eberson, Frederick. **A yeast-agar medium for the meningococcus.** [Abstract.] Absts. Bact. 3: 10. 1919.—"The primary objects of these experiments have been attained, namely to find a cheap and simple medium which would enable us to maintain cultures of a delicate organism such as the meningococcus so that shipment over long distances might be practiced without danger of losing valuable strains."—Preparation of medium: Macerate 10 grams of bakers' or brewers' yeast in 100 cc. of water for 20 minutes; steam for 2 hours at 100°; filter twice through filter paper, or perhaps preferably, clarify by use of glass wool; prepare a 2.5 per cent agar with or without peptone and salt; to each 60 cc. of agar, add 40 cc. of yeast decoction; sterilize in autoclave for 20 to 30 minutes. A semisolid yeast agar (0.5 per cent) "will prolong the viability for beyond the periods observed for the solid medium."—[From abst. of paper read at scientific session, Soc. Amer. Bact.]—*D. Reddick.*

1941. Ferdinandsen, C., and Ö. Winge. **A Phyllachora parasitic on Sargassum.** Mycologia 12: 102–103. *2 fig.* 1920.—*Phyllachora oceanica* is described as a new species. It produces swellings on *Sargassum*.—*H. R. Rosen.*

1942. Gilbert, E. M. **A peculiar entomophthorous fungus.** Trans. Amer. Microsc. Soc. 38: 263–269. *Pl. 27, 28, fig. 1–23.* 1919.—Among the fungi found on fern prothallia grown in water cultures or on moist sphagnum, one appeared from time to time which seemed to be a vigorous parasite. It was isolated and pure cultures were obtained on Thaxter's potato hard-agar plus Lofflund's malt extract. An effort was made to find an insect upon which it would grow; but no infections were secured upon any of the insects of the greenhouse, nor upon vigorous fern prothallia, although it would grow on dying fern prothallia. The fungus seems to be of a decided saprophytic nature. Other investigators have observed a saprophytic condition in certain members of the Entomophthorales. The fungus grows rapidly. No haustoria or rhizoidal growths are found. The hyphae branch and become septate; the cells compare favorably with those of *Empusa*, but differ in many particulars. The shape and size of cells vary greatly. Conidiophores arise usually from terminal cells. No sclerotia are found. Conidiophores, usually simple, are sometimes compound, each branch producing a single conidium. By a process not fully understood, the basidium ruptures and projects the ripened conidium often to a distance of 65 mm. Upon a substratum containing moisture the conidia germinate in from 6 to 12 hours and put forth from one to four germ tubes which develop a typical mycelium. Upon a dry surface the conidia germinate and produce secondary conidia which are discharged like the primary ones, and these may germinate and produce tertiary spores. Primary conidia have diameters of 48 to 60 μ, secondary, 35 to 40 μ, and the tertiary 20 μ. Some conidia do not germinate upon an unfavorable substance; but form a thick wall and appear to be resting spores, although germination has not been observed.—*S. H. Essary.*

1943. Gilkey, Helen M. **Two new truffles.** Mycologia 12: 99–101. *Fig. 1.* 1920.—*Tuber canaliculatum* and *T. unicolor* are described as new species.—*H. R. Rosen.*

1944. Hammer, B. W. **Bacteriological results obtained in practice with vat pasteurization and with one of the final package methods.** Iowa Agric. Exp. Sta. Bull. 190: 151–158. 1919.

1945. Hammer, B. W. **Studies on formation of gas in sweetened condensed milk.** Iowa Agric. Exp. Sta. Res. Bull. 54: 211–220. *2 fig.* 1919.—See Bot. Absts. 5, Entry 2199.

1946. Hammer, B. W., and D. E. Bailey. **The volatile acid production of starters and of organisms isolated from them.** Iowa Agric. Exp. Sta. Res. Bull. 55: 223–246. 1919.—See Bot. Absts. 5, Entry 2172.

1947. Hemmi, Takewo. **Vorläufige Mitteilung ueber eine Anthracnose von Carthamus tinctorius.** [Preliminary report of an anthracnose of Carthamus tinctorius.] Ann. Phytopath. Soc. Japan 1[2]. *11 p., fig. 1–2.* 1919.—See Bot. Absts. 3, Entry 2659.

1948. Herre, Albert C. **Notes on Mexican lichens.** Bryologist 23: 3–4. 1920.

1949. Herre, Albert C. **Hints for lichen studies.** Bryologist 23: 26–27. 1920.—Much valuable work could be done upon the physiology of the growth and luxuriance of lichens, especially in the case of rock- or bark-inhabiting species, without taxonomic knowledge. There are great possibilities in the study of the inheritance of lichen species.—*E. B. Chamberlain.*

1950. KEENE, M. LUCILLE. **Studies of zygospore formation in Phycomyces nitens Kunze.** Trans. Wisconsin Acad. Sci. 19: 1196–1219. *Pl. 16–18, 17 fig.* 1919.—Cytological studies of the plus and minus strains of *Phycomyces nitens* were made but no constant morphological or cytological differences could be determined at any phase of the life cycle. The internal and external changes occurring before and after conjugation are described and illustrated. Following a characteristic grouping of the nuclei, there appears to take place a fusion of nuclei in pairs. The disorganization of part of the nuclei, probably the unfused ones, is followed by the appearance of reserve substances: a large amount of oil and a nucleo-protein-like substance.—*L. K. Bartholomew.*

1951. KLEBAHN, H. **Haupt- und Nebenfruchtformen der Askomyzeten. Erster Teil: Eigene Untersuchungen.** [Perfect and imperfect stages of ascomycetes.] *395 p., 275 fig.* Gebr. Borntráger: Leipzig, 1918.

1952. LATHAM, ROY. **Musci hosts of Cyphella muscigena Fr.** Bryologist 23: 7. 1920.—The author notes that in Southold, New York, the fungus seems to prefer *Thuidium paludosum* as host to the exclusion of other, intimately associated species.—*E. B. Chamberlain.*

1953. LEIDY, JOSEPH. **Modification of Gram's stain for bacteria.** [Abstract.] Absts. Bact. 3: 7. 1919.—"In the course of some experiments in staining bacteria according to Gram's method it was found that any of the metallic iodides soluble in water may be substituted for the potassium iodide in Gram's (Lugol's) solution." [From author's abstract of paper read at scientific session, Soc. Amer. Bact.]—*D. Reddick.*

1954. L'ESTRANGE, W. W., AND R. GREIG-SMITH. **The "springing" of tins of preserved fruit.** Proc. Linnean Soc. New South Wales 43: 409–414. 1918.—Cans of pears and plums, as compared with apricots and peaches, were found to be especially susceptible to "springing." Yeasts chiefly, certain moulds and bacteria sometimes, apparently in an inactive condition, were found. Suggestions for better operating methods are given.—*Eloise Gerry.*

1955. LLOYD, C. G. **Mycological notes. No. 57.** *P. 830–844, fig. 1388–1412.* Cincinnati, Ohio, April, 1919.—A photograph of J. RAMSBOTTOM is accompanied by a brief personal appreciation. The status of the genus *Laschia* is discussed, and about twenty species are cited with annotations. Under the heading "rare or interesting fungi received from correspondents" the following are discussed and in most cases figured: *Clathrus cancellatus, Lentinus rivulosus, Hydnum pulcherrimum, Dacryomitra depallens, Polystictus pinsitus, Dacryopsis nuda, Polystictus felipponei, Polyporus greenii, Polystictus scopulosus, Podocrea xylarioides.*—*H. M. Fitzpatrick.*

1956. LLOYD, C. G. **Mycological notes. No. 58.** *P. 814–828, fig. 1358–1387.* Cincinnati, Ohio, March, 1919.—A short account of ARTHUR LISTER's life and work is accompanied by a photograph of this well known student of the myxomycetes. The following "rare or interesting fungi received from correspondents" are discussed and in most cases figured: *Campanella cucullata, Durogaster brunnea, Rimbachia pezizoidea, Geaster tomentosus, Tremella mellea, Polyporus smaragdinus, Porodiscus rickii, Polystictus hexagonoides, Favolus caespitosus, Pterula fruticum, Daldinia albozonata, Polyporus setiger, Polyporus atrohispidus, Lentinus chordalis, Guepinia elegans, Dacryomyces pallidus, Tremella compacta, Dacryomitra dubia, Stereum corruge, Polyporus pertusus, Lachnocladium brasiliense, Dacryomyces hyalinus.* —*H. M. Fitzpatrick.*

1957. LLOYD, C. G. **Mycological notes. No. 59.** *P. 846–860, fig. 1413–1443.* Cincinnati, Ohio, June, 1919.—A good likeness of GEORGE F. ATKINSON appears on the cover of the pamphlet. A short personal appreciation accompanies it. The genus *Trichoscypha* is discussed and three species are described. These are *T. insititia, T. hindsii,* and *T. Tricholoma.* The following fungi are discussed and in many cases figured: *Trametes heteromorpha, Trametes sepium, Trametes serpens, Tremella candida, Irpex caespitosus, Lenzites betulina, Cata-*

stoma levispora, Trametes truncatospora, Isaria mokanshawii, Polyporus rugosissimus, Isaria ritchiei, Polystictus crocatiformis, Trametes epitephra, Cyphella fuscodisca, Cordyceps lloydii, Polyporus murrillii, Heterochaete gelatinosa, Pseudohydnum guepinoides, Fomes gibbosus, Polyporus suaderis, Lenzites stryacina.—H. M. Fitzpatrick.

1958. Lloyd, C. G. **Mycological notes. No. 60.** *P. 862–876, fig. 1463–1496.* Cincinnati, Ohio, August, 1919.—The cover of this pamphlet bears a good likeness of Charles E. Fairman. A brief statement calls attention to Doctor Fairman's mycological activities. A short review of Kauffman's "Agaricaceae of Michigan" is given. The genus Pterula is discussed and notes and figures are given for twenty-four species. A short note on the genus *Dendrocladium* is appended. Under the heading "tremellaceous plants," notes are given on the following species: *Tremella vesicaria, T. hispanica, T. glaira, T. samoensis, T. sarcoides, Auricularia ornata, A. mesenterica, Exidia janus, Dacryopsis brasiliensis.—H. M. Fitzpatrick.*

1959. Lloyd, C. G. **Mycological notes. No. 61.** *P. 877–903, pl. 124–139.* Cincinnati, Ohio, 1919.—Attention is called to the fact that phalloids and other fleshy forms, when packed in cotton saturated with formalin, can be shipped long distances in good condition. Notes are given on many species of fungi received from correspondents, especially those sent from various countries of the southern hemisphere. These include species of many genera of the higher fungi. New species are described in *Polyporus, Polystictus, Ptychogaster, Hexagona, Stereum, Mitrula, Isaria, Septobasidium, Calocera, Rhizopogon, Calostoma, Lachnocladium, Xerotus, Exidia, Daldinia, Xylaria,* and *Auricularia.* Critical notes are given on many species of *Xylaria.* A discussion is given of the probable identity of *Ceracea* and *Arrhytidia,* and their separation from *Dacryomyces* is questioned. A new genus of the Lycoperdales, *Bovistoidea,* is founded on the species, *B. simplex* n. sp. from South Africa. The genus is characterized by the presence of simple capillitial threads with pointed ends. Attention is directed to several misdeterminations in Baker's "Fungi Malayana." In a discussion of the genus *Septobasidium* it is pointed out that three pileate species are known, and a genus *Rudetum* McGinty is facetiously proposed for these. In the same vein *Pseudothelephora gelatinosa* McGinty is proposed for a gelatinous *Thelephora* received from India. A report of the collection of a species of *Cauloglossum* in the Philippines, *C. saccatum,* is shown to be incorrect, the genus being regarded as still monotypic. Photographs are given for the fungi discussed. Due to the high cost of printing, this number of Mycological Notes is distributed in mimeographed form, and the announcement is made that this policy will be continued.—*H. M. Fitzpatrick.*

1960. MacInnes, L. T., and H. H. Randell. **Dairy produce factory premises and manufacturing processes: the application of scientific methods to their examination.** Agric. Gaz. New South Wales 31: 255–264. *9 fig.* 1920.—See Bot. Absts. 5, Entry 2254.

1961. Merrill, E. D., and H. W. Wade. **The validity of the name Discomyces for the genus of fungi variously called Actinomyces, Streptothrix and Nocardia.** Philippine Jour. Sci. 14: 55–69. 1919.—This is an effort to determine the accurate designation for a group of fungi whose pathogenic members produce various actinomycoses. By the accepted principles of botanical nomenclature, *Streptothrix* Cohn (1875) is invalidated by *Streptothrix* Corda (1839), and *Actinomyces* Harz (1871) by *Actinomyce* Meyen (1827). *Discomyces* Rivolta (1878) would accordingly be valid, *Actinocladothrix* Affanassiew and Schultz (1889) and *Nocardia* Trevisan (1889) are to be regarded as synonyms of *Discomyces. Discomyces* as a generic name is not invalidated by *Discomycetes* as a group name.—Bibliography.—*Albert R. Sweetser.*

1962. Moesz, G. **Mykologiai Közlemények. III. Közlemény.** [**Mycological investigations. III.**] Bot. Közl. 17: 60–78. *11 fig.* 1918. [Summary in German.]—Taxonomic and life history studies of the following. (1) *Herpotrichia nigra* and *Neopeckia coulteri* found on *Pinus pumilio, Juniperus* and *Picea excelsa; Ozonium plica* is connected with latter. Location of these species in herbaria is indicated. (2) *Lizonia emperigonia* (Auersw.) de Not. f. *Baldinii* (Pir.) Moesz on *Polytrichum commune.* (3) *Pachybasidiella microstromoidea* (prior to 1909 as *Gloeosporium*) a saprophyte on capsules of *Catalpa bignonioides*. (4) *P. polyspora* Bub. et Syd. parasitic on leaves of *Acer dasycarpum*. (5) *Leptosphaeria crepini* (Westd.) de Not. on sporophylls of *Lycopodium annotinum* turning them black. (6) *Pyrenochaeta clithridis* n. sp. described from an old fruit body of *Clithris quercina*, *Phoma salsolae* n. sp. from *Salsola kali* and *Aecidium* sp.? from *Rhamnus fallax*. (7) New species of saprophytic fungi described and the host range for old ones extended. [Through abst. by MATOUSCHEK in Zeitschr. Pflanzenkr. 29: 252–253. 1919 (1920).]—*D. Reddick.*

1963. MURRILL, W. A. **A correction.** Mycologia 12: 108–109. 1920.—An error in citation is noted in 25 species of polypores which are found to have been transferred to the genus *Poria* by COOKE two years in advance of SACCARDO's transfers.—*H. R. Rosen.*

1964. MURRILL, W. A. **Daedalea extensa rediscovered.** Mycologia 12: 110–111. 1920.—Specimens collected in Indiana are referred to *D. extensa*; PECK's original description of this species is given.—*H. R. Rosen.*

1965. MURRILL, W. A. **Polyporus excurrens Berk. & Curt.** Mycologia 12: 107–108. 1920.—This species is considered as synonymous with *Trametes rigida* Berk. & Mont., *Polystictus extensus* Cooke, *P. rigens* Sacc. & Cub., *Coriolopsis rigida* (Berk. & Mont.) Murr. Since American specimens referred to *Trametes serpens* are considered distinct from the European material *T. subserpens* is suggested as a new name for American material.—*H. R. Rosen.*

1966. MURRILL, W. A. **Light-colored resupinate polypores—I.** Mycologia 12: 77–92. 1920.—Twenty-seven species of *Poria* are presented including *P. incerta* (Pers.) comb. nov. and the following new species. *P. umbrinescens*, *P. lacticolor*, *P. niveicolor*, *P. cremeicolor*, *P. adpressa*, *P. tenuipora*, *P. Earlei*, *P. corioliformis*, *P. regularis*, *P. polyporicola*, *P. cinereicolor*, *P. subavellanea*, *P. subcorticola*, *P. Amesii*, *P. subcollapsa*, *P. monticola*, *P. lacterata*, *P. rimosa*, and *P. heteromorpha*. "The descriptions included are mainly from dried specimens. Before the hundreds of such specimens in the herbarium here can be intelligently discussed, referred to, or classified, they must be named and more complete descriptions can be prepared later."—*H. R. Rosen.*

1967. MURRILL, W. A. **Illustrations of fungi—XXXII,** Mycologia 12: 59–61. *Pl. 2 (colored).* 1920.—*Boletus luteus*, *Tylopilus alboater* (*Boletus nigrellus*), and *Armillaria nardosmia* are described and illustrated.—*H. R. Rosen.*

1968. NORTHRUP, ZAE. **A new method of preparing cellulose for cellulose agar.** [Abstract.] Absts. Bact. 3: 7. 1919.—"The method is as follows: Melt over a free flame at a low heat 200 grams of ferric chlorid in a porcelain casserole. Add to this completely melted salt a known weight of absorbent cotton, a little at a time (stir with a glass rod), as much as the melted salt will dissolve without making the mixture too thick to be handled readily. When completely dissolved, pour into a large volume of distilled water; a heavy precipitate of finely divided hydrocellulose occurs. Filter by using a Buchner or similar funnel plus suction and wash the precipitate thoroughly on the filter with distilled water. After the thorough washing with distilled water, if any trace of iron chlorid remains, it may be considered as negligible as it is harmless, and may be actually beneficial to the medium. Weigh the moist precipitate to determine the proportion necessary to use per unit weight of original cellulose in making cellulose agar. The weight of hydrocellulose corresponding to 2 grams of absorbent cotton has been found sufficient in Omeliansky's and other cellulose agar media. Pure absorbent cotton dissolves much more satisfactorily than filter paper and gives a more finely divided precipitate, consequently this is the form now employed as a standard in our laboratory." [From author's abst. of paper read at scientific session, Soc. Amer. Bact.]—*D. Reddick.*

1969. NORTHRUP, ZAE. **Agar-liquefying bacteria.** [Abstract.] Absts. Bact. 3: 7. 1919. —Found in anaerobic culture from soil. Pure cultures are to be isolated and studied.—*D. Reddick.*

1970. PAMMEL, L. H. **Perennial mycelium of parasitic fungi.** Proc. Iowa Acad. Sci. 25: 259-263. 1920.—See Bot. Absts. 5, Entry 2082.

1971. PEYRONEL, B. **Sul nerume o marciume nero delle castagne.** [On the blackening or black rot of chestnuts.] Staz. Sperim. Agrarie Italiane 52: 21-41. *Pl. 1-4.* 1919.—See Bot. Absts. 5, Entry 2083.

1972. REINKING, OTTO A. **Phytophthora Faberi Maubl.: The cause of coconut bud rot in the Philippines.** Philippine Jour. Sci. 14: 131-151. *3 pl.* 1919.—See Bot. Absts. 5, Entry 2087.

1973. RETTGER, LEO F., AND C. C. CHEN. **Correlation within the Colon-Aerogenes group.** [Abstract.] Absts. Bact. 3: 1. 1919.—467 cultures isolated from soil, 20 of which were of colon type, and 173 from animals all of which were colon type. Media used were (1) Clark and Lubs medium with Witte's peptone, (2) the same with "Difco" peptone, (3) their synthetic medium. "A total of 3725 individual hydrogen ion concentration determinations and 4632 Voges-Proskauer reactions were made. The P_H was determined by the colorimetric method of Clark and Lubs; the dyes used being brom-thymol blue for the aerogenes group and methyl red for the colon type. Brom-cresol purple was used to check the P_H values of the other two dyes, especially in the range 5.6 to 6.4. The result showed that a three days incubation period was not sufficient for the methyl red test in these media; but an almost perfect correlation between the two types was observed in the synthetic as well as in the Witte's peptone medium (not in the Difco) when the incubation period was prolonged to 5 days.—The results of the Voges-Proskauer tests showed that this test can be made in either of the three media, and that the usual incubation period can be shortened from 5 days to 24 hours (even to ten to fourteen hours). A positive reaction may be obtained by the simple and rapid "shake" method in which the eosin-coloration can be observed for 1 to 3 hours, and its maximum color production from 2 to 8 hours. The method of Levine in which an oxidizing agent is used, and that of Bunker, Tucker and Green in which they expose a thin layer of culture fluid in a Syracuse watch glass both proved either uncertain or too laborious.—With the few exceptions which occurred among the colon strains from soil, the uric acid test of Koser gave very satisfactory correlation with the other reactions when the necessary precautions were taken. —The effect of a mixture of colon-aerogenes types of organisms upon the P_H and upon the Voges-Proskauer test was determined. It was found that the P_H concentration was disturbed between types when such a mixed culture was used, while the Voges-Proskauer reaction proved to be relatively permanent.—The limiting P_H concentration of the colon-aerogenes types of organisms was determined daily in the synthetic medium of Clark and Lubs for a period of 3 weeks. The result showed that the P_H concentration ranged from 4.7 to 7.4 within that period." [From authors' abst. of paper read at scientific section, Soc. Amer. Bact.]—*D. Reddick.*

1974. RETTGER, LEO F., AND MARGARET M. SCOVILLE. **Bacterium anatis, Nov. Spec., an organism of economic importance and a member of the paratyphoid group.** [Abstract.] Absts. Bact. 3: 8. 1919.—An organism resembling very closely *B. paratyphosus B.* was isolated from the internal organs of ducklings which had succumbed. "Indeed so similar were the morphology, cultural characters, etc., of the new organism and different strains of *B. paratyphosus B.* that it has as yet been impossible to differentiate them, although agglutination tests still remain to be made." [From abst. of paper read at scientific session, Soc. Amer. Bact.]—*D. Reddick.*

1975. RIPPEL, AUGUST. **Die chemische Zusammensetzung von Lactaria piperita (Scop.) und Lacteria vellerea (Fries).** [The chemical composition of Lactaria piperita (Scop.) and Lactaria vellerea (Fries.).] Naturw. Zeitschr. f. Forst- u. Landw. 17: 142-146. 1919.—A chemical analysis of the two varieties, which are difficult of distinction to the beginner, is given in two tables. A comparison shows a similar content of phosphoric acid and potassium. Crude fats are slightly more abundant in *vellerea*, and greater in both than in other fungi.

Vellerea also has a higher percentage of crude fiber. The soluble portion of the fiber is not cellulose, but, more than likely, hemicellulose. The chief difference consists in the greater resistance of the cell walls of *vellerea*, which makes it more difficult for digestive juices to attack the nitrogenous constituents and albumen bodies (which are more abundant in *vellerea* than in *piperita*) in this variety. It has not been definitely determined what causes this difficult permeability; it may be chitin. In general, the differences may be traced back with some degree of probability to the tomentose elements of the cap and to the large number of fertile elements resulting from dense-growing lamellae.—*J. Roeser.*

1976. Schøyen, T. H. **Betydningsfulde nyere undersøkelser over furuens blaererust. [Important new investigations on Peridermium pini.]** Tidsskr. Skogbruk 28: 28–29. 1920.

1977. Seaver, F. J. **Notes on North American Hypocreales—IV. Aschersonia and Hypocrella.** Mycologia 12: 93–98. *Pl. 6.* 1920.—*Aschersonia* is considered as the imperfect stage of *Hypocrella.* On this basis a new combination, *Hypocrella turbinata* (Berk.), is made. *H. disjuncta* sp. nov. said to occur on white fly is briefly described and the belief expressed that species of *Hypocrella* may prove to be of economic importance in combating harmful insects. —*H. R. Rosen.*

PALEOBOTANY AND EVOLUTIONARY HISTORY

E. W. Berry, *Editor*

1978. Baccarini, P. **Intorno all'ologenesi. [Concerning ologenesis.]** Nuovo Gior. Bot. Ital. 26: 115–128. 1919.—Daniele Rosa in his recent book "New theory of evolution and the geographic distribution of life," makes an attempt to give on the basis of ologenesis a better explanation of evolution and the distribution of plants and animals than could be had from the theories of Darwin, Lamarck and de Vries. In brief, these are the writer's contentions: (1) The evolution of the specific idioplasm, which is bound up with the phylogeny of the organisms, is predetermined, continuous and independent of external factors. (2) The evolution of the idioplasm is rectilinear up to a certain point when due to increasing complexity the idioplasm divides dichotomously which results in the complete elimination of the mother form and the establishment of two new "species" which in turn develop and then divide. (3) The evolution is not reversible because the products of a dichotomous division have a different constitution since, as was stated above, a certain part of the characters of the mother form have become completely eliminated $A = \frac{B}{C}$. (4) Each new "phyletic species" (the complex of individuals lying between two dichotomous divisions) stands at the end of the genealogical tree and consequently its phylogenetic prospects are much reduced. Furthermore, there is a tendency, as evolution proceeds, for the new forms to become stabilized so that new dichotomous divisions occur only at great intervals. Phyletic and systematic species are not identical. The former has but a limited duration, although throughout its existence it may pass through a number of different stages which would be considered distinct species, or even genera, by the systematist. (5) The large branches of the evolutionary tree are to be sought in the early geological ages when the phylogenetic prospect of the idioplasm was at its prime. The creation of new forms, due to the dichotomous divisions of the idioplasm, does not always find immediate expression because of external conditions. A "mollusk," for example, may have been potentially a mollusk long before climatic and environmental conditions permitted of the existence of mollusks. This indicates why there is such a lack of connections in the evolutionary line, and why there is such an apparent polymorphism. (6) Since the division of the idioplasm of a given form took place simultaneously in all individuals and throughout the entire area occupied by them, it becomes an easy matter to account for the geographic distribution of species and to explain geographic anomalies without having to resort to the migration hypothesis. (7) In the development of the two species of a dichotomous division one form may advance more rapidly and soon reach the apex of its development, while the other,

proceeding slower, gives rise to a greater variety of forms.—When contrasted with the theory of Darwin or de Vries, ologenesis offers a better explanation for the origin of the large evolutionary lines, for the richness of the flora and fauna as far back as the Cretaceous and also for the geographic distribution of species. The chances for the new forms to arise and exist are greater because new forms do not arise as single mutations but simultaneously throughout the entire area occupied by a species which is undergoing division. Of course the explanation of the theory of ologenesis is a teleological one, but it is a teleology which rests on a firm physical and mechanical basis. The author realizes that the hypothetical element in the theory is still large and that it will be necessary to accumulate more evidence before it is accepted altogether.—*E. Artschwager.*

1979. Bancroft, Wilder D. [Rev. of: Jaeger, F. M. **Lectures on the principles of symmetry.** *xii + 333 p.* Elsevier Publishing Co.: Amsterdam, 1917.] Jour. Phys. Chem. **23**: 516. 1919.—See Bot. Absts. 5, Entry 1451.

1980. Baker, Frank C. **The life of the Pleistocene or glacial period.** Univ. Illinois Bull. 17. *vi + 476 p. 57 pl.* 1920.—Essentially geological and zoological, but useful to the botanist in that it contains lists of species of plants and bibliography covering the glaciated and nearby areas of North America.—*E. W. Berry.*

1981. Berry, E. W. [Rev. of: Seward, A. C. **Fossil plants.** Vol. 4. Cambridge Univ. Press: Cambridge, England, 1919.] Plant World **22**: 341–342. (Nov., 1919) March, 1920.

1982. Brown-Blanquet, Josias. **Sur la decouverte du Laurus canariensis Webb et Berth., dans les tufs de Montpellier.** [Discovery of Laurus canariensis in the tuffs of Montpellier.] Compt. Rend. Acad. Sci. Paris **168**: 951–952. 1919.—Description of fragments of leaves determined as *Laurus canariensis* Webb and Berth. The presence of this species in these deposits confirms the oceanic and relatively alpine character of the flora at the time of their formation.—*F. B. Wann.*

1983. Buchholz, John T. **Embryo development and polyembryony in relation to the phylogeny of conifers.** Amer. Jour. Bot. **7**: 125–145. *89 fig.* 1920.—See Bot. Absts. 5, Entry 1883.

1984. Caullery, Maurice. **Parasitism and symbiosis in relation to evolution.** Sci. Amer. Monthly **1**: 399–403. *4 fig.* 1920. [Presidential address delivered before the British Association for the Advancement of Science (the Australian meeting, 1914).]—A criticism of Professor Portier's theory of universal symbiosis.—*Chas. H. Otis.*

1985. Chamberlain, Charles J. **The living cycads and the phylogeny of seed plants.** Amer. Jour. Bot. **7**: 146–153. *Pl. 6.* 1920.—See Bot. Absts. 5, Entry 1885.

1986. Conklin, E. J. **The mechanism of evolution.** Sci. Monthly **10**: 392–403. 1920.—As the chromosomes contain the genes or factors of Mendelian inheritance, many investigators have assumed that the cytoplasm serves only as environment or food for the chromosomes and has nothing to do with heredity. It is true that the spermatozoon is highly differentiated. But the tail of the spermatozoon is either left outside of the egg or its differentiation disappears within the egg. And the yolk of the egg is used up as food.—But there is positive evidence that all cytoplasmic differentiations are not wiped out at this time. Certain cytoplasmic differentiations found in the egg persist in the embryo and adult. Polarity, symmetry, asymmetry, and types of egg organization are of this character.—This egg cytoplasm inheritance is non-Mendelian. Consequently the egg contributes more than the spermatozoon to each generation. This may be somewhat complicated by the fact that the egg has its characters determined by the chromosomes of the cells from which it developed. This would be Mendelian inheritance with its beginnings in the preceding generation. If they are not determined in this way, but are carried from generation to generation in the cytoplasm the inheritance is non-Mendelian. [See also next following Entry, 1987.]—*L. Pace.*

1987. Conklin, E. G. **The mechanism of evolution.** Sci. Monthly 10: 496-515. 1920.—At present there is not sufficient evidence to conclude that modifications of the cytoplasm of the germ cells are ever really inherited or that they are the initial stages in evolution.—Almost all the experimentally produced changes in chromosomes which are known to persist occur during mitoses.—Variations in the volume of chromosomes are dependent upon the volume of the resting nucleus and cytoplasm. These variations have no hereditary or evolutionary value, as is evident from a comparison of the nuclei and chromosomes of the spermatozoa and ova which differ in volume but not in value.—Abnormalities in synapsis, separation and equatorial division of chromosomes are much more important. The two former occur only in the formation of germcells, the latter may occur in any cell.—Changes in the number of chromosomes are known in *Oenothera*, *Ascaris*, and *Drosophila*.—Changes in the constitution of chromosomes by "crossing-over" of sections of homologous chromosomes or by fragmentations or fusions so that a chromosome is not invariably composed of the same chromomeres has been reported.—Experimental modification of chromosomes has produced monstrosities which have not been carried to the next generation. But heat has been shown to increase the number of "cross-overs" in the oocyte of *Drosophila*. These are transmitted. Probably other changes in the constitution of chromosomes may be traced to environmental influences. If so initial stages in evolution may find their causes in such influences.—Genes seem to be subject to all the possibilities just discussed for chromosomes.—In conclusion, it is held that the initial stages in evolution are caused by new combinations of chromosomes, chromomeres, genes, subgenes, and that these new combinations take place in response to stimuli from the external or internal environment.—Germ cells are so complex and so delicately adjusted that they can not usually be greatly changed without rendering them incapable of continued life. The future may show us methods of modifying germ plasm more delicate than those now known. This would make a real experimental evolution possible.—The mystery of mysteries in evolution is how germ plasm ever became so complex. The greatest problem which confronts us is no longer the mechanism of evolution, but the evolution of this mechanism. [See also next preceding Entry, 1986.]—*L. Pace.*

1988. G., A. [Rev. of: Church, A. H. **Thallassiophyta and the subaerial transmigration.** Botanical Memoirs, No. 3. Oxford University Press, 95 p. 1919.] Jour. Botany 58: 59-61. 1920.

1989. Gothan, W., and Nagel, K. **Eine Zechsteinflora (Kupferschieferflora) aus dem untern Zechstein des Niederrheins.** [A flora from the copper shales of the lower Zechstein in the lower Rhine region.] Glückauf 56[6]: 105-107. *1 pl.* Feb., 1920.—Discusses the occurrence of *Ullmannia Bronni* Goeppert, *Ullmannia frumentaria* Goeppert, *Voltzia Liebeana* Goeppert, *Baiera digitata* Heer, *Callipteris Martinsii*, and *Sphenopteris* sp., from the Permian in the vicinity of Wehofen in Western Germany.—*E. W. Berry.*

1990. Groves, J. **A curious fossil Charaphyte fruit.** Geol. Mag. 57: 126-127. *1 fig.* 1920.—Describes specimens of what is probably *Chara merianii* Braun from the Miocene of Locle, Switzerland, showing uniform tubular calcareous hollows on the inner side of the spiral cells that form the oogonium sac.—*E. W. Berry.*

1991. Guppy, H. B. **Fossil botany in the Western World: an appreciation.** Amer. Jour. Sci. 49: 372-374. May, 1920.

1992. Knowlton, F. H. **Evolution of geologic climates.** Bull. Geol. Soc. Amer. 30: 499-566. 1920.—Discusses the factors that might explain the prevailing uniformity of geologic climates, gives an extended summary of the bearing of fossil plants on past climatic conditions, and concludes that the most probable explanation is earth control, the result of internal heat, and not solar control which dominates existing climatic distribution.—*E. W. Berry.*

1993. NEWTON, B. R. **On some freshwater fossils from Central South Africa.** Ann. and Mag. Nat. Hist. 5: 241–249. *Pl. 8.* 1920.—The author describes three specimens of chalcedonized rock found at the base of Kalahari Sand in Matabeleland in Central South Africa. These rocks, representing the first fossils found in this region, contain oogonia and stems of *Chara* and some remains of Gastropods. A more technical description of the *Chara*, including dimensions, is given by Mr. James Groves; but no specific names are mentioned except one oogonium is said to resemble *Chara hipida.* Although the collection is small, the author thinks the combination of *Chara* and Gastropods indicates a correlation between these rocks and the Intertrappean beds of India, and that therefore they belong to the Upper Cretaceous period.—*Harold H. Clum.*

1994. PICQUENARD, CH. **Sur la flore fossile des bassins houillers de Quimper et de Kergogne.** [The fossil flora of the coal beds of Quimper and Kergogne.] Compt. Rend. Acad. Sci. Paris 170: 55–57. 1920.—A list of fossil plants from each of the coal beds named in the title, based on material collected by the author and by others. From the Quimper beds sixteen species are given which in general agree with the flora at the base of the Stephanian stage. Twenty-four species are given from the coal beds of Kergogne, many of which had been previously reported from Blansy and Commentry. Not any new species are described.—*C. H. and W. K. Farr.*

1995. PRINCIPI, P. **Filliti wealdiane della Tripolitania.** [Wealden fossils from Tripoli.] R. Ufficio Geol. Mem. descritt. Carta Geol. d'Italia 18: 71. *2 pl.* 1919.—The engineer ZACCAGNA in a study of the hydrology of western Tripoli in 1914 collected fossil plants in the vicinity of Seck-Scink and Fessato from clay shales of Wealden age including specimens of *Cladophlebis Albertsii* (Dunker) Brongniart, *Dioonites Buchianus* (Ettings.) Bornm., *Sphenolepidium Kurrianum* (Dunker) Heer, somewhat uncertain remains of *Becklesia anomale* Seward, *Cladophlebis zaccagnai* Principi, and *Yuccites* sp. ind. resembling *Yuccites schimperianus* Zigno of the Jurassic of Verona.—*R. Pampanini.*

1996. SERNANDER, R. **Subfossile Flechten.** [Subfossil lichens.] Flora 112: 703–724. *7 fig.* 1918.—The absence of fossil lichens in strata earlier than the Tertiary has been attributed to the rapidity of their decomposition. Observations on *Alectoria jubata* (L.) Ach. in Lapland show that all traces of the plant disappear within a year after it falls on the forest floor. Remains of lichens do not occur in ordinary humus, except as fragments. But a study of post-glacial peat-beds shows recognizable remains of such species as *Cladonia rangiferina* (L.) Web., *Cetraria islandica* (L.) Ach., and *Peltigera canina* (L.) Willd. On partially fossilized tree-trunks, *Opegrapha atra* Pers. is present. Calcareous tufa shows such pitting as is characteristic of *Lecidea immersa* (Web.) Ach. The author believes that these observations prove the possibility of lichens becoming fossilized.—*L. W. Riddle.*

1997. WALCOTT, CHARLES D. **Cambrian geology and paleontology. IV. No. 5. Middle Cambrian algae.** Smithsonian Misc. Coll. 67[5]: 217–260. *Pl. 43–59.* 1919.—*S. F. Blake.*

1998. WIELAND, G. R. **The Tetracentron-Drimys question.** Amer. Jour. Sci. 49: 382–383. May, 1920.—Comments on the question of whether these genera are primitive or reduced, upholding the former view and considering it as conforming to the hypothesis that the Angiosperms are descended from the Cycadeoids through the Ranalian plexus.—*E. W. Berry.*

1999. WIELAND, G. R. **Distribution and relationships of the cycadeoids.** Amer. Jour. Bot. 7: 154–171. *Pl. 7, 5 fig.* 1920.—Author believes that forests of microphyllous and small-stemmed cycadeoids were very numerous in Triassic and Jurassic times. *Williamsoniella* and *Wielandiella* are examples of such plants. They probably shed their leaves with the seasons and were able to thrive in temperate climates. Apparently the climates of the Mesozoic were by no means uniformly tropical. The distribution of cycadophytes in the Mesozoic is briefly considered. Author discusses the relationships which the cycadeoids bear to the cycads, the seed ferns, the cordaites and *Dolerophyllum*, the ginkgos, *Araucaria*, the abietineans, and the Dicotyls and Gnetales. He believes that the cycadeoids gave rise to the angiosperms, and combats the theory of a gentalean origin for the latter group. He suggests that the main plant groups go very far back geologically and have evolved side by side.—*E. W. Sinnott.*

2000. WIELAND, G. R. [Rev. of: SEWARD, A. C. **A text-book for students of botany and geology.** Vol. 4. Price 1£/1s. University Press: Cambridge, 1919.] Amer. Jour. Sci. **49**: 223–224. Mar., 1920.

PATHOLOGY

G. H. COONS, *Editor*
C. W. BENNETT, *Assistant Editor*

2001. ANONYMOUS. **Lime sulphur spray following Bordeaux.** New Zealand Jour. Agric. 19: 371–374. 1919.—It has been reported that lime-sulphur spray following Bordeaux caused russeting of the fruit. Experiments conducted in two orchards indicated that most of the russeting was due to Bordeaux.—*N. J. Giddings.*

2002. ANONYMOUS. **Compatibility of spray mixtures.** New Zealand Jour. Agric. 19: 244–245. 1919.

2003. ANONYMOUS. **Index to American mycological literature.** Mycologia 12: 112–114. 1920.

2004. ANONYMOUS. **Treatment of Armillaria with iron sulphate.** Agric. Gaz. New South Wales 31: 60. 1920.

2005. ANONYMOUS. **Shothole fungi which affect cherry trees.** Jour. Dept. Agric. South Australia 23: 31. 1919.—A brief summary of the results of spray tests with Bordeaux, Burgundy and lime-sulphur mixture for the control of the shothole fungus (*Coccomyces hiemalis*). Bordeaux mixture gave good control, Burgundy mixture fair, while lime-sulphur mixture gave no control.—*Anthony Berg.*

2006. ANONYMOUS. **The skin spot disease of potato tubers (Oospora pustulans).** [Abridged and slightly modified account of: OWEN, Miss M. N. **Skin spot disease of potato tubers.** Kew Bull. Misc. Inf. (London) 1919[8]. 1919.] Jour. Ministry Agric. Great Britain 26: 1245–1250. 1920.

2007. ANONYMOUS. **The Christmas tree. (Nuytsia floribunda.)** Australian Forest. Jour. 3: 10–13. 1920.—This paper discusses the parasitism and root system of *Nuytsia floribunda* which is found always close to banksia or eucalypts, mostly jarrah. The parasite, through the parenchymatous haustoria which develop on the haustoriogen (a continuous fleshy ring encircling the root of the host), obtains an additional supply of organic materials.—*C. F. Korstian.*

2008. ANONYMOUS. **Our botanical immigrants.** Sci. Amer. Monthly 1: 317–319. *5 fig.* 1920.—A popular article on the quarantine regulations of California.—*Chas. H. Otis.*

2009. ANONYMOUS. **Effect of decay on wood pulp.** Sci. Amer. Monthly 1: 247. 1920.

2010. ARNAUD, G. **Sur un mode de traitement de la chlorose.** [**A method for treatment of chlorosis.**] Bull. Soc. Path. Veg. France 6: 136–146. *2 fig.* 1919.—Treatment of chlorosis of pears, poplars, roses, etc., due to an excess of lime, by the injection of sulphate of iron in the trunks and large branches is described. A branch of a chlorotic pear, as a result of this treatment, became green in eight days and is still normal after four years. In some cases slight injury was caused by an excessive dose or the treatment of too small branches. The following formula was used: powdered iron sulphate 35 to 40 grams used with 20 grams olive oil.—*C. L. Shear.*

2011. Averna-Sacca, R. **Molestias da macieria.** [Diseases of apple.] Bol. Agric. [São Paulo] 19: 430–433. *1 fig.* 1918.—Diseases caused by *Ascochyta* sp., *Pleospora herbarum*, and *Sphaerella pomicola.*—*D. Reddick.*

2012. Bancroft, Wilder D. [Rev. of: Peters, Charles A. **The preparation of substances important in agriculture.** 3rd ed. *19×14 cm., vii+81 p.* John Wiley and Sons, Inc.: New York, 1919. $0.80.] Jour. Phys. Chem. 23: 444. 1919.—See Bot. Absts. 5, Entry 1100.

2013. Barker, B. T. P. **Diseases of plants and their treatment.** Jour. Bath and West and South Counties Soc. V, 12: 189–193. 1917–18.—Record of the occurrence of tomato collar rot, a root disease of Belladonna (*Phytophthora*), a bacterial disease of plum trees (hitherto undescribed), a disease of alder (*Fomes igniarius* and *Polyporus sulphureus*), and potato "rust" disease. The following diseases are being investigated: Rhizoctonia of asparagus, apple leaf scorch, apple fruit spot disease and tomato collar rot.—*J. I. Lauritzen.*

2014. Barker, B. T. P., and C. T. Gimingham. **Further experiments on the Rhizoctonia disease of asparagus.** Jour. Bath and West and South Counties Soc. V, 12: 130–134. *1 fig.* 1917–1918.—This is an account of a second series of experiments with soil treatments for *Rhizoctonia violacea*, var. *asparagi* (*R. medicaginis*). For convenience carrots were used in this test also in place of asparagus. The results obtained fully corroborate those of the previous season. The disease was reduced to a mere trace on the plot where bleaching powder was applied (2 ounces per square yard) towards the end of the second week in April, i.e., a few weeks before the test crop was sown. On the check plot the disease was very severe. A considerable importance is attached to the time of application of soil fungicides, that is in the case of *Rhizoctonia* not until the soil temperature begins to rise and the young growth of mycelium makes a good start.—*M. Shapovalov.*

2015. Bastin, S. L. **Some serious potato diseases.** Jour. Bath and West and South Counties Soc. V, 12: 88–106. *2 pl.* 1917–18.—The following diseases of potato are described and control measures suggested: late blight, scab (common), powdery scab, wart disease, Rhizoctonia scab, stalk disease (*Sclerotinia sclerotiorum*), Botrytis disease and black leg.—*J. I. Lauritzen.*

2016. Biers, P. **Le parasitisme probable des Coprins.** [The probable parasitism of Coprinus.] Bull. Soc. Path. Veg. France 6: 159–160. 1919.—*Coprinus domesticus* a close relative of *C. radians* was found associated with a disease of *Broussonetia papyrifera* and is regarded as a probable parasite.—*C. L. Shear.*

2017. Boekker, [—]. **Der Kleekrebs.** [Clover stem-rot (Sclerotinia trifoliorum).] Illustrierte Landw. Zeitg. 39: 402. *Fig. 310.* 1919.

2018. Boyer, G. **Études sur la biologie et la culture des champignons supérieurs.** [Biology and culture of higher fungi.] Mém. Soc. Sci. Phys. Nat. Bordeaux VII, 2: 233–344. *Pl. I–IV, 20 fig.* 1918.—See Bot. Absts. 5, Entry 1931.

2019. Brittlebank, C. C. **The Iceland poppy disease.** Jour. Dept. Agric. Victoria 17: 700–701. 1919.—A brief note discussing the occurrence of a species of *Phytophthora* on Iceland poppy (*Papaver alpinum*).—*J. J. Skinner.*

2020. Bronfenbrenner, J., W. T. Bovie, and Estelle M. Wolff. **A simple arrangement for measuring the rate of heat penetration during sterilization.** [Abstract.] Absts. Bact. 3: 6. 1919.—"A detailed description of the apparatus, with drawings, will appear in the Journal of Industrial and Engineering Chemistry."—*Author.*

2021. Bruner, Esteban. **La pudrición negra del cacao.** [Black rot of the cacao.] Revist. Agric. Com. y Trab. 2: 630. *1 fig.* 1919.—The black rot of the cacao (*Theobroma cacao*) caused by the fungus (*Phytophthora faberi* Maublanc) is reported for the first time from Cuba. The disease is described and methods of control are recommended.—*F. M. Blodgett.*

2022. Bunting, R. H. **Report of the Mycologist.** Rept. Agric. Dept. Gold Coast 1917: 19–21. 1918.—Progress report of work on diseases of cocoa, coffee, para rubber.—*J. I. Lauritzen.*

2023. Butler, E. J. **Report of the Imperial Mycologist.** Sci. Rept. Agric. Res. Inst., Pusa 1918–19: 68–85. 1919.—The report records progress made during the year under report in the study in India of: black band of jute (*Corchorus*) caused by *Diplodia corchori;* diseases of rosaceous plants in the outer Himalayas; various diseases of chili (*Capsicum* spp.); *Pythium* disease of ginger, tobacco (*Nicotiana* spp.), and *Carica papaya;* wilt of *Cajanus indicus;* smut of sugar cane (*Saccharum officinarum*); and wheat rust. Methods of treatment and prevention are recommended.—*Winfield Dudgeon.*

2024. Call, L. E. **Director's Report.** Kansas Agric. Exp. Sta. 1917–18. *63 p.* 1918.—Physiological investigations with sorghum (*Andropogon sorghum*) and corn varieties, showing their comparative drought resistance and water requirements is discussed. Kanred wheat, P1066 and P1068, three hard winter wheats, products of the Kansas Station, have been shown to be very resistant to stem rust, *Puccinia graminis tritici.* The effect of stem rust on the grain of other varieties grown in the same plots with above wheats, is shown in contrast. A new form of stem rust *Puccinia graminis tritici-inficiens* is described. Under corn smut (*Ustilago zeae*) investigations it has been shown that although the smut can be reduced by fungicides, it likewise proportionately reduces the yield. Ecological studies show that infection is local through leaf axils and not systemic. A varietal test of sorghums shows that all are susceptible but milo and feterita. The last named are being studied with a view of discovering what constitutes their resistance. [See also Bot. Absts. 5, Entry 1466.]—*L. E. Melchers.*

2025. Chassignol, F. **La rouille grillagée du poirier (Roestelia cancellata Rebent.) et le Juniperus sabina L.** [The pear rust (Roestelia cancellata Reb.) and Juniperus sabina L.] Bull Soc. Path. Veg. France 6: 133. 1919.—To show the difference in susceptibility of varieties the following case is given. Duchess of Angouleme pear 25 meters from a Juniper had about one-third of its leaves attacked by the *Roestelia,* while an unknown variety only 20 meters from the tree had only four or five leaves affected.—*C. L. Shear.*

2026. Coker, W. C. **A parasitic blue-green alga.** Jour. Elisha Mitchell Sci. Soc. 35: 9. 1919.—Given at the Eighteenth Meeting of the North Carolina Academy of Science, and abstracted in its Proceedings. Oogonia of *Saprolegnia anisospora* were found to be infected by a species of blue-green alga which destroyed the eggs within.—*W. C. Coker.*

2027. Cotton, A. D. **Clover stem-rot (Sclerotinia trifoliorum).** [Rev. of: Amos, A. Clover stem-rot. Jour. Roy. Agric. Soc. England 79: 68–88.] Jour. Ministry Agric. Great Britain 26: 1241–1244. 1920.

2028. Cotton, A. D., and M. N. Owen. **The white rot disease of onion bulbs.** Jour. Ministry Agric. Great Britain 26: 1093–1099. 1920.—The white rot disease of onions, very widespread in England and known to occur in Scotland and Ireland, causes considerable damage to the onion crop, especially in market gardens and allotments, and is caused by *Sclerotium cepivorum.* It attacks both spring and autumn sown onions and is most in evidence from the beginning of June to early August. Few infections appear to take place after that date. In attacked plants the leaves turn yellow, wilt, fall over, and finally the entire plant collapses and is easily pulled from the ground. Under warm, moist conditions a fluffy, white mycelium develops round the base of the bulb which is very characteristic of the White Rot disease and distinguishes it at once from all other diseases of the onion. A little later the surface of the bulb shows the presence of numerous black spherical sclerotia about the size of small poppy seed (0.5 mm. in diameter). The sclerotia appear to persist in the soil at least three or four years and may survive considerably longer. The disease is introduced into new localities by contaminated soil and manure, diseased seedlings and "sets." All common varieties are susceptible. Shallots are usually very resistant as is also true of leeks. The only present known means of control is to keep the infected ground free from onions and allied crops for a number of years. Soil fungicides have not proved effective.—*M. B. McKay.*

2029. CULHAM, A. B. **Report on the agricultural station, Aburi.** Rept. Agric. Dept. Gold Coast 1917: 24–29. 1918.—Includes a note, with table, on distribution of cocoa diseases.—*J. I. Lauritzen.*

2030. DARNELL-SMITH, G. P. **An account of some observations upon the life-history of Phoma citricarpa McAlp. The cause of the "Black Spot" disease in Citrus fruit in New South Wales.** Proc. Linnean Soc. New South Wales 43: 868–882. *Pl. 84–90.* 1918.—The paper first presents a brief historical review of the fungus and the disease. This is followed by a statement of the general symptoms. The disease is serious in New South Wales, producing minute black spots on the foliage throughout the year. On the fruits the spots are rarely seen before the first of August, and vary from $\frac{1}{8}$ to $\frac{1}{2}$ inch or more in diameter. The disease appears almost invariably on the sunny side of the tree and on the side of the fruit exposed to the sun. This has been checked up experimentally and is explained as being due to the lowering of vitality by action of the sun.—Culture data are given and the structure of the mycelium, spores and pycnidia discussed. Two types of spores were found, large viable ones and smaller ones, termed "X" spores, which do not germinate.—The disease can be controlled with Bordeaux.—*C. J. Humphrey.*

2031. DOIDGE, ETHEL M. **The rôle of bacteria in plant diseases.** [Presidential Address, South African Assoc. Adv. Sci., Kingwilliamstown, July, 1919.] South African Jour. Sci. 16: 65–92. 1919.—This is a review of the history of plant bacteriology and a summary of present knoweldge of the rôle of bacteria in plant diseases with special reference to South African conditions and to diseases of plants occurring in South Africa.—*E. M. Doidge.*

2032. DUYSEN, F. **Wurzelbrand im Weizenschlage. [Root-scald in wheat-fields.]** Illustrierte Landw. Zeitg. 39: 372–373. 1919.—The diseases caused by the fungus, *Leptosphaeria culmifraga*, is described and indirect control through increasing the resistance of the host plant by proper fertilization is recommended. Badly diseased fields should be plowed up and replanted with crops other than wheat or rye. Such fields should not be planted with wheat or rye for a term of years.—*John W. Roberts.*

2033. EBERSON, FREDERICK. **A yeast-agar medium for the Meningococcus.** [Abstract.] Absts. Bact. 3: 10. 1919.—See Bot. Absts. 5, Entry 1940.

2034. EKAMBARAM, T. **Suspected parasitism in a moss.** Jour. Indian Bot. 1: 206–211. *6 fig.* 1920.—During the monsoon season a common unidentified moss in Madras is found with its rhizoids and protonemata penetrating colonies of Cyanophyceae. Because the penetrating rhizoids and protonemata are colorless, and become filled with starch coincidentally with the decay of the alga colonies, the author suggests that the moss is parasitic on the algae. Haustorial connections were not observed.—*Winfield Dudgeon.*

2035. ERIKSSON, JAKOB. **Sur l'hètéroecie et la specialisation du Puccinia caricis, Reb. [On heteroecism and specialization in Puccinia caricis Reb.]** Rev. Gén. Bot. 32: 15–18. 1920. —See Bot. Absts. 5, Entry 645.

2036. ERWIN, A. T. **Hot formaldehyde treatment for potato scab.** Potato Mag. 2^9: 14. *1 fig.* 1920.

2037. ERZ, A. A. **The true nature of plant diseases.** Amer. Bot. 26: 20–23. 1920.—The author contends that in favorable situations plants produce substances that render them resistant to disease and that if horticulture is properly conducted the plants will ward off disease by becoming immune.—*W. N. Clute.*

2038. FELT, E. P. **New Philippine gall midges.** Philippine Jour. Sci. 14: 287–294. 1919. —This paper is supplemental to one published in the Philippine Journal of Science for 1918. It describes the gall midges and their food habits, but the appearance of the galls is left for a subsequent paper.—*Albert R. Sweetser.*

2039. FERDINANDSEN, C., AND O. WINGE. **A Phyllachora parasitic on Sargassum.** Mycologia 12: 102–103. *2 fig.* 1920.—See Bot. Absts. 5, Entry 1941.

2040. FLETCHER, J. J., AND C. T. MUSSON. **On certain shoot-bearing tumors of Eucalypts and Angophoras, and their modifying influence on the growth habit of the plants.** Proc. Linnean Soc. New South Wales 43: 191–233. *Pl. 4–26.* 1919.—See Bot. Absts. 5, Entry 1888.

2041. FOEX, ET. **Note sur une maladie du poirier.** [**Note on a pear disease.**] Bull. Soc. Path. Veg. France 6: 102–104. Sept.–Oct., 1919.—A canker on pear branches in France is described and regarded as identical with the disease described by GRIFFON and MAUBLANC. *Diplodia griffoni* Sacc. or *Sphaeropsis pseudo-diplodia* Fckl., the pycnidial form of *Physalospora cydoniae*, was found on the cankers and is regarded as the cause. Cutting out of cankers and spraying with Bordeaux are recommended.—*C. L. Shear.*

2042. FOEX, ET. **Au sujet d'un épi de blé partiellement charbonné.** [**Regarding a partially smutted head of wheat.**] Bull. Soc. Path. Veg. France 6: 105–106. 1919.—A case is reported in which a head of wheat showed the lower spikelets smutted by *Ustilago tritici* and the upper apparently healthy. Three of the unsmutted grains were grown and produced plants free from smut. It is suggested in explanation that the apparently sound spikelets escaped infection or the infection remained dormant. PEGLION is cited as having examined similar cases partially smutted by *Tilletia caries* without finding traces of mycelium in the unsmutted spikelets. A thorough microscopic examination of such cases is necessary in order to determine with certainty whether a partial or undeveloped infection has taken place.—*C. L. Shear.*

2043. FOEX, ET. **Note sur une maladie de l'orge et de l'avoine.** [**Note on a disease of rye and oats.**] Bull. Soc. Path. Veg. France 6: 118–124. Nov.–Dec., 1919.—A disease of oats and rye somewhat resembling foot rot is described. A species of Fusarium was found on the diseased stems. This was compared with *F. rubuginosum* and other species reported on grain but no positive identification made. Soil sterilization and burning of all diseased plants are suggested as control measures.—*C. L. Shear.*

2044. FOEX, ET. **Quelques remarques au sujet de la présence de péritheces de Phyllactinia corylea sur des feuilles de Chêne atteintes de "Blanc."** [**Note on the presence of perithecia of Phyllactinia corylea on oak leaves affected with powdery mildew.**] Bull. Soc. Path. Veg. France 6: 161–166. 1919.—Oak leaves having all the appearance of the mildew attributed to *Microsphaeria quercina* were found to bear perithecia of *Phyllactinia*. Certain peculiarities of the walls of the hyphae of the mildew on the leaves known to occur in *Microsphaeria* but not in *Phyllactinia* lead the author to believe that the perithecia found were not produced on the oak leaves but blown there from some other host.—*C. L. Shear.*

2045. GREENE, LAURENZ, AND I. E. MELHUS. **The effect of crown gall upon a young apple orchard.** Iowa Agric. Exp. Sta. Res. Bull. 59: 147–176. *8 pl., 3 fig.* 1919.—This bulletin deals with the effect of crown gall on a young orchard up until the bearing age. Infected trees were selected and planted on a modified Missouri loess type of soil. The observations extended over a five years period. Crown gall effects were determined by measurements of the trunk diameter, and by consideration of the twigs, their number, length, thickness and weight. The large galls were more injurious than the small ones and those on the stock and union were more harmful than those on the secondary growth.—*I. E. Melhus.*

2046. Grove, O. **Notes on the fruit blossom bacillus. Investigations on diseases of plants and their treatment.** Jour. Path and West and South Counties Soc. 5, 12: 124–128. 1917–18. —The bacillus (specific name not mentioned) which causes disease of pear blossoms was isolated from several samples of soil and is supposed to be common there in April, but not earlier in the year. Cultures made from the roots of various plants yielded apparently the same organism. An experiment was carried on with plants grown in sterilized soil in pots, one set of which was inoculated with cultures of the bacillus. It was found that the latter had a decided beneficial effect upon the growth of the plants. A description is given of morphological, cultural and some biochemical characters of the bacillus.—*M. Shapovalov.*

2047. Hendrick, J. **The use of lime in controlling finger-and-toe in turnips.** Trans. Highl. and Agric. Soc. Scotland V, 30: 137–145. 1918.—The author presents data to show that the application of sufficient lime to neutralize the sourness and leave an excess carbonate of lime in the soil will check or prevent finger-and-toe (*Plasmodiophora brassicae*) in turnips.—*J. I. Lauritzen.*

2048. Hess, E. **Die Mistel auf dem schwarzen Walnussbaum (Juglans nigra). [Mistletoe on the black-walnut (Juglans nigra).]** Schweiz. Zeitschr. Forstw. 71: 1–2. *1 fig.* 1920. —This is the first occurrence of mistletoe on black walnut recorded. It occurred in a park in the village of Champagne, Waadtländer Zura. A possible explanation for its occurrence on this species is the less astringent sap as compared to other nut trees. The mistletoe is supposed to have been disseminated from nearby fruit trees.—*J. V. Hofmann.*

2049. Honnet, G. **Les hybrides en 1919. [1919 hybrids.]** Rev. Vitic. 22: 53–59. 1920.—See Bot. Absts. 5, Entry 1744.

2050. Howard, Albert. **Spike disease of peach trees: an example of unbalanced sap-circulation.** Indian Forester 45: 611–617. 1919.—The characteristics of the spike disease of sandalwood are similar to those of the peach. When the peach is budded on the almond, unless there is close junction between bud-ring and seedling, there is a delayed union and a callus tissue forms until the stock and scion are united. In the former case when the union is perfect, the tree grows normally and vigorously; in the latter case development is slow and the tree becomes "spiked," with the characteristics of form and of mineral and starch content very similar to the sandal. It is suggested that the spike of sandal may be due to the imperfect union of the root haustoria with the host.—*E. N. Munns.*

2051. Hubert, Ernest E. **Disposal of infected slash on timber-scale areas in the northwest.** Jour. Forestry 18: 34–56. 1920.—Factors of available water and food supply, resistance of the host to sporophore production, temperature, humidity and light are most important in the production of sporophores of wood-destroying fungi. These may be present in the slash of cut-over areas and all the destructive wood-rotting fungi can develop on infected slash. These are sources of infection to the remaining trees of the stand. Slash should be burned or charred as far as possible or otherwise dragged into openings where the soil and air is drier and warmer. This is not so important with the yellow pines as with the firs and cedars because of the moister sites occupied by the latter.—*E. N. Munns.*

2052. Kern, Frank D. **Report of the botanist.** Bull. Pennsylvania Dept. Agric. 1[1]: 24–26. 1918.—Attention is called to the greater need for practicing the methods which have already been worked out for the control of crop diseases. Statistics are given on the losses to the oat, potato and apple crop occasioned by plant disease during the season 1917.—*C. R. Orton.*

2053. Klebahn, H. **Haupt- und Nebenfruchtformen der Askomyzeten. Erster Teil: Eigene Untersuchungen. [Perfect and imperfect stages of ascomycetes.]** *395 p., 275 fig.* Gebr. Borntrāger: Leipsig, 1918.

2054. Koerner, W. F. **Auf welche Krankheitsformen ist beim "Durchsehen" und "Aushauen" der zur Saatgewinnung bestimmten Kartoffelfelder besonders zu achten. [What diseases are to be considered especially in going through and thinning out potato fields from which seed potatoes are to be selected.]** Illustrierte Landw. Zeitg. 39: 323–324. *Fig. 252–259.* 1919.

2055. Kornauth, K. **Bericht der K. K. landwirtschaftlich-bakteriologischen und Pflanzenschutzstation in Wien für das Jahr 1917. [Report for 1917, of the Vienna institute for agriculture, bacteriology and plant protection.]** Zeitschr. landw. Versuchsw. Österr. 21: 377–393. 1918.—Occurrence of potato black leg and an early severe outbreak of blight (*Phytophthora infestans*), tomato rot caused by *Phytobacter lycopersicum*, core rot of apple caused by *Fusarium putrefaciens* and a disease of *Picea pugens* caused by *Cucurbitaria piceae.*—Seeds of cucumber, onion and bean were tested for tolerance to a variety of proprietary disinfectants.—"Bosnapasta is a satisfactory preventive of cucumber mildew (*P. cubensis*) and scab (*Cladosporium*). [Through abst. by Matouschek in Zeitschr. Pflanzenkr. 29: 241–242. 1919 (1920).]—*D. Reddick.*

2056. Lee, H. Atherton, and Harry S. Yates. **Pink disease of citrus.** Philippine Jour. Sci. 14: 657–671. *7 pl., 2 fig.* 1919.—The disease is caused by *Corticium salmonicolor* B. & Br. At present localized in a small area, hence the importance of a description of the disease and the method of eradication to prevent further spread. The method of dissemination is studied and recommendations are made for its treatment with lime sulphur spray.—*Albert R. Sweetser.*

2057. Lees, A. H. **"Reversion" of black currants.** Jour. Bath and West and South Counties Soc. 5, 12: 134–135. 1917–1918.—An explanation is given as to the probable causes of an abnormal lateral growth in currants, known as big bud or reversion. It is said to be due to a check in terminal growth of which two cases were observed: the mite-checked terminal and the formation of a terminal flower. The latter was found to occur on shoots that were making a comparatively weak growth.—*M. Shapovalov.*

2058. Lees, A. H. **Further experiments on big bud mite.** Jour. Bath and West and South Counties Soc. 5, 12: 137–139. 1917–1918.—Experiments were conducted to determine the number of sprays necessary for the control of the big bud mite, and the best time for their application. It was found that 2 applications give better results than one and possibly 3 are necessary. The following months were selected: (a) beginning of December, (b) beginning of January and (c) end of February. A satisfactory control was obtained with a mixture containing 10 per cent of soap and 5 per cent of crude carbolic acid.—*M. Shapovalov.*

2059. Lees, A. H. **Copper stearate.** Jour. Bath and West and South Counties Soc. 5, 12: 139–142. 1917–1918.—A proper combination of soap and copper sulphate, called for convenience copper stearate, possesses high wetting and spreading properties. Ordinarily both Burgundy and Bordeaux mixtures alone are deficient in these qualities. The wetting powers of the copper soap mixture may be greatly increased by combining it with a 2 per cent paraffin emulsion [kerosene].—*M. Shapovalov.*

2060. Levy, E. Bruce. **Investigation of dry-rot in swedes.** New Zealand Jour. Agric. 19: 223–228. 1919.—A dry rot disease of swede turnips (*Brassica campestris*) is serious in certain sections and is frequently followed by soft rot. The article deals only with direct control measures. The effects of various fertiliser combinations were tried and a superphosphate-guano mixture seemed to give a slight improvement. Seed from different sources gave little variation in the amount of disease. A large number of varieties were tested and some were found to be slightly resistant. Selection of resistant plants for seed is to be practiced.—*N. J. Giddings.*

2061. Lewis, A. C. **Annual Report of the State Entomologist for 1919.** Georgia State Bd. Entomol. Bull. 55. *31 p. Fig. 2.* 1920.—Contains a statement of the work conducted by the Georgia State Board of Entomology, one of the main lines having been the dusting of peaches (*Amygdalus persica*) against diseases and insects. In the garden and truck work, spraying against the Mosaic disease of peppers (*Capsicum annuum*) was undertaken. Black Leaf 40 was used against the plant lice in an effort to prevent spread of the trouble. Experiments appeared successful, but it was found hard to control the lice. This work will be continued.—The breeding of cotton (*Gossypium hirsutum*), the testing of varieties and the growing of wilt resistant strains were part of the activities of the Board during 1918.—Two new insects were reported in Georgia, one being a species of Margarodes and the other the Chrysanthemum Midge (*Diarthronomyia hypogaea*). The latter part of the report contains a list of the Georgia nurseries inspected for 1918–1919.—*T. H. McHatton.*

2062. Lloyd, C. G. **Mycological notes, No. 61.** *P. 877–993, pl. 124–139.* Cincinnati, Ohio, 1919.—See Bot. Absts. 5, Entry 1959.

2063. Matz, Julius. **Algunas enfermedades del follaje en las plantas. [Foliage diseases.]** Revist. Agric. Com. y Trab. 2: 624–625. 1919.—Reprinted from Revist. Agric. Puerto Rico.

2064. McKay, M. B. **Verticillium wilt of potatoes in Oregon.** Potato Mag. 2[4]: 10–11, 38, 42. *5 fig.* 1919.—*V. alboatrum* may be present in apparently healthy tubers and absent in tubers with discolored strands. It may survive a winter in either end of a tuber from a diseased hill, or in trash from diseased plants in the soil. The fungus first attacks the small roots. It spreads through the soil along the row. Infection in 90 per cent of the hills reduced the yield by 32.5 per cent.—*Donald Folsom.*

2065. McRae, W. **Administration report of the Government Mycologist for the year 1917–18.** Rept. Dept. Agric. Madras 1917-18: 77–80. 1918.—A progress report of the work being done on miscellaneous diseases is given.—*J. I. Lauritzen.*

2066. Melhus, I. E., and L. W. Durrell. **Cereal rusts of small grains.** Iowa Agric. Exp. Sta. Circ. 62. *15 p., 11 fig.* 1919.—The five different rusts commonly attacking the small grains are described in a popular manner. The time of appearance of stem rust (*Puccinia graminis*) in the spring and its spread from the common barberry (*Berberis vulgaris*) is shown in tabular and graphic form. During the past two years (1917 and 1918) a great many barberry bushes have been found in the state growing as hedges in the country and town or as clump plantings on public and private grounds in the cities. Previous to 1917, all of the nurseries in the state carried extensive plantings for distribution. In some cases these plantings covered five acres. In addition to being domesticated, this shrub is at present tending to run wild in some localities, 20 such places having been found. Data at hand show that in 1917 before the barberry eradication movement was begun, there were in Iowa at least a million bushes. Their distribution was general over the state, and they were found in every county, although the largest numbers were found in the larger cities. The relation of crown rust to the various species of buckthorn (*Rhamnus*) in the state is explained. There are three species of buckthorn in Iowa. Two of them have been introduced from Europe and are sold by nurserymen for ornamental and hedge purposes. These are *Rhamnus cathartica* and *R. frangula.* The latter species is very resistant to the alternate stage of crown rust.—*I. E. Melhus.*

2067. Melhus, I. E., and L. W. Durrell. **Studies on the crown rust of oats.** Iowa Agric. Exp. Sta. Res. Bull. 49: 115–144. *6 fig.* 1919.—A progress report dealing largely with factors influencing the growth and reaction of crown rust on oats and different species of *Rhamnus.* The minimum, optimum and maximum temperatures for urediniospore germination are given. The per cent of germination of urediniospores produced in the greenhouse is variable. Urediniospores must be in direct contact with water in order to germinate. Vaseline and paraffine oil in contact with water acted as stimulants. The special form of crown rust on oats uses *Rhamnus cathartica* and *R. lanceolata* as alternate hosts. *R. frangula, R. caroliniana* and *R. alnifolia,* according to the data presented, do not harbor the aecial stage of crown rust of oats.—*I. E. Melhus.*

2068. Miović and Anderlić. **Über Tomatener krankungen.** [Tomato diseases.] Zeitschr. landw. Versuchsw. Österr. 21: 407–415. 1918.—*Phytophthora infestans* and *Gloeosporium phomoides* attacked tomatoes in Dalmatia. The latter fungus attacked only the variety Ficarazzi causing wrinkled, unmarketable fruit. The diseases were controlled by 4 applications of 1 per cent Bordeaux mixture the first application being made in the hot-bed. [Through abst. by Matouschek in Zeitschr. Pflanzenkr. 29: 253–254. 1919 (1920).]—*D. Reddick.*

2069. Mirande, Robert. **Sur une maladie de la Coque de Noix.** [A disease of the shell of walnut (Juglans regia).] Bull. Soc. Path. Veg. France 6: 134–136. *Pl. 1, 6 fig.* Nov.–Dec., 1919.—Nuts of *Juglans* (cultivated) from certain trees show thin places or irregular lesions penetrating the shell. No insect or fungus was associated with the trouble and it is regarded as a physiological disorder or degeneration. It is confined to certain trees which show the disease each year.—*C. L. Shear.*

2070. Moesz, G. **Mykologiai Közlemények. III. Közlemény.** [Mycological investigations. III.] Bot. Közl. 17: 60–78. *11 fig.* 1918.—See Bot. Absts. 5, Entry 1962.

2071. Moesz, G. **Megjegyzés Schilbersky K.—nak a fekete á gabonarozsda tárgyában tett javaslatához.** [Remarks on Schilbersky's lecture on black rust of cereals.] Bot. Közl. 17: 49–51. 1918.—Review of facts concerning overwintering of *Puccinia graminis* and rôle of barberry in its perpetuation and dissemination. Suggests that critical study be made before restrictive measures of the more northerly countries are adopted in Hungary. [Through abst. by Matouschek in Zeitschr. Pflanzenkr. 29: 255-256. 1919 (1920).]—*D. Reddick.*

2072. Munn, M. T. **The seed analyst's responsibility with reference to seed-borne plant diseases.** Proc. Assoc. Official Seed Analysts of North America 1919: 31–35. 1919.

2073. Nicholls, H. M. **Annual report of the Government Microbiologist.** Tasmania Agric. and Stock Dept. Rept. 1916–17: 20–23. 1917. [Appeared, 1918.]—"Owing to the phenomenally wet season, fungous diseases of all kinds were very common in fruit and other crops." Apple scab, "powdery mildew or fire blight," black rot (*Sphaeropsis malorum*) abundant and destructive on apples. *Puccinia pruni* injured stone fruits, generally, including apricots; *Coryneum beyerincki* also was injurious to stone fruits causing shot-hole. Potato blight [*Phytophthora*] was widespread and losses ranged up to 100 per cent. Experiments for the control of a pea disease, caused by *Rhizoctonia*, are reported but were practically without result owing to wet weather. *Peronospora viciae* does some damage to peas.—Iron sulfid spray gave satisfactory control of apple mildew. [See also next following Entries, 2074, 2075.]—*D. Reddick.*

2074. Nicholls, H. M. **Annual report of the Government Microbiologist.** Tasmania Agric. and Stock Dept. Rept. 1917–18: 13–16. 1918.—Diseases much less prevalent than previous year on account of dry season. In addition to notes on apple diseases and potato blight, *Fusarium solani* is reported as the cause of a destructive potato wilt. [See also next preceding and next following Entries 2073, 2075.]—*D. Reddick.*

2075. Nicholls, H. M. **Annual report of the Government Microbiologist.** Tasmania Agric. and Stock Dept. Rept. 1918–19: 20–23. 1919.—*Oidium lactis* has been found to cause rancidity in butter. Slow pasteurisation of cream is effective in prevention.—Potato tubers were subjected to a temperature of 125°F. for 4 hours to kill *Phytophthora*. When planted, they sprouted sooner and more evenly than untreated tubers and made a better crop. Owing to the dry season late blight did not develop in the field. Early blight (*Macrosporium*), wilt (*Fusarium*), scab (*Rhizoctonia*), scurf (*Spondylocladium*), potato moth, and eel-worm were prevalent on potato. *Rhizoctonia* of potato also injures field pea. Fruit diseases occurred as in 1916–17 (see second preceding entry) but were not so serious owing to dry season.—Young apple trees which suddenly wilt and die were found affected with a fungus "identical in every respect with *Fusarium vasinfectum*." Action of fungus seems to be purely mechanical (thrombotic). Cold wet springs are favorable to the disease. Trees are very susceptible up to the eighth year. Indications are that fungus gains entrance at time of budding or grafting.—Somewhat similar disease of apricots is said to be caused by *Nectria cinnabarina*.—Differential stain for mycelium of these two organisms in wood is: very weak solution Delafield's haematoxylin, 24 hours, differentiated in ammoniated distilled water. [See also next preceding Entries, 2073, 2074.]—*D. Reddick.*

2076. Nicholson, C. G. **Some vegetable parasites.** Sci. Amer. 122: 87–97. *4 fig.* 1920.—A popular article on flowering plants and fungi that derive nourishment from other plants.—*Chas. H. Otis.*

2077. Osborne, T. G. B. **Black leg disease of cabbages.** Jour. Dept. Agric. South Australia 23: 107–110. *1 fig.* 1919.—The article contains a brief summary of the history of the disease in South Australia. A detailed description of the symptoms, and remedial measures based upon K. P. Henderson's work (Phytopathology 7: *379–431.* 1918) is given.—*Anthony Berg.*

2078. Osborne, T. G. B. **Two serious new wilt diseases.** Jour. Dept. Agric. South Australia 23: 437. 1919.—Two serious wilt diseases hitherto unrecorded in the state have come to the attention of the author. The one a spotted wilt of tomato which develops on the young leaves, leaf stalk and stems in irregular, brown spots and within a few days the whole plant wilts from above downward. The other is a strawberry wilt. Apparently healthy plants wilt within a few hours in hot weather; though seldom killed outright the first season the plants fail to make thrifty growth or to bear fruit. The disease can be spread by planting offshoots from diseased plants. Healthy plants set out in beds that had a diseased crop the previous season become affected.—*Anthony Berg.*

2079. Osmaston, A. E. **Observations on some effects of fires and on lightning-struck trees in the chir forests of the North Garhwall Division.** Indian Forester 46: 125–131. 1920.—Chir forests were badly burned in 1916 and the trees apparently have not been killed by heat directly but through the subsequent action of insects, especially bark beetles, and fungi. Similar action is seen in trees struck by lightning, the infection spreading to surrounding trees in the group. This may be due to electrical disturbances and action on the cambium as well as to external agencies.—*E. N. Munns.*

2080. Osterwalder. **Vom Apfelmehltau.** [**Apple mildew.**] Schweis. Zeitschr. Obst. u. Weinbau 1918: 161. 1918.—Sulfur and lime-sulfur solution are worthless for control. Best control is early careful cutting and burning of infected twigs. The following varieties are very susceptible: Parkers Pepping, Orleans- and Landsberger-Reinette, Goldreinette von Blenheim, Boiken. [Through abst. by Matouschek in Zeitschr. Pflanzenkr. 29: 261–262. 1919 (1920).]—*D. Reddick.*

2081. Paine, S. G., and C. M. Haenseler. **Decay in potato clamps due to "black-leg."** Jour. Ministry Agric. Great Britain 27: 78–80, 1920.—Cultural studies indicate that some of the trouble from the rotting of potatoes in out-door storage in Britain during the winter of 1918–19 was due to the "black-leg" organism (*Bacillus atrosepticus*). It is not certain whether it was responsible for the initial injury or whether its presence was general in rotting potatoes throughout the country.—*M. B. McKay.*

2082. Pammel, L. H. **Perennial mycelium of parasitic fungi.** Proc. Iowa Acad. Sci. 25: 259–263. 1920.—The author enumerates many species of fungus with perennial mycelium. Of *Ustilago striaeformis* he states "The purpose of this note is to call attention to the fact that the same stool of timothy will produce the smut for years."—*H. S. Conard.*

2083. Peyronel, B. **Sul nerume o marciume nero delle castagne.** [**On the blackening or black rot of chestnuts.**] Staz. Sperim. Agrarie Italiane 52: 21–41. *4 pl.* 1919.—A study of the black rot of chestnuts, a condition distinctly recognisable in the ripe fruit but which, according to the author, is conveyed to the flower at the time of flowering. The causal organism is carefully described and studied in its natural and cultural environments. It is found that the optimum temperature lies between 14°C. and 16°C., while a temperature of 10°C. below 0°C. is not injurious to the organism, although growth is checked during the time of exposure. Higher temperatures than the optimum bring about a luxuriant growth which does not last more than a very few days. On relatively dry media there is the formation of sclerotic tissues that are considered by the author as the adaptations for the tiding over of dry periods rather than cold periods. Microscopically the fungus causing the rot resembles closely the one described by Peglion and by Bainier and with a few differences of a minor importance the incomplete descriptions of the above authors are suited for the description of the present form. Systematically the causal organism has been placed in a newly formed genus under the name *Rhacodiella castaneae* (Banier) Peyronel. Asphyxiation of the fungus which is an obligate aerobe, by means of CO_2 or simple soaking in water for a few days, may prove beneficial if care is then taken to spread the chestnuts to dry in a thin layer in a cool and dry place. Sulphur fumigation was of no avail in the treatment of the fruit. Infected chestnuts being of less specific gravity allows separation from sound nuts by flotation methods.—*A. Bonazzi.*

2084. Pridham, J. T. **An obscure disease in wheat.** Agric. Gaz. New South Wales 31: 229–231. *2 fig.* 1920.—A non-technical description of a wheat trouble is given. Abnormal conditions appear at heading time. Heads have a faded dull appearance, are constricted, and contain shrunken grain. The characters of the disease do not indicate take-all, *Ophiobolus graminis.* Disease not amenable to seed treatments used. Disease has been noticed at points in New South Wales since 1911 but nearly absent several years.—*L. R. Waldron.*

2085. Rambousek. **Über die praktische Anwendung des Sulfins gegen Schimmelpilze und Schädlinge.** [On the applicability of Sulfin for fungous diseases and insect pests.] Zeitschr. Zuckerind. Böhmen 42: 649. 1918.—Sulfin is a new proprietary powder containing sodium bisulfate and gypsum. Results secured thus far are satisfactory and the material is worthy of further test. [Through abstract by Matouschek in Zeitschr. Pflanzenkr. 29: 280. 1919 (1920).]—*D. Reddick.*

2086. Ravaz, L. **Traitement de l'Anthracnose.** [Control of the anthracnose.] Prog. Agric. et Vitic. 74: 103–104. 1920.

2087. Reinking, Otto A. **Phytophthora Faberi Maubl.: The cause of coconut bud rot in the Philippines.** Philippine Jour. Sci. 14: 131–151. *3 pl.* 1919.—The history, distribution, and nature of the disease are outlined, followed by detailed description of field and laboratory studies of the disease. The indications pointed to bacterial agency and an organism resembling *Bacillus coli* was isolated. Inoculations with pure cultures of *Bacillus coli* produced many symptoms of the disease. Bacterial causation, however, was deemed insufficient to account for the rapid dissemination. Trees inoculated with Phytophthora isolated from Cacao proved positive in a large percentage of cases. Later the same fungus was isolated from the woody tissue of the coconut and reinfections proved it to be the cause of bud rot. A taxonomic study revealed the presence of several species of the fungus. Methods of treatment recommended and a bibliography is appended.—*Albert R. Sweetser.*

2088. Rosen, H. R. **The mosaic disease of sweet potatoes.** Arkansas Agric. Exp. Sta. Bull. 167. *16 p., 5 pl.* 1920.—The mosaic disease of sweet potatoes was first identified by the author in 1918 and has since been the subject of study. Isolation and infectivity studies have been carried on, so far with negative results. The disease is classified as a non-infectious, heritable chlorosis. The appearance of the disease is described and illustrated. Roguing diseased plants is recommended as a control for the disease.—*John A. Elliott.*

2089. Salmon, S. C. **Establishing kanred wheat in Kansas.** Kansas Agric. Exp. Sta. Circ. 74. *16 p.* 1919.—See Bot. Absts. 5, Entry 1205.

2090. Sanders, J. G. **A handbook of common garden pests.** Bull. Pennsylvania Dept. Agric. 1[3]: 1–24. *20 fig.* 1918.

2091. Sanders, J. G., and L. H. Wible. **List of owners of commercial orchards and licensed nurserymen in Pennsylvania including list of registered dealers in nursery stock.** Bull. Pennsylvania Dept. Agric. 1[10]: 1–56. 1918.—*C. R. Orton.*

2092. Schellenberg. **Versuche zur Bekämpfung der Peronospora. [Investigations of control of grape downy mildew.]** Schweiz. Zeitschr. Obst- u. Gartenbau **1918**: 81. 1918.—Best mixture is 1 per cent copper sulfate, 1 per cent iron sulfate and 1 per cent hydrated lime. Of proprietary mixtures, Martini mixture is preferable to Bordola paste. [Through abst. by Matouschek in Zeitschr. Pflanzenkr. **29**: 254–255. 1919 (1920).]—*D. Reddick.*

2093. Schilberszky, K. **Hipertrófos paraszemölesök almagyümölcsökön. [Hypertrophied lenticels on fruit of apple.]** Bot. Közlemények **17**: 93. 1918.—The condition is thought to be caused by excessive amount of water in soil. Tissue underlying hypertrophied area appears water soaked. [Through abst. by Matouschek in Zeitschr. Pflanzenkr. **29**: 249. 1919 (1920).]—*D. Reddick.*

2094. Schilberszky, K. **Javaslat a fekete gabonarozsda tárgyában. [A lecture on black rust of cereals.]** Bot. Közlemények **17**: 43–48. 1918.—Summary in German.

2095. Schönfeld, Leo. **Beizen des Hirsesaatgutes. [Disinfecting millet seeds.]** Wiener landw. Zeitg. **68**: 257. 1918.—In Hungary, millet seed is poured through the flame of burning straw to free it from smut. Five per cent copper sulfate is effective but a solution of this strength injures those seeds which are broken in threshing. [Through abst. by Matouschek in Zeitschr. Pflanzenkr. **29**: 255. 1919 (1920).]—*D. Reddick.*

2096. Schøyen, T. H. **Betydningsfulde nyere undersøkelser over furuens blaererust. [Important new investigations on Peridermium pini.]** Tidsskr. Skogbruk **28**: 28–29. 1920.

2097. Škola, Vlad. **Über die Zusammensetzung der durch Rhizoctonia zersetzten Rübe. [Composition of sugar beets destroyed by R.]** Zeitschr. Zuckerind. Böhmen **42**: 135–138. 1918. —Affected tissue contains invert sugar but no saccharose. [Through abst. by Matouschek in Zeitschr. Pflanzenkr. **29**: 263. 1919 (1920).]—*D. Reddick.*

2098. Spiekermann. **Der falsche Kartoffelkrebs. [False potato wart.]** Illustr. landw. Zeitg. **1918**: 153. 1918.—Lesions have the appearance of true wart. Microscopic examination necessary for diagnosis. Cause of false wart not stated. [Through abst. by Matouschek in Zeitschr. Pflanzenkr. **29**: 252. 1919 (1920).]—*D. Reddick.*

2099. Spinks, G. T. **Damping-off and collar rot of tomatoes.** Jour. Bath and West and South Counties Soc. **5, 12**: 128–130. 1917–1918.—Both damping-off and collar rot of tomatoes are ascribed to a fungus placed in the genus *Phytophthora*, but the actual species has not been yet identified. From the results of certain studies it is concluded that the fungus is most active and causes most damage in the first 3 or 4 months of the year and that the infection may be carried on from year to year in the soil.—*M. Shapovalov.*

2100. Stevens, F. L. **Foot-rot diseases of wheat—historical and bibliographic.** Bull. Illinois Nat. Hist. Surv. **13**: 259–286. 1919.—The recent discovery of a foot-rot disease of wheat in southwestern Illinois (Madison County) and the lack of agreement among American and European pathologists as to the cause of this and similar diseases are the reasons given by the author for presenting this preliminary statement. A brief historical review is given in which attention is called to a wide variance among investigators as to symptoms and causes of foot-rot. These points of disagreement are summarized. The body of the publication consists of a bibliography of 188 titles. In some cases brief abstracts are given.—*H. W. Anderson.*

2101. STUART, G. A. D. **Mycology and operations against disease.** Rept. Dept. Agric. Madras 1917–18: 17–20. 1918.—An account of the progress in the study and control of: secondary leaf fall of *Hevea;* a disease of paddy, variety korangu samba (caused by *Piricularia oryzae*); bleeding disease of coconuts; rot of stored potatoes; Palmyra disease; and Mahali disease on Areca palm nuts.—The cause of secondary leaf fall in *Hevea* is *Phytophthora meadii*, which differs from a somewhat similar fungus reported brom Ceylon.—*J. I. Lauritzen.*

2102. STUCKEY, H. P., AND B. B. HIGGINS. **Spraying peaches.** Georgia Agric. Exp. Sta. Bull. 135: 91–101. 1920.—The bulletin discusses briefly peach diseases and peach insects and recommends formulae for controlling sprays. The effects of commercial lime-sulphur upon peach foliage is discussed, the results being obtained from experiments with six plats of Elberta peaches. Solutions of the following densities were used: 1.003, 1.004, 1.005, 1.006, 1.007, 1.008 specific gravity. Five days after application, the following conditions were found: (1) sprayed with lime sulphur of specific gravities 1.003 to 1.004, not injured; (2) 1.005 burned about 5 per cent of the leaves and these fell from the trees; (3) 1.006 approximately 10 per cent of the leaves injured and fallen; (4) 1.007 and 1.008 did not cause the leaves to fall but burned holes where the droplets of spray collected.—*T. H. McHatton.*

2103. THOMAS, P. H. **Annual report of the Assistant Fruit and Forestry Expert.** Tasmania Agric. and Stock Dept. Rept. 1918–19: 19–20. 1919.—See Bot. Absts. 5, Entry 1777.

2104. THOMAS, ROY C. **A new lettuce disease.** Monthly Bull. Ohio Agric. Exp. Sta. 5: 24–25. 1920.—A brief note is given of the discovery of a disease of lettuce new to Ohio observed in lettuce grown under glass. The causal organism is a bacterium which attacks the roots of the plants gaining entrance when they are young seedlings, or when unfavorable cultural conditions result in a checking of growth. Preliminary investigations indicate that the disease is similar to one previously reported from South Carolina.—*R. C. Thomas.*

2105. UICHANCO, LEOPOLD B. **A biological and systematic study of Philippine plant gall.** Philippine Jour. Sci. 14: 527–554. *15 pl.* 1919.—In the present paper only the galls caused by the action of animals, and known as zoocecidia are taken into consideration, which may be caused by insects and arachnida, as practically no work has been done on the galls in the Philippines. This was a virgin field. Fifty-seven species of galls were described and drawn or photographed and the insects were reared from them.—*Albert R. Sweetser.*

2106. VÖCHTING, HERMANN. **Untersuchungen zur experimentellen Anatomie und Pathologie des Pflanzenkörpers. II. Die Polarität der Gewächse. [Experimental anatomy and pathology of the plant body. II. Polarity.]** *vi + 333 p., 12 pl., 113 fig.* Tübingen, 1918.—Review by O. VON K[IRCHNER] in Zeitschr. Pflanzenkr. 29: 242–249. 1919 (1920).

2107. VON K[IRCHNER], O. [Rev. of: VOCHTING, HERMANN. **Untersuchungen zur experimentellen Anatomie und Pathologie des Pflanzenkörpers. II. Die Polarität der Gewächse. (Experimental anatomy and pathology of the plant body. II. Polarity.)** *vi + 333 p., 12 pl., 113 fig.* Tübingen, 1918.] Zeitschr. Pflanzenkr. 29: 242–249. 1919 (1920).

2108. WATERHOUSE, W. L. **A note on the over-summering of wheat rust in Australia.** Agric. Gaz. New South Wales 31: 165–166. 1920.—Observations indicated that volunteer wheat plants probably serve as an important medium in carrying over the rust *Puccinia graminis*. Uredinia were formed at intervals during the summer months.—*L. R. Waldron.*

2109. WECK, R. **Saatgutbehandlung der Wintergerste. [Seed treatment of winter barley.]** Illustrierte Landw. Zeitg. 39: 315. 1919.

2110. WHITEHOUSE, W. E. **Cold storage for Iowa apples. (Third progress report.)** Iowa Agric. Exp. Sta. Bull. 192: 181–216. *14 fig.* 1919.—See Bot. Absts. 5, Entry 1787.

2111. Wilcox, E. Mead. **The nature and classification of plant diseases.** Publ. Nebraska Acad. Sci. 10: 5-14. 1920.—We may recognise four great bases for the classification of plant diseases: taxonomy, etiology, morphology, physiology. The paper closes with a two page classification of plant diseases, with examples, under the captions Ontopathology and Phylopathology, relating respectively to functions having to do with the maintenance of life and those concerned with the perpetuation of the species.—*H. S. Conard.*

SUGAR CANE DISEASES

2112. Ashby, S. F. **Mottling or yellow-stripe disease of sugar cane.** Jour. Jamaica Agric. Soc. 23: 344-347. 1919.—A compiled account covering damage caused, distribution, symptoms, varieties attacked, and control measures of the mottling or yellow-stripe disease of sugar cane, now prevalent in Porto Rico and the southern United States. The disease has not been found to date in Jamaica.—*John A. Stevenson.*

2113. Cross, W. E. **The Kavangire cane.** Louisiana Planter and Sugar Manufacturer 63: 397-399. *1 fig.* 1919.—A discussion of the desirable and undesirable qualities of the Kavangire cane, the variety that has been proved to be immune to the mosaic disease, is given. It is a cane very susceptible to frost and drought injury and its small size also makes it expensive to handle.—*C. W. Edgerton.*

2114. Earle, F. S. **The mosaic or new sugar cane disease.** Louisiana Planter and Sugar Manufacturer 63: 167. 1919.—In a criticism of the article of R. M. Grey (Louisiana Planter and Sugar Manuf. 63: 90. 1919), the behavior of the mosaic disease is stated as being often contradictory yet in the main it is capable of causing an immense loss. A cane stalk once affected with the disease never recovers. It is probable that Grey confused the mosaic with other sugar-cane troubles.—*C. W. Edgerton.*

2115. Edgerton, C. W. **Mosaic or mottling disease of sugar cane.** Louisiana State Univ. Div. Agric. Ext. Circ. 32: 1-6. *1 fig.* 1919.—A popular discussion of the mosaic disease of sugar cane, including a description of the disease, varietal susceptibility, distribution and methods of control.—*C. W. Edgerton.*

2116. Edgerton, C. W., and C. C. Moreland. **Effect of fungi on the germination of sugar cane.** Louisiana Agric. Exp. Sta. Bull. 169. *40 p., 9 pl., 2 fig.* 1920.—The average germination of the buds of sugar cane in Louisiana is around 20 per cent. Among the many factors instrumental in causing this low germination is that of the action of several fungi. The common or serious fungi found on deteriorating seed cane in Louisiana, include *Colletotrichum falcatum*, *Melanconium sacchari*, *Gnomonia iliau*, *Marasmius plicatus*, *Thielaviopsis paradoxa* and species of *Fusarium* and *Scopularia*. Of these, *C. falcatum* seems to cause the most loss in Louisiana. Stalks of seed cane inoculated with this fungus at planting time show an average deterioration of about 50 per cent. Stalks that have a heavy infection of the red rot disease, caused by *C. falcatum*, before cutting, do not deteriorate so rapidly when used for seed as stalks that are inoculated after cutting. The other fungi, with the possible exception of a *Fusarium*, are of little economic importance in Louisiana as far as the germination of the buds is concerned. Preliminary tests in "seed" treatment using corrosive sublimate and formaldehyde have given encouraging results.—*C. W. Edgerton.*

2117. Edgerton, C. W., and others. **The mosaic disease.** Louisiana Planter and Sugar Manufacturer 63: 253-255, 350. 1919.—A stenographic report of a discussion at a meeting of the Louisiana Sugar Planters' Association on the mosaic disease of sugar cane.—*C. W. Edgerton.*

2118. Fawcett, G. L. **The yellow-stripe or mosaic disease in the Argentine.** Louisiana Planter and Sugar Manufacturer 64: 41. 1920.—The mosaic disease has been in Argentina for at least fifteen years. In all the sugar provinces except one, it is impossible to find a plant of a susceptible variety that is free from the disease. The bad effects of this disease seem to be comparatively small as these susceptible varieties have been grown successfully for years. The mosaic disease is not curable and it does not seem to be influenced by the root disease or by fertilization and cultivation.—*C. W. Edgerton.*

2119. GREY, ROBERT M. **The mosaic or mottling disease.** Louisiana Planter and Sugar Manufacturer 63: 199. 1919.—An answer to the communication of F. S. EARLE (Louisiana Planter and Sugar Manuf. 63: 167. 1919). Sugar cane plants affected with the mosaic, and so identified by authorities of the United States Department of Agriculture, recovered from the trouble in 116 days.—*C. W. Edgerton.*

2120. GREY, R. M. **The new cane disease in Cuba.** Louisiana Planter and Sugar Manufacturer 63: 90. 1919.—The mosaic or mottling disease has been in Cuba for a number of years. From observations made at the Harvard Experiment Station, Central Soledad, Cienfuegos, Cuba, the claim is made that the disease causes little or no loss and that stalks will frequently ougtrow the trouble. It is believed that the prevalency of the disease is influenced by such weather conditions as rainfall.—*C. W. Edgerton.*

2121. JOHNSTON, JOHN R. **The new cane disease in Cuba.** Louisiana Planter and Sugar Manufacturer 63: 43. 1919.—The mosaic, yellow-stripe, or mottling disease of sugar cane exists in at least three provinces of Cuba. The disease tends to stunt the growth of the cane, causing a decrease in tonnage. The history of the disease in other countries is discussed and the author considers that cane should be prohibited from entering Cuba from the other countries.—*C. W. Edgerton.*

2122. ZENO, RAFAEL DEL VALLE. **"Mottling" or "Yellow Stripe" disease of sugar cane. (Some facts relative to the importance of the discovery of the "morbid" cause.)** Published privately with two colored plates by author. New York, 1919.—Symptoms of the disease are given as a general yellowing of the leaves, which by close inspection is seen to be caused by interrupted streaks, elongated more or less in the direction of the midrib, of a pale green color. Growth of the plants is slow and "closing" of the rows retarded. Development of the canes is more puny than in the healthy plants; the internodes are spindle shaped. Terminal roots are destroyed and the plant can not obtain sufficient nutritive elements from the soil. Good cultivation has no effect on the course of the disease.—"Not because of greater merit than that of my predecessors, but by the chance of having been guided to the right road I can offer today to my country and to all those who have cane plantations the solution of this vital problem, having discovered the cause of 'mottling' and practical methods for raising plantations completely free from this disease and saving the sugar world millions of dollars."—"Cost of the treatment will vary with the class of labor in each locality, method of application (manual or mechanical), number of cuttings per acre, etc., but it is an insignificant sum, possible to be reckoned always as an ordinary expense in raising plantations of cane. Before any sugar planter need pay for the revelation of the secret of this discovery, a series of experiments demonstrating the truth and efficacy of the treatment will be made before a committee composed of competent agronomists and interested planters."—"The committee, composed of Srs. Georgetty, Benítez, D. E. Colon, Wale & Veve, has stipulated certain conditions to be fulfilled."—The writer makes some general remarks on other diseases, states that the pulling out of diseased stools has no scientific basis and proposes to reveal his secret for a prize. Appended to the paper are credentials consisting of letters of introduction from the governor of Porto Rico, other officials and prominent sugar planters and extracts from statistical reports of the Insular Department of Agriculture, showing decreases in production of sugar from 1916 to 1919.—*E. D. Brandes.*

PHARMACEUTICAL BOTANY AND PHARMACOGNOSY

Heber W. Youngken, *Editor*
E. N. Gathercoal, *Assistant Editor*

2123. Albes, E. **Scented soap from Paraguay oranges.** Sci. Amer. Supplem. 88: 382–383. *5 fig.* 1919. [From the *Pan American Union.*]—Concerns the distillation of oil of petit grain, used for scenting toilet soaps, from the leaves of the bitter orange or bigarrade (*Citrus bigaradia*). There are between 30 and 40 factories operating in Paraguay, employing rather primitive stills. From 500 to 600 pounds of leaves are required to produce about a quart of the ordinary oil of petit grain. The average still will produce about 4 quarts per day. In 1913, the amount of oil exported was 71,322 pounds.—*Chas. H. Otis.*

2124. Anonymous. **A new source of vegetable oil.** Sci. Amer. 122: 399. 1920. [Extract from the Bull. Imp. Inst. United Kingdom Great Britain.]—Note on a semi-siccative oil from the seeds of *Lactuca scariola*, var. *oleifera*.—*Chas. H. Otis.*

2125. Bargellini, G. **Sul 1-2-3-triossiflavone. Contributo alla conoscenza della costituzione della Scutellareina. [On the 1-2-3-trioxy-flavone. Contribution to the knowledge of the constitution of Scutellarein.]** Gaz. Chim. Italiana 49: 47–63. 1919.

2126. Bargellini, G., and E. Peratoner. **Sul 1-3-2′ triossi-flavonolo. Ricerche per la sintesi della Datiscetina. [On 1-3-2′ trioxy-flavonol. Researches on the synthesis of Datiscetin.]** Gaz. Chim. Italiana 49: 64–69. 1919.—See Bot. Absts. 5, Entry 2160.

2127. Cauda, A. **Contenuto in essenza dei semi di senape. [Essence content of mustard seeds.]** Staz. Sperim. Agrarie Italiane 52: 122. 1919.—A short note on the total content of essence in seeds of different species and of the same species cultivated in different regions. *Brassica alba*, *B. nigra* and *B. carinata* were studied and the determination made by bromine oxydation in a paraffin bath and subsequent weighing as sulphate. *B. nigra* seeds were found to contain a higher percentage of essence than *B. alba* and *B. carinata* while seeds from plants grown in northern localities contained greater percentages than the seeds from plants grown in southern regions. Size of seed seems also to have an influence, the smaller having a higher percentage than the larger.—*A. Bonazzi.*

2128. Cohn, Edwin J., Joseph Gross, and Omer C. Johnson. **The isolectric points of the proteins in certain vegetable juices.** Jour. Gen. Physiol. 2: 145–160. *5 tables, 3 fig.* 1919.

2129. Cusmano, G. **Sui principi ipotensivi del Viscum album. [Hypotensive compounds of Viscum album.]** Gaz. Chim. Italiana 49: 225–228. 1919.—The author prepares a solution of the substances found in *Viscum* by dialyzing a decoction of fresh leaves with water. The hypotensive components pass through the membrane, and their solution thus obtained is concentrated on a water bath and extracted with alcohol (96 per cent). At first there is the formation of a homogeneous mixture, but on standing two strata are separated and the lower one is discarded. The supernatant liquid is again concentrated and again extracted with alcohol. As a guide for the separation of the hypotensive compounds the author used the method of injection in the blood stream of the dog.—*A. Bonazzi.*

2130. Dodd, Sydney. **St. John's wort and its effects on live stock.** Agric. Gaz. New South Wales 21: 265–272. 1920.—Deals with the effect of a plant, probably *Hypericum perforatum*, upon the different classes of live stock. Sensitized areas appear upon the body, especially where pigment is deficient. Develops mainly under conditions of insolation. Feeding experiments are described.—*L. R. Waldron.*

2131. McAtee, W. L. **Notes on the flora of Church's Island, North Carolina.** Jour. Elisha Mitchell Sci. Soc. 35: 61–75. 1919.—See also Bot. Absts. 5, Entry 2419.

2132. MOFFAT, C. B. **Some notes on Oenanthe crocata: its character as a poisonous plant.** Irish Nat. 29: 13-18. Feb., 1920.—The "Water hemlock-Dropwort" is notoriously deadly. Many fatal cases are known from eating the plant. DR. CHRISTISON, however, made the discovery that in the vicinity of Edinburgh this species is, for some unknown reason, devoid of toxic properties. In County Wexford the author observed three herds of cows feeding on the plant by preference with no injurious effects. Cases are on record of cattle in other parts of Ireland killed by this poison. Some suggestions are made but no explanation offered.—*W. E. Praeger.*

2133. SALEEBY, N. M. **The treatment of human beriberi with autolyzed yeast extract.** Philippine Jour. Sci. 14: 11-14. 1919.—The extract was prepared by the Bureau of Science, from brewers yeast obtained in Manila, by incubating at 35°C. for 48 hours, then filtering and concentrating to one third the volume in partial vacuum below 60°C. About forty acute cases were treated. The dose for adults was 15-40 cc. and children 2-4 cc. Marked results were noted in less than three days and full relief in a week. This extract seemed to behave much the same as hydrolysed extract of rice polishings, only weaker.—*Albert R. Sweetser.*

2134. SCHÜLER, D. B. **Vergiftungen durch Herbstzeitlose und deren Bekämpfung. [Poisoning by meadow saffron (Colchicum autumnale) and its control.]** Illustrierte Landw. Zeitg. 39: 457. *Fig. 361-363.* 1919.

2135. WELLS, A. H. **The physiological active constituents of certain Philippine medicinal plants. III.** Philippine Jour. Sci. 14: 1-7. *1 pl.* 1919.—As a result of chemical analyses, made in the chemical laboratory of the Bureau of Science, Manila, *Arcangelica flava* (Linn.) Merr. gave 4.8 per cent berberine; *Cassia siamea* Lam., an undetermined alkaloid; and the rhizome of *Geodorum nutans* Ames., 14 per cent of a water soluble adhesive; and *Coriaria intermedia* Mats., a poisonous glucoside, in its leaves and fruit. A bibliography is appended.—*Albert R. Sweetser.*

PHYSIOLOGY

B. M. DUGGAR, *Editor*

CARROLL W. DODGE, *Assistant Editor*

GENERAL

2136. BANCROFT, WILDER D. [Rev. of: HALDANE, J. S. **The new physiology.** *22 x 14 cm., viii+156 p.* J. B. Lippincott Company: Philadelphia, 1919.] Jour. Phys. Chem. 23: 586-587. 1919.

DIFFUSION, PERMEABILITY

2137. BUSCALIONI, L. **Nuove osservazione sulle cellule artificiali. [Further observations on artificial cells.]** Malpighia 28: 403-434. *Pl. 11-12.* 1919.—See Bot. Absts. 5, Entry 1267.

2138. COUPIN, H. **Sur le lieu d'absorption de l'eau par la racine. [Absorption of water by roots.]** Compt. Rend. Acad. Sci. Paris 168: 1005-1008. 1919.—The roots of pea, bean, sunflower, pumpkin, pine, corn, and rice grew more rapidly and produced more laterals when merely the tip was suspended in water than when the whole root was immersed. Growth was extremely slow when corn roots were moistened in the region of the root hairs only. The author concludes that roots absorb water exclusively by the tip and not by the root hairs; the latter protect the root against too rapid drying out and attach the root firmly to soil particles.—*F. B. Wann.*

2139. CURTIS, OTIS F. **The upward translocation of foods in woody plants. I. Tissues concerned in translocation.** Amer. Jour. Bot. 7: 101-124. *4 fig.* 1920.—Attention is called by the author to the general belief that in woody plants food stored in the lower part of the trunk and in the roots passes upward in the spring through the xylem. He brings forward evidence, derived from ringed stems, that this is not the case but that the food travels upward chiefly in the phloem.—If a ring of tissues extending to the cambium is removed at the base of a growing twig, growth above the ring is reduced even if the leaves remain, and practically ceases if the leaves are removed. This check to growth is probably due primarily to a lack of food necessary for energy or for building material. If the leaves are left above the ring, enough food is ordinarily manufactured by them to allow of considerable growth. The author suggests that in some cases, especially where tissues above the ring tend to wilt, the check caused by ringing may be due to an inability of the stem to carry up above the ring such osmotically active substances as carbohydrates, and to a consequent inability to draw up water osmotically. He finds that the osmotic concentration of the sap of a twig above a ring is reduced, and is very markedly so if the twig is also defoliated. He suggests the importance of the distribution of osmotically active substances as a factor in causing polarity.—Ringing of the stem below a fruit was found to check the growth of the fruit.—Ringing of dormant twigs was found to decrease greatly the growth of shoots coming from buds above the ring. Such growth as took place was evidently at the expense of starch stored above the ring and proportional to its amount, for at the cessation of growth this starch had quite disappeared. In several species two rings, separated by from 15 to 107 cm., were cut out from dormant twigs in early April, and the twigs examined for starch and sugar about a month later. In all cases starch was found to be practically absent above the upper ring, very abundant between the rings, and considerably less abundant below the lower ring and throughout a similar twig which was unringed. Tests for sugar above, between, and below the rings gave essentially similar results, sugar being much more abundant between the rings than elsewhere.—From these facts the author concludes that although large amounts of carbohydrates are stored in the xylem, there is no appreciable longitudinal transfer of sugars through this tissue, but that to be translocated the stored food must pass radially into the phloem, where it may readily be carried upward or downward. The author also suggests that at least some of the mineral nutrients from the soil may move primarily through the phloem.—*E. W. Sinnott.*

2140. Kofler, Johanna. **Der Dipmorhismus der Spaltöffnungen bei Pandanus. [Dimorphism of the stomata in Pandanus.]** Oesterreich. Bot. Zeitschr. 67: 186–196. *3 fig.* 1918.

2141. LeFevre, Edwin. **Brine tolerance in certain rot organisms.** [Abstract.] Absts. Bact. 3: 3–4. 1919.—Softening of cucumbers in brine is caused by a wide range of bacteria, among them being organisms causing soft rots, those destroying cellulose, and spore-bearing aerobes. *Bacillus vulgatus* is probably the cause of much of the spoilage, since it has the highest sodium chlorid tolerance and fourth highest acid tolerance of 50 organisms tested. The concentration of salt for preserving cucumbers is between 7 and 8 per cent. [From author's abst. of paper read at scientific session, Soc. Amer. Bact.]—*D. Reddick.*

2142. Loeb, Jacques. **Influence of the concentration of electrolytes on the electrification and the rate of diffusion of water through collodion membranes.** Jour. Gen. Physiol. 2: 173–200. *16 fig.* 1919.—Solutions of electrolytes when separated from pure water by a collodion membrane affect the diffusion through the membrane in a way different from that of non-electrolytes. The latter influence the initial rate of diffusion of water approximately in direct proportion to their concentration, which the writer calls the gas effect, as it follows the laws of gas pressure. This effect of the diffusion of water under the conditions of the experiments was noticeable at concentrations above M/64 or M/32. Solutions of electrolytes may also show this gas pressure effect upon the initial rate of water diffusion, but it commences only at higher concentrations, usually at M/16 or higher. With weaker solutions of electrolytes, the gas effect is not evident, but the rate and direction of diffusion of water is determined more by the electrical charge of water, by the nature of the ions and the charges borne by them. Two rules for the sign of the charge of the water were previously given (Bot. Abst., vol. 3, Entry 1203). With an increase in concentrations of electrolytes up to about M/256 or above, the rate of diffusion of water towards the solution is rapidly increased, which is apparently due to increased attraction for the water by the ions bearing a charge opposite to that borne by water. With a further increase in concentration from M/256 to about M/16. depending somewhat upon the nature of the electrolyte, the rate of the diffusion of water towards the solution is less than that at weaker concentrations, which is apparently due to a more rapid increase in the repelling action of that ion bearing the same charge as the water particles. In fact, this repelling action may become so dominant as to develop negative osmosis when diffusion takes place from the solution toward the pure water decreasing the volume of the solution. Therefore, within the range above stated, the reverse of what would be expected from van't Hoff's law is observed; that is, with an increase in concentration of the electrolyte, the attraction for water diminishes. This was demonstrated with a number of solutions, in some cases when water behaved as if positively charged and repelled by the cations, and in others when it behaved as if negatively charged and repelled by anions, especially those with higher valences. When experimenting to determine the effects of solutions on the diffusion of negatively charged water, it was necessary to use membranes previously treated with gelatin.—*Otis F. Curtis.*

2143. MacDougal, D. T., and H. A. Spoehr. **The solution and fixation accompanying swelling and drying of biocolloids and plant tissues.** Plant World 22: 129–137. 1919.—Desiccated slices of *Opuntia discata* showed vigorous swelling in water, dilute acids, alkalies, and salt solutions; but on being dried after the first swelling, they exhibited a greatly reduced power of swelling. Substances giving the sections their high imbibition capacity are believed to be extracted during the first swelling. The loss during the first swelling was about 7 per cent of the total solids, and mainly amino-acids, hexoses, malates, and salts. Biocolloids like agar and gelatine-agar show similar losses during swelling, about 15 per cent being extracted. Reduced swelling after extraction and drying may also be related to changes in the colloidal mesh, aggregations, or coagulations which cannot be reversed by simple hydration.—*Charles A. Shull.*

2144. Paterno, E. **Origini e sviluppo della crioscopa.** [**Origin and development of cryoscopy.**] Gaz. Chim. Italiana 49: 381–411. 1919.—A historical study and digest of the literature on the subject of cryoscopic methods, and measurements, chiefly considered from the standpoint of pure chemistry.—*A. Bonazzi.*

WATER RELATIONS

2145. Harding, S. T. **Relation of the moisture equivalent of soils to the moisture properties under field conditions of irrigation.** Soil Sci. 8: 303–312. *6 fig.* 1919.—See Bot. Absts. 5, Entry 2320.

2146. Hill, Leonard, and Hargood-Ash, D. **On the cooling and evaporative powers of the atmosphere, as determined by the Kata-thermometer.** Proc. Roy. Soc. London 90B: 438–447. 1919.—Data are presented endorsing the efficiency and applicability of the Kata-thermometer as an instrument for determining the cooling and evaporative powers of the atmosphere.—*R. W. Webb.*

2147. Middleton, Howard E. **The moisture equivalent in relation to the mechanical analysis of soils.** Soil Sci. 9: 159–167. *1 fig.* 1920.—See Bot. Absts. 5, Entry 2331.

2148. Sayre, J. D. **The relation of hairy leaf coverings to the resistance of leaves to transpiration.** Ohio Jour. Sci. 20: 55–75. *7 fig.* 1920.—Mullein (*Verbascum thapsus*) leaves offer greater resistance to water loss in darkness than in light and less in wind than in still air, when compared to tobacco (*Nicotiana sp.*) leaves, and they respond as much or more to environmental changes. Removal of hairs does not alter resistance of mullein leaves in still air and light; but slightly decreases resistance in wind and light, and greatly decreases resistance in still air and darkness, because the cuticular surface is more exposed. In darkness stomata are closed and water loss is cuticular. Removal of hairs increases cuticular water loss only. As water loss from surface of mesophyll cells is 20 to 40 times cuticular water loss, leaf hairs may be disregarded as protection against ordinary wind and light.—*H. D. Hooker, Jr.*

MINERAL NUTRIENTS

2149. Ames, J. W., and C. J. Schollenberger. **Calcium and magnesium content of virgin and cultivated soils.** Soil Sci. 8: 323–335. 1919.—See Bot. Absts. 5, Entry 2293.

2150. De Turk, Ernest. **Potassium-bearing minerals as a source of potassium for plant growth.** Soil Sci. 8: 269–301. 1919.—See Bot. Absts. 5, Entry 2290.

2151. Howard, L. P. **The relation of certain acidic to basic constituents of the soil affected by ammonium sulfate and nitrate of soda.** Soil Sci. 8: 313–321. 1919.—See Bot. Absts. 5, Entry 2261.

2152. Lamprov, E. **Les engrais radioactifs. [Radioactive fertilizers.]** Rev. Hortic. [Paris] 91: 393–394. 1919.—See Bot. Absts. 6, Entry 123.

2153. Rudolfs, W. **Influence of sodium chloride upon the physiological changes of living trees.** Soil Sci. 8: 397–425. *7 pl.* 1919.—The application of 1 to 10 pounds of sodium chloride to oak, birch, and maple trees shows a favorable effect in the smaller amounts and a toxic action in the larger. Maple is most easily affected, followed by birch and oaks. The higher trees are more resistant than the lower ones of the same species.—*W. J. Robbins.*

2154. Shive, John W. **The influence of sand upon the concentration and reaction of a nutrient solution for plants.** Soil Sci. 9: 169–179. 1920.—A nutrient solution, consisting of potassium dihydrogen phosphate, calcium nitrate, and magnesium sulphate was added to washed or unwashed sea sand and after longer or shorter intervals of contact, the solution was drawn off and the freezing point and hydrogen-ion concentration determined. With washed sand no adsorptive effect was noted. The unwashed sand during the first 24 hour period reduced the freezing point of the solution 8.5 per cent but did not affect the reaction. By renewing the solution, the adsorptive effect of the washed sand was eliminated.—*W. J. Robbins.*

2155. Winterstein, E. **Über das Vorkommen von Jod in Pflanzen. [The occurrence of iodine in plants.]** Zeitschr. Physiol. Chem., **104**: 54–58. 1919.

PHOTOSYNTHESIS

2156. Anonymous. **Starch formation in leaves, and photographic prints.** Sci. Amer. Monthly 1: 416. 1920.

METABOLISM (GENERAL)

2157. Allen, Paul W. **"Rope" producing organisms in the manufacture of bread.** [Abstract.] Absts. Bact. 3: 4. 1919.—*Bacillus subtilis* and 14 other very similar spore-bearing organisms produced "rope" in bread during the first 30 hours when bread was stored at 25°. *Bacillus bulgaricus*, *B. aerogenes viscosus*, and *Bact. lactis viscosus* failed to produce "rope" under similar conditions.—In a commercial bread oven the internal temperature of a loaf did not reach 100° although the oven was held uniformly at 204°. [From author's abstract of paper read at scientific session, Soc. Amer. Bact.]—*D. Reddick.*

2158. Allen, Paul W. **The manufacture of starch and other corn products as affected by "rope" producing organisms.** [Abstract.] Absts. Bact. 3: 4. 1919.—"In a wet process of the manufacture of products from corn, 'rope' production often develops during hot weather, causing serious difficulties in the operation of the reels and cutting down the yield of starch per bushel of corn.—*B. bulgaricus* was repeatedly isolated from viscous starch and gluten liquors. This organism was also isolated from the corn as it arrived in the cars. Normal starch and gluten liquors became exceedingly viscous when inoculated with it and held at 37°C. for twenty-four hours."—[Author's abst. of paper read at scientific session, Soc. Amer. Bact.]

2159. BARGELLINI, G. **Sul 1-2-3-triossiflavone. Contributo alla conoscenza della costituzione della Scutellareina. [On 1-2-3-trioxy-flavone. Contribution to the knowledge of the constitution of scutellarein.]** Gaz. Chim. Italiana 49: 47-63. 1919.

2160. BARGELLINI, G., AND E. PERATONER. **Sul 1-3-2. triossi-flavonolo. Ricerche per la sintesi della Datiscetina. [On 1-3-2. trixoy-flavonol. Researches on the synthesis of Datiscetin.]** Gaz. Chim. Italiana 49: 64-69. 1919.—A theoretical study of the chemical constitution of the derivatives of the glucoside of *Datisca cannabina* and of the synthetic preparation of the following compounds: 2'oxy-4'-6'-2-trimethoxy-calcone, 1-3-2' tri-methoxy-flavonone, 1-3-2' trimethoxy-isonitrous-flavonone and of 1-3-2" trimethoxy-flavonol.—*A. Bonazzi.*

2161. BUNKER, JOHN W. M. **Some factors influencing diphtheria toxin production.** [Abstract.] Absts. Bact. 3: 8-9. 1919.—"Toxin production depends upon growth, but growth alone does not assure toxin. By controlling conditions which affect growth, toxin production can in turn be influenced." The initial hydrogen-ion concentration of the medium (optimum P_H 7 to 7.5), the final hydrogen-ion concentration (range bounded by P_H 7.8 to 8.25), and the presence of suitable polypetids in the medium are among the controllable factors which influence toxin production by *Bacterium diphtheriae.* [From author's abst. of paper read at scientific session, Soc. Amer. Bact.]—*D. Reddick.*

2162. COHN, EDWIN J., JOSEPH GROSS, AND OMER C. JOHNSON. **The isoelectric points of the proteins in certain vegetable juices.** Jour. Gen. Physiol. 2: 145-160. *3 fig.* 1919.

2163. DE BESTEIRO, D. C., AND M. MICHEL-DURAND. **Influence de la lumière sur l'absorption des matières organique du sol par les plantes. [The influence of light on the absorption by plants of the organic materials of the soil.]** Compt. Rend. Acad. Sci. Paris 168: 467-470. 1919.—The pea, a heliophile plant which cannot adapt its assimilation of CO_2 by the green leaves to a condition of feeble light, is likewise incapable of increasing the absorptive power of the roots whereby it might draw upon the soil for a larger quantity of organic carbon. There is for this plant no parallelism or compensation between the absorption of CO_2 by the leaves and the absorption of organic carbon by the roots.—*G. M. Armstrong.*

2164. DONK, P. J. **Some organisms causing spoilage in canned foods, with special reference to flat sours.** [Abstract.] Absts. Bact. 3: 4. 1919.—"A thermophilic organism was isolated from cans of 'flat sour' corn. This is a large aerobic, facultative anaerobic bacterium, Gram negative, spore-bearing and non-motile, with minimum, optimum and maximum temperatures of 45°, 60° and 76°C. respectively. It grows well on all ordinary culture media and does not produce gas when grown in any of the standard sugar-broths. Pure culture introduced into sterile cans of a variety of canned foods (corn, peas, string beans, pumpkins, and tomatoes) produced the same characteristic 'flat sour.'"—Twenty other organisms were identified from various sources. Critical conditions are being determined especially with reference to temperature and acidity, for both vegetative and spore forms. [From author's abst. of paper read at scientific session, Soc. Amer. Bact.]—*D. Reddick.*

2165. DRUMMOND, JACK CECIL. **Researches on the fat-soluble accessory substance. I. Observations upon its nature and properties.** Biochem. Jour. 13[1]: 81-94. 1919.—Temperature, rather than oxidation or hydrolysis, appears to be the chief agent in the inactivation of fat-soluble A of natural animal fats. Destruction occurs at temperatures ranging from 100° to 37°, the severity varying with the temperature. Destruction at relatively low temperatures suggests that the fat-soluble A may be an ill-defined and labile substance. The substance may be extracted with alcohol, but not with acid or water, and it has not been identified with any of the recognized components of fat.—*R. W. Webb.*

2166. Drummond, Jack Cecil. **Researches on the fat-soluble accessory substance. II. Observations on its rôle in nutrition and influence on fat metabolism.** Biochem. Jour. 13[1]: 95–102. 1919.—The presence of fat soluble A in the diet of adult rats is essential to their health, while the absence of this substance increases their susceptibility to bacterial diseases. A deficiency of fat-soluble A causes no characteristic pathological lesion in adult rats; does not directly influence the absorption of fats, and appears to play no important part in the absorption of fatty acids nor in their synthesis into fats.—*R. W. Webb.*

2167. Durbin, H. E., and M. J. Lewi. **The preparation of a stable vitamine product and its value in nutrition.** Amer. Jour. Med. Sci. 159: 264–286. 1920.—Following a review of the literature on the relation of vitamines to growth in animals the authors describe a method of preparing a stable vitamine from corn, autolyzed yeast, and orange juice, the final product being a grayish, non-hygroscopic powder which retains its effectiveness for 5 months or longer. Experiments showing the efficiency of this vitamine in treating malnutrition in children, pigeons, and guinea pigs are described.—*Harris M. Benedict.*

2168. Eddy, Walter H. **The vitamine.** Absts. Bact. 3: 313–330. 1919.—This is a bibliographic review dealing with the following: historical, methods of preparation, sources, structure, function, and organisms requiring vitamines for development. The bibliography contains 236 titles.—*D. Reddick.*

2169. Gillespie, L. J. **Colorimetric determination of hydrogen-ion concentration without buffer mixtures, with especial reference to soils.** Soil Sci. 9: 115–136. *1 fig.* 1920.—See Bot. Absts. 5, Entry 2324.

2170. Grace, L. G., and F. Highberger. **Variations in the hydrogen ion concentration in uninoculated culture medium.** Jour. Infect. Diseases 26: 457–462. 1920.—A medium consisting of Liebig's Beef Extract 0.3 per cent, Difco Peptone 1 per cent, NaCl 0.5 per cent, glucose 1 per cent, and adjusted to a reaction of P_H 6.4, 6.8, 7.2, 7.6, and 8.0, was found to change in reaction not only on autoclaving, but also on allowing the control medium to incubate. Plain broth, free from glucose, did not give as great variations in reaction as the glucose broth. It is suggested that the acid is formed in the medium by the breaking up of the glucose and perhaps also by the formation of amino acids from the peptone.—*Selman A. Waksman.*

2171. Hägglund, Erik. **Beiträge zur Kenntnis des Lignins.** [Lignin.] Arkiv. Kemi, Min., Geol. 7[6]: 1–20. 1918–19.

2172. Hammer, B. W., and D. E. Bailey. **The volatile acid production of starters and of organisms isolated from them.** Iowa Agric. Exp. Sta. Res. Bull. 55: 223–246. 1919.—A study of a number of "starters" of good quality showed that more than one organism was present. Experimental data showed that the high volatile acid content of starters is not altogether due to the action of *Bacterium lactis acidi*.—*Florence Willey.*

2173. Harrington, Geo. T. **Comparative chemical analyses of Johnson grass seeds and Sudan grass seeds.** Proc. Assoc. Official Seed Analysts of North America 1919: 58–64. 1919. —See Bot. Absts. 5, Entry 1148.

2174. Hess, Alfred F., and Lesser J. Unger. **The effect of heat, age, and reaction on the antiscorbutic potency of vegetables.** Proc. Soc. Exp. Biol. and Med. 16: 52–53. 1919.—Results obtained from experiments with guinea pigs show that the antiscorbutic value of vegetables decreases with increase in age of the vegetables and also with their subjection to high temperature. Their efficacy remains the same for both acid and alkaline reactions. However, the effect of alkalinization or of heat is greatly influenced by the time-factor.—*R. W. Webb.*

2175. Lucius, Franz. **Über die Trennung von Glykose und Fructose.** [Separation of glucose and fructose.] Zeitschr. Untersuch. Nahrungs-u. Genussmittel 38: 177–185. 1919.

2176. Mellanby, John. **The composition of starch. I. Precipitation by colloidal iron. II. Precipitation by iodine and electrolytes.** Biochem. Jour. 13[2]: 28-36. 1919.—A detailed account is given of the effects produced by colloidal iron and by iodine, in the presence and absence of electrolytes, on a solution of potato starch in water. The results indicate that, while starch grains are composed chiefly of amylogranulose, they contain various polymers ranging in complexity from amylodextrin to amylocellulose; however, the relative quantities of the dextrin and the cellulose compounds are small.—*R. W. Webb.*

2177. Molliard, Marin. **Influence de certaines conditions sur la consommation comparée du glucose et du lévulose par le Sterigmatocystis nigra a partir du saccharose. [The influence of certain conditions on the comparative consumption of glucose and levulose (derived from inversion of saccharose) by Sterigmatocystis nigra.]** Compt. Rend. Acad. Sci. Paris 167: 1043-1046. 1918.—The ratio of consumption of glucose and levulose in a modified Raulins' solution varies upon the addition of different quantities of HCl and with changes in the nitrogen ratio, the glucose being used more rapidly. The utilization of the two sugars appears to depend on a function of the mycelium and not on the differential diffusion of the sugars. —*G. M. Armstrong.*

2178. Northrup, Zae. **Agar-liquefying bacteria.** [Abstract.] Absts. Bact. 3: 7. 1919. See Bot. Absts. 5, Entry 1969.

2179. Osborne, Thomas B., and Lafayette B. Mendel. **The extraction of "fat-soluble vitamine" from green foods.** Proc. Soc. Exp. Biol. and Med. 16: 98-99. 1919.—Contrary to the statements of several investigators, the writers experimentally demonstrate that it is both possible and practicable to obtain "fat-soluble" vitamine from green foods by means of ether extraction.—*R. W. Webb.*

2180. Rivière, G. **De la progression de la maturation dans les poires a couteau. [Progression of ripening in table pears.]** Jour. Soc. Nation. Hortic. France 20: 306-307. 1919.— See Bot. Absts. 5, Entry 1770.

2181. Schowalter, E. **Zur Titration von Zuckerarten. [Titration of sugars.]** Zeitschr. Untersuch. Nahrungs- u. Genussmittel 38: 221-227. 1919.

2182. Tasaki, Buhachirô, and Ushio Tanaka. **On the toxic constituents in the bark of Robinia pseudacacia L.** Jour. Coll. Agric. Tokyo Imp. Univ. 3: 337-356. *2 fig.* 1918.—The toxic constituent proved to be a glucoside and has been named "Robitin." It amounts to about 1 per cent of the fresh bark. The symptoms of intoxication in animals are discussed.—*B. M. Duggar.*

2183. Waksman, Selman A. **On the metabolism of actinomycetes.** [Abstract.] Absts. Bact. 3: 2-3. 1919.

2184. Waksman, Selman A., and Jacob S. Joffe. **Studies in the metabolism of actinomycetes. IV. Changes in reaction as a result of the growth of actinomycetes upon culture media.** Jour. Bact. 5: 31-48. 1920.—The hydrogen-ion concentration of various media was tested before and after the growth of various forms of Actinomyces with a view to determine the changes in the media due to the different substances added as sources of carbon and nitrogen. It was found that no appreciable amount of acid was formed from the carbohydrates studied which included glucose, lactose, sucrose, maltose, mannitol, glycerol, starch, inulin, and sodium acetate. When sodium nitrate was added to the medium with the different carbohydrates, an alkaline reaction resulted; if sodium nitrite was added instead of the nitrate an acid was produced. When ammonium salts of strong acids are present as the only source of nitrogen, the medium tends to become distinctly acid; with proteins and amino acids the reaction may be unchanged or may become either acid or alkaline depending on the species, source of carbon, and the hydrogen-ion concentration of the medium.—*Chester A. Darling.*

2185. Zellner, J. **Über die chemische Zusammensetzung der Agave americana L. nebst Bemerkungen über die Chemie der Succulenten im allgemeinen. [Chemical composition of Agave americana and the chemistry of succulents in general.]** Zeitschr. Physiol. Chem. **104:** 2–10. 1919.

METABOLISM (NITROGEN RELATIONS)

2186. Bokorny, T. **Notizen über Harnstoff und einige andere N-Quellen der grünen Pflanzen. [Urea and a few other sources of nitrogen for green plants.]** Pflüger's Arch. Physiol. 172: 466–496. 1918.

2187. Conn, H. J., and R. S. Breed. **The use of the nitrate-reduction test in characterizing bacteria.** New York Agric. Exp. Sta. [Geneva] Tech. Bull. 73. *31 p.* 1919.—This is a reprint of an article in: Jour. Bact. 4: 267–290. 1919.—*Abstractor.*

2188. Gibbs, W. M. **The isolation and study of nitrifying bacteria.** Soil Sci. 8: 427–481. *4 pl., 1 fig.* 1919.—Pure cultures of *Nitrosomonas* and *Nitrobacter* isolated from the soil were grown on washed agar or silicic acid gel containing suitable nutrient salts. On plates the colonies were extremely small and required a microscope for their study. Pure cultures of *Nitrosomonas* and *Nitrobacter* did not produce visible growth when inoculated into bouillon. Pure cultures of these organisms were maintained in a liquid medium indefinitely. Sodium chloride at a concentration of 1 per cent was very toxic for *Nitrosomonas.* The soil extract used to prepare nutrient solutions for these organisms did not prove toxic. The thermal death point for *Nitrobacter* was 56–58°C. and for *Nitrosomonas,* 53–55°. At 28°C. *Nitrobacter* in pure culture produced a maximum of 527 mgm. of nitrogen as nitrates per 100 cc. of solution. *Nitrosomonas* at 28°C. in pure culture produced a maximum of 218.9 mgm. of nitrogen as nitrites per 100 cc. of solution.—*W. J. Robbins.*

2189. Mayer, A., and G. Schaeffer. **Extension aux cas des microbes de la notion d'acides aminés indispensables. Rôle de l'arginine et de l'histidine dans la culture du bacille de Koch sur milieux chimiquement définis. [The indispensable amino acids for microörganisms. The rôle of arginine and of histidine in the culture of Koch's bacillus on synthetic media.]** Compt. Rend. Soc. Biol. 82: 113–115. 1919.

2190. Meisenheimer, Jakob. **Die stickstoffhaltigen Bestandteile der Hefe. [The nitrogen constituents of yeast.]** Zeitschr. Physiol. Chem. **104:** 229–283. 1919.

2191. Saillard, Émile. **Balance de l'azote pendant la fabrication du sucre. Précipitation des matières albuminoides de la betterave par l'acide sulfureux, les bisulfites et les hydrosulfites. [The balance of nitrogen during the refining of sugar. Precipitation of the albuminoids of the beet by sulphurous acid, bisulphites and hydrosulphites.]** Compt. Rend. Acad. Sci. Paris 170: 129–130. 1920.—The determination of the relative amounts of nitrogenous compounds present in the sugary extract of the beet at the various steps in the commercial refining of sugar is given. There is also included the effect of various reagents used in the processes in precipitating these nitrogenous compounds.—*C. H. and W. K. Farr.*

2192. Waksman, Selman A. **Studies in the metabolism of actinomycetes. III. Nitrogen metabolism.** Jour. Bact. 5: 1–30. 1920.—The utilization of different nitrogenous compounds by several different species of Actinomycetes and the transformation of these substances due to the action of the organisms are considered. Various nitrogenous compounds were tested, and glycerol or glucose was used principally as the source of carbon. The conclusions reached are: the Actinomycetes do not utilize atmospheric nitrogen; proteins and amino acids furnish the best sources of nitrogen, amides being utilized to a limited extent; nitrates and nitrites are utilized fairly well; ammonium salts are poor sources of nitrogen if glycerol is used as a source of carbon, but if glucose is used these salts are readily utilized; the production of ammonia from proteins and amino acids is not characteristic of the group, although some may be produced on continued incubation. Pigments are produced by many species when grown in media containing proteins and amino acids. [See also Bot. Absts. 3, Entries 2860, 2883.]—*Chester A. Darling.*

METABOLISM (ENZYMES, FERMENTATION)

2193. André, G. **Sur l'inversion du sucre de canne pendant la conservation des oranges. [The inversion of sucrose in oranges during storage.]** Compt. Rend. Acad. Sci. Paris 170: 126–128. 1920.—Oranges were cut in two, one-half being analyzed at once for the amounts of citric acid, sucrose, and invert sugar present. The other half was deposited in a container in which was also placed a small vessel of toluene, and an analysis was made of this portion after an interval of 4 or 5 months. From 11.65 to 57.33 per cent of the sucrose originally present is changed during this period to invert sugar. The amount of citric acid remains about the same during the interval, although specimens differ in the original amount. The rate of inversion is more rapid at certain times during this period than at others. The rate of inversion of sucrose by citric acid was also determined *in vitro* at the concentrations obtaining in the expressed orange juice. 94.08 per cent of the sucrose is inverted in 78 days. The rate is thus faster *in vitro* than in the orange tissue. The variations in the rate of inversion within the tissue are explicable on the basis of lack of homogeneity. It does not appear that the rate of inversion is affected by the amount of acid present.—*C. H. and W. K. Farr.*

2194. Anonymous. **Fresh information concerning yeast.** Sci. Amer. Monthly 1: 417–420. 1920.—Certain investigations on yeast in progress at the Berlin Institute of Fermentation and at the Mellon Institute at Pittsburgh are described.—*Chas. H. Otis.*

2195. Euler, H. v., and E. Moberg. **Invertase und Gärungsenzyme in einer Oberhefe. [Invertase and ferment enzymes in surface yeast.]** Arkiv Kemi, Min., Geol. 7[12]: 1–17. 1918–19.

2196. Euler, Hans v., and Olof Svanberg. **Enzymchemische Studien. [Enzyme chemistry.]** Arkiv Kemi, Min., Geol. 7[11]: 1918–19.

2197. Giaja, J. **La levure vivante provoque-t-elle la fermentation du sucre uniquement par sa zymase? [Does the living yeast cell induce fermentation merely by zymase?]** Compt. Rend. Soc. Biol. 82: 804–806. 1919.

2198. Grigaut, A., F. Guérin, and Mme. Pommay-Michaux. **Sur le mesure de la protéolyse microbienne. [Estimation of microbic proteolysis.]** Compt. Rend. Soc. Biol. 82: 66–70. 1919.

2199. Hammer, B. W. **Studies on formation of gas in sweetened condensed milk.** Iowa Agric. Exp. Sta. Res. Bull. 54: 211–220. *2 fig.* 1919.—Gas formation in sweetened condensed milk was found to be due to a budding organism *Torula lactis-condensi*. There was a variation in different brands of condensed milk in their susceptibility to fermentation with the yeast studied. The milk solids may retard the growth, since the yeast may grow in a saturated sucrose solution.—*Florence Willey.*

2200. Harvey, R. B. **Apparatus for measurement of oxidase and catalase activity.** Jour. Gen. Physiol. 2: 253–254. 1920.

2201. Hérissey, H. **Sur la conservation du ferment oxydant des champignons. Preservation of the oxidizing ferment of fungi.]** Compt. Rend. Soc. Biol. 82: 798–800. 1919.

2202. Kopeloff, Nicholas, S. Byall, and Lillian Kopeloff. **The effect of concentration on the deteriorative activity of mold spores in sugar.** Louisiana Planter and Sugar Manufacturer 64: 270–271. 1920.—Spores of *Aspergillus sydowi*, *Aspergillus niger*, and *Penicillium expansum* are responsible for some of the deterioration of sugar and sugar products. This deterioration increases with a decreased concentration of the molasses or of the films around the sugar crystals.—*C. W. Edgerton.*

2203. Kopeloff, Nicholas, and Lillian Kopeloff. **The deterioration of manufactured sugar by molds.** Louisiana Planter and Sugar Manufacturer 63: 202–206. 1919.—The data given in this article have been abstracted from another source (Kopeloff, Nicholas, and Lillian Kopeloff. The deterioration of cane sugar by fungi. Louisiana Agric. Exp. Sta. Bull. 166. *72 p. Pl. 1-2, fig. 1.* 1919.)—*C. W. Edgerton.*

2204. Lemoigne. **Fermentation butylèneglycolique du saccharose par les bactéries du groupe du Bacillus prodigiosus. [Butylèneglycolic fermentation of saccharose by bacteria of the group Bacillus prodigiosus.]** Compt. Rend. Soc. Biol. 82: 234-236. 1919.

2205. Lemoigne. **Réaction spécifique du 2-3-butylèneglycol et de l'acétylméthylcarbinol, produits de la fermentation butylèneglycolique. [The specific reaction of 2-3-butyleneglycol and of acetylmethylcarbinol as products of butyleneglycolic fermentation.]** Compt. Rend. Acad. Sci. Paris 170: 131-132. 1920.—The group of bacteria including *Bacillus lactis aerogenes* and *B. coli* which accomplish the fermentation of butyleneglycol is found capable of very accurate detection by oxidising the products of this fermentation with ferric chlorid and the treatment of the compound thus formed with a nickel salt. The reaction is highly sensitive and specific.—*C. H. and W. K. Farr.*

2206. McGuire, Grace, and K. George Falk. **Studies on enzyme action. XVIII. The saccharogenic actions of potato juice.** Jour. Gen. Physiol. 2: 215-227. 1920.—A study was made to determine the effect of saccharogenic enzymes of potato juice on carbohydrates added as well as those contained in the juice. Amylase was present and was most active both upon the starch of the juice and upon added starch at a hydrogen ion concentration of P_H 6 to 7, which corresponded to that of the normal juice. Sucrase was present and was most active upon the sucrose (or raffinose) present in the juice, as well as upon added sucrose at a hydrogen ion concentration of P_H 4 to 5. No maltase was detected.—*Otis F. Curtis.*

2207. Oelsner, Alice, and A. Koch. **Über den abweichenden Verlauf der Alkoholgärung in alkalischen Medien. [Irregular course of alcoholic fermentation in alkaline media.]** Zeitschr. Physiol. Chem. 104: 175-181. 1919.

2208. Prinsen Geerligs, H. C. **Manufacture of glycerin from molasses.** Louisiana Planter and Sugar Manufacturer 63: 268-269. 1919. [Translated from: De Suikerindustrie 19: 195-202, by F. W. Zerban.]—An account of the fermentation process involved in the manufacture of glycerin.—*C. W. Edgerton.*

2209. Went, F. A. F. C. **On the course of the formation of diastase by Aspergillus niger.** Proc. K. Akad. van Wetenschappen te Amsterdam 21: 479-493. *3 fig.* 1919.—The fungus was grown on a liquid medium using glucose and NH_4NO_3 as sources of C and N. The fungus mats were ground with kieselguhr and extracted with the culture fluid. The quantity of diastase was determined by following the time interval required for the disappearance of starch from a starch solution of known strength, using a dilute iodine solution as indicator. Destruction of the enzyme in the mycelium takes place from the beginning, but this is negligible at first in comparison with the production of the enzyme, A maximum of production is reached in about 5 days from the commencement of germination, after which the total quantity declines rapidly. The nutrient fluid never shows more than a small part of the total enzyme, and this perhaps from dead cells.—*C. R. Hursh.*

METABOLISM (RESPIRATION)

2210. Anonymous. **How age affects the respiration of leaf cells.** Sci. Amer. Monthly 1: 310. 1920.—A brief report of several investigations of respiratory phenomena, and especially those of M. Nicholas in: Revue Générale de Botanique 30, No. 335, 1918.—*Chas. H. Otis.*

2211. Linhart, George A. **The free energy of biological processes. Preliminary paper.** Jour. Gen. Physiol. 2: 247-251. 1920.—This is a brief statement of a problem which is being started to determine by thermodynamic calculations the efficiency in the use of energy from the carbohydrate of a culture solution during the process of nitrogen fixation by *Azotobacter.* —*Otis F. Curtis.*

2212. Nicolas, G. **Contribution a l'étude des relations qui existent dans les feuilles, entre la respiration et la présence de l'anthocyane. [Relations which exist in the leaves between respiration and the presence of anthocyanin.]** Rev. Gén. Bot. 31: 161–178. 1919.—Comparative studies of the respiration of red and green leaves of the same species were made. It was found that leaves which become red as a result of some external influence (for example, light intensity, low temperature, or attacks of parasites) and those leaves which are red when young, becoming green later in their development, show an ihtensity of respiration greater than the green leaves of the same species. This is especially true with regard to the amount of oxygen absorbed. The leaves which are normally red, that is, turn red in old age, have a much lower respiratory intensity than the green leaves of the same species. The influence of old age furnishes sufficient explanation for this lower value. The respiratory quotient (CO_2/O_2) is, with one exception, always lower in the cases of the red leaves. These results indicate a greater fixation of oxygen in the red leaves than in the green leaves. Analyses for acidity showed in every case a greater acidity in the red leaves. The author states that in the leaves accidentally reddened there is a greater accumulation of soluble carbohydrates. He thinks that the greater acidity of the red leaves is due to the presence of these compounds, resulting in a greater fixation of oxygen and a consequent lowering of the respiratory quotient. —*R. S. Nanz.*

2213. Peirce, G. J. **Testing seeds with a thermometer.** Sci. Amer. Monthly 1: 259. 1920.—The vitality, germinating and growing power, cleanness and soundness of seeds can be determined, according to the kind of seed, by their temperature behavior when placed in sterile water in Dewar flasks or thermos bottles for 2 days.—*Chas. H. Otis.*

ORGANISM AS A WHOLE

2214. Boyer, G. **Études sur la biologie et la culture des champignons supérieurs. [Biology and culture of higher fungi.]** Mem. Soc. Sci. Phys. Nat. Bordeau xVII. 2: 233–344. *IV pl., 20 fig.* 1918.—See Bot. Absts. 5, Entry 1931.

2215. Brenchly, Winifred E. **Some factors in plant competition.** Ann. Appl. Biol. 6: 142–170. *Pl. 5, 10 fig.* 1919.

2216. C. A. H. [Rev. of: Lumière, Anguste. **Le mythe des symbiotes. (The myth of symbiosis.)** *xi+205 p.* 8°. Masson: Paris, 1919.] Jour. Botany 58: 26. 1920.

2217. Jivanna Rao, P. S. **The formation of leaf-bladders in Eichornia speciosa, Kunth (Water hyacinth).** Jour. Indian Bot. 1: 219–225. *5 fig.* 1920.—See Bot. Absts. 5, Entry 1893.

2218. Manaresi, A. **Sulla biologia florale del pesco. 2 nota. [Floral biology of the peach. 2nd note.]** Staz. Sperim. Agrarie Italiane 52: 42–67. 1919.—See Bot. Absts. 5, Entry 1757.

2219. Shreve, Forrest. **Physiology of the mangrove.** [Rev. of: Bowman, H. H. M. **Ecology and physiology of the red mangrove.** Proc. Amer. Phil. Soc. 56: 589–672. *Pl. 4–9.* 1917.] Plant World 22: 146–147. 1919.

GROWTH, DEVELOPMENT, REPRODUCTION

2220. Calkins, Gary N. **The effect of conjugation.** Proc. Soc. Exp. Biol. and Med. 16: 57–60. 1919.—From a study of *Uroleptus mobilis*, the writer presents data showing that the absence of conjugation promotes a noticeable physiological weakness ultimately ending in natural death, while the presence of conjugation promotes a rejuvenescence of the protoplasm. —*R. W. Webb.*

2221. Chambers, Mary H. **The effect of some food hormones and glandular products on the rate of growth of Paramecium caudatum.** Biol. Bull. [Woods Hole] 36: 82–91. 1919.—As a food hormone potato extract has little effect on the division rate. The influence of yeast is evident in the resulting increase of the division rate. Contrasting results were obtained with pituitary solution added to the basis fluid. Suprarenal extract caused an increase in the rate of division.—*C. R. Hursh.*

2222. Linossier, G. **Sur le développment de l'Oidium lactis en milieux artificiels. Influence de la quantité de semence sur le poids de la récolte.** [**The development of Oidium lactis in artificial media. Influence of the quantity of inoculum on the weight of the fungous product resulting.**] Compt. Rend. Soc. Biol. 82: 240–242. 1919.

2223. MacDougal, D. T. **Hydration and growth.** Carnegie Inst. Wash. Publ. 297. *17 x 25 cm. V+176 p., 52 fig.* 1920.—The author prepared biocolloids by mixing proteins, usually of plant origin, such as bean or oat protein, with agar, gum arabic, prosopis gum, tragacanth, or opuntia mucilage. The colloidal suspension of these mixtures in water was partially dried in thin plates and the hydration (that is, the amount of water taken up when sections of these plates were immersed in solutions) was measured by means of an auxograph developed especially for this purpose. Solutions of acids, alkalis, and salts were employed and a rather close parallelism was shown between the swelling of these biocolloids and cell masses, such as sections of joints of opuntia, cotyledons of beans, and leaves of various plants. In this connection the point is brought out that vegetative cell masses, such as are responsible for growth, are composed of colloids predominantly of a carbohydrate character, frequently of pentosan nature. These pentosans do not dissociate and their swelling capacity is less in electrolytes than in pure water. The hydration of carbohydrates is retarded by hydrogen ions.—Biocolloids behave in much the same way as do cell massses, in nutrient solutions and in bog and swamp waters. Under fluctuating or alternating hydration effects, the basis of xerophily and succulence, the writer details experiments in which biocolloids were subjected to alternate treatments of acids and alkalis in solution. As a result of this treatment, an alternate swelling and shrinking of the biocolloid was brought about. He considers these phenomena as related to the structural variation of leaves of *Castilleia latifolia;* these leaves being thin and highly acid when growing under mesophytic conditions while succulent and less acid leaves in arid locations. Temperature effects and water deficit, or unsatisfied hydration capacity, both in biocolloids and cell masses, are discussed.—Growth of tissues consists of two fundamental features, hydration of the colloidal material of the plasma and the arrangement of additional colloidal material in colloidal structures with entailed additional capacity for absorbing water. The character of the hydration depends upon the character of the cell colloids, proteinaceous colloids showing increases of hydration capacity with acidity, while when the colloidal material is more largely carbohydrate—such as pentosans—the reverse is apparently the case. Nutrient salts always modify hydration capacity. The author is directing his studies toward an analysis of the phenomena of plant growth based on the physicochemical properties of colloid gels, especially with reference to imbibition and swelling.—*Lon A. Hawkins.*

2224. Seifriz, William. **The length of the life cycle of a climbing bamboo. A striking case of sexual periodicity in Chusquea abietifolia Griseb.** Amer. Jour. Bot. 7: 83–94. *5 fig.* 1920.—The author notes the fact that several species of bamboo display sexual periodicity, flowering at intervals of a definite number of years. *Chusquea abietifolia*, of the Blue Mountains of Jamaica, went through such a flowering period in 1918, during which practically all individuals blossomed, produced seed and died. The next year the species was represented only by seedlings, except for one small area discovered by the author in an unusually arid situation where the plants were still thriving and flowerless. The only previous flowering period recorded for this species was in 1885, thus establishing a cycle of 33 years, very similar to that of the Indian *Bambusa arundinacea*, which is 32 years.—The author discusses possible factors which may cause such a periodicity and shows that seasonal differences, particularly in moisture, are probably insufficient to explain them, and suggests that the problem may be of the same nature as that of puberty and senility in organisms. No sufficient explanation is as yet forthcoming for the remarkable fact that fully 98 per cent of the individuals of the species come into flower simultaneously over a great stretch of country.—*E. W. Sinnott.*

2225. Sieglinger, John B. **Temporary roots of the sorghums.** Jour. Amer. Soc. Agron. 12: 143–145. 1920.—Under greenhouse conditions the radicle is the only temporary root developed in sorghums. Shortly after germination the first node develops below the surface and from this node the first permanent roots develop.—*F. M. Schertz.*

MOVEMENTS OF GROWTH AND TURGOR CHANGES

2226. Bremekamp, C. E. B. **Theorie des Phototropismus. [The theory of phototropism.]** Recueil Trav. Bot. Néerland. 15: 123–184. *Fig. 1–14.* 1918.

2227. Jivanna Rao, P. S. **Note on the geotropic curvature of the inflorescence in Eichornia speciosa Kunth (water hyacinth).** Jour. Indian Bot. 1: 217–218. *1 fig.* 1920.—Bending of the floral axis begins immediately after the flowers close, and results in complete submergence of the inflorescence. The reaction is geotropic rather than hydrotropic.—*Winfield Dudgeon.*

GERMINATION, RENEWAL OF ACTIVITY

2228. Bastin, S. L. **Colored glass for seed germination.** Sci. Amer. 122: 165. *1 fig.* 1920.

2229. Duysen, F. **Ueber die Keimkraftdauer einiger landwirtschaftlich Wichtiger Samen. [The vitality of certain agriculturally important seeds.]** Illustrierte Landw. Zeitg. 39: 282–283. 1919.—See Bot. Absts. 5, Entry 1132.

2230. Martin, J. N., and L. E. Yocum. **A study of the pollen and pistils of apples in relation to the germination of the pollen.** Proc. Iowa Acad. Sci. 25: 391–410. *Fig. 163–166.* 1920. —See Bot. Absts. 5, Entry 1759.

TEMPERATURE RELATIONS

2231. Bancroft, Wilder D. [Rev. of: Griffeths, Ezer. **Methods of measuring temperature.** *23 x 17 cm., xi + 174 p.* Philadelphia: J. B. Lippincott Company, 1918.] Jour. Phys. Chem. 23: 286–288. 1919.—The review is chiefly concerned with methods for measuring temperatures above the boiling point of water.—*H. E. Pulling.*

2232. Bronfenbrenner, J., W. T. Bovie, and Estelle M. Wolff. **A simple arrangement for measuring the rate of heat penetration during sterilization.** [Abstract.] Absts. Bact. 3: 6. 1919.—A detailed description of the apparatus, with drawings, will appear in the Journal of Industrial and Engineering Chemistry.—*Authors.*

2233. Crocker, William. **Optimum temperatures for the after-ripening of seeds.** Proc. Assoc. Official Seed Analysts of North America 1919: 46–48. 1919.—See Bot. Absts. 5, Entry 1123.

2234. Shreve, Edith Bellamy. **The rôle of temperature in the determination of the transpiring power of leaves by hygrometric paper.** Plant World 22: 172–180. *1 fig.* 1919.—Thermoelectric measurement of the temperature of the cobalt chloride slip used in determining the index of transpiring power in plants shows that the temperature of the slip varies so little from that of the air temperature that the latter may be used in calculating the indices. Similarly, in standardizing the cobalt slips over a porous evaporating surface in a small closed room, the air temperature may be used instead of the temperature of the slip without significant error.—*Charles A. Shull.*

RADIANT ENERGY RELATIONS

2235. Dubois, Raphael. **Luminous living creatures.** Sci. Amer. Monthly 1: 9-12. *7 fig.* 1920. [Translated from Science et la Vie (Paris).]—Devoted mainly to a discussion of luminous animal life; but briefly considers luminous fungi and certain photobacteria.—*Chas. H. Otis.*

2236. Pulling, Howard E. **Sunlight and its measurement.** Plant World 22: 151–171, 187–209. *5 fig.* 1919.—The author presents a general discussion of the nature, distribution, and variability in amount of solar radiation reaching the earth, as modified by extra-terrestrial influences, and by atmospheric conditions. Three general methods of measuring radiation are discussed: radiometry, photometry, and actinometry. The difficulties involved in each method, their limitations, the precautions to be observed in manipulating the instruments, and the interpretations of measurements are considered. An extensive bibliography accompanies the text.—*Charles A. Shull.*

2237. Raunkiaer, C. **Über das biologische Normalspektrum.** [The biological "normal spectrum."] Kgl. Danske Vidensk. Selskab. Biol. Meddel. 1[4]: 1–18. 1918.

2238. Schanz, Fritz. **The effects of light on plants.** Sci. Amer. Monthly 1: 12–16. 1920. [Translated from the *Biologisches Centralblatt* (Berlin).]—Some of the topics considered are: how light affects the albumens of plants; substances which act as catalyzers; the meaning of colors in flowers; and effect on plants of varying intensity of light.—*Chas. H. Otis.*

TOXIC AGENTS

2239. Breasola, M. **Le devitalizzazione dei semi di Cuscuta.** [The killing of Cuscuta seeds.] Staz. Sperim. Agrarie Italiane 52: 193–207. 1919.—See Bot. Absts. 5, Entry 1112.

2240. Ciamician, G., and C. Ravenna. **Sul contegno di alcune sostanze organishe nei vegetali. Nota XI.** [On the behavior of certain organic substances in plants. XIth contribution.] Gaz. Chim. Italiana 49: 83–126. *Pl. 1–2, fig. 1–20.* 1919.—The present contribution is divided in two parts. Part I. The authors study the effect on the growth of beans (germinated in cotton and distilled water) of repeated doses of one per thousand solutions of the substances investigated. In nearly every case when galvanized iron containers were used instead of glass, there was a distinct reduction in toxicity of the compounds studied. The results may be summarized as follows: Mono-methyl-amine was slightly toxic while di-methyl-amine and tri-methyl-amine were more toxic in the order named. Ammoniacal salts, urea, pyridine and uric acid show no toxic action in the conditions studied while tetra-methyl-ammonium tartrate and tetra-ethyl-ammonium tartrate, piperidine, nicotine, and theobromine are very slightly, if at all, toxic. The function of the methyl group in toxicity is brought out very plainly by the fact that potassium salicylate is very slightly toxic while methyl salicylate is distinctly toxic. A list is given of the substances found to be toxic under the conditions mentioned. Part II. This section is given to the study of the oxidative changes undergone by some organic compounds when incubated with spinach pulp in the presence of adequate oxygen and of small amounts of toluol as an antiseptic. Attention is also given to the inoculation of some compounds into living maize and to the changes undergone by these in the living organism. Two examples will indicate the direction of the results. Succinic acid, which by the action of light is transformed to acetic aldehyde, acetic and propionic acids, also glyoxal, is changed by plant enzymes into acetic aldehyde and a compound decomposed by emulsion. Lactic acid in the light yields acetic acid and acetic aldehyde, while only the latter compound results when acted upon by enzymes. In respect to the above the general conclusions is that the enzymes of spinach leaves have a selective oxidizing function which in some cases does not equal the action of light, though surpassing it in other cases. With respect to the behavior of organic compounds inoculated into maize and tobacco the results obtained point to the fact that compounds very resistant to oxidation, such as pyridine and benzoic acid, are only found in very small amounts in the extract of the plats after inoculation. The strong oxidizing power of plants and especially of living plants may not be due to the ordinary oxydases, but more probably to protoplasmic enzymes insoluble in water and apparently also in glycerin.—*A. Bonazzi.*

2241. Malisoff, William, and Gustav Egloff. **Ethylene.** Jour. Phys. Chem. 23: 65–138. 1919.—This is a collection "on a logically convenient basis" of the physical and chemical data on ethylene, including references to its effects on plants. A bibliography of 324 citations is appended.—*H. E. Pulling.*

2242. MAQUENNE, L., AND E. DEMOUSSY. **Sur la distribution et la migration du cuivre dans les tissus des plantes vertes.** [The occurrence and translocation of copper in the tissues of green plants.] Compt. Rend. Acad. Sci. Paris 170: 87–93. 1920.—Chemical analyses were made of various parts of 27 types of cultivated herbaceous and woody plants and in some cases of the expressed sap of such parts with a view to determining the amount of copper present. The cupro-zinc-ferrocyanid method was employed, 3 grams of dry vegetable matter being used for each test. Copper is found to be present in all plants tested and in all the parts which were analyzed. The amount varies from 0.25 mgm. per liter of centrifuged expressed sap of potato to 40 mgm. per kilogram of dry leaf substance of lettuce. Copper is found in greatest abundance in cells which are active in growth or metabolism, hence the authors conclude that its translocation is controlled by nutritive processes or processes accompanying metabolism.—*C. H. and W. K. Farr.*

2243. WINSLOW, C.-E. A., AND DOROTHY F. HOLLAND. **The disinfectant action of glycerol in varying concentrations.** Proc. Soc. Exp. Biol. and Med. 16: 90–92. 1919.—Glycerol in 9 per cent solution exerts no appreciable effect upon the viability of *Bacillus coli*, but in strengths of 28–100 per cent there is a progressively increasing "disinfecting" action, nine-tenths of the bacteria being killed in 3 hours at 100 per cent.—*R. W. Webb.*

2244. WOGLUM, R. S. **Is it safe to fumigate while trees are in bloom?** California Citrograph 5: 190. *1 fig.* 1920.—See Bot. Absts. 5, Entry 1788.

MISCELLANEOUS

2245. BANCROFT, WILDER D. **The colors of colloids. II. Reflection and refraction.** Jour. Phys. Chem. 23: 1–35. 1919. **III. Reflection and visibility.** *Ibid.* 23: 154–185. 1919. **IV. Interference and diffraction.** *Ibid.* 23: 253–282. 1919. **V. Metallic and vitreous lustre.** *Ibid.* 23: 289–347. 1919. **VI. Blue eyes.** *Ibid.* 23: 356–361. 1919. **VII. Bluefeathers.** *Ibid.* 23: 365–414. 1919. **VIII. Metallic colors.** *Ibid.* 23: 445–468. 1919. **IX. Colloidal metals.** *Ibid.* 23: 554–571. 1919. **X. Glasses and glazes.** *Ibid.* 23: 603–633. 1919. **XI. Gems.** *Ibid.* 23: 640–644. 1919.—This is a collection of excerpts and abstracts, which includes numerous examples, some biological, chiefly from standard works, on the physical optics of the phenomena incompletely indicated by the sub-titles.—*H. E. Pulling.*

2246. BANCROFT, WILDER D. [Rev. of: ALEXANDER, JEROME. **Colloid chemistry. An introduction with some practical applications.** *17 x 12 cm., vi+90 p.* D. Van Nostrand Co.: New York, 1919.] Jour. Phys. Chem. 23: 441–442. 1919.

2247. BANCROFT, WILDER D. [Rev. of: BECHHOLD, H. **Colloids in biology and medicine.** Translated by J. G. M. BULLOWA. *24 x 16 cm., xiv+464 p.* D. Van Nostrand Co.: New York, 1919.] Jour. Phys. Chem. 23: 513–515. 1919.—"It is a great pleasure to welcome an English translation of this excellent book."—*Reviewer's summary.*

2248. BANCROFT, WILDER D. [Rev. of: OSTWALD, WOLFGANG. **A handbook of colloid chemistry.** (Translated by M. H. FISCHER with notes added by EMIL HATSCHEK.) *2nd ed., 14 x 17 cm., xvi+284 p.* P. Blakiston's Son & Co.: Philadelphia, 1919.] Jour. Phys. Chem. 23: 364. 1919.—With a few exceptions, chiefly notes on the viscosity of colloids, the volume is the same as the first edition and does not represent the present knowledge of the subject.—*H. E. Pulling.*

2249. BANCROFT, WILDER D. [Rev. of: PRIDEAUX, E. B. R. **The theory and use of indicators.** *22 x 15 cm., ix+375 p.* D. Van Nostrand & Co.: New York, 1918.] Jour. Phys. Chem. 23: 203–204. 1919.

2250. BANCROFT, WILDER D. [Rev. of: WILLOWS, R. S., AND E. HATSCHEK. **Surface tension and surface energy.** *2nd ed., 19 x 13 cm., viii+114 p.* P. Blakiston's Son & Co.: Philadelphia, 1919.] Jour. Phys. Chem. 23: 443. 1919.—"Books like these are interesting and worth while, but condensation seems to lead more often than necessary to inaccuracy of statement."—*Reviewer's summary.*

2251. Carles, P. **La prune d'ente et les pruneaux d'Agen: Explication scientifique de leur preparation et des moyen de les conserver temporairement pour l'Europe et de facon indéfinie pour l'exportation mondiale. [A scientific account of methods used in preparing "prunes of Agen" for foreign and domestic consumption.]** Mém. Soc. Sci. Phys. Nat. Bordeaux VII. 2: 219–232. 1918.—See Bot. Absts. 5, Entry 1866.

2252. Kopeloff, Nicholas. **Micro-organisms in the sugar factory.** Louisiana Planter and Sugar Manufacturer 64: 14–15. 1920.—This is in continuation of the experiments published in Louisiana Agric. Exp. Sta. Bull. 166. 1919. The results obtained in 1919 agree with those of the previous year. In the sugar factory, the greatest number of molds and bacteria is found in the raw juice. The clarification process reduces the number in the other sugar products.—*C. W. Edgerton.*

2253. Laborde, J. **Recherches sur le vieillissement du vin. [Aging of wine.]** Mém. Soc. Sci. Phys. Nat. Bordeaux VII. 2: 37–75. 1918.

2254. MacInnes, L. T., and H. H. Randell. **Diary produce, factory premises and manufacturing processes: The application of scientific methods to their examination.** Agric. Gaz. New South Wales 31: 255–264. *9 fig.* 1920.—The authors give the results of an investigation relative to the bacterial flora of dairy products at various stages of manufacture and of the various substances with which the products come in contact, including the air of the butter factory. Not only are plat counts given of the bacteria, yeasts, and molds, but a classification is made relative to the physiological action of the various organisms. Suggestions are also presented in regard to creamery methods.—*L. R. Waldron.*

2255. Murray, Benjamin L. **Standards and tests for reagent chemicals.** *400 p.* Van Nostrand Co.: New York, 1920.

2256. Seidell, Atherton. **Solubilities of inorganic and organic compounds.** *2nd ed., 867 p.* Van Nostrand Co.: New York, 1920.

SOIL SCIENCE

J. J. Skinner, *Editor*
F. M. Schertz, *Assistant Editor*

ACIDITY AND LIMING

2257. Bancroft, Wilder D. [Rev. of: Brideaux, E. B. R. **The theory and use of indicators.** *22 x 13 cm. ix + 375 p.* D. Van Nostrand & Co.: New York, 1917. $5.00.] Jour. Phys. Chem. 23: 203–204. 1919.

2258. Corson, Geo. E. **The use of lime on Iowa soils.** Iowa Agric. Exp. Sta. Circ. 58. *7 p.* 1919.

2259. Fippin, Elmer O. **The status of lime in soil improvement.** Jour. Amer. Soc. Agron. 12: 117–124. 1920.—A general discussion of liming of soils.—*F. M. Schertz.*

2260. Howard, L. P. **The reaction of soil as influenced by the decomposition of green manure.** Soil Sci. 9: 27–39. 1920.—The lime requirements of land on which corn has grown since 1894 but a part of which has for about 25 years grown rye or legumes shows that no acidity has developed from the use of rye as a cover crop. The legumes, however, have during the same time considerably increased the lime requirement. In plot experiments, with the same soil, green rye increased the lime requirement twice as much as an equal weight of green clover.—*W. J. Robbins.*

2261. Howard, L. P. **The relation of certain acidic to basic constituents of the soil affected by ammonium sulfate and nitrate of soda.** Soil Sci. 8: 313–321. 1919.—Studies made on limed and unlimed plots which have been treated with ammonium sulfate or sodium nitrate show that the hydrogen ion concentration in the unlimed ammonium sulfate treated plot is very similar (about P_H 4) to that produced by even quite large additions of aluminium salts to buffer solutions. Extractions with potassium chloride solution and 0.2 normal hydrochloric acid solution remove relatively large amounts of aluminum and iron from the soil of the unlimed ammonium sulfate treated plot.—*W. J. Robbins.*

2262. Lipman, J. G., and A. W. Blair. **The lime factor in permanant soil improvement. 1. Rotation without legumes.** Soil Sci. 9: 83–90. 1920. **2. Rotation with legumes.** *Ibid.* 9: 91–114. 1920. A 5-year rotation of corn, oats, wheat and 2 years of timothy was grown on plots which were unlimed or which received 1 ton of lime as carbonate per acre for the first 5 years and 2 tons of lime per acre for the second 5 years. The total yields of dry matter and of nitrogen for the 10-year period for the limed and unlimed plots were essentially the same. Analyses of the soil at the beginning of the experiment and after each 5-year period showed a loss of nitrogen from both limed and unlimed plots but a greater loss from the limed plots. Four 5-year rotations each containing a leguminous crop were carried out on plots which were unlimed or which received 1000, 2000 or 4000 pounds per acre of calcium or magnesium limestone. During a 10-year period, the limed plots yielded distinctly larger crops and more total nitrogen than the unlimed. Analyses of the soil show in most cases an amount of nitrogen in the limed plots equal to or greater than that in the unlimed. The magnesium limestone was slightly superior to the calcium limestone.—*W. J. Robbins.*

2263. MacIntire, W. H. **The liberation of native soil potassium induced by different calcic and magnesic materials.** Soil Sci. 8: 337–395. *Pl. 1. 19 fig.* 1919.—The results of five years experiments show that practical or economical applications of burnt calcareous limestone, burnt dolomitic limestone, ground calcareous limestone or ground dolomitic limestone will not effect a direct chemical liberation of native soil potassium.—*W. J. Robbins.*

2264. Stutzer, A. **Beiträge zur Dungekalkfrage. [A contribution to the calcium fertilizer problem.]** Illustrierte Landw. Zeitg. 39: 333–334. 1919.

2265. [Tansley, A. G.] **Investigations on soil.** [Rev. of: Hartwell, B. L., F. R. Pember and L. P. Howard. **Lime requirement as determined by the plant and the chemist.** Soil Sci. 279–282. 1919.] Jour. Ecol. 7: 214. 1919.

2266. Walker, Seth S. **The effect of aeration and other factors on the lime requirement of a muck soil.** Soil Sci. 9: 77–81. 1920.—Air-drying a black muck soil increases the lime requirements. The increase in lime requirements was less in a stirred moist portion than in a water covered undisturbed portion. The lime requirement of stored moist samples increased but that of stored dry samples decreased. Soil neutralized with calcium carbonate and stored moist showed a greater increase in lime requirement than unneutralized soil.—*W. J. Robbins.*

FERTILIZATION

2267. Beckwith, Charles C. **The effect of certain nitrogenous and phosphatic fertilizers on the yield of cranberries.** Soil Sci. 8: 483–490. 1919.—See Bot. Absts. 5, Entry 1723.

2268. Blair, A. W. **Barium phosphate experiments.** Amer. Fert. 52: 142–144. 1920.—Experiment was made comparing barium phosphate and other phosphate materials. Beans and corn were grown. Practically no increased crop production was secured from the use of barium phosphate.—*J. J. Skinner.*

2269. Forman, L. W. **Reclaiming Iowa's "push" soils.** Iowa Agric. Exp. Sta. Bull. 191: 162–176. *5 fig.* 1919.

2270. Frear, William. **Some notes of fertilizers and the war.** Bull. Pennsylvania Dept. Agric. 1[1]: 29–33. 1918.—A brief summation of the past and present sources of supply of potash, nitrogen and phosphoric acid with remarks concerning the difficulties which are being encountered among the domestic manufactures of fertilizers.—*C. R. Orton.*

2271. Harrison, W. H. **Report of the Imperial Agricultural Chemist.** Sci. Rept. Agric. Res. Inst. Pusa 1918–19: 35–45. 1919.—A summary of the work carried on during the year at the Agricultural Research Institute, Pusa, India, and a program for 1919–20. From studies in the method of retention of superphosphate in soil, it is concluded that the phosphate is held in non-calcareous soils by absorption, and in calcareous soils by chemical combination, and therefore the range of application and method of employment of superphosphate as fertiliser must be different in the two types of soil.—Sugar cane (*Saccharum officinarum*) stored in windrows in the North-West Frontier Province shows increasing content of both glucose and sucrose, but other changes render the final sucrose yield nearly constant with continued storage. Immediately following heavy rainfall there is rapid deterioration of the cane.—In fertilizer experiments with rice (*Oryza sativa*), green manure combined with ammonium sulphate gave an increase in yield almost exactly proportional to that given by sulphate alone. —*Winfield Dudgeon.*

2272. Jacob, A. **Beeinträchtigung der Bodenstruktur durch Kochsalz-Düngung.** [Injury of the soil structure through applications of sodium chloride.] Illustrierte Landw. Zeitg. 39: 420–421. 1919.

2273. Jordan, W. H., and G. W. Churchill. **An experience in crop production.** New York Agric. Exp. Sta. [Geneva] Bull. 465. *20 p.* 1919.—See Bot. Absts. 5, Entry 1164.

2274. Mitscherlich, Eilh. Alfred. **Zum Gehalt der Haferpflanze an Phosphorsäure und seinen Beziehungen zu der durch eine Nährstoffzufuhr bedingten Ertragserhohung.** [On the phosphoric acid content of the oat plant and its relation to the increased yield resulting from the addition of nutrients.] Jour. Landw. 67: 171–176. *1 fig.* 1919.

2275. Münter. **Pflanzenanalyse und Düngerbedürfnis des Bodens.** [Plant analysis and fertilizer requirement of the soil.] Jour. Landw. 67: 229–266. 1919.—The following results reported were obtained on the Lauchstedt loessal loam soil with winter wheat when fertilised with different materials: Fertilising with potassium and phosphoric acid increased the silicic acid content of the straw, fertilising with nitrogen decreased it.—Fertilising with potassium and phosphoric acid decreased the nitrogen, calcium and magnesium content of the straw; nitrogen increased it.—The nitrogen content in the grain was decreased by potassium and increased by phosphoric acid.—The chemical analysis of the wheat plants of a fertilised plat gave no sure indication of the fertiliser need of the soil.—The better the growing season, the more does nitrogen control the formation of organic substance, especially in the grain, therewith the total calcium, magnesium, potassium and phosphoric acid taken up. In poorer growing seasons potassium influences more the plant production. Phosphoric acid is apparently indifferent.—Nitrogen, potassium or phosphoric acid used alone first influences the straw.—The weather condition of any year exerts a strong influence upon the taking up of nitrogenous matter, sometimes even more than the fertiliser applied, thereby rendering the percentages of nitrogen resulting from incomplete fertiliser applications unreliable in indicating fertiliser needs of the soil.—The nitrogen requirement of the Lauchstedt soil may be determined by the quantities of N, CaO, and MgO in the wheat plant. When the sum of N, CaO and MgO in grain and straw for 1 hectar amounts to more than 90 kgm., or in grain more than 60 kgm., or in straw more than 30 kgm., then there is sufficient nitrogen present in the soil.—If after subtracting the sum of the N+CaO+MgO percentages from the potassium percentage the result is positive, the potassium content of the soil is sufficient for plant production, if it is negative, potassium is lacking.—The plants from the plats without fertilizer and with full fertilizer usually contain the same percentages of N and P_2O_5. Only the potassium content of the straw is higher in the fully fertilized plot than in the unfertilized plot. A comparison of the plant analysis of unfertilized and fully fertilized plots gives no information as to the plant food in a soil. The fertilizer requirement of a soil becomes evident if the plants of two incompletely fertilized plots are investigated, e.g., plots receiving (1) N, and (2) $P_2O_5+K_2O$ applications of fertilizer. If thereupon the ratio of $N:K_2O$ is less than 100:200 potassium is lacking; if it is wider, then sufficient potassium is present. If the ratio $N:P_2O_5$ from the nitrogen plot is wider than 100:35 it lacks in phosphoric acid; if less, then no lack exists. If from the $P_2O_5+K_2O$-plot the ratio of $N:P_2O_5$ is less than 100:60 it lacks in nitrogen. If the ratio of $SiO_2:N$ is wider than 100:6 there is not sufficient N present; if less, the N content is sufficient for wheat growth. If the N percentage in the wheat straw found for the N-plot is considered as 100, then enough N is present in the soil of the $P_2O_5+K_2O$-plot when the ratio of the 2 percentages is less than 100:60.—*C. E. Leighty.*

2276. Reimer, F. C., and H. V. Tartar. **Sulfur as a fertilizer for alfalfa in Southern Oregon.** Oregon Agric. Exp. Sta. Bull. 163. *40 p. 9 fig.* 1919.—Various fertilizers containing sulfur, such as flowers of sulfur, superphosphate, gypsum, iron sulfate, ammonium sulfate, potassium sulfate, magnesium sulfate and sodium sulfate, on various types of soil generally increased the yields of clover and alfalfa very greatly. Most of the soils experimented with were well supplied with potassium, calcium, magnesium, and iron but contained only limited amounts of sulfur. None of them were acid, and none contained noticeable amounts of alkali. Analyses of the alfalfa plants which had received applications of sulfate fertilizers showed that they had larger root systems with more nodules on them and that they contained much more sulfur, more protein, and more nitrogen. In the hay from the sulfur fertilized plats from 71 to 79 per cent of the sulfur was in the organic form, the remainder in the sulfate form, while from the unfertilized plats it was all in the organic form. Up to the present time the returns from the use of superphosphate have not been greater than those from calcium sulfate alone. Flowers of sulfur produce as marked results as does calcium sulfate but a somewhat longer period is required since it must first be changed to the sulfate form before it can be utilized by the plants. On soils deficient in lime, flowers of sulfur should be used only in conjunction with liberal quantities of lime or rock phosphate to avoid conditions of acidity.—*E. J. Kraus.*

SOIL BIOLOGY

2277. Bornebusch, C. H. **Bedømmelse om Skovjordens Godhed ved Hjaelp af Bundfloraen.** [Judging the quality of soil by the flora.] Dansk Skovforenings Tidsskr. 5: 37–50. 1920.

2278. Fellers, C. R., and F. E. Allison. **The protozoan fauna of the soils of New Jersey.** Soil Sci. 9: 1–25. *Pl. 1–4.* 1920.—Protozoa were found in all soils examined, the number of species ranging from 2 to 28. About 5000 per gram of soil were found. It is believed that in normal New Jersey soils, the protozoa exist mainly in a nontrophic state.—*W. J. Robbins.*

2279. Geilmann, [—]. **Untersuchung des Bakteriennährpräparates der Superphosphatfabrik Nordenham.** [Investigation of the bacterial food preparation of the Nordenham superphosphate factory.] Jour. Landw. 67: 209–227. 1919.—The superphosphate factory at Nordenham has introduced a peat preparation which is designed to furnish food material to soil bacteria and to stimulate them to greater activity. The preparation itself is not supposed to act as a fertilizer, but only to bring about nitrogenous fertilization through increased bacterial activity. Better physical condition and higher productive power of the soil and prevention of lodging of grain crops should then result. These investigations have shown: (1) the absolute ineffectiveness of the preparation; (2) that an increase in nitrogen content of the soil does not result from use of the preparation; (3) that it does not act in the least as nitrogenous fertilizer; and (4) that it does not result in increased bacterial activity either in the soil or in nutrient solutions, but that any good results are due to the $CaCO_3$ content.—*C. E. Leighty.*

2280. GIBBS, W. M. **The isolation and study of nitrifying bacteria.** Soil Sci. 8: 412–481. *4 pl., 1 fig.* 1919.—See Bot. Absts. 5, Entry 2188.

2281. GREIG-SMITH, R. **Contributions to our knowledge of soil-fertility. No. XVI. The search for toxin-producers.** Proc. Linnean Soc. New South Wales 34: 142–190. 1918.—This paper is one of a series on the subject of soil toxins. In the earlier papers it was shown that soil extracts sometimes contain bacterio-toxic substances. Investigations on the possibility that these toxic substances are formed by bacteria, moulds and amoebae are reported. These organisms were grown in various media and under varying conditions; and in all cases, the signs of toxicity to the test organism *Bacillus prodigiosus* which became manifest could be attributed to an alteration in the reaction of the media. This toxic effect was found to be of a different order from that previously noted with soil extracts.—*E. Truog.*

2282. HUTCHINSON, C. M. **Report of the Imperial Agricultural Bacteriologist.** Sci. Rept. Agric. Res. Inst. Pusa 1918–19: 106–114. 1919.—The report summarises investigations in progress during the year under report in nitrification; nitrogen fixation; green manuring; biological analyses of soils; indigo manufacture; pebrine disease of the silkworm; and sterilisation of water.—*Winfield Dudgeon.*

2283. LYON, T. L., J. A. BIZZELL, AND B. D. WILSON. **The formation of nitrates in a soil following the growth of red clover and timothy.** Soil Sci. 9: 53–64. 1920.—Cylinders of soil treated with dried blood, acid phosphate, potassium chloride, and ground limestone and planted to timothy or clover were leached with distilled water during the period of the growth of the crops and a 7 months fallow period thereafter. Twice as much nitrogen was present in the drainage water from the clover pots as the timothy pots. There was little difference in the quantities of nitrogen leached from the timothy and clover soils during the growth of those crops but during the first two months of fallowing, ten times as much nitrogen was leached from the clover soil as from the timothy soil. Corn and oats planted after one month fallowing yielded twice as much in the clover soil as in the timothy soil. The total nitrogen in the drainage water and in the corn and oats was over twice as much in the case of the clover soil as in the timothy soil.—*W. J. Robbins.*

2284. MIEGE, E. **La desinfection du sol.** [The disinfection of the soil.] Prog. Agric. et Vitic. 74: 133–140. 1920.—A discussion of results obtained by the use of a number of antiseptic substances on the yields of various plants. Generally, most of these substances have increased very markedly the yields of these plants. Sulfur and copper sulfate have been very efficacious on potatoes; lysol and formaldehyde were very favorable on carrots. Toluol, charcoal, potassium permanganate and calcium hypochlorite have also given good results on truck crops.—*L. Bonnet.*

2285. SMITH, T. A. J. **Manures and fertilizers for tobacco.** Jour. Dept. Agric. Victoria 17: 674–675. 1919.—The need of phosphoric acid for Victorian soils is shown. The soils are naturally rich in potash, and nitrogen is secured by growing leguminous crops. Acid phosphate is recommended, applying at the rate of 100 to 200 pounds per acre. A crop of tobacco yielding 1875 pounds per acre removes 65 pounds of nitrogen, 89 pounds potash and 8 pounds of phosphoric acid.—*J. J. Skinner.*

2286. WAKSMAN, SELMAN A. **Microbiological studies on the cranberry bog soils. I. The effect of liming upon the microbial population of the cranberry soil.** [Abstract.] Absts. Bact. 3: 2. 1919.—"The addition of ground limestone, at the rate of 8000 pounds per acre, to a Savannah bottom cranberry bog resulted in a distinct change in soil reaction and microbial flora, accompanied by a twofold increase in the crop for the four years after the lime had been applied. This study was made on the fourth year after the application of lime.—The hydrogen ion concentration of the unlimed soil was $P_H = 5.2$ to 5.4; the P_H of the limed soil was equal to 6.2 to 6.4. Ammonia was found in traces in both soils. The limed soil contained nitrites and a trace of nitrates, while the unlimed soil had no nitrates and practically no nitrites, indicating a more active nitrification resulting from the change of reaction. On adding the two soils to nitrifying solutions, nitrification was found to be more active in the limed than in the unlimed soil. The aerobic nitrogen-fixing organisms, *Azotobacter*, were found in the limed soil, but not in the unlimed soil. The unlimed soil contains 6000 bacteria and 5000 molds (spores and pieces of mycelium) per gram, while the limed soil contained 20,000 bacteria and 1500 molds per gram, showing the decrease in acidity resulted in an increase in the bacterial and a decrease in the mold flora." [Author's abstract of paper read at scientific session, Soc. Amer. Bact.]—*D. Reddick.*

2287. WHITING, ALBERT L., AND WARREN R. SCHOONOVER. **The comparative rate of decomposition of green and cured clover tops in soil.** Soil Sci. 9: 137–149. 1920.—Green clover at the rate of 50 tons per acre or cured clover in equivalent amounts was mixed with a brown silt and incubated in tumblers or 1 gallon pots. Under aerobic conditions the green and cured clover underwent the same type of decomposition but the curing retarded the decomposition as measured by ammonification, nitrification and loss of carbon. Under anaerobic conditions, the types of decomposition of green and cured clover were very different.—*W. J. Robbins.*

FERTILIZER RESOURCES

2288. ANONYMOUS. **German potash production.** Amer. Fertilizer 52: 70. 1920.—During January, 1920, the potash production in Germany was 550,000 tons.—*J. J. Skinner.*

2289. BANCROFT, WILDER D. [Rev. of: LLOYD, STRAUSS L. **Mining and manufacture of fertilizing materials and their relation to soils.** *19 x 14 cm., vi+153 p.* D. Van Nostrand Co.: New York, 1918. $2.00.] Jour. Phys. Chem. 23: 442. 1919.

2290. DE TURK, ERNEST. **Potassium-bearing minerals as a source of potassium for plant growth.** Soil Sci. 8: 269–301. 1919.—Applications of 2 tons per acre of orthoclase, microline, leucite and alunite to limed peat soil increased the yield of buckwheat from 20 to 35 per cent. Lepidolide was detrimental probably due to an excess of soluble lithium. The potassium in dune sand crushed to pass a 100 mesh sieve (100 meshes to an inch) will produce 0.114 pound of soluble potassium.—*W. J. Robbins.*

2291. FROST, A. C. **The phosphate production in Algeria.** Amer. Fertilizer 52: 70. 1920. —There were 201,013 tons of phosphate produced in Algeria for the first three quarters of 1919. —*J. J. Skinner.*

2292. SMITH, T. A. J. **The importance of lime in agriculture.** Jour. Dept. Agric. 17: 682–683. 1919.—The forms of lime are described. Large deposits of limestone are found in Northern, Northeastern, Western and Gippsland Districts of Victoria.—*J. J. Skinner.*

SOIL ANALYSIS

2293. AMES, J. W., AND C. J. SCHOLLENBERGER. **Calcium and magnesium content of virgin and cultivated soils.** Soil Sci. 8: 323–335. 1919.—Determinations of the total calcium and magnesium, the calcium and magnesium soluble in 0.2 normal nitric acid, the carbonates and the reaction of virgin and cultivated soils from 23 locations in Ohio show that there is a concentration of readily soluble calcium and magnesium at the surface in most virgin soils. When the proportion of the total bases which is soluble is high the soil is likely to contain more carbonate and to be more basic to tests.—*W. J. Robbins.*

2294. [TANSLEY, A. G.] **Investigations on soil.** [Rev. of: HIBBARD, P. L. **Changes in composition of the soil and of the water extract of the soil following the addition of manure.** Soil Sci. 7: 259–272. 1919.] Jour. Ecol. 7: 214–215. 1919.

SOIL CLASSIFICATION

2295. Beck, M. W., M. Y. Longacre, and others. **Soil survey of Howard County, Arkansas.** Advance sheets, Field Operations Bur. Soils, U. S. Dept. Agric. 1917: 5-57. *1 fig., 1 map (colored).* 1919.—For character of report see Bot. Absts. 5, Entry 2316.

2296. Carter, W. T., J. M. Snyder, and O. C. Bruce. **Soil survey of Baltimore County, Maryland.** Advance sheets, Field Operations Bur. Soils, U. S. Dept. Agric. 1917: 5-40. *1 fig., 1 map (colored).* 1919.—For character of report see Bot. Absts. 5, Entry 2298.

2297. Cobb, W. B., E. S. Vanatta, L. L. Brinkley, S. F. Davidson, and F. N. McDowell. **Soil survey of Beaufort County, North Carolina.** Advance sheets, Field Operations Bur. Soils, U. S. Dept. Agric. 1917: 7-39. *1 fig., 1 map (colored).* 1919.—For character of report see Bot. Absts. 5, Entry 2316.

2298. Davis, L. Vincent, and H. W. Warner. **Soil survey of Buena Vista County, Iowa.** Advance sheets, Field Operations Bur. Soils, U. S. Dept. Agric. 1917: 5-36. *Fig. 1, 1 map (colored).* 1919.—Buena Vista County is situated in the northwestern part of Iowa in a prairie region. The topography is flat to gently rolling. Morainic deposits contributed to the more rolling topography. The ruling elevation of the county is 1537 feet above sea level.—The Missouri-Mississippi river drainage divide passes through the county in a general north and south direction. The incipient drainage systems arise in poorly drained areas. Artificial drainage is generally necessary for satisfactory cropping.—Transportation facilities are furnished by five railroads.—The mean annual precipitation is 29.80 inches; and is distributed favorably for crops. The mean annual temperature is 46.30°F. The average growing season is 151 days. Numerous low-lying areas are particularly subject to early frost in fall.—Agriculture which is the principal industry in Buena Vista County consists mainly in the production of corn, oats and hay and the raising and feeding of hogs, cattle, horses and sheep. Corn is the principal crop.—The soils of the county are mainly of glacial origin. The soils are predominantly dark-colored. In the poorly drained areas the lime content is often high. Alluvial soils are found on the terraces along the Little Sioux River and on the first bottoms of those natural drainage ways of sufficient size to have developed flood plains. Several areas of Muck and Peat are found in the county.—Steep slopes of the glacial soils frequently are forested, principally with bur oak, soft maple, elm, basswood and red oak. In the muck and peat areas water loving flora are still to be found in various stages of decomposition.—Eighty-five per cent of the population is rural. Artificial drainage has permitted the extension of the limits of arable land.—*F. B. Howe.*

2299. Deeter, E. B., and F. H. Cohn. **Soil survey of Faulkner County, Arkansas.** Advance sheets, Field Operations Bur. Soils, U. S. Dept. Agric. 1917: 5-33. *1 fig., 1 map (colored).* 1919.—For character of report see Bot. Absts. 5, Entry 2316.

2300. Eckmann, E. C., and A. T. Strahorn. **Soil survey of Anaheim Area, California.** Advance sheets, Field Operations Bur. Soils, U. S. Dept. Agric. 1916: 5-77. *1 fig., 1 map (colored).* 1919.—For character of report see Bot. Absts. 5, Entry 2316.

2301. Goodman, A. L., A. H. Meyer, R. W. McClure, and B. H. Hendrickson. **Soil survey of Amite County, Mississippi.** Advance sheets, Field Operations Bur. Soils, U. S. Dept. Agric. 1917: 5-37. *1 fig., 1 map (colored).* 1919.—For character of report see Bot. Absts. 5, Entry 2316.

2302. Hall, E. C., and E. I. Angell. **Soil survey of Wapello County, Iowa.** Advance sheets, Field Operations Bur. Soils, U. S. Dept. Agric. 1917: 5-42. *1 fig., 1 map (colored).* 1919.—For character of report see Bot. Absts. 5, Entry 2316.

2303. JONES, E. M., AND A. T. SWEET. **Soil survey of Covington County, Mississippi.** Advance sheets, Field Operations Bur. Soils, U. S. Dept. Agric. 1917: 5–39. *1 fig., 1 map (colored).* 1919.—For character of report see Bot. Absts. 5, Entry 2316.

2304. KRUSEKOPF, H. H., J. H. AGEE, AND R. H. HALL. **Soil survey of Callaway County, Missouri.** Advance sheets, Field Operations Bur. Soils, U. S. Dept. Agric. 1916: 5–37. *1 fig., 1 map (colored).* 1919.—For character of report see Bot. Absts. 5, Entry 2316.

2305. MAXSON, E. T., C. E. DEARDORFF, W. A. ROCKIE AND J. M. SNYDER. **Soil survey of Burke County, Georgia.** Advance sheets, Field Operations Bur. Soils, U. S. Dept. Agric. 1917: 5–29. *1 fig., 1 map (colored).* 1919.—For character of report see Bot. Absts. 5, Entry 2316.

2306. MEYERS, A. H., AND T. H. BENTON. **Soil survey of Henry County, Iowa.** Advance sheets, Field Operations Bur. Soils, U. S. Dept. Agric. 1917: 5–31. *1 fig., 1 map (colored).* 1919.—For character of report see Bot. Absts. 5, Entry 2316.

2307. MEYER, A. H., AND B. H. HENDRICKSON. **Soil survey of St. Martin Parish, Louisiana.** Advance sheets, Field Operations Bur. Soils, U. S. Dept. Agric. 1917: 5–31. *1 fig., 1 map (colored).* 1919.—For character of report see Bot. Absts. 5, Entry 2316.

2308. NELSON, J. W., C. J. ZINN, AND OTHERS. **Soil survey of the Los Angeles Area, California.** Advance sheets, Field Operations Bur. Soils, U. S. Dept. Agric. 1916: 5–76. *3 pl., 1 fig., 1 map (colored).* 1919.—For character of report see Bot. Absts. 5, Entry 2316.

2309. ROGERS, R. F., AND W. G. SMITH. **Soil survey of Calhoun County, Michigan.** Advance sheets, Field Operations Bur. Soils, U. S. Dept. Agric. 1916: 5–52. *1 fig., 2 maps (colored).* 1919.—For character of report see Bot. Absts. 5, Entry 2316.

2310. ROGERS, R. F., AND L. A. WOLFANGER. **Soil survey of Chase County, Nebraska.** Advance sheets, Field Operations Bur. Soils, U. S. Dept. Agric. 1917: 5–64. *1 fig., 1 map (colored).* 1919.—For character of report see Bot. Absts. 5, Entry 2316.

2311. SMIES, E. H. **Soil survey of Canadian County, Oklahoma.** Advance sheets, Field Operations Bur. Soils, U. S. Dept. Agric. 1917: 5–58. *1 fig., 1 map (colored).* 1919.—Canadian County, Oklahoma, is situated in the Great Plains region and consists of undulating to rolling uplands with a ruling elevation of 1375 feet above sea level. The area is thoroughly drained by four of the parallel streams that cross western Oklahoma in a southeastward direction.—Grain farming is the important industry of the county with the raising and fattening of livestock as the coördinate industry. The principal farm crops are corn, oats, wheat, grain sorghums, alfalfa, hay and cotton. Fruit growing is developed to some extent in part of the county. Railroad facilities are good.—The mean annual rainfall is about 32 inches. The highest rainfall occurs during the growing season while the winter months are comparatively dry. The lowest annual rainfall recorded is 17.27 inches. The mean annual temperature is 58.6°F. Hot, dry winds from the south sometimes cause considerable damage to crops.—The upland soils of the county are classed into two general divisions, residual prairie soils and soils largely of wind blown origin. The residual prairie soils are derived from the underlying red sandstones and shales, which form a part of the Permian Red Beds. They are usually calcareous. The wind blown soils are composed for the most part of material blown up over the uplands from the near-by alluvial flood plains. The alluvial bottom-land soils are divided into two general divisions, terrace or second-bottom soils, and the more recent alluvial or first-bottom soils.—The principal native grasses of the upland soils consisted chiefly of blue stem, buffalo grass, grama, mesquite and a variety of bunch grasses. Blue stem disappears after being pastured for a few years and the principal growth is mesquite. Timber belts lie along most of the drainage ways in the more rolling sections. The trees are chiefly elm, hackberry, black walnut, cottonwood and oak. Red cedar was once abundant.—The farms in the vicinity of the larger streams and on the prairie soils are fairly well improved.—*F. B. Howe.*

2312. Tartar, H. V., and F. C. Reimer. **The soils of Jackson County.** Oregon Agric. Exp. Sta. Bull. 164. *62 p. 1 map.* 1920.—An area of approximately 544 square miles of valley and adjacent hill and mountain land in the central part of Jackson County were studied. The soil types are numerous and fall principally into two classes, residual and alluvial, ranging from fine sandy loam to clay adobe. Results of chemical analyses of the most important soil types showed, that there is an abundant supply of potassium, calcium and magnesium, that none are acid, that the phosphorus supply is only fair to low, that the nitrogen content (also organic matter) is prevailingly low, and that sulfur is present in most of the soils in quantities so small that it is one of the limiting factors in the growth of crops making large demands for that plant food. Irrigation and drainage are needed in some places.—*E. J. Kraus.*

2313. Thorp, W. E., and H. J. Harper. **Soil survey of Blackhawk County, Iowa.** Advance sheets, Field Operations Bur. Soils, U. S. Dept. Agric. 1917: 7–43. *1 fig., 2 pl., 1 map (colored).* 1919.—For character report see Bot. Absts. 5, Entry 2316.

2314. Tillman, B. W., F. A. Hayes, and F. Z. Hutton. **Soil survey of Drew County, Arkansas.** Advance sheets, Field Operations Bur. Soils, U. S. Dept. Agric. 1917: 5–46. *1 fig., 1 map (colored).* 1919.—For character of report see Bot. Absts. 5, Entry 2316.

2315. Tillman, B. W., and B. F. Hensel. **Soil survey of Phelps County, Nebraska.** Advance sheets, Field Operations Bur. Soils, U. S. Dept. Agric. 1917: 5–40. *1 fig., 1 map (colored).* 1919.—For character of report see Bot. Absts. 5, Entry 2316.

2316. Tillman, B. W., and B. F. Hensel. **Soil survey of Wayne County, Nebraska.** Advance sheets, Field Operations Bur. Soils, U. S. Dept. Agric. 1917: 5–47. *1 fig., 1 map (colored).* 1919.—Situated in northeastern Nebraska, Wayne County covers about 450 square miles. The topography is uneven, ranging from hills to level areas. Three-fourths of the county is upland, one-eighth bottom land and the remainder terrace. The bottom areas lie at about 1500 feet above sea level, while the hills are 160 feet higher.—The climate is suited to general farming, with an annual precipitation of 28 inches and a mean annual temperature of 48°F. The growing season of 144 days receives about one-half of the annual rainfall.—The upland soils, comprising 76 per cent of the county, are loess of the *Marshall* and *Knox* series. The former is a black soil while the latter is light brown. Both are silt loams, and quite productive. The sedimentary soils, covering 17.5 per cent of the county are the most productive although the terrace areas, ranking with the loess in fertility, are excellent.—The main industry of the county is agriculture. The principal crops are corn, oats, alfalfa, clover, timothy, wheat and hay. Wheat is about the only cash crop. Over one-third of the crop acreage every year is corn. Stock raising is constantly receiving greater attention.—Progressive farmers follow systematic crop rotation. Drainage, especially on the bottom lands, is being rapidly developed. Good crops are obtained in all parts of the county although the production is below what it should be for soils of such high natural fertility. The cropping systems in vogue are not keeping up the productiveness of the land. More attention should be paid to green manures and legumes.—*H. O. Buckman.*

2317. Van Duyne, C., L. R. Schoenmann, and S. D. Averitt. **Soil survey of Shelby County, Kentucky.** Advance sheets, Field Operations Bur. Soils, U. S. Dept. Agric. 1916: 5–64. *1 fig., 1 pl., 1 map (colored).* 1919.—For character of report see Bot. Absts. 5, Entry 2316.

2318. Van Duyne, C., W. E. McLendon, W. J. Latimer, and I. M. Morrison. **Soil survey of Marlboro County, South Carolina.** Advance sheets, Field Operations Bur. Soils, U. S. Dept. Agric. 1917: 5–72. *2 fig., 1 map (colored).* 1919.—Marlboro County occupies a belt in northeastern South Carolina extending from the crest of the Sandhill region down into the lower Coastal Plain. The elevations range from 140 to 300 feet. The area is in part undulating and in part flat and poorly drained. Drainage is into the Pee Dee River.—The winters are short and mild while the summers are long and hot. Two-thirds of the 47 inches of rain fall during the summer months. The growing season is about 216 days. A great variety of crops may be grown.—Marlboro County is one of the best developed counties agriculturally of the state. Many different soil types occur, those of the coastal plain being extensively farmed and mostly to cotton. While the terrace soils along the Pee Dee River are cropped, the bottom lands yet remain to be developed. Corn, cowpeas, wheat and oats do well. Peanuts yield splendidly on all soils. The first bottoms are fine grass lands and offer splendid opportunities for cattle raising.—Crops are not very often grown in rotation and the land is running down. Constantly increasing amounts of fertilizer are necessary. Complete mixed fertilizers are most generally purchased. Some nitrate of soda is used as a top dressing. Lime although needed has not come into general use.—*H. O. Buckman.*

2319. WATKINS, W. I., E. D. FOWLER, H. I. COHN, J. A. MACKLIS, AND H. H. KRUSEKOPF. **Soil survey of Texas County, Missouri.** Advance sheets, Field Operations Bur. Soils, U. S. Dept. Agric. 1917: 5-36. *1 fig., 1 map (colored).* 1919.—For character of report see Bot. Absts. 5, Entry 2316.

MOISTURE RELATIONS

2320. HARDING, S. T. **Relation of the moisture equivalent of soils to the moisture properties under field conditions of irrigation.** Soil Sci. 8: 303-312. *6 fig.* 1919.—A comparison was made of the moisture equivalent with the critical moisture points of soils under actual field conditions of irrigation practice. The results include over 9000 individual moisture determinations and 136 determinations of moisture equivalent varying from 4.1 to 37.6. The maximum field capacity, the normal field capacity, soil moisture before irrigation, and soil moisture at permanent wilting of the crop were studied. Expressed as per cent of the moisture equivalent the moisture at the time of permanent wilting alone shows a linear relationship with the moisture equivalent. This for the surface foot is about 15 per cent less than that given by the formula of BRIGGS and SHANTZ.—*W. J. Robbins.*

2321. KNAPP, GEORGE S. **Winter irrigation for western Kansas.** Kansas Agric. Exp. Sta. Circ. 72. *8 p.* Jan., 1919.

METHODS

2322. BEAR, FIRMAN E., AND GEORGE M. MCCLURE. **Sampling soil plots.** Soil Sci. 9: 65-75. *4 fig.* 1920.—The composite from a one-twentieth acre plot should be made up of 20 samples, each 12 inches in depth and uniformly distributed over the plot.—*W. J. Robbins.*

2323. GARDNER, WILLARD. **A new soil elutriator.** Soil Sci. 9: 191-197. *2 fig. Pl. 1.* 1920.—An elutriator for the mechanical analysis of soil is described and figured.—*W. J. Robbins.*

2324. GILLESPIE, L. J. **Colorimetric determination of hydrogen-ion concentration without buffer mixtures, with especial reference to soils.** Soil Sci. 9: 115-136. *1 fig.* 1920.—A simple method is described for the colorimetric determination of the hydrogen-ion exponent without the use of buffer mixtures. The method also provides for the elimination of errors due to the turbidity of the solution in which the determination is made. Each color standard consists of two test tubes, one tube containing 5 cc. of dilute acid, the other 5 cc. of dilute alkali. The tubes together contain 10 drops of indicator solution, the 10 being divided between the alkaline and acid tubes in various "drop ratios." To 10 cc. of the unknown solution, 10 drops of the indicator solution are added and compared with the two color standards by means of a simple comparator. A table is given of the pH for each drop ratio of the indicators used which cover a range of P_H 3.1 to P_H 9.75. Soil extracts, water clear, were prepared by the use of colloidal iron solution as a precipitant and pH measurements of the water extracts of nine soils prepared by this method gave the same results as were obtained by the usual methods.—*W. J. Robbins.*

2325. Hurst, C. T., and J. E. Greaves. **Some factors influencing the quantitative determination of chlorides in soil.** Soil Sci. 9: 41–51. 1920.—A soil extract is obtained by filtering through a Pasteur-Chamberland filter or by the use of alum and the chlorides determined by the method given in detail.—*W. J. Robbins.*

2326. Robinson, R. H. **Concerning the effect of heat on the reaction between lime-water and acid soils.** Soil Sci. 9: 151–157. 1920.—The length of time of heating and the temperature used during the process of evaporation affects the lime requirement of acid soils as determined by the Veitch method. Variations in the lime requirement of a soil from 1300 pounds per acre when evaporation occurred in 2.5 hours at 70° to 4600 pounds per acre where evaporation occurred at 110° in 8 hours were found.—*W. J. Robbins.*

MISCELLANEOUS

2327. Call, L. E. **Director's report.** Kansas Agric. Exp. Sta. 1917–18. *63 p.* 1918.—See Bot. Absts. 5, Entries 1466, 2024.

2328. Jovino, S. **Osservazioni sull'aridocoltura italiana.** [**Observations upon dry farming in Italy.**] Staz. Sperim. Agrarie Italiane 52: 69–121. 125–192. 1919.—A lengthy study of the subject divided in the following way: (1) the climate of the arid regions of Italy, (2) the soil of the arid regions of Italy, (3) biological characteristics of Italian dry farming, (4) the function of fallowing in Italy, (5) the critical period in the spring, (6) the summer critical period, (7) means of favoring the evolution of the present cultural conditions. In this paper are studied the adaptations of plants to the conditions of the arid regions: low soil-water content, high temperature and strong illumination. A lengthy abstract of this paper with special emphasis on the technical side is to be found in Monthly Bull. Internation. Instit. Agric. Rome 10[6]: 522–526. 1919. (English edition.)—*A. Bonazzi.*

2329. Hodsoll, H. E. P. **The care of the soil.** Jour. Roy. Hortic Soc. 45: 22–28. 1919. —*J. K. Shaw.*

2330. Howard, A., and G. L. C. **Report of the Imperial Economic Botanists.** Sci. Rept. Agric. Res. Inst. Pusa 1918-19: 46–67. *Pl. 5 and 6.* 1919.—See Bot. Absts. 5, Entry 1159.

2331. Middleton, Howard E. **The moisture equivalent in relation to the mechanical analysis of soils.** Soil Sci. 9: 159–167. *1 fig.* 1920.—The maximum percentage of water which a soil can retain in opposition to a force equal to 1000 times that of gravity (the moisture equivalent) was compared with the mechanical analyses. The relation between the percentage of sand, silt and clay and the moisture equivalent was found to be 0.063 sand+0.291 silt+0.426 clay = moisture equivalent. The presence of considerable organic matter increases the moisture equivalent and disturbs the above relation.—*W. J. Robbins.*

2332. Powers, W. L. **Duty of water in irrigation.** Oregon Agric. Exp. Sta. Bull. 161. *20 p., 1 fig.* 1920.—Proper economical irrigation is necessary to permanent irrigative agriculture. By saving 50 per cent of the water now used in many places, it will be possible to double the crop producing area. The economical use and duty of irrigation water depend upon a wide variety of conditions of culture, method of distributing and handling of the water, types of crops produced, and environment. Soil fertility is one of the most important factors affecting irrigation requirements, for it is frequently possible to double the returns from each unit of water supplied by applying needed simple fertilizers. At times one ton of manure may equal 100 tons of water in securing returns. Irrigation farming reaches its highest development in connection with intensive farming. In general it is better economy to provide only a moderate allowance of water with reasonably priced structures than to provide a liberal supply at a great expense and invite additional drainage assessments later.—*E. J. Kraus.*

2333. POWERS, W. L., AND W. W. JOHNSTON. **The improvement and irrigation requirement of wild meadow and tule land.** Oregon Agric. Exp. Sta. Bull. 167. *44 p., 25 fig.* 1920.—See Bot. Absts. 5, Entry 1198.

2334. WHERRY, EDGAR T. **Soil tests of Ericaceae and other reaction-sensitive families in northern Vermont and New Hampshire.** Rhodora 22: 33–49. 1920.

2335. WITTMACK, L. **Die Bonitierung des Bodens nach der Unkrautpflanzen. [The rating of soils according to the weeds growing on them.]** Illustrierte Landw. Zeitg. 39: 391–392. 1919.

TAXONOMY OF VASCULAR PLANTS

J. M. GREENMAN, *Editor*
E. B. PAYSON, *Assistant Editor*

GENERAL

2336. ANONYMOUS. [Rev. of: WILLIAM MANSFIELD. **Squibb's atlas of the official drugs.** *686 p., illustrated.* 1919.] Druggists Circ. 63: 243. 1919.—See Bot. Absts. 3, Entry 1691.

2337. B. D. **Quelques plantes nouvelles. [Some new plants.]** Rev. Hortic. [Paris] 91: 260–262. *Fig. 84–85.* Apr., 1919.

2338. BOLUS, HARRIET M. L. **Elementary lessons in systematic botany.** Based on familiar species of the South African Flora, with an introduction and eight summaries. Illustrated by MARY M. PAGE. *96 p., 24 fig.* 1919.

2339. BROWN, WILLIAM H., AND ARTHUR F. FISCHER. **Philippine mangrove swamps.** Bur. Forestry Dept. Agric. and Nat. Resources [Manila] Bull. 17. *132 p. Pl. 1–47.* 1918.—About 30 species are listed as mangrove-swamp plants in the Philippine Islands; these belong to 17 families. A key to the genera is given, the species are described and their local names recorded. The paper is copiously illustrated by reproductions from photographs.—*J. M. Greenman.*

2340. BUSWELL, W. M. **Familiar wildflowers of Florida.** Amer. Bot. 25: 90–93. 1919.

2341. CHEVALIER, AUG. **Catalogue des plantes du jardin botanique de Saigon. [Catalogue of plants in the Botanical Garden of Saigon.]** *68 p.* 1919.—The introductory matter includes an interesting historical sketch of the Botanical Garden. In appendix II is included a number of changes in nomenclature, the new binomials proposed being necessitated by the determination of the exact status of some of LOUREIRO's hitherto imperfectly known species. —*E. D. Merrill.*

2342. CREMATA, MERLINO. **Cercas, alambradas y setos en Cuba. [Fences and hedges in Cuba.]** Revist. Agric. Com. y Trab. 2: 259–272. *29 fig.* 1919.—See Bot. Absts. 3, Entry 527.

2343. EWART, A. J. **Contributions to the flora of Australia.** No. 26. Proc. Roy. Soc. Victoria (N. S.) 30: 173–177. 1918.

2344. GERTZ, OTTO. **Christopher Rostii herbarium vivum. Ein deutsches herbar vom jahre 1610. [The herbarium of Christopher Rostius. A German herbarium of the year 1610.]** Oesterr. Bot. Zeitschr. 67: 369–382. 1918.—This collection consists of 363 specimens of plants, chiefly of central Europe and the Mediterranean region, mounted in a bound volume 20 x 16.5 cm. in size. The original author is unknown, but a history of the herbarium is in part recorded. A list of the original names accompanying the specimens is given with their present binomial equivalents.—*J. M. Greenman.*

2345. HALLIER, HANS. **Ueber Gaertner'sche Gattungen und Arten unsicherer Stellung, einige Rubiaceen, Sapotaceen, Cornaceen und über versunkene Querverbindungen der Tropenländer. [Horticultural genera and species of uncertain position, some Rubiaceae, Sapotaceae, Cornaceae; submerged land-connections in the tropics.]** Recueil Trav. Bot. Néerlandais 15: 27–122. 1918.

2346. HEMSLEY, W. B., AND OTHERS. **Flora of Aldabra: with notes on the flora of the neighboring islands.** Kew Bull. Misc. Inf. [London] 1919: 108–153. 1919.—See Bot. Absts. 4, Entry 339.

2347. KOPS, JAN, F. W. VAN EEDEN, AND L. VUYCK. **Flora Batava. Afbeelding en Beschrijving der Nederlandische Gewassen. [Flora of Batavia. Illustrations and descriptions of plants of Holland.]** Aflevering 396e–399e. *Pl. 1977–1992.* Martinus Nijhoff's, Gravenhage. 1919.—The present parts contain descriptions and colored illustrations of the following vascular plants: *Carex Kneuckeriana* Zahn, *Cyperus vegetus* Willd., *Glyceria plicata* Fr., *Veronica praecox*, All., *Solanum nitidibaccatum* Bitter, *Rubus humifusus* Weihe & Ness, *R. pyramidalis* Kaltenb., *R. caesius* var. *aquaticus* Weihe & Ness, *Rumex odontocarpus* Sandor, and *Lathyrus cicera* L. The non-vascular plants included are: *Hydnum violaceum* Thore, *H. nigrum* Fr., *Psathyrella disseminata* P., *Peziza hemisphaerica* Hoff., *Clavaria aurea* Schaeff., *Mycena epipterygia* Scop., *Amanita porphyria* Fr., and *Hygrophorus pratensis* Fr.—*J. M. Greenman.*

2348. LANE-POOLE, C. E. **Report of the Woods and Forests Department for the half-year ended 30th of June, 1918.** Semi-Ann. Progress Rept. Woods and Forests Dept. Western Australia. *17 p.* 1919.—See Bot. Absts. 4, Entry 443.

2349. MOLA, PASQUALE. **Flora delle acque Sarde. Contributo delle Piante idrofite ed igrofite della Sardegna. [Flora of the Sardinian waters. Hydrophytes and hygrophytes of Sardinia.]** Atti R. Accad. Sci. Torino 54: 478–502. 1918–1919.—See Bot. Absts. 4, Entry 1025.

2350. NELSON, JAMES C. **A comparison of the flora of southern British Columbia with that of the State of Washington, as illustrated by the floras of Henry and Piper. [Rev. of: HENRY, JOSEPH KAYE. Flora of Southern British Columbia and Vancouver Island.** *363 p.* W. J. Gage & Co.: Toronto, 1915.] Torreya 19: 174–184. 1919.—HENRY's Flora, although covering a territory at least twice as large as the State of Washington, and extending to the eastward so as to include the Rocky Mountain flora, mentions only 2359 named forms as compared with 2511 in PIPER's Flora of Washington. Of these 1517, or about 60 per cent, are common to both manuals. Assuming equal thoroughness on the part of both authors, two conclusions seem to be justified. (1) That Washington is a region of more marked endemism than British Columbia. (2) That the 49th parallel seems to come very near to a line marking the extreme northward dominance of the Californian flora on the one hand, and the extreme southern extension of the Alaskan flora on the other. In Henry's Flora there are 764 forms not mentioned by PIPER; in PIPER's 928 not mentioned by Henry. These species are arranged by groups to show distribution and degree of endemism. A table of discrepancies in the case of 18 of the larger genera is presented. Prof. Henry displays a commendable conservatism in his conception of taxonomic relations. The book is marred by many inaccuracies in capitalization, grammatical agreement, orthography, abbreviation, citation and etymology, but on the whole is a valuable effort to contribute to the fuller knowledge of the Northwest Flora. —*J. C. Nelson.*

2351. PHILLIPS, EDWIN PERCY. **Some notes on a collecting trip to French Hoek.** South African Jour. Sci. 15: 450–478. 1919.—See Bot. Absts. 4, Entry 298.

2352. QUER, P. FONT. **Plantas de Tetuán. [Plants of Tetuán.]** Bol. R. Soc. Española Hist. Nat. 19: 93–95. 1919.—List of eighty-four species of plants collected in the vicinity of Tetuan, northern Morocco, by Manual Pando in April, 1916. Proposed as new are *Cistus* *salviifolius* var. *Pandoanus*, *Linum strictum α cymosum* f. *scaberrimum*, *Trifolium campestre* var. *Pandoi*, *Cerinthe oranensis* f. *parviflora*. New combinations appear to be *Lathyrus Clymenium* race *articulatus* (L.), and *Convolvulus tricolor* race *pseudotricolor* (Bert.).—*O. E. Jennings*.

2353. Quer, P. Font. **Adiciones a la flora de Menorca.** [**Additions to the flora of Minorca.**] Bol. R. Soc. Espanola Hist. Nat. 19: 268–273. 1919.—This is an annotated list with localities and other information relating to 69 species, varieties, or forms. Former workers on this flora are referred to and the following new species or varieties are published: *Fumaria muralis* Sond. var. *longipes* Pau, *Calycotome spinosa* Link race *villosa* Link var. *Fontqueri* Pau, *Lotus fallax* Quer, *Cotyledon umbilicus* L. var. *minoricensis* Pau, and *Avellinia Michelii* Parl. var. *longiaristata* Quer. Nine of these plants are new to the flora of the Balearic Islands.—*O. E. Jennings*.

2354. Salisbury, F. S. **Naturalized plants of Albany and Bathurst.** Rec. Albany Mus. [Grahamstown, South Africa] 3: 161–177. 1919.

2355. Stone, Herbert. **Les bois utiles de la Guyane Française.** [**The useful woods of French Guiana.**] Ann. Mus. Colonial, Marseille III, 6 : 1–68. 1918.—The present article continues the author's enumeration of the useful woods of French Guiana and includes well known species of the following families: Combretaceae, Myrtaceae, Melastomaceae, Samydaceae, Passifloraceae, Araliaceae, Rubiaceae, Sapotaceae, Ebenaceae, Styricaceae, Oleaceae, Apocynaceae, Borraginaceae, Bignoniaceae, Myoporaceae, Verbenaceae, and Polygonaceae.—*J. M. Greenman*.

2356. Turrill, W. B. **Contributions to the flora of Macedonia.** Kew Bull. Misc. Inf. [London] 1919: 105–108. 1919.—See Bot. Absts. 4, Entry 368.

2357. Vuijk, L. **Verslag der excursie gehouden te 's-Hertogenbosch 26 Juli 1918 en volgende dagen.** [**Report of the excursion held in Hertogenbosch, Holland, etc.**] Nederland. Kruidkundig Arch. 1918: 19–30. May, 1919.—A rather complete enumeration of the plants found by the members of the society on the trip. A six page list with additions to the flora is given. —*J. A. Nieuwland*.

2358. Waby, J. F. **Notes on a collection of preserved fruits and seeds (Part 1).** Jour. Bd. Agric. British Guiana 12: 2–6. 1919.—Descriptions of a very large collection of tropical fruits and seeds preserved in glass jars in the Herbarium of the Botanic Garden of Georgetown. In this part are given descriptions of plants, flowers, fruits and seeds of *Entada scandens*, *E. polystachya*, *Poinciana regia*, *Cassia grandis*, *C. fistula*, *C. javanica*, *Pterocarpus guianensis*, and *Platymiscium polystachyum*.—*J. B. Rorer*.

2359. Waby, J. F. **Notes on a collection of dried fruit and seeds** (continued). Jour. Bd. Agric. British Guiana 12: 102–111. 1919.—Descriptions of seeds and fruits, together with common names, many interesting notes and superstitions, of the following plants: *Eperua falcata*, *E. Schomburghii*, *E. Jenmani*, *Bauhinia Vahlii*, *Enterolobium cyclocarpum*, *E. Timboüva*, *Caesalpinia Bonducella*, *Macrolobium acaciaefolium*, *M. hymaenoides*, *Caesalpinia Sappan*, *Peltophoroum ferrugineum*, *Caesalpinia bijuga*, *C. feriea*, *C. coriaria*, *Piscidia Erythrina*, *Acacia arabica*, *Detarium senegalense*, *Flemingia strobilifera*, *Drepanocarpus lunatus*, *Ormosia dasycarpa*, *O. jamaicensis*, *Copaifera officinalis*, *Myrospermum Pereirae*, *Mucuna urens*, *M. pruriens*, *M. Fawcettii*. *Stizolobium altissimum*, *Adenanthera Pavonina*, *Erythrina corallo dendron*. *E. indica*, *Psophocarpus tetraagonolobus*, and *Trachylobium Hornemannianum*.—*J. B. Rorer*.

2360. Williams, Frederic N. **Pulteney's references to the Flora Londinensis.** Jour. Botany 57: 100. 1919.—Notes on the so-called "MS. of Pulteney," and on the confusion of plates, and chronological puzzles in the above flora.—*K. M. Wiegand*.

PTERIDOPHYTES

2361. Barnola, Joaquin Ma. de. **Las Licopodiales de la península Ibérica, citas y notas críticas.** [Catalogue of Iberian Lycopodiales.] Broteria Ser. Bot. 17: 17-27. 1919.—The author lists the species and varieties of *Lycopodium*, *Selaginella*, and *Isoetes* which grow in Spain or Portugal, with keys, detailed citation of localities, some critical notes on distribution, and a bibliography of 22 titles; no new forms are described.—*Edward B. Chamberlain.*

2362. Beck, G. **Einige Bemerkungen über heimische Farne.** [Some observations on native ferns.] Oesterr. Bot. Zeitschr. 67: 52-63, 113-123. 1918.—The author gives an annotated list of ferns of south-central Europe and records particularly the spore characters of several species and forms.—*J. M. Greenman.*

2363. Benedict, R. C. **The simplest fern in existence.** Amer. Fern Jour. 9: 48-50. *Pl. 3, 7 fig.* 1919.

2364. Graves, E. W. **The Botrychiums of Mobile County, Alabama.** Amer. Fern Jour. 9: 56-58. 1919.—*Botrychium obliquum*, *B. biternatum* (Lam.) Underw. and *B. alabamense* Maxon are found growing together in this county. *B. alabamense* may be distinguished from *B. biternatum* by the manner in which it holds its sterile fronds and also by the time of fruiting. The former holds its sterile fronds three to ten inches above the ground and completes fruiting by October 15, while the latter holds its sterile fronds not more than an inch above the ground and matures its fruit about March 1.—*F. C. Anderson.*

2365. Maxon, William R. **Ferns of the District of Columbia.** Amer. Fern Jour. 9: 38-48. 1919.—After briefly describing the area adopted for the "District flora," the author lists 56 species, distributed among 25 genera. The occurrence and habitat of each species is discussed.—*F. C. Anderson.*

2366. Palmer, Ernest J. **Texas Pteridophyta—II.** Amer. Fern Jour. 9: 50-56. 1919.—The author continues the enumeration of the Pteridophytes of Texas, listing 17 species with habitat and localities. A reduced form of *Botrychium obliquum* Muhl. may represent a distinct and undescribed variety.—*F. C. Anderson.*

2367. Weatherby, C. A. **Changes in the nomenclature of the Gray's Manual ferns.** Rhodora 21: 173-179. 1919.—A discussion of the changes which have been accepted in the nomenclature of the *Polypodiaceae* and the *Osmundaceae* of Gray's Manual since the publication of the seventh edition and an explanation of these changes. The summary gives a list of thirty changes, in each case giving the Manual name, the later name and authority, and the synonyms.—*James P. Poole.*

2368. Woynar, H. **Betrachtungen über Polypodium austriacum Jacquin.** [Considerations on Polypodium austriacum Jacquin.] Oesterr. Bot. Zeitschr. 67: 267-275. 1918.—The author presents a discussion of this fern particularly with reference to the nomenclatorial status of the specific name.—*J. M. Greenman.*

SPERMATOPHYTES

2369. Beringer, G. M. [Rev. of: Maiden, J. H. **A critical revision of the genus Eucalyptus.** Vol. IV, Part 6. Published by the Government of the State of New South Wales.] Amer. Jour. Pharm. 91: 328-329. 1919.—*Anton Hogstad, Jr.*

2370. Blake. S. F. **The genus Homalium in America.** Contrib. U. S. Nation. Herb. 20: 221-235. 1919.—Nineteen species are recognized, in addition to one doubtful one (*H. senarium* Moc. & Sessé). The following are new: *H. nicaraguense*, *H. mollicellum*, *H. pleiandrum*. *H. leiogynum*, *H. hemisystylum*, *H. racemosum* subsp. *barbellatum*, *H. Pittieri*, *H. trichocladum*, *H. eleutherostylum*, *H. columbianum*, *H. stenosepalum*, *H. eurypetalum*.—*S. F. Blake.*

2371. Blake, S. F. **New South American spermatophytes collected by H. M. Curran.** Contrib. U. S. Nation. Herb. 20: 237–245. 1919.—The following new species and new names occur: *Dorstenia anthuriifolia, Coussapoa Curranii, Coccoloba cyclophylla, Ruprechtia oxyphylla, R. coriacea* (Karst.) Blake, *Triplaris euryphylla, T. laxa, Schizolobium parahybum* (Vell.) Blake, *Guarea racemiformis, Trichilia alta, T. Curranii, T. microdonta, T. triphylla, Fischeria blepharopetala, Macroscepis barbata.*—*S. F. Blake.*

2372. Blom, Carl. **Lepidium bonariense L., Lepidium neglectum Thell., samt Rumex salicifolius L. funna i Sverge.** Bot. Notiser 1919: 181. 1919.—The first and the last of these are recorded from ballast at Malmö, and the second one from Borås and Stockholm.—*P. A. Rydberg.*

2373. Chevalier, A. **Quelques légumineuses d'Extreme-Orient utiles à répandre. [Some legumes of Indo-China worthy of wider use.]** Bull. Agric. Inst. Sci. Saigon 1: 87–92. 1919.—Contains the new combination *Mucuna cochinchinensis* (Lour.) A. Chev. based on *Marcanthus cochinchinensis* Lour., the oldest valid name for *Mucuna nivea* W. & A.—*E. D. Merrill.*

2374. Chevalier, A. **Le pommier à cidre des hauts plateaux de l'Indochine. [The cider apple of the high plateaus of Indo-China.]** Bull. Agric. Inst. Sci. Saigon 1: 142–150. 1919.—The utilisation of the fruits of *Pyrus Doumeri* Bois is discussed and the species redescribed.—*E. D. Merrill.*

2375. Chevalier, A. **Une nouvelle variété de palmier Elaeis. [A new variety of the Elaeis palm.]** Bull. Agric. Inst. Sci. Saigon 1: 154, 155. 1919.—Reduces *Elaeis Poissoni* Annet to *E. guineensis* Aubl. as var. *Poissoni* (Annet) A. Chev.—*E. D. Merrill.*

2376. Clute, Willard N. **Phlox nomenclature.** Amer. Bot. 25: 100, 101. *Fig. 1.* 1919.—Eastern and western forms of *Phlox divaricata* appear to differ in the shape and size of the flowers. The differences were noted long ago by Alphonso Wood who called the western variety, *Laphamii.* The differences in the two forms have been ignored by systematists but it is suggested that the western form be called *Phlox Laphamii* (Wood).—*W. N. Clute.*

2377. Cremata, Merlino. **Plantas meliferas. [Melliferous plants.]** Revist. Agric. Com. y Trab. 2: 140–152. *10 fig.* 1919.—See Bot. Absts. 4, Entry 215.

2378. De Candolle, Casimir. **Begoniaceae Centrali-Americanae et Ecuadorenses. [Begoniaceae of Central America and Ecuador.]** Smithsonian Misc. Collections 68[12]: 1–10. 1919.—The following new species and new names appear, with Latin descriptions: *Begonia Kellermanii* (Guatemala), *B. fissurarum* (*B. leptophylla* C.DC. 1908, not Taub. 1896), *B. stenoptera* (Costa Rica), *B. garagarana, B. brevicyma, B. mucronistipula, B. uvana, B. mameiana, B. villipetiola, B. cilibracteola, B. leptopoda, B. pubipedicella, B. serratifolia, B. chiriquina, B. chepoensis, B. caudilimba, B. udisilvestris, B. parcifolia* (Ecudaor). With the exceptions noted, all these are described from Panama.—*S. F. Blake.*

2379. Fernald, M. L. **I. The unity of the genus Arenaria. II. The type of the genus Alsine. III. The earlier names for Alsinopsis. IV. The American representatives of Arenaria sajanensis. V. The specific identity of Arenaria groenlandica and A. glabra. VI. The American variations of Arenaria verna** [Contrib. Gray Herb. Harvard Univ. New Series.—No. LVII]. Rhodora 21: 1–22. 1919.—The subject-matter under the six separate titles deals with the genus *Arenaria* which the author maintains in its broad sense. The following new combinations, new names, and new species are published: *Arenaria arenarioides* (*Cerastium arenarioides* Crantz), *A. bryophylla* (*Ar. musciformis* Edgew. & Hook. f., not Triana & Planch.), *A. Funkii* (*Alsine Funkii* Jord.), *A. cymifera* (*Alsine cymifera* Rouy & Fouc.), *A. iberica* (*Minuartia dichotoma* L., not *Ar. dichotoma* Krock), *A. caucasica* (*Alsine caucasica* Boiss.), *A. anatolica* (*Alsine anatolica* Boiss.), *A. Thevenaei* (*Alsine Thevenaei* Reut.), *A. attica* (*Alsine attica* Boiss.), *A. sphagnoides* (*Sabulina sphagnoides* Froel.), *A. aizoides* (*Alsine aizoides* Boiss.),

A. decipiens (*Alsine decipiens* Fenzl), *A. dianthifolia* (*Alsine dianthifolia* Boiss.), *A. intermedia* (*Alsine intermedia* Boiss.), *A. leucocephala* (*Alsine leucocephala* Boiss.), *A. pulvinaris* (*Alsine pulvinaris* Boiss.), *A. makmelensis* (*Alsine libanotica* Boiss., not *Ar. libanotica* Kotschy), *A. rimarum* (*Alsine rimarum* Boiss. & Balansa), *A. Schimperii* (*Alsine Schimperii* Hochst.), *A. stellata* (*Cherleria stellata* Clarke), *A. diversifolia* (*Moehringia diversifolia* Dolliner), *A. Grisebachii* (*Moehringia Grisebachii* Janka), *A. Jankae* (*Moehringia Jankae* Griseb.), *A. dasyphylla* (*Moehringia dasyphylla* Bruno), *A. dasyphylla* var. *sedoides* (*Moehringia mucosa* β *sedoides* Cumino), *A. Tommasinii* (*Moehringia Tommasinii* Marches), *A. glaucovirens* (*Moehringia glaucovirens* Bertol.), *A. polygonoides* Wulf. var. *obtusa* (*A. obtusa* All.), *A. papulosa* (*Moehringia papulosa* Bertol.), *A. platysperma* (*Moehringia platysperma* Maxim.), *A. Cossoniana* (*Moehringia stellarioides* Coss., not *Ar. stellarioides* Willd.), *A. octandra* (*Cherleria octandra* Sieb.), *A. obtusiloba* (*Alsinopsis obtusiloba* Rydb.), *A. marcescens*, *A. groenlandica* (Retz.) Spreng. var. *glabra* (*A. glabra* Michx.), *A. verna* L. var. *pubescens* (*A. hirta* β *pubescens* Cham. & Schlecht.), and *A. verna* var. *pubescens* forma *epilis* (*A. verna* var. *propinqua* forma *epila* Fernald).—*James P. Poole.*

2380. GIROLA, CARLOS D. **Maices argentinos y aclimatodos: Variedades de Maiz cultivadis en Argentina.** *169 p. 35 pl.* Buenos Aires. 1919.—See Bot. Absts. 4, Entry 71.

2381. GLEASON, HENRY ALLAN. **Taxonomic studies in Vernonia and related genera.** Bull. Torrey Bot. Club 46: 235–252. 1919.—The following species and varieties of *Vernonia* are discussed: *V. borinquensis* Urban, *V. borinquensis* var. *Stahlii* Urban, *V. sericea* L. C. Rich., *V. gnaphaliifolia* Rich., *V. icosantha* DC., *V. racemosa* Delp., *V. rigida* Sw., *V. mollis* HBK., *V. missurica* Raf., *V. altissima* var. *pubescens* (Morris) Daniels. Descriptions of new species appear as follows: *V. Shaferi*, *V*, *morelana*, *V. salamana*, *V. ctenophora*, *V. aborigina*, *V. jucunda.* The following new varieties are given: *V. borinquensis* var. *resinosa*, *V. borinquensis* var. *hirsuta*, *V. gnaphaliifolia* var. *platyphylla*, *V. Sagraeana* var. *angusticeps* (Ekman), *V. missurica* var. *austroriparia*, *V. fasciculata* var. *nebraskensis*, *V. altissima* var. *brevipappa*, *V. altissima* var. *laxa*, *V. flaccidifolia* var. *angustifolia*, and *V. ovalifolia* var. *purpurea.* A new genus **Ekmania** is created for *E. lepidota* (Griseb.); *Vernonia Milleri* Johnston is referred to the genus *Oliganthes*; and *Piptocoma rufescens* var. *latifolia* and *Elephantopus elatus* var. *intermedius* are described as new varieties.—*P. A. Munz.*

2382. GOURLAY, W. BALFOUR, AND G. M. VEVERS. **Vaccinium intermedium Ruthe.** Jour. Botany 57: 259–260. 1919.—See Bot. Absts. 3, Entry 2128.

2383. LORENZ, ANNIE. **Nardus stricta in the White Mountains.** Rhodora 21: 22–23. 1919.—Reporting new station for *Nardus stricta* at Waterville, New Hampshire. Description of habitat and list of stations in U. S. A. previously reported.—*James P. Poole.*

2384. MATOUSCHEK. [Rev. of: HOLMBERG, O. **Orobanche caryophyllacea Sm. tagen i Sverige. (Orobanche caryophyllacea in Schweden entdeckt.) (Orobanche caryophyllacea discovered in Sweden.)**] Bot. Notiser 1917: 193–195. *1 fig.* 1917.] Zeitschr. Pflanzenkrankh. 29: 59. 1919.

2385. MILLER, W. DEW. **A distinction between two Carices.** Rhodora 21: 23–24. 1919. —An additional character distinguishing *Carex laxiculmnis* Schweinitz and *C. digitalis* Willd. One to three staminate flowers at the base in most of the pistillate spikes of the former, but in the latter all staminate flowers are at the tip of the spike.—*James P. Poole.*

2386. NAKAI, TAKENOSHIN. **Genus novum Oleacearum in Corea media inventum. [New genus of the Oleaceae found in central Corea.]** Bot. Mag. Tôkyô 33: 153–154. 1919.—Latin diagnoses of the new genus **Abeliophyllum** Nakai and the new species *Abeliophyllum distichum* Nakai.—*L. L. Burlingame.*

2387. NELSON, JAMES C. **The grasses of Salem, Oregon, and vicinity.** Torreya 19: 216–227. 1919.—See Bot. Absts. 4, Entry 357.

2388. [NORDSTEDT, C. F. O.] [Swedish rev. of: ALMQUIST. **Sveriges Rosae.** (Swedish roses.) *50 p.*, 1919.] Bot. Notiser 1919: 168. 1919.—*P. A. Rydberg.*

2389. [NORDSTEDT, C. F. O.] [Swedish rev. of: JÖRGENSEN, E. **Die Eyphrasia-Arten Norwegens.** (The species of Euphrasia of Norway.) Bergens Mus. Aarsbok, 1916–1917. *337 p., 11 maps, 14 pl., 54 fig.*] Bot. Notiser 1919: 182. 1919.—*P. A. Rydberg.*

2390. PENNELL, FRANCIS W. **A brief conspectus of the species of Kneiffia with the characterization of a new allied genus.** Bull. Torrey Bot. Club 46: 363–373. 1919.—A key is presented for the species of *Kneiffia* with descriptions of the following new species: *K. sessilis*, *K. brevistipata*, *K. semiglandulosa*, and *K. velutina*. The following new combinations are made: *K. fruticosa humifusa* (Allen), *K. tetragona* (Roth), *K. tetragona hybrida* (Michx.), and *K. perennis* (L.); while *K. tetragona* var. *longistipata* is offered as a new variety. A new allied genus **Peniophyllum** is made for *P. linifolium* (Nutt.) Pennell, comb. nov.—*P. A. Munz.*

2391. PENNELL, FRANCIS W. **Scrophulariaceae of the local flora. I.** Torreya 19: 107–119. 1919.—The area concerned is that included within the local flora range of the Torrey Botanical Club and the Philadelphia Botanical Club. The author has personally collected material of each species and made descriptions of fresh corollas. The object of the study is (1) to present detailed keys to the genera and species included in our flora, (2) to confirm the nomenclature, by stating the type-species and tracing the later history, (3) to give preliminary observations on distribution. Detailed keys for the entire family are presented, representing 8 tribes and 21 genera. The genera and species are then taken up in detail; the present installment discusses the tribes *Verbasceae* and *Cheloneae*, including the genera *Verbascum* (4 species), *Pentstemon* (5 species), *Chelone* (1 species), and *Scrophularia* (2 species). One new combination is proposed, *Chelone glabra* L. forma *tomentosa* (Raf.) Pennell. The study will be continued.—*J. C. Nelson.*

2392. PENNELL, FRANCIS W. **Scrophulariaceae of the local flora, III.** Torreya 19: 161–171. 1919.—This installment takes up the tribe *Digitaleae*, containing the genera *Veronicastrum* (1 species) and *Veronica* (15 species, 1 variety). A detailed key to the species of *Veronica* is presented. Two new species are described: *Veronica Brittonii* Porter, Columbia University, the type from Marble Hill, Phillipsburg, New Jersey; and *V. glandifera* Pennell, from Suffolk, Nansemond County, Virginia. One new combination is made: *Veronica xalapensis* HBK. is reduced to a variety of *V. peregrina* L. *V. humifusa* Dickson of GRAY'S Manual, Ed. 7, is identified with *V. ruderalis* Vahl.—*J. C. Nelson.*

2393. PENNELL, FRANCIS W. **Scrophulariaceae of the local flora. IV.** Torreya 19: 205–216. 1919.—This installment takes up the tribe *Buchnereae*, containing the genera *Aureolaria* (4 species, 2 varieties), *Agalinis* (8 species) and *Otophylla* (1 species). One new variety is described, namely, *Aureolaria pedicularia* (L.) Raf. var. *intercedens*, collected at Mt. Arlington, Morris Co., New Jersey by K. K. Mackensie, Aug. 26, 1906. Detailed notes on synonymy and distribution are continued.—*J. C. Nelson.*

2394. ROGERS, R. S. **Chiloglottis Pescottiana sp. nov.** Proc. Roy. Soc. Victoria (N. S.) 30²: 139–141. *Pl. 25.* 1918. [Contains papers read Sept. to Dec. 1917.] A description of this new species from specimens from Tallangatta, Victoria, is given. This is accompanied by an analytical table presenting data which differentiate this species from the six other Australian members of the genus.—*Eloise Gerry.*

2395. ROLFE, R. A. **The true mahogonies.** Kew Bull. Misc. Inf. [London] 1919: 201–207. 1919.—See Bot. Absts. 3, Entry 2050.

2396. SALMON, C. E. **A hybrid Stachys.** Jour. Linnean Soc. Bot. London 44: 357–362. *1 fig.* 1919.—An account of the natural origin in a garden of a hybrid between *Stachys germanica* and *S. alpina*. The mistaken identity of this plant with Aiton's *S. intermedia* of North

America is discussed. This hybrid apparently arises frequently in European gardens, occurring in somewhat varying forms. The puzzling synonymy of this plant is worked out, and the characteristics of the hybrid and its parents tabulated in detail. [See Bot. Absts. 3, Entry 2188.]—*A. J. Eames.*

2397. SMALL, JAMES. **The origin and development of the Compositae.** New Phytol. 18: 65–89. *Fig. 41–55.* 1919.

2398. STANDLEY, PAUL C. **Studies of tropical American phanerogams.—No. 3.** Contrib. U. S. Nat. Herb. 20.: 173–220. 1919.—This paper contains revisions of the Mexican species of *Ateleia*, the Mexican and Central American species of *Erythrina*, and the Panamanian species of *Leiphaimos*, together with descriptions of many new species of woody plants, chiefly Leguminosae and Rubiaceae. The following new names appear: *Ateleia Arsenii*, *A. insularis*, *Erythrina cochleata*, *E. montana* Rose & Standl., *E. occidentalis*, *E. Goldmanii*, *Capparis discolor*, *Forchammeria macrocarpa*, *F. lanceolata*, *Steriphoma macrantha*, *Acacia polypodioides*, *A. leucothrix*, *A. laevis*, *A. penicillata*, *A. Conzattii*, *A. sororia*, *A. Rosei*, *A. vernicosa*, *Calliandra Conzattii*, *Leucaena cuspidata*, *L. plurijuga*, *Pithecolobium leiocalyx*, *P. calostachys*, *P. macrosiphon*, *P. confine*, *Calophyllum Rekoi*, *C. chiapense*, *Maba nicaraguensis*, *M. Rekoi*, *Diospyros oaxacana*, *Leiphaimos truncatus*, *L. stellatus*, *L. Pittieri*, *L. albus*, *L. thalesioides*, *L. pulcherrimus*, *L. simplex* (Griseb.) Standl., *Randia cinerea* (Fernald) Standl., *R. lasiantha* (*Basanacantha lasiantha* Standl.). *R. Pittieri* (*B. Pittieri* Standl.), *R. portoricensis* (Urban) Standl., *R. spinifex* (Roem. & Schult.) Standl., *R. subcordata* (*Basanacantha subcordata* Standl.), *R. calycosa*, *R. laevigata*, *R. pleiomeris*, *R. guatamalensis*, *R. malacocarpa*, *R. Rosei*, *Hoffmannia rotundata*, *H. uniflora*, *H. panamensis*, *H. Tonduzii*, *H. orizabensis*, *H. decurrens*, *H. confertiflora*, *H. angustifolia*, *H. chiapensis*, *Hamelia costaricensis*, *H. panamensis*, *Cassia jacquinioides* (Griseb.) Standl., *Duroia costaricensis*, *Phialanthus macrostemon*, *Machaonia Coulteri* (Hook. f.) Standl., *Chiococca pubescens*, *Guettarda Deamii*, *G. filipes*, *Brosimum Conzattii*, *Coussapoa Rekoi*, *Struthanthus densiflorus* (Benth.) Standl., *S. diversifolius* (Benth.) Standl., *S. Grahami* (Benth.) Standl., *S. Haenkeanus* (Presl) Standl., *S. Hartwegi* (Benth.) Standl., *S. inconspicuus* (Benth.) Standl., *S. inornus* (Robins. & Greenm.) Standl., *Phrygilanthus sonorae* (S. Wats.) Standl., *Ximenia pubescens*, *Platanus chiapensis*, *P. oaxacana*, *Prunus prionophylla*, *Caesalpinia acapulcensis*, *C. caladenia*, *C. sclerocarpa*, *Cassia chiapensis*, *C. Tonduzii*, *Indigofera sphinctosperma*, *Cracca Brandegei*, *C. tepicana*, *Andira Galeottiana*, *Picramnia pistaciaefolia* Blake & Standl., *Rhus Barclayi* (Hemsl.) Standl., *R. jaliscana*, *Marcgravia guatemalensis*.—*S. F. Blake.*

2399. TAYLOR, NORMAN. **Rock's Lobelioideae of Hawaii.** [Rev. of: ROCK, J. F. **A monographic study of the Hawaiian species of the tribe Lobelioideae, family Campanulaceae.** *394 p. 217 pl.*. Honolulu, Feb. 20, 1919.] Torreya 19: 228–230. 1919.—The flora of the Hawaiian Islands has been long noted for its extreme endemism. The tribe *Lobelioideae*, synonymous with the family Lobeliaceae, is discussed with reference to its affinities with its nearest relatives. The genus *Cyanea* is regarded as still in process of evolution. Seven genera, containing 149 species and varieties, are included. Four of the endemic genera are related to American genera. The species are fully described and illustrated. The book is truly a monograph in the best sense of the word.—*J. C. Nelson.*

2400. WABY, J. F. **Some interesting species of palms.** Jour. Bd. Agric. British Guiana 12: 49–55. 1919.—Descriptions of *Oreodoxa regia*, *O. regia* var. *Jenmanii*, *O. oleracea*, *Euterpe edulis*, *E. stenophylla*, *E. ventricosa*, *E. acuminata*, *E. Jenmanii*, and *E. utilis*.—*J. B. Rorer.*

2401. WABY, J. F. **Some interesting species of palms.** Jour. Bd. Agric. British Guiana 12: 112–115. 1919.—Gives descriptions, common names and interesting facts about the following palms—*Mauritia flexuosa*, *Chrysalidocarpus lutescens*, *Cystostachys renda*, *Desmoncus* sp., *Nipa fruticans*, and *Hyphaene thebaica*.—*J. B. Rorer.*

2402. WARD, MARTHA E. **Galax aphylla introduced in Massachusetts.** Rhodora 21: 24. 1919.—Few plants of *Galax aphylla* found in the woods in Swampscott, Mass., where previously reported by Fernald.—*James P. Poole.*

MISCELLANEOUS, UNCLASSIFIED PUBLICATIONS

B. E. LIVINGSTON, *Editor*

2403. ANONYMOUS. **Sea-grass fibre as a packing material.** Sci. Amer. Monthly 1: 96. 1920. [Abstract from La Nature, Oct. 11, 1919, in Technical Rev.]

2404. ANONYMOUS. **Substitutes for pollen and nectar.** Agric. Gaz. New South Wales 31: 116. 1920.—Discusses rye flour as a substitute for pollen and nectar as food for bees.—*L. R. Waldron.*

2405. ANONYMOUS. **Putting flax on a modern basis.** Sci. Amer. 122: 166, 175-176. *4 fig.* 1920.—Pertains to manufacturing processes.—*Chas. H. Otis.*

2406. BOYER, G. **Sur l'inclusion de brins d'herbes par les champignons. [Concerning the inclusion of bits of plants by certain fungi.]** Actes Soc. Linn. Bordeaux (Proces-verbaux) 69: 49-50. 1915-16.—Stems and leaves of grasses remain living after their inclusion by growth of polyporous fungi.—*W. H. Emig.*

2407. CARDOT, J. **A letter from M. Cardot to the Sullivant Moss Society.** Bryologist 23: 7. 1920.

2408. CHALMERS, ALBERT J. **Sadd dermatitis.** Jour. Tropical Med. and Hygiene 23: 57-59. *7 fig.* 1920.—The stiff hairs of *Panicum pyramidale* Lam., one of the chief grasses forming the floating and rooted masses of vegetation which sometimes block the White Nile, are shown to cause a dermatitis in human beings, by their mechanical action.—*E. A. Bessey.*

2409. CHEEL, E., AND DUCKWORTH, A. C. **The cultivation of native plants.** Australian Nat. 4: 131-133. 1920.

2410. CLAUDY, C. H. **The fruits of scientific farming.** Sci. Amer. 122: 216. 1920.—A popular article on some of the activities of the United States Department of Agriculture.—*Chas. H. Otis.*

2411. DEBORD, GEO. G. **Comments on the examination of canned salmon.** [Abstract.] Absts. Bact. 4: 11. 1920.—Twelve hundred and eighty-three cans were examined bacteriologically of which 34 per cent were not sterile. The organisms found were aerobic, sporulating bacteria. There was no correlation between the sterility and the odor of the can. [From author's abst. of paper read at scientific session, soc. Amer. Bact.]—*D. Reddick.*

2412. DODD, SYDNEY. **Infestation of the skin, etc., of sheep by grass seeds.** Jour. Comparative Path. and Therap. 22: 90-95. 1919.—In many parts of Australia much injury, sometimes death, results in sheep from the penetration of the skin or eyes by seeds of various grasses, chiefly of the genera *Stipa* and *Aristida* and also *Hordeum murinum*, *Festuca bromoides* and possibly species of *Andropogon.*—*E. A. Bessey.*

2413. DUNHAM, ELIZABETH M. **Mounting mosses for exhibition purposes.** Bryologist 23: 6. 1920.—The author describes how specimens may be mounted on cardboard and protected against dust and breakage by sheets of celluloid.—*E. B. Chamberlain.*

2414. EDMONDSON, RUTH B., GEO. G. DEBORD, AND CHARLES THOM. **Botulism from canned ripe olives.** [Abstract.] Absts. Bact. 4: 10. 1920.—All cans which were swelled or "off" in odor showed living organisms. Twenty-seven cans from a "batch" which had caused poisoning cases were tested for *B. botulinus* and the organism was isolated from 7 cans. [From author's abst. of paper read at scientific session, Soc. Amer. Bact.]—*D. Reddick.*

2415. ESTY, J. R., AND C. C. WILLIAMS. **Resistant bacteria causing spoilage in canned foods.** [Abstract.] Absts. Bact. 4: 11. 1920.—The organisms causing this spoilage were facultative and obligate anaerobes and were classified according to the range in temperature where growth occurred. The facultative anaerobes fell between (1) 42° and 80°C.; (2) 22° and 80°C.; (3) 37° and 80°C.; (4) 22° and 55°C.; (5) 37° and 55°C.; (6) 37° and 65°C.; (7) 22° and 45°C. All the five obligate anaerobes isolated were vigorous gas formers and fell into four groups according to the above classification. (1) 45° and 80°C.; (2) 30° and 65°C.; (3) 42° and 65°C.; (4) 22° and 45°C. [From author's abst. of paper read at scientific session, Soc. Amer. Bact.]—*D. Reddick.*

2416. HAMMER, B. W., AND L. R. SANDERS. **A bacteriological study of the method of pasteurizing and homogenizing the ice cream mix.** Iowa Agric. Exp. Sta. Bull. 186: 19–26. 1919.

2417. KING, ALBERT E. W. **The mechanical properties of Philippine bast-fiber ropes.** Philippine Jour. Sci. 14: 561–655. *5 pl., 2 fig.* 1919.—These investigations were undertaken to secure quantitative results on the mechanical properties of Philippine bast-fiber ropes. Thirty-two kinds of fibrous material were obtained from bast-plant species, and seven from those that gave no bast-fibers. These were compared with abaca and maguey. The plain stripping process of obtaining fiber was compared with the water-retting process, to the advantage of the latter. The circumference and cross sectional area were calculated and the tensile strength was determined, the results being collected in a series of tables. The individual species of the fiber plants are described.—*Albert R. Sweetser.*

2418. MAGGIORA, A., AND CARBONE, D. **Sull'impiego del Bacillus felsineus per la macerazione industriale della canapa.** [**The utilization of B. felsineus in the retting of hemp on an industrial scale.**] Stas. Sper. Agrarie Ital. 52: 449–462. 1919.—The present investigation aims at the study of the commercial application of a biological method for retting hemp (*Cannabis*). The material is introduced into masonry tanks containing water sufficient to cover it and maintained at 37°C. by means of steam pipes. Inoculation of the mass with cultures of *B. felsineus* and *Saccharomyces ellipsoideus* in relatively moderate amounts brings about retting of the fiber in 60–90 hours. The quality of the product is "perfect" in terms of commercial standards. The quantity retted varied in the experiments from 100 kgm. to 492 kgm. and this is considered by the authors as an indication that the method is applicable to larger lots on a commercial scale. Slight variations in the technic may be introduced in the procedure as a result of scientific investigations.—*A. Bonazzi.*

2419. MCATEE, W. L. **Some local names of plants, III.** Torreya 20: 17–27. 1920.—A list of 150 local names, applied to 104 species of American plants belonging to 59 families, is presented. The locality is cited wherever possible, and the source from which the name was obtained is indicted. [Previous installments appeared in: Torreya 13: 225–236. 1913. *Ibid.* 16: 235–242. 1916.]—*J. C. Nelson.*

2420. MURRILL, W. A. **Plant growths that shed light.** Sci. Amer. 122: 427, 440. *4 fig.* 1920.—Popular description of certain luminous fungi.—*Chas. H. Otis.*

2421. MUTCH, NATHAN. **The isolation of a single bacterial cell.** Jour. Roy. Microsc. Soc. London 1919: 221–225. *1 fig.* 1919.—The organism to be studied is grown upon a solid medium for six or eight hours, and the resulting growth emulsified in sterile broth or normal saline solution. One or two narrow rings of filter paper are then placed in the hanging drop cell and moistened with saline solution. The rim of the cell is prepared with vaseline. A clean cover slip is flamed and when cooled a micro-drop of emulsion of bacterium is placed in its center by means of a very small loop of platinum wire. The slip is immediately placed in position over the moist chamber. A series of such drops can be prepared and examined rapidly and the dilution of the original emulsion adjusted until a drop containing a solitary organism is found. The cover slip is raised from the cell, a large drop of suitable medium is placed close to the micro-drop and the slip is tilted until the two coalesce. The slip is then placed on another moist cell, incubated for 24 hours, and again examined. When a solid medium is employed, if original observation was correct, one colony only will have developed. In working with delicate organisms the process must be carried out at body temperature on a warm stage, and the filter paper ring must be replaced by a small drop of saline solution, only two or three times as large as the micro-drop. When the observation is complete, the large drop of medium is added, the slip is transferred to a moist cell containing paper ring, and the preparation is incubated as before.—The advantages of this method are that no special skill or practice is called for, no special preparation is needed, and the work can be performed with the ordinary apparatus found on a bacteriological bench; the time required is only one or two hours.—*Julia Mossel Haber.*

2422. RUDOLFS, W. **Experiments on the value of common rock salt and sulfur for killing stumps.** Soil Sci. 9: 181-189. *Pl. 1-2.* 1920.—Sulphur applied to high or low brush stimulated the growth of the live tree stumps. Rock salt up to 2.5 tons per acre did little harm while 0.5 to 1 ton per acre acted as a fertilizer. Applications of 2 to 3 tons per acre of sodium chloride to stumps cut in the winter killed or severely injured them. The salt should be applied in the spring just before the leaves appear.—*W. J. Robbins.*

2423. SAGASPE, M. J. **Sur la Digitale (Digitalis purpurea).** [A note concerning Digitalis purpurea.] Actes Soc. Linn. Bordeaux. Proces-verbaux. 68: 64-65. 1914.—Three monstrosities of *Digitalis purpurea* are briefly described.—*W. H. Emig.*

2424. TROWBRIDGE, P. F. **Report of the director, July 1, 1917, to June 30, 1919.** North Dakota Agric. Exp. Sta. Bull. 136. *23 p., 3 fig.* 1920.—This embraces the annual report of the station for two years. A brief summary is given of the experimental work including a limited amount of data.—*L. R. Waldron.*

✓ 2425. WAKSMAN, SELMAN A. **The industrial application of enzymes of Aspergillus oryzae.** [Abstract.] Absts. Bact. 4: 7. 1920.—The enzymes of *A. oryzae* hydrolyze starch completely whereas malt diastase does not; and the quantity of starch hydrolyzed is 4 to 6 times greater. The enzymes are useful in the textile industry for removing "size," in clearing fruit extracts which contain some starch, and in the manufacture of various starch derivatives. [From author's abst. of paper read at scientific session, Soc. Amer. Bact.]—*D. Reddick.*

2426. WYANT, ZAE NORTHRUP. **Experiments in silage inoculation.** [Abstract.] Absts. Bact. 4: 6. 1920.—Various strains of lactic acid producing bacteria were used to inoculate ensilage. After fermentation for 5 weeks the ensilage proved very palatable to calves.—Platings were made from the interior to determine whether the types introduced predominated or not. From the first pair of inoculations with *Bact. lactis acidi* and *Bact. bulgaricum* the first organism was recovered without difficulty, the latter not at all. The organisms which predominated in each silo were short rods in pairs which resemble *Bact. lactis acidi* in morphology, spore-forming rods, and a few yeasts. [From author's abst. of paper read at scientific session, Soc. Amer. Bact.]—*D. Reddick.*

INDEX TO AUTHORS' NAMES IN VOLUME V

(References are to Entry numbers; an asterisk before a number signifies that the entry referred to is by citation alone—no abstract.)

Vol. V AUGUST, 1920 No. 1

ENTRIES 1–1085

BOTANICAL ABSTRACTS

A monthly serial furnishing abstracts and citations of publications in the international field of botany in its broadest sense

WILLIAMS & WILKINS COMPANY
BALTIMORE, U. S. A.
THE CAMBRIDGE UNIVERSITY PRESS
FETTER LANE, LONDON, E. C.

Entered as second-class matter, November 9, 1918, at the post office at Baltimore, Maryland, under the Act of March 3, 1879

Price, net postpaid for two volumes: $6.00 United States, Mexico, Cuba; $6.25 Canada; $6.50 Other countries

1919 Volumes: I and II
1920 Volumes: III, IV, V and VI

CONTENTS

BOARD OF EDITORS FOR 1920 AND ASSISTANT EDITORS

EDITORS FOR SECTIONS

BIBLIOGRAPHY COMMITTEE FOR 1920

INFORMATION CONCERNING BOTANICAL ABSTRACTS

The purpose of BOTANICAL ABSTRACTS is to supply complete citations and analytical abstracts of all papers dealing with botanical subjects, wherever published, just as soon as possible after they appear. Every effort is made to present complete and correct citations with abstracts of original work, of all papers and reviews, appearing after January 1, 1919. As an adequate index of progress, BOTANICAL ABSTRACTS is of use to the intelligent grower, field agent and inspector, extension worker, teacher and investigator. The international scope of the work should appeal especially to those workers who have restricted library facilities. It is hoped that the classification by subjects will prove to be a great aid even to those having access to large libraries, while the topical index should serve a most useful purpose to every one interested in plants.

The service of BOTANICAL ABSTRACTS is planned for botanists and all workers with plants, throughout the world. The services of all the botanical workers who are connected with BOTANICAL ABSTRACTS in any way, are given without any compensation except the satisfaction of participation in such a great coöperation toward the advancement of science. It is hoped that all students of plants will feel that BOTANICAL ABSTRACTS is their journal. Although the physical exigencies of the enterprise have made it practically necessary that the actual work of preparing the issues be largely done within a relatively short distance from the place of publication, yet this does not imply that the coöperation is not needed of residents of other countries than the United States and Canada. Many collaborators and abstractors reside in other countries, but the aim has not been to distribute the actual work throughout the world; rather has it seemed best to distribute the work so as to give prompt and efficient service, without reference to the particular countries in which the workers reside. It is physically necessary that the burden of the work and the finding of funds for clerical assistance, etc., should rest largely on North American workers, but the field covered is international and the results are available to all.

The Board of Control of Botanical Abstracts, Incorporated, has charge of publication. The board is a democratic organization made up of members elected from many societies, as is shown on the first cover page. Each society elects, in its own way, two representatives, each for a period of four years. One new member is elected each biennium (beginning January 1, 1921) to replace the representative who retires. In the list on the first cover page, the member first named in each group is to serve till January 1, 1923; the second member in each group is to serve till January 1, 1921. Members are not eligible for immediate reëlection.

The Executive Committee of the Board of Control of BOTANICAL ABSTRACTS consists of five members, elected annually by the Board. It has charge of *ad interim* affairs not involving matters of general policy. Its membership is shown by the asterisks in the list on the first cover page. The chairman of the Committee for 1920 is Donald Reddick, Cornell University, Ithaca, New York.

The Board of Editors of BOTANICAL ABSTRACTS consists of an Editor-in-Chief and Editors for Sections, as shown on the second cover page. The Editors are elected annually by the Board of Control. Assistant Editors are appointed by the Editors. Editors for Sections, with the aid of Assistant Editors for Sections, are responsible for editing the material of their respective sections as this is supplied by the Bibliography Committee (from the Collaborators and other Abstractors), and also for citations and abstracts of non-periodical literature. They also supply abbreviated titles for the author index of each volume and subject-index entries (for the occasional subject indexes) pertaining to their respective sections. The Editor-in-Chief, with the help of the Associate Editor-in-Chief and with the approval of the Board of Editors, is responsible for the general make-up of the issues, for the final compilations of the author and subject indexes, and for such other details as are left to him by the Editors for Sections.

The Bibliography Committee of BOTANICAL ABSTRACTS, the membership of which is shown on the second cover page, is appointed annually by the Executive Committee of the Board of Control. The Bibliography Committee is charged with the responsibility of arranging for the prompt citing and abstracting of serial botanical literature. In performing this function, the Committee assigns to individual Collaborators the complete responsibility for furnishing the abstracts of all botanical papers in a specified serial publication, or in a limited number of serials. The Committee is further charged with the duty of maintaining an accurate record, through a system of reports furnished currently by the Collaborators, of the state of abstracting for each serial publication. This record enables the Committee to detect and correct delinquencies in the work of abstracting and to keep the work up to date. The number of assigned serials will eventually exceed 2000, for each of which a record of the state of abstracting will be maintained in the office of the Bibliography Committee. Readers are earnestly requested to aid the Bibliography Committee by bringing to its attention any serial publications that are not being properly represented in BOTANICAL ABSTRACTS. The chairman of the Committee for 1920 is J. R. Schramm, Cornell University, Ithaca, New York.

Collaborators of BOTANICAL ABSTRACTS. A large number of botanical workers in all parts of the world have volunteered to assume complete responsibility for securing citations and abstracts from one or more serial publications as assigned to them by the Bibliography Committee. This corps of voluntary workers (called Collaborators) really constitute the basis of the service rendered by BOTANICAL ABSTRACTS. Through their work it is made certain that all serial publications are promptly entered. A list of the names of Collaborators is published in each volume of BOTANICAL ABSTRACTS. It is desirable that a considerable reserve list of collaborators be maintained, in order to allow for necessary changes, and additional collaborators are therefore earnestly solicited.

Abstractors for BOTANICAL ABSTRACTS. Collaborators frequently prepare abstracts themselves, and are thus Abstractors, but they also arrange for others to prepare them. Every abstract is signed by the Abstractor who prepared it, but entries by citation alone are not signed. The Collaborators are responsible for these citations. A list of Abstractors is published for each volume of BOTANICAL ABSTRACTS. It includes many names of voluntary contributors to the enterprise, besides those of the Collaborators.

The Printing and Circulation of BOTANICAL ABSTRACTS is in the hands of the Publishers, according to the terms of a definite contract between them and the Board of Control. All other matters are directly in the hands of the Board of Control. Correspondence concerning subscriptions should be addressed to the Publishers or their agents; other matters should be referred to the Chairman of the Board of Control, to the Chairman of the Bibliography Committee, or to the Editor-in-Chief.

Readers of BOTANICAL ABSTRACTS are earnestly requested to make careful note of any errors that occur in the journal, with their corrections, and to send these notes to the Editor-in-Chief. If all will coöperate in this it will be possible to supply a page of corrigenda with each volume. These notes should be on sheets about 22 × 28 cm. (8½ × 11 inches).

BOTANICAL ABSTRACTS is published monthly, two, three, or four volumes being issued each year at present. Each volume contains about 300 pages. The current (1920) volumes are III, IV, V and VI. Subscriptions are accepted for Vols. III and IV, (January–July, incl.), and V and VI, (August–December, incl.). Volumes I and II can no longer be furnished by the publishers. The price for two volumes is $6.00, for the United States and its dependencies, Mexico and Cuba; $6.25, for Canada; $6.50, for other countries. Prices are net postpaid. No claims are allowed for copies lost in the mails unless such claims are received within 30 days (90 days for places outside of the United States and Canada) of the date of issue.

Subscriptions are received at the following addresses, for the respective countries:

United States of North America and dependencies; Mexico; Cuba: Williams & Wilkins Company, Mount Royal and Guilford Avenues, Baltimore.

Argentina and Uruguay: Beutelspacher y Cia., Sarmiento 815, Buenos Aires.

Australia: Stirling & Co., 317 Collins St., Melbourne.

Belgium: Henri Lamertin, 58 Rue Coudenberg, Bruxelles.

The British Empire, except Australia and Canada: The Cambridge University Press, C. F. Clay, Manager, Fetter Lane, London, E. C. British subscribers are requested to make checks and money orders payable to Mr. C. F. Clay, Manager, at the London Address.

Canada: Wm. Dawson & Sons, Ltd., 87 Queen Street, East Toronto.

Denmark: H. Hagerup's Boghandel, Gothersgade 30, Kjöbenhavn.

France: Emile Bougault, 48, Rue des Ecoles, Paris.

Germany: R. Friedlander & Sohn, Carlstrasse 11, Berlin N. W., 6.

Holland: Scheltema & Holkema, Rokin 74–74, Amsterdam.

Italy: Ulrico Hoepli, Milano.

Japan and Korea: Maruzen Company, Ltd. (Maruzen-Kabushiki-Kaisha), 11–16 Nihonbashi Tori-Sanchome, Tokyo; Fukuoka, Osaka, Kyoto, and Sendai, Japan.

Spain: Ruiz Hermanos, Plaza de Santa Ana 13, Madrid.

Switzerland: Georg & Cie., Freistrasse 10, Bâle.